# Springer Optimization and Its Applications

Volume 146

*Managing Editor*
Panos M. Pardalos (University of Florida)

AF620850

*Honorary Editor*
Ding-Zhu Du (University of Texas at Dallas)

*Advisory Editors*
J. Birge, Booth School of Business (University of Chicago)
S. Butenko, Department of Industrial and Systems Engineering (Texas A & M University)
F. Giannessi, Dipto. Matematica (University of Pisa)
S. Rebennack, Institute of Operations Research (Karlsruhe Institute of Technology)
T. Terlaky, Department of Industrial and Systems Engineering (Lehigh University)
Y. Ye, Department of Engineering (Stanford University)

*Aims and Scope*

Optimization has been expanding in all directions at an astonishing rate during the last few decades. New algorithmic and theoretical techniques have been developed, the diffusion into other disciplines has proceeded at a rapid pace, and our knowledge of all aspects of the field has grown even more profound. At the same time, one of the most striking trends in optimization is the constantly increasing emphasis on the interdisciplinary nature of the field. Optimization has been a basic tool in all areas of applied mathematics, engineering, medicine, economics and other sciences.

The series *Springer Optimization and Its Applications* publishes undergraduate and graduate textbooks, monographs and state-of-the-art expository works that focus on algorithms for solving optimization problems and also study applications involving such problems. Some of the topics covered include nonlinear optimization (convex and nonconvex), network flow problems, stochastic optimization, optimal control, discrete optimization, multi-objective programming, description of software packages, approximation techniques and heuristic approaches.

More information about this series at http://www.springer.com/series/7393

Themistocles M. Rassias ·
Valentin A. Zagrebnov
Editors

# Analysis and Operator Theory

## Dedicated in Memory of Tosio Kato's 100th Birthday

Foreword by Barry Simon, IBM Professor of Mathematics and Theoretical Physics, Emeritus; California Institute of Technology, Pasadena, CA 91125, USA

Springer

*Editors*
Themistocles M. Rassias
National Technical University of Athens
Zografou Campus
Athens, Greece

Valentin A. Zagrebnov
Institute of Mathematics
Aix-Marseille University
Marseille, France

ISSN 1931-6828 ISSN 1931-6836 (electronic)
Springer Optimization and Its Applications
ISBN 978-3-030-12663-6 ISBN 978-3-030-12661-2 (eBook)
https://doi.org/10.1007/978-3-030-12661-2

Library of Congress Control Number: 2019930373

Mathematics Subject Classification (2010): 46L80, 47A58, 46Txx, 00A79

© Springer Nature Switzerland AG 2019
This work is subject to copyright. All rights are reserved by the Publisher, whether the whole or part of the material is concerned, specifically the rights of translation, reprinting, reuse of illustrations, recitation, broadcasting, reproduction on microfilms or in any other physical way, and transmission or information storage and retrieval, electronic adaptation, computer software, or by similar or dissimilar methodology now known or hereafter developed.
The use of general descriptive names, registered names, trademarks, service marks, etc. in this publication does not imply, even in the absence of a specific statement, that such names are exempt from the relevant protective laws and regulations and therefore free for general use.
The publisher, the authors and the editors are safe to assume that the advice and information in this book are believed to be true and accurate at the date of publication. Neither the publisher nor the authors or the editors give a warranty, expressed or implied, with respect to the material contained herein or for any errors or omissions that may have been made. The publisher remains neutral with regard to jurisdictional claims in published maps and institutional affiliations.

This Springer imprint is published by the registered company Springer Nature Switzerland AG
The registered company address is: Gewerbestrasse 11, 6330 Cham, Switzerland

# Foreword

Tosio Kato (1917–1999) profoundly influenced the science of his time. He was the founder of a large field, the theory of Schrödinger operators (which describes the mathematics of the fundamental objects of the physics of ordinary matter at small scale). Indeed, one could say collection of fields—for example, shortly after of writing this, I will be attending a conference on almost periodic and random Schrödinger operators.

Kato not only founded the field with his 1951 paper on self-adjointness. He continued with influential contributions for the rest of his life, especially the next 30 years, on a variety of aspects of the theory including perturbation and scattering theory, spectral analysis, properties of eigenfunctions, and adiabatic theory. His work is marked by depth, beauty, and elegance.

Like Newton who also made pivotal contribution to Coulomb systems, Kato found his great self-adjointness result while evacuated from his University and its library. Because of the War, he spent the middle part of the 1940s in the Japanese countryside where he also found his basic results on perturbation theory. This disruption explains why Kato only received his doctorate when he was over 30 years of age.

Like many of the other founders of modern mathematical physics (e.g., Jost, Thirring, Wightman), Kato's formal training was in physics at the University of Tokyo, where he served as a Professor of Physics, but he learned to exploit rigorous proof and spent the latter half of his career as a Professor of Mathematics at University of California, Berkeley. Kato's opus wasn't limited to Schrödinger operators. He also made seminal contributions to the theory of nonlinear PDEs where he was an early pioneer. Besides these two topics, he has worked in semi-group theory and to a variety of parts of functional analysis.

This book, written by a group of admirers of Kato who were all influenced by his work, was begun during the centennial year of Kato's birth and is dedicated to his memory. Its breadth mirrors his.

Speaking for myself, I have been impacted by Kato's work throughout my career up until today. As a graduate student 50 years ago, I learned an enormous amount from his great book on Perturbation Theory which was central to my first major work on the quantum anharmonic oscillator. During the 1970s, we exchanged numerous letters which stimulated research on both our parts.

I am very glad to be able to honor Kato with my contribution here and would like to thank the editors for putting this project together, for inviting me to participate and asking me to write this foreword.

Los Angeles, USA
October 2018

Barry Simon
IBM Professor of Mathematics and
Theoretical Physics, Emeritus;
California Institute of Technology
Pasadena, CA 91125, USA
e-mail: bsimon@caltech.edu

# Preface

*Analysis and Operator Theory—Dedicated in Memory of Tosio Kato's 100th Birthday* features a collection of carefully selected research as well as survey articles devoted to a broad spectrum of subjects of Mathematical Analysis and Mathematical Physics, to which Kato contributed monumental results.

The breadth and full depth of Kato's legacy is impossible to exhaust by even more than one volume dedicated to him. A brilliant mathematician, world-renowned specialist in functional analysis and operator theory, he was also an outstanding teacher. His great book "Perturbation Theory for Linear Operators" first published by Springer in 1966 is an encyclopedia for mathematicians and mathematical physicists. Many of his papers, e.g., on operator theory, evolution equations, or spectral analysis, are so remarkably written that they are still great classics to be read in the original.

The book chapters presented here have been written by a number of experts in the various areas of Kato's interest. They are aiming to present the actual state of the art in the topics, which were developed by Tosio Kato in various branches of analysis and operator theory. Some of his achievements, such as Kato's inequality, the Kato type matrix limit theorem, the Howland–Kato commutator problem, the Kato class of potentials, or the Trotter–Kato product formulae, are directly manifesting in the titles of the chapters of this volume. Among others, there are related to or inspired by certain ideas of Kato.

A special article presents a report of Tosio Kato's work on nonrelativistic Quantum Mechanics, and one of the chapters is dedicated to an unpublished paper of his.

It is hoped that this publication provides an extensive account of research results which will be of usefulness for a wide readership, from graduate students to established researchers in the corresponding domains.

We would like to express our gratitude to all the scientists who contributed valuable works in this book as well as to Barry Simon who wrote the foreword. Additionally, we would like to acknowledge the superb assistance of the staff of Springer for the preparation of this publication.

Athens, Greece — Themistocles M. Rassias
Marseilles, France — Valentin A. Zagrebnov

# Contents

# Complementarity and Stochastic Independence

**Luigi Accardi and Yun-Gang Lu**

**Abstract** A mathematical approach to the notion of complementarity in quantum physics is described and its historical development is shortly reviewed. After that, the notion of n-complementarity is introduced as a natural extension of complementarity and at the same time as weak form of stochastic independence. Several examples in which $n$-complementarity is realized but not independence are produced. The construction of these examples is based on the structure of Interacting Fock Space (IFS) that is strictly related to the classical theory of orthogonal polynomials. A brief description of both this notion and this connection is included to make the paper self-contained.

## 1 Introduction

The history of the early development of the notion of complementarity is vividly described in Chapter 7.2 of M. Jammer's monograph [17] on the development of Quantum Mechanics. N. Bohr first introduced the term *complementarity* during the Como conference in 1927 as a candidate to unify De Broglie's idea of *wave–particle duality* with Heisenberg indeterminacy principle and to capture in great generality the physical essence of both statements. In the same Chapter Jammer recalls the comment, attributed by Wigner to von Neumann: *Well, there are many things that do not commute and you can easily find three operators that do not commute*, and explains that this comment should be interpreted as an implicit critique to Bohr's idea that complementarity refers to *pairs* of attributes (or observables).

L. Accardi
Centro Vito Volterra, Università di Roma Tor Vergata, 00133 Rome, Italy
e-mail: accardi@volterra.mat.uniroma2.it

Y.-G. Lu (✉)
Department of Mathematics, University of Bari Aldo Moro, Campus of University Ernesto Quagliariello, Via Edoardo Orabona 4, 70126 Bari, Italy
e-mail: accardi@volterra.mat.uniroma2.it

© Springer Nature Switzerland AG 2019

T. M. Rassias and V. A. Zagrebnov (eds.), *Analysis and Operator Theory*,
Springer Optimization and Its Applications 146,
https://doi.org/10.1007/978-3-030-12661-2_1

In quantum theory the non-commutativity of two Hermitean linear operators implies a lower bound on the product of the variances of these operators, with respect to any state, which is zero on every state if and only if the two operators commute. The fact that there exist states such that this lower bound can be strictly positive is interpreted as a *statistical formulation of Heisenberg indeterminacy principle* because any attempt to experimentally decrease the variance of one of the two operators implies an increase in the variance of the other one. In terms of the physical observables represented by these Hermitean operators, this means that any increase in the precision of a measurement of one of these observables implies a larger imprecision in the measurement of the other one.

Pairs of observables satisfying Heisenberg commutation relations are characterized by the property that this lower bound is strictly positive and independent of the state. This fact suggests a natural way to defend Bohr's contention on complementarity from von Neumann critique by selecting, among pairs of incompatible observables, the complementary ones as those which *are in some sense maximally incompatible*. The problem is then reduced to give a mathematical formulation of the notion of *maximal incompatibility*.

On the other hand, Heisenberg commutation relations are a model-dependent statement without a direct physical interpretation while, as explained by von Neumann [29], all statements about quantum mechanics that can be directly compared with experiments can be reduced to the calculation of expectation values or equivalently to values of observables and their transition (conditional) probabilities.

The program to find an intrinsic, model-independent property that expresses the maximal incompatibility of two observables was formulated in [1] where it was shown that it leads to a natural problem in operator theory which is still open nowadays even in the finite-dimensional case.

## 1.1 An Open Problem in Operator Theory

The starting point of [1] was the idea that *two observables should be considered maximally incompatible if the exact knowledge of the value of one of them gives zero information on the values of the other one.*

If the possible values of these observables (spectrum of the corresponding operators) have a finite cardinality, the natural way to express the *zero information condition* on these values is to say that, with the given information, all the values of the given observable are equiprobable. If the two observables have spectrum with the same cardinality, the only case considered in [1], the situation is perfectly symmetric.

**Definition 1** A pair $A$, $B$ of $N$–valued observables ($N < +\infty$) is called **complementary** if, denoting $(a_\alpha)$, $(b_\beta)$ their values, their transition probability matrix corresponds to the maximum indeterminacy, i.e.,

$$P(A = a_\alpha | B = b_\beta) = \frac{1}{N}\,; \quad \alpha, \beta = 1, \ldots, N \tag{1}$$

we assume that the values $(a_\alpha)$, $(b_\beta)$ are mutually different (non-degenerate spectrum).

In the quantum model the two observables $A$, $B$ are represented by Hermitean operators on a complex $N$-dimensional Hilbert space $\mathcal{H} \equiv \mathbb{C}^N$ and we will use the same symbol for an observable and the corresponding operator. Denoting $(\psi_\alpha^A)$, $(\psi_\beta^B)$ their eigenvectors, the corresponding transition probability matrix is given by $\left(|\langle\psi_\alpha^A, \psi_\beta^B\rangle|^2\right)$. This motivates the following definition:

**Definition 2** Two orthonormal bases in $\mathcal{H}$, $(\varphi_\alpha)$, $(\psi_\beta)$ are called **complementary** if

$$|\langle\varphi_\alpha, \psi_\beta\rangle|^2 = \frac{1}{N} \quad ; \quad \forall \alpha, \beta \tag{2}$$

*Remark* If $A$, $B$ satisfy the finite version of the CCR [27], namely

$$e^{ihA}e^{ikB} = e^{ihk}\, e^{ikB}e^{ihA}\; ; \quad h, k = 1, \ldots, N \tag{3}$$

then they are complementary (for a proof see the end of section 1.2).

Condition (2) is equivalent to

$$\tau(E_A(I) \cdot E_B(J)) = c\lambda(I) \cdot \lambda(J) \tag{4}$$

where $I$, $J$ are sub-sets of $\{1, \ldots, N\}$, $E_A(\cdot)$, $E_B(\cdot)$ the spectral projectors of $A$, $B$ respectively, $\lambda$ is the uniform measure on $\{1, 2, \ldots, N\}$ (i.e., $\lambda \equiv (1, 1, \ldots, 1)$), $\tau$ the non-normalized trace on $\mathcal{B}(\mathcal{H}) \equiv N \times N$ $\mathbb{C}$-matrices and $c = 1/N$. The formulation (4) of the complementarity condition is better suited than (2) for infinite-dimensional extension.

**Definition 3** Let $\mathcal{A}$ be a semi-finite von Neumann algebra acting on a Hilbert space $\mathcal{H}$ and with a trace $\tau$. A pair of self-adjoint operators $A$, $B$ on $\mathcal{H}$, affiliated to $\mathcal{A}$ is called $(\tau, \lambda)$-**complementary** where here and in the following $\lambda$ denotes the Lebesgue measure on $\mathbb{R}$ if, for any pair $I$, $J$ of bounded Borel subsets of $\mathbb{R}$, condition (4) is satisfied with $c$ a strictly positive constant.

*Remark* The identity

$$\tau(|E_A(I)E_B(J)|^2) = \tau(E_B(J)E_A(I)E_B(J)) = \tau(E_A(I)E_B(J)) \tag{5}$$

implies that $E_Q(I)E_P(J)$ is trace class if and only if it is Hilbert-Schmidt and in this case its trace coincides with its Hilbert–Schmidt norm.

The above definition of complementarity was given in [1] and one of the problems posed there (Problem (I.f)) was

**Problem (I.f)**. Classify the complementary pairs in $\mathcal{B}(\mathcal{H})$ (or more generally, in a semi-finite von Neumann algebra) up to the natural unitary equivalence.

Later developments (see section 2 and the following ones) have shown that Definition 3 has to be extended in order to include all examples available now. Furthermore, these examples show that there is more than one **natural equivalence relation** among complementary pairs. To explain this issue it is convenient to give a third equivalent formulation of the relation (4) introducing the abelian von Neumann sub-algebras $\mathscr{A}_A$, $\mathscr{A}_B$ of $\mathscr{B}(\mathscr{H})$ generated by the spectral projections of $A$, $B$ respectively. The condition of non-degeneracy of the spectrum becomes in this case the requirement that the algebras $\mathscr{A}_A$ and $\mathscr{A}_B$ are maximal abelian sub-algebras of $\mathscr{A}$. The integral with respect to $\lambda$, defines a weight on $L^\infty(\mathbb{R}, \lambda)$ which is finite for $f \in L^\infty \cap L^1(\mathbb{R}, \lambda)$.

Using the same symbol for a measure and the associated integral, the identities

$$\lambda_A(f(A)) := \lambda(f) =: \lambda_B(f(B))$$

show that $\lambda$ induces weights $\lambda_A$, $\lambda_B$ respectively on $\mathscr{A}_A$ and $\mathscr{A}_B$.

With these notations (4) can be re-written in the equivalent form

$$\tau(f(A) \cdot g(B)) = c\lambda_A(f(A))\lambda_B(g(B)) = c\lambda(f) \cdot \lambda(g)\,;\; f, g \in L^\infty \cap L^1(\mathbb{R}) \quad (6)$$

The identity (6) suggests a natural way to construct new complementary pairs out of a given one. It is natural to consider *equivalent* two pairs obtained one from the other from this construction. As already mentioned in the infinite-dimensional case, there are several variants of this definition that can also be considered, but we will not discuss them here.

**Definition 4** Two complementary pairs $(A, B)$, $(A', B')$ of self-adjoint operators on $\mathscr{H}$, affiliated to $\mathscr{A}$ are said to be **equivalent** if there exist automorphisms $u_A$, $u_B$ of $\mathscr{A}$ satisfying

$$u_A(\mathscr{A}_A) = \mathscr{A}_A \quad ; \quad u_B(\mathscr{A}_B) = \mathscr{A}_B \quad (7)$$

$$\lambda_B \circ u_A = \lambda_A \quad ; \quad \lambda_B \circ u_B = \lambda_B \quad (8)$$

$$A' = u_A(A) \quad ; \quad B' = u_B(B) \quad (9)$$

where we use the same symbols $u_A$, $u_B$ to denote their restrictions on $\mathscr{A}_A$ and $\mathscr{A}_B$ respectively and $u_A(A)$, $u_B(B)$ are defined in terms of their spectral decompositions.

If $u_A$, $u_B$ satisfy conditions (7), (8) and $(A', B')$ are defined by (9) then, for any $f, g \in L^\infty \cap L^1(\mathbb{R})$, one has the identity

$$\tau(f(A') \cdot g(B')) = \tau(f(u_A(A)) \cdot g(u_B(B))) = \tau(u_A(f(A)) \cdot u_B(g(B)))$$

But, due to (7), $u_A(f(A))$ and $u_B(g(B))$ are still in $\mathscr{A}_A$ and $\mathscr{A}_B$ respectively. Therefore one can apply (6) and from (8) one has

$$\tau(f(A') \cdot g(B')) = c\lambda_A(u_A(f(A))\lambda_B(u_B(g(B))) = c\lambda_A(f(A)\lambda_B(g(B))) = c\lambda(f) \cdot \lambda(g)$$

i.e. $(A', B')$ is a complementary pair.

If $\mathscr{A} = \mathscr{B}(\mathscr{H})$ and $A, B$ have pure point spectrum, then by maximal abelianity $\mathscr{A}_A$ and $\mathscr{A}_B$ are generated by rank-one projectors of the form $\varphi_\alpha \varphi_\alpha^*$, $\psi_\beta \psi_\beta^*$ where $(\varphi_\alpha)$, $(\psi_\beta)$ are two orthonormal bases of $\mathscr{H}$ and

$$\psi_\beta \psi_\beta^*(\xi) := \langle \psi_\beta^*, \xi \rangle \psi_\beta \quad ; \quad \xi \in \mathscr{H}$$

In this case, $u_A$ and $u_B$ act as permutations on the indexes $\alpha$ and $\beta$ and the complementarity of $(A', B')$ is simply expression of the invariance of (2) under two separate permutations of these indexes. In the continuous case permutations are replaced by $\lambda$-preserving invertible bi-measurable maps.

Problem (I.f) above, i.e., the classification of complementary pairs up to the equivalence relation introduced in Definition 4, arises for complementary pairs a problem analogue to that solved by von Neumann on the uniqueness up to isomorphism and irreducibility of continuous representations of the CCR in bounded form for systems with finitely many degrees of freedom. The analogue of the irreducibility condition here is that the spectral projections of the complementary pair generate the whole von Neumann algebra $\mathscr{A}$. It was already clear in [1] that, contrarily to the CCR case, in finite dimensions the problem has a negative solution, i.e., there exist inequivalent complementary pairs.

In the infinite-dimensional case there is almost no literature on the equivalence problem with the exception of a paper by Cassinelli and Varadarajan [13]. As shown in section 1.2 below, the position and momentum operators on $L^2(\mathbb{R}, \lambda)$ form a complementary pair. The problem then is if there exist other complementary pairs in $L^2(\mathbb{R}, \lambda)$ inequivalent to it in the sense of Definition 4. This problem is an open challenge for experts in operator theory.

In [1] it was conjectured that the answer to this problem is negative because the requirements of Definition 4 are effectively very weak as one can see from the finite-dimensional case.

Since the Lebesgue measure is not a probability measure, in the continuous case one cannot interpret the left-hand side of (4) as a joint probability. However, following Renyi [26], for a $\sigma$-finite measure $\mu$ on $\mathbb{R} \times \mathbb{R}$ a compatible system of conditional probabilities defined by

$$P([A \in I_0] \cap [B \in J_0] | [A \in I] \cap [B \in J]) := \frac{\mu(I_0 \cap J_0)}{\mu(I \cap J)} \quad ; \quad I_0 \subseteq I \, , \, J_0 \subseteq J$$

where $I, J$ are sets of finite $\mu$-measure. With this definition (4) becomes equivalent to

$$P([A \in I_0] \cap [B \in J] | [A \in I] \cap [B \in J]) = \frac{\lambda(I_0)}{\lambda(I)} \quad ; \quad I_0 \subseteq I \, , \, \forall J \, , \, \lambda(J) < +\infty$$

and this is the analogue of the zero information condition (1) for infinite measures.

## 1.2 Complementarity of Position and Momentum

If, in the finite version of the CCR (3), one introduces the rescalings $h \mapsto h/\sqrt{n}, k \mapsto k/\sqrt{n}$ and lets $n \to \infty$, so that $h/\sqrt{n} \to t$, $k/\sqrt{n} \to s$ and moreover the equality $e^{i(h/\sqrt{n})A}e^{i(k/\sqrt{n})B} = e^{i\frac{hk}{n}}\, e^{i(k/\sqrt{n})B}\, e^{i(h/\sqrt{n})A}$ becomes

$$e^{itA}e^{isB} = e^{ist}\, e^{isB}e^{itA}$$

By von Neumann uniqueness theorem one can then identify

$$A \equiv Q \text{ Position} \quad ; \quad B \equiv P \text{ Momentum}$$

There are at least 2 ways to make these heuristic considerations rigorous: one is based on contractions of Lie algebras (see [10]) and another one on quantum central limit theorems (see [2]). Therefore we know that position and momentum operators in $L^2(\mathbb{R}, \lambda)$ are approximated by complementary pairs. Intuitively this suggests that they too should be a complementary pair. The following theorem shows that this intuition is correct:

**Theorem 1** *Position and momentum operators in $L^2(\mathbb{R}, \lambda)$ are a complementary pair with constant $c = 1/(2\pi)$.*

*Proof* Let $Q$, $P$ denote position and momentum operators in $L^2(\mathbb{R}, \lambda)$. Denote $E_P(B)$ (resp. $E_Q(B)$) their spectral projection on the Borel set $B$ and $\mathscr{F}$ Fourier transform

$$\hat{f}(k) := \mathscr{F}f(k) := \frac{1}{\sqrt{2\pi}}\int_{\mathbb{R}} e^{-ikx} f(x)dx$$

Then for $\psi$ in the space $\mathscr{S}(\mathbb{R})$

$$E_P(B)\psi(p) = \mathscr{F}\mathscr{F}^{-1}E_Q(B)\psi(p) = \mathscr{F}E_Q(B)\mathscr{F}^{-1}\psi(p)$$

$$= \frac{1}{\sqrt{2\pi}}\int e^{-ipq}[E_Q(B)\mathscr{F}^{-1}\psi](q)dq = \frac{1}{2\pi}\int e^{-ipq}dq\ \chi_B(q)\int e^{ip'q}\psi(p')dp'$$

$$= \frac{1}{2\pi}\int dq\int dp' e^{-i(p-p')q}\chi_B(q)\psi(p') = \frac{1}{\sqrt{2\pi}}\int \psi(p')dp'\frac{1}{\sqrt{2\pi}}\int dq e^{-i(p-p')q}\chi_B(q)$$

$$= \frac{1}{\sqrt{2\pi}}\int dp'\psi(p')\hat{\chi}_B(p-p')$$

Therefore

$$E_Q(I)E_P(J)\psi(q) = \chi_I(q)\frac{1}{\sqrt{2\pi}}\int dq'\psi(q')\hat{\chi}_J(q-q') = \frac{1}{\sqrt{2\pi}}\int dq'\left(\chi_I(q)\hat{\chi}_J(q-q')\right)\psi(q')$$

so that $E_Q(I)E_P(J)$ has an integral kernel given by

$$E_Q(I)E_P(J)(q,q') = \frac{1}{\sqrt{2\pi}}\chi_I(q)\hat{\chi}_J(q-q')$$

For bounded Borel sets $I$, $J$ and with $\|\cdot\|$ denoting the $L^2$-norm, one has

$$\int\int dqdq'|E_Q(I)E_P(J)(q,q')|^2 = \int\int dqdq'\frac{1}{2\pi}\chi_I(q)|\hat{\chi}_J(q-q')|^2$$
$$= \frac{1}{2\pi}\int dq\chi_I(q)\int dq'|\hat{\chi}_J(q-q')|^2 = \frac{1}{2\pi}\int dq\chi_I(q)\|\hat{\chi}_J\|^2 = \frac{1}{2\pi}|I|\,\|\chi_J\|^2 = \frac{1}{2\pi}|I|\,|J| < +\infty$$

Having a square-integrable kernel, $E_Q(I)E_P(J)$ is an Hilbert–Schmidt operator and its Hilbert–Schmidt norm coincides with the $L^2$-norm of its kernel (see [18], Chap. V, sect. 2.4). Finally from (5) we conclude that $E_Q(I)E_P(J)$ is trace class and that its trace coincides with the its Hilbert–Schmidt norm. □

*Remark* Theorem 1 and the Stone–von Neumann theorem imply that, also in the infinite dimensional case, any pair $A$, $B$ of self-adjoint operators satisfying the CCR

$$e^{ihA}e^{ikB} = e^{ihk}e^{ikB}e^{ihA} \quad ; \quad h,k\in\mathbb{R} \tag{10}$$

and such that the maps $t\mapsto e^{itA}$, $t\mapsto e^{itB}$ are continuous in the weak operator topology, is a complementary pair.

A less computational and quicker proof of Theorem 1, which however does not give the explicit form of the constant $c$, is the following.

Using the above-mentioned result that, for bounded Borel sets $I, J$ of $\mathbb{R}^d$, $E_Q(I)E_P(J)$ is trace class, and the properties of the trace on $\mathscr{B}(L^2(\mathbb{R}^d))$ one verifies that the map

$$I\times J\subset\mathbb{R}^{2d}\mapsto\tau(E_Q(I)E_P(J)) \tag{11}$$

defines a positive $\sigma$-finite countably additive measure on $\mathbb{R}^{2d}$ finite on bounded Borel sets. Similarly, for $N\in\mathbb{N}$ and $Q$, $P$ replacing $A$, $B$ in (3), one has the measure

$$I\times J\subset\{1,\ldots,N\}\times\{1,\ldots,N\}\mapsto\tau(E_Q(I)E_P(J))$$

From the CCR (10) (resp. (3)) one deduces that, with addition interpreted as addition modulo $N$ in the $N$-dimensional case,

$$\tau(E_Q(I+s)E_P(J+t)) = \tau(e^{-isp}E_Q(I)e^{isp}e^{itq}E_P(J)e^{-itq}) \tag{12}$$
$$= \tau(E_Q(I)e^{isp}e^{itq}E_P(J)(e^{isp}e^{itq})^*) = \tau(E_Q(I)e^{itq}e^{isp}E_P(J)(e^{itq}e^{isp})^*)$$
$$= \tau(e^{-itq}E_Q(I)e^{itq}e^{isp}E_P(J)e^{-isp}) = \tau(E_Q(I)E_P(J))$$

Therefore the measure (11) is translation invariant and it is known that any such a measure is a multiple of the Lebesgue measure in the infinite-dimensional case, or a multiple of the uniform measure in the $N$-dimensional case, i.e., for some $c > 0$ one must have

$$\tau(E_Q(I)E_P(J)) = c\lambda(I)\lambda(J)$$

where $\lambda$ is the Lebesgue measure in the infinite-dimensional case and the counting measure in the $N$-dimensional case.

## 1.3 Further Developments of the Theory of Complementary Pairs

After 1982 a large literature has accumulated on complementary pairs.

Popa [25] studied complementary pairs in finite von Neumann algebras and used them to solve a problem posed by Kadison. Busch and Lahti [12] simplified the proof of Theorem 1 (their proof is used here).

Kraus [19] adopted Definition 2 of complementarity and begun a systematic study of the equivalence problem in the finite-dimensional case. He discussed, always in the finite-dimensional case, entropic uncertainty principles, which can be described as follows. In addition to the Boltzmann entropy, many numerical indexes associated to probability measures on $\mathbb{R}$ have been introduced and often the name **entropy** is used also for them. If such an index $S$ is calculated on the probability distribution of an observable $A$ in a state $\varphi$, the resulting number is denoted $S_{\varphi}(A)$. An entropic uncertainty principle for two observables $(A, B)$ is a lower bound on the sum $S_{\varphi}(A) + S_{\varphi}(B)$. Kraus conjectured an optimal lower bound in the case when $S$ is the Boltzmann entropy and explicitly calculated it for complementary pairs. This conjecture was proved by Maassen and Uffink [21] for a large family of entropic-type indexes including the original Boltzmann entropy. An excellent survey on complementary pairs and entropic uncertainty principles is in Ohya and Petz's monograph [22] which also contains additional bibliography.

The use of complementarity in state estimation was discussed by Parthasarathy in [23] and Petz and Ruppert in [24]. Wootters and Fields [30] discovered that complementary bases in the sense of Definition 2 play an important role in quantum information and re-baptized them *mutually unbiased bases*, a term now widely used in quantum information where the associated unitary matrices are called *complex Hadamard matrices* in view of the fact that Hadamard used some matrices with entries $-1, 0, 1$ whose columns satisfy condition (2) up to normalization (see [28] for a survey). The fact that the eigenvectors of a pair of Hermitean matrices $A$, $B$ satisfying the finite version of the CCR (3) also satisfies (2) was noted by Schwinger [27] without mentioning neither complementarity nor probabilistic or information theoretical implications, nor the continuous case.

In connection with state estimation, Ivonovic [16] extended Schwinger's result proving that in $\mathbb{C}^N$ one can construct $N + 1$ ortho-normal bases any pair of which satisfies (2). Considering, for each of these bases, the algebra generated by the rank-one projectors on its vectors, one obtains $N + 1$ sub-algebras of the algebra $M_N(\mathbb{C})$ of $N \times N$ complex matrices that are 2-complementary, with respect to the non-normalized trace, in the sense of Definition 5 in section 2. The problem of finding the maximal number of sub-algebras of $M_N(\mathbb{C})$ with this property was studied by Björck [11]. Haagerup [15] determined this number for the subclass of circulant Hadamard matrices (i.e., those whose $(i, j)$-entry depends on $i - j$) in the case $N = 2, 3, 5, 7$ and gave an upper estimate for general prime $N$.

In the infinite-dimensional case, with the exception of the above mentioned paper of Cassinelli and Varadarajan [13] and Chapter 16 in the monograph of Ohya and Petz, there is essentially nothing on the equivalence problem.

## 2 $n$-Complementarity

Definition 5 in the present section includes all notions of complementarity discussed up to now and extends them in three directions

- inclusion of the infinite-dimensional case,
- factorization of expectation values of $n$-tuples rather than pairs,
- consideration of the above two properties for non-tracial states.

The motivation of this extension comes both from the theory of orthogonal polynomials and from quantum probability. In fact it is clear from (6) that complementarity is a weak form of stochastic independence, more precisely tensor independence, but it differs from it in two important items. First, the weights appearing in the right-hand side are not the restrictions of the weight on the left-hand side on the two sub-algebras (which, in the case of (6) are identically infinite). Second, the factorization of the expectation value in the left-hand side takes place only for pairs and not for general $n$-tuples as required by tensor independence. This creates a hybrid situation in which some expectations values factorize according to the rule of tensor independence, but others may factorize according to the rule of other notions of independence. The present paper is a first step in the investigation of this phenomenon, not known up to now in quantum probability.

It is easy to construct examples of tensor independence, just by taking tensor products of algebras and of states. So the problem is to construct examples of algebras and states which satisfy the complementarity condition, or its extension given in Definition 5 below, but are not tensor independent.

The notion of Interacting Fock Space (IFS) (see section 3 below) was motivated by the stochastic limit of quantum electro-dynamics without dipole approximation and it was immediately clear that it provided a natural tool to construct examples of stochastic independence (which are in fact forms of stochastic dependence easily coded in some algebraic rules, but difficult to read from the joint expectations, see [9] for a discussion of this issue).

In view of the connections between independence and complementarity, it is natural to ask oneself if the framework of IFS can also be used to produce new nontrivial examples of complementary pairs. We will see in the following that this is indeed the case.

The theory of IFS (see section 3) includes the theory of orthogonal polynomials, and in fact is identified with it in the case of polynomials in one variable (see [3] for bibliography and more details). One of the consequences of this inclusion is the notion of **quantum decomposition of a classical random field** (or random variable) with all moments which allows to associate, in a canonical way, to every classical random field a non-commutative algebra of operators acting on an IFS. This allows to extend classical random fields to some typically quantum notions such as complementarity.

Experience coming from the stochastic limit of quantum theory, where this structure was first deduced, shows that in physics nontrivial (i.e., non-tensor product type) IFS arise naturally in connection with **nonlinear interactions**. This has been verified in a number of models such as: particles interacting with a Boson field (QED, polaron model, ...), Anderson model, tri-linear field–field interaction, Stochastic bosonization of Fermi system in dimensions $\geq 3, \ldots$. These models are described in Part II of the monograph [8] (see also [6, 7, 9, 20]).

The connection between IFS and nonlinear interactions is confirmed by the fact that the assignment of a family of orthogonal polynomials, in any even infinite dimensions, is equivalent to the assignment of a special class of symmetric IFS (characterized in [3]) and the property that the associated IFS is of tensor product type (like usual Bose or Fermi fields) **characterizes product measures** [5].

Motivated by the examples discussed in the previous sections and by those to be discussed in the following we introduce the following definition.

**Definition 5** Let $\mathscr{A}$ be a $*$-algebra, in the following, unless otherwise specified, and let $(\mathscr{A}_j)_{j\in I}$ ($I$ a set) be a family of $*$-sub-algebras of $\mathscr{A}$. Given a weight $\varphi$ on $\mathscr{A}$ and $n \in \mathbb{N}$, the family $(\mathscr{A}_j)_{j\in I}$ of $*$-sub-algebras of $\mathscr{A}$ is called $n$**–complementary** with respect to $\varphi$, or $\varphi$-$n$–complementary, if there exist weights $\varphi_j$ on $\mathscr{A}_j$ ($j \in I$) such that, for any $k \in \{1, 2, \ldots, n\}$ and any **injective** function $j : h \in \{1, \ldots, k\} \to j_h \in I$ one has:

$$\varphi\left(Y_{j_1} \cdots Y_{j_k}\right) = \prod_{h=1}^{k} \varphi_{j_h}\left(Y_{j_h}\right) \quad ; \quad Y_{j_h} \in \mathscr{A}_{j_h} \, , \; h \in \{1, \cdots, k\}$$

in the sense that, if either side of the identity exists and is finite, the same is true for the other one and the identity holds.

2–complementarity is simply called **complementarity**.

If the family $(\mathscr{A}_j)_{j\in I}$ is $\varphi$–$n$–complementary for any $n \in \mathbb{N}$ with $n \leq |I|$ (cardinality of $I$), it is called **completely complementary**.

*Remark* In Definition 12 below we introduce the notion of **strong complementarity** which includes the additional condition that

$$\varphi_j := \varphi\Big|_{\mathscr{A}_j} \quad , \quad j \in I \tag{13}$$

where $\Big|$ denotes restriction. This condition allows to relate in a natural way the examples discussed in section 3 below with the basic original example of position and momentum discussed in section 1.2. In fact, in a general Interacting Fock Space (IFS) one deals with creation and annihilation operators $a^{\pm}$ rather than position and momentum. In any IFS one can define a **generalized position operator** using the standard relation between field operators and creation/annihilation operators valid in usual quantum mechanics, i. e. $q := (a^{+} + a)/\sqrt{2}$ and a **generalized momentum operator** $p := (a^{+} - a)/i\sqrt{2}$. In IFS coming from orthogonal polynomials and for vacuum–symmetric random variables (i.e., with vanishing odd vacuum moments) one can prove that the generalized position operator coincides with the usual multiplication operator and that the commutation relations between generalized position and momentum operators extend in a natural way the Heisenberg commutation relations between position and momentum. Since condition (13) involves algebras rather than pairs of self–adjoint operators, one can exploit the linear relation expressing generalized position and momentum operators as linear combinations of creation/annihilation operators and conclude that, if the algebras are $\varphi$-strongly complementary then the corresponding generalized position and momentum operators satisfy condition (6) with $\tau$, $\lambda_A$, $\lambda_B$ replaced by $\varphi$. In this case $\varphi$ is a state, hence the normalization condition

$$\varphi(1) = c\varphi(1)\varphi(1) \iff c = 1$$

uniquely fixes the constant $c$ to be 1. Note that the analogue condition for usual position and momentum operators, with $\tau = Tr$ and $\lambda_A = \lambda_B =$Lebesgue measure, would be difficult to formulate because creation/annihilation operators are not self-adjoint and to our knowledge for them there is no analogue of the trace formula of Theorem 1.

# 3 Interacting Fock Spaces

## *3.1 Tensor Algebras Over Vector Spaces*

Let $H_1$ be a vector space (always complex, unless otherwise specified) and, for each $n \in \mathbb{N}$, denote $H_1^{\otimes n}$ (resp. $H_1^{\widehat{\otimes} n}$) the vector space algebraic tensor product (resp. symmetric tensor product) of $n$ copies of $H_1$.

The tensor algebra (resp. symmetric tensor algebra) over $H_1$ is defined by

$$\text{Tens}(H_1) := \dot{\sum}_{n\in\mathbb{N}} H_1^{\otimes n} \quad ; \quad \text{Tens}_{sym}(H_1) := \dot{\sum}_{n\in\mathbb{N}} H_1^{\widehat{\otimes} n} \tag{14}$$

where $\dot{\sum}$ denotes vector space direct sum and, in the right hand sides of (14), by definition

$$H_1^{\otimes 0} = H_1^{\widehat{\otimes} 0} := \mathbb{C} \cdot \Phi \tag{15}$$

where $\Phi$ is a norm-one vector. The spaces $\left(H_n\right)_{n \in \mathbb{N}}$ are called the $n$**–particle spaces**, if $n = 0$ one speaks of the vacuum space.
If $K_1$ is another vector space, a **linear morphism** from $\text{Tens}(H_1)$ to $\text{Tens}(K_1)$ is a linear map $A : \text{Tens}(H_1) \to \text{Tens}(K_1)$ preserving the vector space gradation such that

$$A\Phi_{H_1} = \Phi_{K_1}$$

## 3.2 *Actions of $H_1$ and $H_1^*$ on $Tens(H_1)$ and $Tens_{sym}(H_1)$*

For a general vector space $K$, we denote $\mathscr{L}(K)$ the algebra of linear operators of $K$ into itself. For $n \in \mathbb{N}$, $f, g_n, \ldots, g_1 \in H_1$, the **creation operator** $\ell_f^* : \text{Tens}(H_1) \to \text{Tens}(H_1)$ is algebraically defined

$$\ell_f^*(g_n \otimes \cdots \otimes g_2 \otimes g_1) := f \otimes g_n \otimes \cdots g_2 \otimes g_1 \tag{16}$$

Similarly one defines the **symmetric creation operators** $B_f^+ : \text{Tens}_{sym}(H_1) \to \text{Tens}_{sym}(H_1)$ by

$$B_f^+(g_n \widehat{\otimes} \cdots \widehat{\otimes} g_2 \widehat{\otimes} g_1) := f \widehat{\otimes} g_n \widehat{\otimes} \cdots \widehat{\otimes} g_2 \widehat{\otimes} g_1 \tag{17}$$

Both maps $f \mapsto \ell_f^*, B_f^+$ are linear.

By polarization, $\text{Tens}_{sym}(H_1)$ is linearly generated by the vectors of the form $g^{\otimes n}$. On these vectors $B_f^+$ takes the form

$$B_f^+ g^{\otimes n} := f \hat{\otimes} g^{\widehat{\otimes} n} := \frac{1}{n+1} \sum_{k=0}^{n} g^{\otimes k} \otimes f \otimes g^{\otimes (n-k)} \tag{18}$$

Denote $\langle H_1^*, H_1 \rangle$ any duality between $H_1$ and its algebraic dual $H_1^*$.

For $n \in \mathbb{N}$, $g_n, \ldots, g_1 \in H_1$ and $f^* \in H_1^*$, the algebraically defined operator $\ell_{f^*} : \text{Tens}(H_1) \to \text{Tens}(H_1)$

$$\ell_{f^*}(g_n \otimes \cdots \otimes g_2 \otimes g_1) := \langle f^*, g_n \rangle g_{n-1} \otimes \cdots g_2 \otimes g_1 \quad ; \quad \ell_{f^*}\Phi := 0 \tag{19}$$

is called **full annihilation operators** (or full annihilator) and the algebraically defined operator $B_{f^*} : \text{Tens}_{sym}(H_1) \to \text{Tens}_{sym}(H_1)$

$$B_{f^*}(g_n \hat{\otimes} \cdots \hat{\otimes} g_2 \hat{\otimes} g_1) := \sum_{k=1}^{n} \langle f^*, g_k \rangle g_n \hat{\otimes} \cdots \hat{\otimes} g_{k+1} \hat{\otimes} g_{k-1} \cdots \hat{\otimes} g_1 \; ; \; B_{f^*}\Phi := 0 \tag{20}$$

This implies

$$B_{f^*}h^{\otimes(n+1)} = (n+1)\langle f^*, h\rangle h^{\otimes n} \quad ; \quad h^{\otimes 0} := \Phi\ ,\ B_{f^*}\Phi = 0 \tag{21}$$

is called **annihilation operator** (or annihilator). Whenever it is given an anti-linear embedding $f \in H_1 \mapsto f^* \in H_1^*$, for example induced by a semi-scalar product on $H_1$ (see, e.g., [14]) we will write $\ell_f$ for $\ell_{f^*}$ and $B_f$ for $B_{f^*}$. With these notations the annihilation operators are anti-linear in $f \in H_1$.

The **number operator** $\Lambda$ is also algebraically defined on both $\mathrm{Tens}(H_1)$ and $\mathrm{Tens}_{sym}(H_1)$ by the following prescription: for any function $F : \mathbb{N} \to \mathbb{C}$

$$F(\Lambda)\Big|_{H_1^{\otimes n}} = F(\Lambda)\Big|_{H_1^{\widehat{\otimes} n}} := \text{multiplication by } F(n) \tag{22}$$

## 3.3 Commutations Relations

In this section, we suppose that it is given an anti-linear embedding $f \in H_1 \mapsto f^* \in H_1^*$ and we use the notations $\ell_f$ and $B_f$ for the annihilation operators.

For $A \in \mathscr{L}(H_1)$ denote

$$\Gamma_1(A) : \mathrm{Tens}(H_1) \to \mathrm{Tens}(H_1)$$

the linear operator defined by linear extension of

$$\Gamma_1(A)(g_n \otimes \cdots \otimes g_2 \otimes g_1) := (Ag_n) \otimes \cdots \otimes g_2 \otimes g_1 \tag{23}$$

Similarly denote

$$\Gamma(A) : \mathrm{Tens}(H_1) \to \mathrm{Tens}(H_1)$$

the linear operator defined by linear extension of

$$\Gamma(A)g^{\otimes n} := (Ag)^{\otimes n} \tag{24}$$

One can prove that $\Gamma(A)$ is well defined. From (19), (21), (16), (17) it follows that, for $n \in \mathbb{N}$, $g, g_n, \ldots, g_1 \in H_1$ and $f^* \in H_1^*$, one has

$$\ell_f \ell_g^*(g_n \otimes \cdots \otimes g_2 \otimes g_1) = \ell_f(g \otimes g_n \otimes \cdots \otimes g_2 \otimes g_1) = \langle f^*, g\rangle g_n \otimes \cdots \otimes g_2 \otimes g_1$$

$$\ell_g^* \ell_f(g_n \otimes \cdots \otimes g_2 \otimes g_1) = \langle f^*, g_n\rangle \ell_g^* g_{n-1} \otimes \cdots \otimes g_2 \otimes g_1 = \langle f^*, g_n\rangle (g \otimes g_{n-1} \otimes \cdots \otimes g_2 \otimes g_1)$$

$$= ((gf^*)g_n \otimes g_{n-1} \otimes \cdots \otimes g_2 \otimes g_1) = \Gamma_1(gf^*)(g_n \otimes g_{n-1} \otimes \cdots \otimes g_2 \otimes g_1)$$

where $\Gamma_1$ is defined by (23) and

$$g \otimes f^* \in H_1 \otimes H_1^* \mapsto gf^* \in \mathscr{L}(H_1) := \{\text{linear operators on } H_1\}$$

is the natural embedding defined by

$$gf^*(h) := \langle f^*, h \rangle g$$

In conclusion in the full case one has the following multiplication table:

$$\ell_f \ell_g^* = \langle f^*, g \rangle \cdot 1 \quad ; \quad \ell_g^* \ell_f = \Gamma_1(gf^*) \tag{25}$$

In the symmetric case (20), i.e.,

$$B_f(g_n \hat{\otimes} \cdots \hat{\otimes} g_2 \hat{\otimes} g_1) := \sum_{k=1}^{n} \langle f^*, g_k \rangle g_n \hat{\otimes} \cdots \hat{\otimes} \widehat{g_k} \hat{\otimes} \cdots \hat{\otimes} g_2 \hat{\otimes} g_1$$

implies that, denoting $g_{n+1} := g$ and $g_k := h$ for any $k \in \{1, \cdots, n\}$,

$$B_f B_g^+ h^{\otimes n} = B_f g_{n+1} \hat{\otimes} \cdots \hat{\otimes} g_1 = \sum_{k=1}^{n+1} \langle f^*, g_k \rangle g_{n+1} \hat{\otimes} \cdots \hat{\otimes} g_{k+1} \hat{\otimes} g_{k-1} \hat{\otimes} \cdots \hat{\otimes} g_1 \tag{26}$$

$$= \langle f^*, g \rangle h^{\otimes n} + \sum_{k=1}^{n} \langle f^*, g_k \rangle g_n \hat{\otimes} \cdots \hat{\otimes} g_{k+1} \hat{\otimes} g_{k-1} \hat{\otimes} \cdots \hat{\otimes} g_1 = \langle f^*, g \rangle h^{\otimes n} + \langle f^*, h \rangle \sum_{k=1}^{n} g \hat{\otimes} h^{\otimes n}$$

$$= \langle f^*, g \rangle h^{\otimes n} + n \langle f^*, h \rangle g \hat{\otimes} h^{\otimes n}$$

Similarly

$$B_g^+ B_f h^{\otimes n} = n \langle f^*, h \rangle B_g^+ h^{\otimes (n-1)} = n \langle f^*, h \rangle g \hat{\otimes} h^{\otimes k}$$

So that

$$B_f B_g^+ - B_g^+ B_f = \langle f^*, g \rangle \tag{27}$$

## 3.4 *Interacting Fock Spaces Over a Semi-Hilbert Space*

We choose the framework of semi-Hilbert spaces rather than the more usual one of Hilbert spaces because it is useful to have a common space for all the operators involved. This choice is familiar in the theory of orthogonal polynomials where it is used to compare operators associated to mutually singular measures. (see [3] for a discussion of this choice).

**Definition 6** A semi-scalar product $\langle \cdot , \cdot \rangle$ on a vector space $H_1$ is a positive sesquilinear form on $H_1$. This is equivalent to give an anti-linear embedding

$$f \in H_1 \to f^* \in H_1^*$$

satisfying

$$\langle f^*, f \rangle \geq 0 \quad ; \quad \forall f \in H_1$$

where $\langle \cdot , \cdot \rangle$ is the duality between $H_1$ and its algebraic dual $H_1^*$ (see (19)).

A semi-Hilbert space is a pair $(H_1, \langle \cdot , \cdot \rangle)$ where $H_1$ is a vector space and $\langle \cdot , \cdot \rangle$ semi-scalar product on $H_1$. A zero-norm vector is a vector $f \in (H_1, \langle \cdot , \cdot \rangle)$ such that $\|f\| = 0$ (notice that $\| \cdot \|$ is a semi-norm). These vectors are orthogonal to $H_1$. If $(K_1, \langle \cdot , \cdot \rangle_1)$ is another semi-Hilbert space, a semi-Hilbert space linear operator

$$A : (H_1, \langle \cdot , \cdot \rangle) \to (K_1, \langle \cdot , \cdot \rangle_1)$$

is a linear operator mapping zero norm vectors into zero norm vectors. Such an operator is called **adjointable** if there exists another linear operator

$$A^* : (K_1, \langle \cdot , \cdot \rangle_1) \to (H_1, \langle \cdot , \cdot \rangle)$$

satisfying

$$\langle A\xi, \eta \rangle_1 = \langle \xi, A^*\eta \rangle_1 \quad ; \quad \forall \xi \in H_1 \ , \ \forall \eta \in K_1$$

A **semi-Hilbert space isometry** is an adjointable operator preserving the scalar products. Semi-Hilbert space direct sums are defined as orthogonal sums without completion.

*Remark* Unless otherwise specified we will always suppose in the following that the topology induced by a semi-scalar product is separable.

**Definition 7** Let $H_1$ be a vector space. An **interacting Fock space over** $H_1$ is defined by

(i) A semi-scalar product $\langle \cdot , \cdot \rangle$ on $\mathrm{Tens}(H_1)$ such that the vector space direct sums in (14) become semi-Hilbert space orthogonal sums.
(ii) $\Phi$, defined by (15) is a unit vector in $(\mathrm{Tens}(H_1) \ , \ \langle \cdot , \cdot \rangle)$.
(iii) For each $f \in H_1$, the creation operator $\ell_f^*$, defined by (16), has an adjoint $a_f$ with respect to the semi-scalar product $\langle \cdot , \cdot \rangle$ on $\mathrm{Tens}(H_1)$ called the **annihilation operator on** $(\mathrm{Tens}(H_1) \ , \ \langle \cdot , \cdot \rangle)$ **with test function** $f$ .

If $K_1$ is another vector space, a **morphism**

$$V : (\mathrm{Tens}(H_1) \ , \ \langle \cdot , \cdot \rangle) \to (\mathrm{Tens}(K_1) \ , \ \langle \cdot , \cdot \rangle_1)$$

is a linear isometry preserving the vector space gradation and such that $V\Phi = \Phi_1$.

Replacing everywhere in the above definition the tensor products by symmetric tensor products, $\mathrm{Tens}(H_1)$ becomes $\mathrm{Tens}_{sym}(H_1)$ and, with the definition of creator given by (17), one obtains the definition of **symmetric interacting Fock space over** $H_1$, $\big(\mathrm{Tens}_{sym}(H_1)\ ,\ \langle\,\cdot\ ,\ \cdot\,\rangle_{sym}\big)$, whose creator (resp. annihilator) is denoted $B^+_f$ (resp. $B_f$)

**Notations**. When, for each $n \in \mathbb{N}$, the restriction of the semi-scalar product on $(\mathrm{Tens}(H_1)\ ,\ \langle\,\cdot\ ,\ \cdot\,\rangle)$ (resp. $\big(\mathrm{Tens}_{sym}(H_1)\ ,\ \langle\,\cdot\ ,\ \cdot\,\rangle_{sym}\big)$) coincides with the $n$-th tensor power (resp. symmetric $n$-th tensor power) of the semi-scalar product $\langle\,\cdot\ ,\ \cdot\,\rangle_1$, for the pair $(\ell^*_f, a_f)$ (resp. $(B^+_f, a_f)$) we use the notation $(\ell^*_f, \ell_f)$ (resp. $(B^+_f, B_f)$). Otherwise we use the notation $(a^+_f, a_f)$ in both cases. This is motivated by the fact that many interesting IFS arise as *deformations* of the two special cases described above. Finally, vacuum expectation values will be denoted indifferently

$$\varphi(\,\cdot\,) \equiv \langle\Phi, (\,\cdot\,)\Phi\rangle \equiv \langle(\,\cdot\,)\rangle$$

*Remark* There is a more intrinsic definition of IFS (resp. symmetric IFS) from which the concrete representations of Definition 7 can be deduced (see [3]).

### 3.5 1-*Mode Type IFS*

1-mode type IFS were introduced in [4]. Here we recall their definition and some basic properties.

**Definition 8** Let $H_1$ be a vector space. Fix a principal Jacobi sequence $(\omega_n)$, i.e., a sequence satisfying

$$\omega_0 = 0 \quad ; \quad \omega_n = 0 \Rightarrow \omega_{n+k} = 0 \qquad ; \qquad \forall k \in \mathbb{N}^*$$

Define the sequence $\lambda \equiv (\lambda_n)$ by:

$$\lambda_n := \omega_n! := \omega_1\omega_2\cdots\omega_n$$

The 1-**mode type IFS** (1-MTIFS) over $H_1$ **with Jacobi sequence** $\omega$, denoted $\Gamma_{\omega,full}(H_1)$, is defined as the IFS with creator defined by

$$a^+_f := \ell^*_f \tag{28}$$

where $\ell^*_f$ is given by (16) and semi-scalar product uniquely determined by the conditions

$$\langle f_n\otimes\cdots\otimes f_1, g_n\otimes\cdots\otimes g_1\rangle_n := \lambda_n\langle f_n\otimes\cdots\otimes f_1, g_n\otimes\cdots\otimes g_1\rangle_{\otimes n} \tag{29}$$

$$= \lambda_n \prod_{j=1}^{n} \langle f_j, g_j \rangle_{H_1} \quad ; \quad \forall n \in \mathbb{N}$$

**Definition 9** The **full Fock space over** $H_1$, denoted $\Gamma_{full}(H_1)$, is the 1-mode type IFS over $H_1$ with principal Jacobi sequence

$$\omega_n := 1 \quad ; \quad \forall n \in \mathbb{N}$$

In this section we prove that the structure of a general 1-MTIFS $\Gamma_{\omega,full}(H_1)$ is obtained as a deformation of the structure of $\Gamma_{full}(H_1)$. Recall that the creation (annihilation) operators on $\Gamma_{full}(H_1)$ are denoted $\ell_f^*$, $\ell_f$ and those on $\Gamma_{\omega,full}(H_1)$, $a_f^{\pm}$.

From (28) and (29) one deduces the action of the annihilator

$$\langle a_f(f_n \otimes \cdots \otimes f_2 \otimes f_1), g_{n-1} \otimes \cdots \otimes g_1 \rangle_{n-1} = \langle f_n \otimes \cdots \otimes f_2 \otimes f_1, a_f^+(g_{n-1} \otimes \cdots \otimes g_1) \rangle_n$$

$$= \lambda_n \langle f_n \otimes \cdots \otimes f_1, g_n \otimes \cdots \otimes g_1 \rangle_{\otimes n} = \lambda_n \langle f_n, f \rangle \prod_{j=1}^{n-1} \langle f_j, g_j \rangle_{H_1}$$

$$= \frac{\lambda_n}{\lambda_{n-1}} \langle f_n, f \rangle \left( \lambda_{n-1} \langle f_{n-1} \otimes \cdots \otimes f_1, g_{n-1} \otimes \cdots \otimes g_1 \rangle_{\otimes n} \right)$$

$$= \omega_n \langle f_n, f \rangle \langle f_{n-1} \otimes \cdots \otimes f_1, g_{n-1} \otimes \cdots \otimes g_1 \rangle_{n-1} = \langle \omega_n \langle f, f_n \rangle f_{n-1} \otimes \cdots \otimes f_1, g_{n-1} \otimes \cdots \otimes g_1 \rangle_{n-1}$$

and since $n \in \mathbb{N}$ and $f, f_1, \ldots, f_n, g_1, \ldots, g_{n-1} \in H_1$ are arbitrary, this is equivalent to

$$a_f(f_n \otimes \cdots \otimes f_2 \otimes f_1) = \omega_n \langle f, f_n \rangle f_{n-1} \otimes \cdots \otimes f_1 \quad ; \quad \forall n \in \mathbb{N} \tag{30}$$

Comparing (19) with (30) and recalling the notation (22) one finds

$$a_f = \ell_f \omega_\Lambda \quad ; \quad \forall f \in H_1 \tag{31}$$

Notice that

$$a_f \Phi = B_f \omega_\Lambda \Phi = 0 \tag{32}$$

and, due to (28) and (31), the operators $\ell_g \ell_f^*$ and $\ell_f^* \ell_g$ preserve the gradation built with the operators $a_f^{\pm}$.

### 3.6 1-*Mode Type Symmetric IFS*

**Definition 10** Let $H_1$ be a vector space and let $(\omega_n)$ and $(\lambda_n)$ be as in Definition 8. Fix a principal Jacobi sequence $(\omega_n)$, i.e. a sequence satisfying The 1-**symmetric 1-MTIFS** over $H_1$ **with Jacobi sequence** $\omega$, denoted $\Gamma_\omega(H_1)$, is defined as the symmetric IFS over $H_1$ with creator defined by

$$a_f^+ := \ell_f^* \tag{33}$$

where $\ell_f^*$ is given by (16) and semi-scalar product uniquely determined by the conditions

$$\langle f_n\widehat{\otimes}\cdots\widehat{\otimes}f_1, g_n\widehat{\otimes}\cdots\widehat{\otimes}g_1\rangle_n := \lambda_n\langle f_n\widehat{\otimes}\cdots\widehat{\otimes}f_1, g_n\widehat{\otimes}\cdots\widehat{\otimes}g_1\rangle_{\widehat{\otimes}n} \tag{34}$$

By polarization, in the symmetric case, (29) is equivalent to

$$\langle f^{\otimes n}, g^{\otimes n}\rangle_n := \lambda_n\langle f^{\otimes n}, g^{\otimes n}\rangle_{\widehat{\otimes}n} = \lambda_n\langle f, g\rangle_{H_1}^n \quad ; \quad \forall n \in \mathbb{N} \tag{35}$$

**Definition 11** The **Boson Fock space over** $H_1$, denoted $\Gamma(H_1)$, is the 1-mode type symmetric IFS over $H_1$ with principal Jacobi sequence

$$\omega_n := 1 \quad ; \quad \forall n \in \mathbb{N}$$

In this section we prove that the structure of a general symmetric 1-MTIFS $\Gamma_\omega(H_1)$ is obtained as a deformation of the structure of $\Gamma(H_1)$ and we show how this deformation affects the commutation relations. Recall that the creation (annihilation) operators on $\Gamma(H_1)$ are denoted $B_f^\pm$ those on $\Gamma_\omega(H_1)$, $a_f^\pm$.

With these notation, one has

$$\langle f^{\otimes n}, g^{\otimes n}\rangle_n = \langle f, g\rangle_{H_1}\langle f^{\otimes(n-1)}, g^{\otimes(n-1)}\rangle_{n-1}$$

it follows that

$$\langle f^{\otimes n}, g^{\otimes n}\rangle_n = \langle f, g\rangle\frac{\lambda_n}{\lambda_{n-1}}\left(\lambda_{n-1}\langle f, g\rangle_{H_1}^{n-1}\right) = \omega_n\langle f, g\rangle\langle f^{\otimes(n-1)}, g^{\otimes(n-1)}\rangle_{n-1}$$

so that

$$\|f^{\otimes n}\|_n^2 = \langle f^{\otimes n}, f^{\otimes n}\rangle_n = \omega_n!\langle f^{\otimes n}, f^{\otimes n}\rangle_{\otimes n} = \lambda_n\langle f, f\rangle_{H_1}^n = \lambda_n\|f^{\otimes n}n\|_{\widehat{\otimes}n}^2$$

We know that as vector spaces $\Gamma(H_1) = \Gamma_\omega(H_1) = \mathrm{Tens}_{sym}(H_1)$, and that one has

$$a_f^+ = B_f^+ \quad ; \quad \forall f \in H_1 \tag{36}$$

The adjoint of $a_f^+$ is computed as follows:

$$\langle a_f^+ g^{\otimes n}, h^{\otimes(n+1)}\rangle_{n+1} = \lambda_{n+1}\langle B_f^+ g^{\otimes n}, h^{\otimes(n+1)}\rangle_{\otimes(n+1)} = \lambda_{n+1}\langle f, h\rangle\langle g^{\otimes n}, h^{\otimes n}\rangle_{\otimes n}$$

$$= \omega_{n+1}\langle f, h\rangle\lambda_n\langle g^{\otimes n}, h^{\otimes n}\rangle_{\otimes n} = \omega_{n+1}\langle f, h\rangle\langle g^{\otimes n}, h^{\otimes n}\rangle_n = \langle g^{\otimes n}, a_f h^{\otimes(n+1)}\rangle_n$$

Therefore, recalling the definition (22) of the number operator $\Lambda$ on $\mathrm{Tens}_{sym}(H_1)$, one has

$$a_f h^{\otimes(n+1)} = \omega_{n+1}\langle f, h\rangle h^{\otimes n} = \frac{\omega_{n+1}}{n+1}(n+1)\langle f, h\rangle h^{\otimes n} = \frac{\omega_{\Lambda+1}}{\Lambda+1} B_f h^{\otimes(n+1)} \quad (37)$$

$$= B_f \frac{\omega_\Lambda}{\Lambda} h^{\otimes(n+1)}$$

Equivalently

$$a_f = B_f \frac{\omega_\Lambda}{\Lambda} \quad (38)$$

Notice that, with the convention $0/0 := 1$, one has

$$a_f \Phi = B_f \omega_\Lambda \Phi = 0 \quad (39)$$

and, due to (36) and (39), the operators $B_g B_f^+$ and $B_f^+ B_g$ preserve the gradation built with the operators $a_f^\pm$.

**Lemma 1**

$$a_f F_\Lambda = F_{\Lambda+1} a_f \quad ; \quad F_\Lambda a_f^+ = a_f^+ F_{\Lambda+1} \quad (40)$$

$$[a_f, a_g^+] = B_f\,(\omega_\Lambda - \omega_{\Lambda-1})\,B_g^+ + \langle f, g\rangle\omega_\Lambda = B_f B_g^+\,(\omega_{\Lambda+1} - \omega_\Lambda) + \langle f, g\rangle\omega_\Lambda \quad (41)$$

*Proof* For any function $F : \mathbb{N} \to \mathbb{C}$ one has

$$a_f F_\Lambda h^{\otimes n} = F_n a_f h^{\otimes n} = F_{\Lambda+1} a_f h^{\otimes n}$$

Therefore

$$a_f F_\Lambda = F_{\Lambda+1} a_f$$

which is the first of (40). Taking adjoints and denoting $\bar{F}$ the pointwise conjugate of $F$ one finds

$$\bar{F}_\Lambda a_f^+ = a_f \bar{F}_{\Lambda+1}$$

which, since $F$ is arbitrary is equivalent to the second of (40).

Finally, from (36) and (39) it follows that

$$[a_f, a_g^+] = [B_f\omega_\Lambda, B_g^+] = B_f[\omega_\Lambda, B_g^+] + [B_f, B_g^+]\omega_\Lambda \quad (42)$$

and, since

$$\omega_\Lambda B_g^+ h^{\otimes n} = \omega_\Lambda g \widehat{\otimes} h^{\otimes n} = \omega_{n+1} g \widehat{\otimes} h^{\otimes n} = \omega_{n+1} B_g^+ h^{\otimes n}$$

$$B_g^+ \omega_\Lambda h^{\otimes n} = \omega_n B_g^+ h^{\otimes n}$$

it follows that

$$[\omega_\Lambda, B_g^+]h^{\otimes n} = \omega_{n+1} B_g^+ h^{\otimes n} - \omega_n B_g^+ h^{\otimes n} = (\omega_{n+1} - \omega_n)\, B_g^+ h^{\otimes n} = (\omega_\Lambda - \omega_{\Lambda-1})\, B_g^+ h^{\otimes n}$$

i.e.

$$[\omega_\Lambda, B_g^+] = (\omega_\Lambda - \omega_{\Lambda-1})\, B_g^+ = B_g^+ \,(\omega_{\Lambda+1} - \omega_\Lambda)$$

where in the last identity we have used (40). Therefore (42) becomes

$$[a_f, a_g^+] = B_f\,(\omega_\Lambda - \omega_{\Lambda-1})\, B_g^+ + \langle f, g\rangle \omega_\Lambda = B_f B_g^+\,(\omega_{\Lambda+1} - \omega_\Lambda) + \langle f, g\rangle \omega_\Lambda$$

and this proves (41). □

## 4 Complementarity in IFS

For any semi-Hilbert space $\left(H_1\ ,\ \langle\cdot,\cdot\rangle_{H_1}\right)$ and any IFS space over $H_1$

$$\left(\Gamma_I(H_1) := \bigoplus_{n\in\mathbb{N}} \left(H_1^{\widehat{\otimes} n}\ ,\ \langle\cdot,\cdot\rangle_n\right)_{n\in\mathbb{N}}\ ,\ a^+\ ,\ \Phi\right)$$

we denote

$$\mathscr{P}(a^+, a) := \text{polynomial} * \text{-algebra in the variables } \{a_h^+, a_h\ :\ h \in H_1\}$$

$$\varphi(\,\cdot\,) := \langle \Phi, (\,\cdot\,)\Phi\rangle\Big|_{\mathscr{P}(a^+,a)} \in \mathscr{S}(\mathscr{P}(a^+, a))$$

the restriction of the vector state $\langle \Phi, (\,\cdot\,)\Phi\rangle$ on $\mathscr{P}(a^+, a)$ induced by the vacuum state in $\Gamma_I(H_1)$ and, for $f \in H_1$

$$\mathscr{P}(a_f^+, a_f) := \text{polynomial } * \text{–algebra in the variables } \{a_f^+, a_f\}$$

**Definition 12** An IFS $\Gamma_I(H_1)$ is said to enjoy the **strong complementarity property** if, for any family $(f_j)_{j\in J}$ $(J \subseteq \mathbb{N})$ of mutually orthogonal vectors, the $*$-algebras $\left(\mathscr{P}(a_{f_j}^+, a_{f_j})\right)_{j\in J}$ are strongly complementary with respect to the vacuum state $\varphi$ in the sense of Definition 5, with

$$\mathscr{A} := \mathscr{P}(a^+, a)\ \ ;\ \ \mathscr{A}_j := \mathscr{P}(a_{f_j}^+, a_{f_j})\ \ ;\ \ \varphi_j := \varphi\Big|_{\mathscr{P}(a_{f_j}^+, a_{f_j})}\ ,\ \ j \in J \qquad (43)$$

The following theorem provides a useful tool to construct examples of complementary algebras acting on IFS.

**Theorem 2** *If the IFS $\Gamma_I(H_1)$ satisfies the following two properties:*

***Property (1)***
*For any $m \in \mathbb{N}$, any $\varepsilon := (\varepsilon_n, \dots, \varepsilon_1) \in \{-1, +1\}^n$ and any $f \in H_1$, the vector*

$$a_f^{\varepsilon_m} \cdots a_f^{\varepsilon_1} \Phi$$

*belongs to*

$$lin\text{-}span\text{–}\{(a_f^+)^n \Phi \ : \ n \in \mathbb{N}\} \quad ; \quad (a_f^+)^0 := 1$$

*where 1 denotes the identity in $\mathscr{P}(a^+, a)$.*
***Property (2)***
*For any $N, n \in \mathbb{N}$, any $\varepsilon := (\varepsilon_N, \dots, \varepsilon_2, +1) \in \{-1, +1\}^N$*

$$\varphi\left(a_{g_N}^{\varepsilon_N} \cdots a_{g_1}^{\varepsilon_1} (a_f^+)^n\right) = 0 \tag{44}$$

*whenever $f, g_1, \dots, g_N \in H_1$ are such that $g_j \perp f$ for all $j$ and*

$$j \neq k \ \Rightarrow \ g_j \perp g_k$$

*Then $\Gamma_I(H_1)$ enjoys the strong complementarity property in the sense of Definition 12.*

*Proof* We have to prove that, for any $n \in \mathbb{N}$ and any $f_1, \dots, f_n \in H_1$ such that $f_j \perp f_k$ for $j \neq k$, the algebras $\mathscr{P}(a_{f_j}^+, a_{f_j})$ $(j \in \{1, \dots, n\})$ are $\varphi$–$n$–complementary in the sense of Definition 5, with $\varphi_j$ given by (43), i.e.,

$$\varphi(Y_{f_1} \cdots Y_{f_j}) = \prod_{j=1}^{n} \varphi(Y_{f_j}) \quad ; \quad \forall Y_j \in \mathscr{P}(a_j^+, a_j) \tag{45}$$

By linearity (45) is equivalent to

$$\varphi\left((a_{f_n}^{\varepsilon_{n,m_n}} \cdots a_{f_n}^{\varepsilon_{n,1}}) \cdots (a_{f_1}^{\varepsilon_{1,m_1}} \cdots a_{f_1}^{\varepsilon_{1,1}})\right) = \prod_{j=1}^{n} \varphi\left(a_{f_j}^{\varepsilon_{j,m_j}} \cdots a_{f_j}^{\varepsilon_{j,1}}\right) \tag{46}$$

where the $\varepsilon_{j,h} \in \{-1, +1\}$ are arbitrary. Re-labeling the indexes we write

$$(a_{f_n}^{\varepsilon_{n,m_n}} \cdots a_{f_n}^{\varepsilon_{n,1}}) \cdots (a_{f_2}^{\varepsilon_{2,m_2}} \cdots a_{f_2}^{\varepsilon_{2,1}}) =: a_{g_N}^{\varepsilon_N} \cdots a_{g_1}^{\varepsilon_1}$$

So that the left-hand side of (46) becomes

$$\varphi\left(a_{g_N}^{\varepsilon_N} \cdots a_{g_1}^{\varepsilon_1} (a_{f_1}^{\varepsilon_{1,m_1}} \cdots a_{f_1}^{\varepsilon_{1,1}})\right)$$

By Property (1) there exists a finite sub-set $F \subset \mathbb{N} \setminus \{0\}$ such that

$$a_{f_1}^{\varepsilon_{1,m_1}} \cdots a_{f_1}^{\varepsilon_{1,1}} \Phi = c_0(f_1) + \sum_{n \in F} c_n(f_1)(a_{f_1}^+)^n \Phi$$

for some complex constants $c_0(f_1)$, $c_n(f_1)$ $(n \in F)$. By the Fock property

$$c_0(f_1) = \varphi\left(a_{f_1}^{\varepsilon_{1,m_1}} \cdots a_{f_1}^{\varepsilon_{1,1}}\right)$$

Therefore the left-hand side of (46) can be written in the form

$$\varphi\left(a_{g_N}^{\varepsilon_N} \cdots a_{g_1}^{\varepsilon_1}\right) c_0(f_1) \; + \; \sum_{n \in F} c_n(f_1) \varphi\left(a_{g_N}^{\varepsilon_N} \cdots a_{g_1}^{\varepsilon_1} (a_{f_1}^+)^n\right)$$

Since $f_1$ and the $g_j$ satisfy the conditions of Property (2), it follows that

$$\varphi\left(a_{g_N}^{\varepsilon_N} \cdots a_{g_1}^{\varepsilon_1} (a_{f_1}^+)^n\right) = 0 \quad ; \quad \forall n \in F$$

Therefore the left-hand side of (46) is equal to

$$\varphi\left((a_{f_n}^{\varepsilon_{n,m_n}} \cdots a_{f_n}^{\varepsilon_{n,1}}) \cdots (a_{f_2}^{\varepsilon_{2,m_2}} \cdots a_{f_2}^{\varepsilon_{2,1}})\right) \varphi\left(a_{f_1}^{\varepsilon_{1,m_1}} \cdots a_{f_1}^{\varepsilon_{1,1}}\right)$$

and the thesis follows by induction. □

## 4.1 All 1MTIFS Enjoy the Complementarity Property

In this section we prove that the quantum polynomial algebras associated to 1MTIFS satisfy Properties (1) and (2) of Theorem 2. This implies that these algebras enjoy the strong complementarity property with respect to the vacuum state in the sense of Definition 12.

**Theorem 3** *Every 1-mode type IFS or symmetric 1-mode type IFS satisfies Property (1) and Property (2) of Theorem 2.*

*Proof* **Property (1)**. We have to prove that for any $m \in \mathbb{N}$, $\varepsilon := (\varepsilon_n, \ldots, \varepsilon_1) \in \{-1, +1\}^n$ and $f \in H_1$, the vector

$$a_f^{\varepsilon_m} \cdots a_f^{\varepsilon_1} \Phi \tag{47}$$

belongs to

$$\text{lin-span–}\{(a_f^+)^n \Phi \; : \; n \in \mathbb{N}\} \quad ; \quad (a_f^+)^0 := 1 \tag{48}$$

where 1 denotes the identity in $\mathscr{P}(a^+, a)$. This is clear if either $m = 1$ or $a_f^{\varepsilon_1} = a_f$ because the zero vector belongs to the space (48). So we can suppose that $\varepsilon_1 = +1$. Suppose by induction that the statement is true up to $m - 1$ and define

$$h := \max\{j \le m \;:\; \varepsilon_r = +1 \,\forall\, r \in \{1, \cdots, j\}\}$$

then $h \ge 1$ because of $\varepsilon_1 = +1$. If $h = m$, then the vector (47) is equal to $(a_f^+)^m \Phi$ which belongs to the space (48) by definition. Otherwise $\varepsilon_{h+1} = -1$ and the vector (47) has the form

$$a_f^{\varepsilon_m} \cdots a_f (a_f^+)^h \Phi \tag{49}$$

If $h = 1$ this expression becomes, both in the symmetric and the full case

$$\langle f, f\rangle \omega_1 a_f^{\varepsilon_m} \cdots a_f^{\varepsilon_3} \Phi$$

and the thesis follows by induction on $m$.

If $h > 1$, in the full case, by (30), (49) is equal to

$$\omega_m \langle f, f\rangle a_f^{\varepsilon_m} \cdots a_f^{\varepsilon_{h+2}} (a_f^+)^{h-1} \Phi$$

and again the thesis follows by induction on $m$. In the symmetric case, by (41), (49) is equal to

$$\sum_{j=0}^{h-1} a_f^{\varepsilon_m} \cdots (a_f^+)^j [a_f, a_f^+] (a_f^+)^{h-j-1} \Phi$$

$$= \sum_{j=0}^{h-1} a_f^{\varepsilon_m} \cdots (a_f^+)^j \left( B_f B_f^+ (\omega_{\Lambda+1} - \omega_\Lambda) + \langle f, f\rangle \omega_\Lambda \right) (a_f^+)^{h-j-1} \Phi$$

$$= \sum_{j=0}^{h-1} a_f^{\varepsilon_m} \cdots (a_f^+)^j B_f B_f^+ (\omega_{\Lambda+1} - \omega_\Lambda) (a_f^+)^{h-j-1} \Phi + \langle f, f\rangle \sum_{j=0}^{h-1} a_f^{\varepsilon_m} \cdots (a_f^+)^j \omega_\Lambda (a_f^+)^{h-j-1} \Phi$$

$$= \sum_{j=0}^{h-1} (\omega_{h-j} - \omega_{h-j-1}) a_f^{\varepsilon_m} \cdots (a_f^+)^j B_f B_f^+ (a_f^+)^{h-j-1} \Phi + \langle f, f\rangle \left( \sum_{j=0}^{h-1} \omega_{h-j-1} \right) a_f^{\varepsilon_m} \cdots (a_f^+)^{h-1} \Phi$$

$$= \omega_h \langle f, f\rangle a_f^{\varepsilon_m} \cdots (a_f^+)^{h-1} B_f B_f^+ \Phi + \langle f, f\rangle \left( \sum_{j=0}^{h-1} \omega_{h-j-1} \right) a_f^{\varepsilon_m} \cdots (a_f^+)^{h-1} \Phi$$

and also in this case the thesis follows by induction on $m$.

**Property (2)**. We have to prove that, for any $N, n \in \mathbb{N}$, any $\varepsilon := (\varepsilon_N, \ldots, \varepsilon_2, +1) \in \{-1, +1\}^N$

$$\varphi \left( a_{g_N}^{\varepsilon_N} \cdots a_{g_1}^{\varepsilon_1} (a_f^+)^n \right) = 0 \tag{50}$$

for any $f, g_1, \ldots, g_N \in H_1$ such that $g_j \perp f$ for all $j$ and $j \neq k \Rightarrow g_j \perp g_k$.

The identity (50) is clear if $\varepsilon_j = +1$ for any $j \in \{1, \cdots, N\}$, so one can suppose that the set of $j \in \{1, \ldots, N\}$ such that $\varepsilon_j = -1$ is non-empty. Denote

$$h := \min\{j \in \{1, \ldots, N\} \,:\, \varepsilon_j = -1\}$$

Then the right hand side of (50) has the form

$$\varphi\left(a_{g_N}^{\varepsilon_N} \cdots a_{g_{h+1}}^{\varepsilon_{h+1}} a_{g_h} a_{g_{h-1}}^+ \cdots a_{g_1}^+ (a_f^+)^n\right) \tag{51}$$

and the identity (50) is true if $h - 1 = N = 0$ by the Fock property. Otherwise in the full case, by (30), (51) is equal to

$$\begin{cases} \omega_{n+h-1}\langle g_h, g_{h-1}\rangle \varphi\left(a_{g_N}^{\varepsilon_N} \cdots a_{g_{h+1}}^{\varepsilon_{h+1}} a_{g_{h-2}}^+ \cdots a_{g_1}^+ \cdots (a_f^+)^n\right), & \text{if } h > 1 \\ \omega_n\langle g_1, f\rangle \varphi\left(a_{g_N}^{\varepsilon_N} \cdots a_{g_2}^{\varepsilon_2} (a_f^+)^{n-1}\right), & \text{if } h = 1 \end{cases}$$

and in both cases this quantity is 0 because by assumption $g_j \perp f$ for all $j$ and $g_{h-1} \perp g_h$. Thus the thesis follows by induction.

Under the same assumptions, in the symmetric case using (38), (51) becomes

$$\langle \Phi, a_{g_N}^{\varepsilon_N} \cdots a_{g_{h+1}}^{\varepsilon_{h+1}} B_{g_h} \omega_\Lambda \omega_\Lambda B_{g_{h-1}}^+ \cdots \omega_\Lambda B_{g_1}^+ (\omega_\Lambda B_f^+)^n \Phi\rangle$$

$$= \omega_{n+h-1}(\omega_{n+h-1}!)\langle \Phi, a_{g_N}^{\varepsilon_N} \cdots a_{g_{h+1}}^{\varepsilon_{h+1}} B_{g_h} B_{g_{h-1}}^+ \cdots B_{g_1}^+ (B_f^+)^n \Phi\rangle$$

and by the Fock property this is equal to

$$= \omega_{n+h-1}(\omega_{n+h-1}!)\langle \Phi, a_{g_N}^{\varepsilon_N} \cdots a_{g_{h+1}}^{\varepsilon_{h+1}} [B_{g_h}, B_{g_{h-1}}^+ \cdots B_{g_1}^+ (B_f^+)^n] \Phi\rangle$$

$$= \omega_{n+h-1}(\omega_{n+h-1}!) \sum_{j=1}^{h-1} \langle \Phi, a_{g_N}^{\varepsilon_N} \cdots a_{g_{h+1}}^{\varepsilon_{h+1}} B_{g_{h-1}}^+ \cdots B_{g_{j+1}}^+ [B_{g_h}, B_{g_j}^+] B_{g_{j-1}}^+ \cdots B_{g_1}^+ (B_f^+)^n \Phi\rangle$$

$$+ \omega_{n+h-1}(\omega_{n+h-1}!) \sum_{j=0}^{n-1} \langle \Phi, a_{g_N}^{\varepsilon_N} \cdots a_{g_{h+1}}^{\varepsilon_{h+1}} B_{g_{h-1}}^+ \cdots B_{g_1}^+ (B_f^+)^j [B_{g_h}, B_f^+] (B_f^+)^{n-j-1} \Phi\rangle$$

with the obvious conventions on the extreme values of the sums. Using (27) with the embedding $f \in H_1 \mapsto f^* \in H_1^*$ given by the semi-scalar product, this becomes

$$\omega_{n+h-1}(\omega_{n+h-1}!) \sum_{j=1}^{h-1} \langle \Phi, a_{g_N}^{\varepsilon_N} \cdots a_{g_{h+1}}^{\varepsilon_{h+1}} B_{g_{h-1}}^+ \cdots B_{g_{j+1}}^+ \langle g_h, g_j\rangle B_{g_{j-1}}^+ \cdots B_{g_1}^+ (B_f^+)^n \Phi\rangle$$

$$+ n \, \langle g_h, f\rangle \, \omega_{n+h-1}(\omega_{n+h-1}!)\langle \Phi, a_{g_N}^{\varepsilon_N} \cdots a_{g_{h+1}}^{\varepsilon_{h+1}} B_{g_{h-1}}^+ \cdots B_{g_1}^+ (B_f^+)^{n-1} \Phi\rangle$$

which is zero by the orthogonality assumption. Thus again the thesis follows by induction. □

## 4.2 *Complementarity Without Independence*

In this sub-section, we prove that in 1-MTIFS, in products with repeated test functions, the factorization property of vacuum expectation values in general does not hold.

This proves that $n$-complementarity is a strictly weaker property than independence.

Since $n$-complementarity implies complementarity for each pair of algebras involved, it is sufficient to consider the case of complementarity.

In all IFS the field operators are defined by

$$Q_f \equiv X_f = a_f^+ + a_f \in \mathscr{P}(a_f^+, a_f) \tag{52}$$

The vacuum distribution of $X_f$ defines a classical $\varphi$-symmetric random variable, i.e.,

$$\langle \Phi, X_f^{2n+1} \Phi \rangle = \langle X_f^{2n+1} \rangle = 0$$

whose variance is

$$\langle X_f^2 \rangle = \langle a_f a_f^+ \rangle = \omega_1 \| f \|^2$$

Now consider

$$X_f (a_f^+)^n \Phi = (a_f^+ + a_f)(a_f^+)^n \Phi = (a_f^+)^{n+1} \Phi + a_f (a_f^+)^n \Phi$$

$$= (a_f^+)^{n+1} \Phi + \sum_{k=0}^{n-1} (a_f^+)^k B_f \left[ a_f, a_f^+ \right] B_f^+ (a_f^+)^{n-k-1} \Phi$$

$$= (a_f^+)^{n+1} \Phi + \sum_{k=0}^{n-1} (a_f^+)^k B_f \left( \omega_{\Lambda+2} - \omega_{\Lambda+1} \right) B_f^+ (a_f^+)^{n-k-1} \Phi$$

$$= (a_f^+)^{n+1} \Phi + \left( \sum_{k=0}^{n-1} (\omega_{n-k+1} - \omega_{n-k}) \right) (a_f^+)^{n-1} \Phi$$

Since $\omega_0 = 0$, it follows that

$$X_f (a_f^+)^n \Phi = (a_f^+)^{n+1} \Phi + \omega_n (a_f^+)^{n-1} \Phi \tag{53}$$

which is the monic Jacobi relation for the symmetric vacuum random variable $X_f$, where the $(a_f^+)^n\Phi$ is its $n$-th orthogonal polynomial. Since the 1MTIFS are of type I in the sense of [6], it follows that the right hand side of (53) is isomorphic to the **canonical quantum decomposition** of $X_f$ (see [6] for the definition) and that in the isomorphism

$$\bigotimes_{n\in\mathbb{N}} \mathbb{C}\cdot(a_f^+)^n\Phi \equiv L^2_{\text{pol}}(\mathbb{R},\mu) := \text{closure of the polynomials in } L^2(\mathbb{R},\mu)$$

$X_f$ goes into the multiplication operator and the $(a_f^+)^n\Phi$ ($n\in\mathbb{N}$) into the orthogonal polynomials of $X_f$. In the language of [6] this means that the quantum decomposition (52) is isomorphic to the canonical quantum decomposition of $X_f$.

**Lemma 2** *Suppose that $f\perp g$. Then*

$$[X_f, X_g] = \left(B_g^+B_f - B_f^+B_g\right)(\omega_{\Lambda+1}-\omega_\Lambda) \tag{54}$$

*Proof* Using (41) and $f\perp g$, one finds

$$[X_f, X_g] = [a_f^+ + a_f, a_g^+ + a_g] = [a_f^+, a_g] + [a_f, a_g^+] = [a_f, a_g^+] - [a_g, a_f^+]$$

$$= \left(B_g^+B_f - B_f^+B_g\right)(\omega_{\Lambda+1}-\omega_\Lambda)$$

which is (54). □

**Theorem 4** *Let $f, g\in H_1$ be such that $f\perp g$. If $\omega_1\neq 0$ (otherwise one has a trivial IFS), then the operator random variables $X_f$, $X_g$ are not $\varphi$-tensor-independent in $\Gamma_{\omega,full}(H_1)$. If $\omega_1\neq\omega_2$ they are also not $\varphi$-tensor-independent in $\Gamma_\omega(H_1)$. In particular, the algebras $\mathscr{P}(a_f^+, a_f)$ and $\mathscr{P}(a_g^+, a_g)$ in both cases are $\varphi$-complementary but not $\varphi$-tensor-independent*

*Proof* In the full case one has

$$\langle X_fX_gX_fX_g\rangle = \langle X_f\Phi, X_gX_fX_g\Phi\rangle = \langle a_f^+\Phi, (a_g^+ + a_g)(a_f^+ + a_f)a_g^+\Phi\rangle$$

$$= \langle a_g^+a_f^+\Phi, a_f^+a_g^+\Phi\rangle = \langle a_f^+\Phi, a_ga_f^+a_g^+\Phi\rangle = 0$$

but, using (31) one finds

$$\langle X_f^2\rangle\langle X_g^2\rangle = \langle a_f^+\Phi, a_f^+\Phi\rangle\langle a_g^+\Phi, a_g^+\Phi\rangle = \langle\omega_\Lambda\ell_f^*\Phi, \omega_\Lambda\ell_f^*\Phi\rangle\langle\omega_\Lambda\ell_g^*\Phi, \omega_\Lambda\ell_g^*\Phi\rangle$$

$$= \omega_1^2\langle\ell_f^*\Phi, \ell_f^*\Phi\rangle\langle\ell_g^*\Phi, \ell_g^*\Phi\rangle = \omega_1^2\|f\|^2\|g\|^2 \neq 0$$

In the symmetric case one has from the formula (54)

$$\langle X_f X_g X_f X_g \rangle = \langle X_f [X_g, X_f] X_g \rangle + \langle X_f^2 X_g^2 \rangle$$

$$= \langle X_f \left( B_g^+ B_f - B_f^+ B_g \right) (\omega_{\Lambda+1} - \omega_\Lambda) X_g \rangle + \langle X_f^2 X_g^2 \rangle$$

$$= \langle a_f \left( B_g^+ B_f - B_f^+ B_g \right) (\omega_{\Lambda+1} - \omega_\Lambda) a_g^+ \rangle + \langle X_f^2 X_g^2 \rangle$$

and $\langle X_f^2 X_g^2 \rangle = \langle X_f^2 \rangle \langle X_g^2 \rangle$ *because of* $f \perp g$

Using (38) and $\omega_0 = 0$ this becomes

$$(\omega_2 - \omega_1) \langle a_f^+ \Phi, \left( B_g^+ B_f - B_f^+ B_g \right) a_g^+ \Phi \rangle + \langle X_f^2 X_g^2 \rangle$$

$$= (\omega_2 - \omega_1) \langle \omega_\Lambda B_f^+ \Phi, \left( B_g^+ B_f - B_f^+ B_g \right) \omega_\Lambda B_g^+ \Phi \rangle + \langle X_f^2 X_g^2 \rangle$$

$$= \omega_1^2 (\omega_2 - \omega_1) \langle B_f^+ \Phi, B_g^+ B_f B_g^+ \Phi \rangle - \omega_1^2 (\omega_2 - \omega_1) \langle B_f^+ \Phi, B_f^+ B_g B_g^+ \Phi \rangle + \langle X_f^2 X_g^2 \rangle$$

$$= \omega_1^2 (\omega_2 - \omega_1) \langle B_g B_f^+ \Phi, B_f B_g^+ \Phi \rangle - \omega_1^2 (\omega_2 - \omega_1) \langle B_f B_f^+ \Phi, B_g B_g^+ \Phi \rangle + \langle X_f^2 X_g^2 \rangle$$

$$= \omega_1^2 (\omega_2 - \omega_1) \langle f, g \rangle^2 - \omega_1^2 (\omega_2 - \omega_1) \| f \|^2 \| g \|^2 = -\omega_1^2 (\omega_2 - \omega_1) \| f \|^2 \| g \|^2 + \langle X_f^2 X_g^2 \rangle$$

Thus from Theorem 3 we conclude that

$$\langle X_f X_g X_f X_g \rangle = -(\omega_2 - \omega_1) \| f \|^2 \| g \|^2 + \langle X_f^2 \rangle \langle X_g^2 \rangle \neq \langle X_f^2 \rangle \langle X_g^2 \rangle$$

if $\omega_2 \neq \omega_1$. □

## 5 Weak Complementarity in Monotone IFS

Monotone IFS arise from functional central limit theorems of monotone-independent operator random variables.
In this section, we prove that monotone IFS do not enjoy the strong complementarity property in the sense of Definition 12.

Then we prove that they enjoy a weak form of the strong complementarity property.

### *5.1 The Monotone IFS*

**Theorem 5** *There exists a unique IFS over* $L^2(\mathbb{R})$ *with the following properties*

$$a_f^+ F_n := f \otimes F_n \tag{55}$$

$$a_f a^+_{f_1} \cdots a^+_{f_n} \Phi = \int (\bar{f} f_1)(t) a^+_{f_2 \chi_{[0,t]}} \cdots a^+_{f_n \chi_{[0,t]}} \Phi \, dt \tag{56}$$

$$a_f \Phi = 0 \tag{57}$$

*Proof* We prove that the 3 conditions above uniquely determine the semi-scalar product on $\mathrm{Tens}(H_1)$. For each $n \in \mathbb{N}$, define the sesqui-linear form

$$\kappa_n(f_1 \otimes \cdots \otimes f_n \, ; \, g_1 \otimes \cdots \otimes g_n) := \langle a^+_{f_1} \cdots a^+_{f_n} \Phi, a^+_{g_1} \cdots a^+_{g_n} \Phi \rangle =$$

$$= \int dt (\bar{f}_1 g_1)(t) \langle a^+_{f_2 \chi_{[0,t]}} \cdots a^+_{f_n \chi_{[0,t]}} \Phi \, , \, a^+_{g_2 \chi_{[0,t]}} \cdots a^+_{g_n \chi_{[0,t]}} \rangle$$

$$= \int dt (\bar{f}_1 g_1)(t) \kappa_{n-1} \left( f_2 \chi_{[0,t]} \otimes \cdots \otimes f_n \chi_{[0,t]} \, ; \, g_2 \chi_{[0,t]} \otimes \cdots \otimes g_n \chi_{[0,t]} \right)$$

Clearly $\kappa_1$ is positive definite. Suppose by induction that $\kappa_{n-1}$ is positive definite. Then since, for each $t \in \mathbb{R}$ the kernel

$$\begin{aligned} &(f_1 \otimes f_2 \otimes \cdots \otimes f_n \, , \, g_1 \otimes g_2 \otimes \cdots \otimes g_n) \\ &\mapsto (\bar{f}_1 g_1)(t) \kappa_{n-1} \left( f_2 \chi_{[0,t]} \otimes \cdots \otimes f_n \chi_{[0,t]} \, ; \, g_2 \chi_{[0,t]} \otimes \cdots \otimes g_n \chi_{[0,t]} \right) \end{aligned}$$

is positive definite by the induction assumption and Schur's Lemma, $\kappa_n$ is positive definite as an integral of kernels with this property with respect to a positive measure. Uniqueness of the IFS follows from the fact that the scalar product is uniquely determined. □

## 5.2 Weak Complementarity in Monotone IFS

The following Lemma shows that monotone IFSs do not enjoy the strong complementarity property.

**Lemma 3** *In the monotone Fock space over $L^2([0, 1])$:*

*(i) Property (2) of Theorem 2 is not satisfied,*
*(ii) There exist orthogonal test functions $f_1, f_2 \in L^2([0, T])$ such that the associated polynomial algebras $\mathcal{P}(a^+_{f_1}, a_{f_1})$, $\mathcal{P}(a^+_{f_2}, a_{f_2})$ are not complementary with respect to the vacuum state.*

*Proof* Consider the functions

$$f_1 = \chi_{[0, \frac{1}{2}]} - \chi_{[\frac{1}{2}, 1]} \quad , \quad f_2 = \chi_{[0,1]}$$

They satisfy $\langle f_1, f_2 \rangle = 0$ and

$$\langle a_{f_1}\rangle = \langle a_{f_1} a^{+2}_{f_2}\rangle = 0$$

However

$$\langle a_{f_1} a_{f_2} a^{+2}_{f_2}\rangle = \int_0^1 |f_2|^2(t)\langle a_{f_1} a^+_{f_2\chi_{[0,t]}}\rangle dt = \int_0^1 f_1(s)(1-s)ds$$

$$= \int_0^{\frac{1}{2}}(1-s)ds - \int_{\frac{1}{2}}^1 (1-s)ds = \frac{1}{2} - \frac{1}{2}s^2\Big|_0^{\frac{1}{2}} - \frac{1}{2} + \frac{1}{2}s^2\Big|_{\frac{1}{2}}^1 = \frac{1}{2}1^2 - \frac{1}{2}\left(\frac{1}{2}\right)^2 - \frac{1}{2}\left(\frac{1}{2}\right)^2 = \frac{1}{4} \neq 0$$

i.e. even if $\langle f_1, f_2\rangle = 0$, one has:

$$\frac{1}{4} = \langle a_{f_1}\left(a_{f_2} a^{+2}_{f_2}\right)\rangle \neq \langle a_{f_1}\rangle\langle a_{f_2} a^{+2}_{f_2}\rangle = 0$$

From this both statements of the theorem follow. □

**Lemma 4** *On the monotone Fock space over* $L^2([0,T])$ *with* $T \in [1,+\infty]$ *(*$[0,+\infty]$ *understood as* $[0,+\infty)$*), for any* $n \geq 0$, $m \geq 1$ *and* $\varepsilon \in \{-1,1\}^n$, *for any* $\{f_k\}_{k=1}^n$ *and* $\{g_h\}_{h=1}^m$ *such that*

$$f_k g_h =_{a.s.} 0 \quad ; \quad \forall k, h$$

*one has*

$$\langle a^{\varepsilon(1)}_{f_1}\cdots a^{\varepsilon(n)}_{f_n} a^+_{g_1}\cdots a^+_{g_m}\rangle = 0 \tag{58}$$

*Proof* (58) is trivial for $n = 0$. For $n = 1$, (58) is trivial if $\varepsilon(1) = +1$ and, if $\varepsilon(1) = -1$, then

$$a^{\varepsilon(1)}_{f_1} a^+_{g_1}\cdots a^+_{g_m}\Psi = a_{f_1} a^+_{g_1}\cdots a^+_{g_m}\Psi = \int \bar{f}_1 g_1(t) a^+_{g_2\chi_{[0,t]}}\cdots a^+_{g_m\chi_{[0,t]}}\Phi$$

which is equal to zero thanks to the fact that $f_1 g_1 =_{a.s.} 0$. Thus (58) holds for $n = 0, 1$. Suppose that the thesis is true for all $n \leq N$ and consider the case $n = N+1$.

If $\varepsilon(k) = +1$ for any $k$, (58) is trivial. So we examine (58) for $\varepsilon$ such that $\varepsilon^{-1}(\{-1\}) \neq \emptyset$. If $\varepsilon(N+1) = -1$ , then

$$a^{\varepsilon(N+1)}_{f_{N+1}} a^+_{g_1}\cdots a^+_{g_m}\Phi = a_{f_{N+1}} a^+_{g_1}\cdots a^+_{g_m}\Phi$$

which is equal to zero in virtue of the fact

$$f_{N+1} g_1 =_{a.s.} 0$$

So we consider such $\varepsilon$ that $\varepsilon^{-1}(\{-1\}) \neq \emptyset$ and

$$k := \max \varepsilon^{-1}(\{-1\}) < N+1$$

In this case

$$\langle a_{f_1}^{\varepsilon(1)} \cdots a_{f_{N+1}}^{\varepsilon(N+1)} a_{g_1}^+ \cdots a_{g_m}^+ \rangle = \langle a_{f_1}^{\varepsilon(1)} \cdots a_{f_{k-1}}^{\varepsilon(k-1)} a_{f_k} a_{f_{k+1}}^+ \cdots a_{f_{N+1}}^+ a_{g_1}^+ \cdots a_{g_m}^+ \rangle$$

$$= \int \bar{f}_k f_{k+1}(t) \langle a_{f_1}^{\varepsilon(1)} \cdots a_{f_{k-1}}^{\varepsilon(k-1)} a_{f_{k+2}\chi_{[0,t]}}^+ \cdots a_{f_{N+1}\chi_{[0,t]}}^+ a_{g_1\chi_{[0,t]}}^+ \cdots a_{g_m\chi_{[0,t]}}^+ \rangle$$

Now notice that for any $r \in \{1, \cdots, k-1\} \cup \{k+2, \cdots, N+1\}$ and for any $h \in \{1, 2, \cdots, m\}$, it follows from the fact $g_h f_r =_{a.s.} 0$ that

$$g_h \chi_{[0,t]} f_r = (g_h \chi_{[0,t]})(f_r \chi_{[0,t]}) =_{a.s.} 0$$

So the induction assumption gives

$$\langle a_{f_1}^{\varepsilon(1)} \cdots a_{f_{k-1}}^{\varepsilon(k-1)} a_{f_{k+2}\chi_{[0,t]}}^+ \cdots a_{f_{N+1}\chi_{[0,t]}}^+ a_{g_1\chi_{[0,t]}}^+ \cdots a_{g_m\chi_{[0,t]}}^+ \rangle = 0 \quad ; \quad \forall t$$

hence

$$\langle a_{f_1}^{\varepsilon(1)} \cdots a_{f_{N+1}}^{\varepsilon(N+1)} a_{g_1}^+ \cdots a_{g_m}^+ \rangle = 0$$

Therefore the thesis follows by induction. □

Analysis of the proof of Lemma 3, suggests the following weaker form of Definition 12.

**Definition 13** Let $H_1$ be a semi-Hilbert space of functions defined on a measure space $(S, \mathscr{B}, \mu)$ with any scalar product such that any two functions with disjoint support are orthogonal.

An IFS $\Gamma_I(H_1)$ is said to enjoy the **weak complementarity property** if any family $(f_j)_{j\in J}$ $(J \subseteq \mathbb{N})$ of vectors in $H_1$ with mutually disjoint supports is such that the algebras $\left(\mathscr{P}(a_{f_j}^+, a_{f_j})\right)_{j\in I}$ are complementary with respect to the vacuum state $\varphi$ in the sense of Definition 5, with

$$\varphi_j := \varphi\Big|_{\mathscr{P}(a_{f_j}^+, a_{f_j})} \quad ; \quad j \in I$$

**Theorem 6** *The monotone Fock space over $L^2([0, T])$ enjoys the weak complementarity property in the sense of Definition 13.*

*Proof* It is sufficient to prove that, for any $n \geq 0$, $m \geq 1$ and $\{f_k\}_{k=1}^n$, $\{g_h\}_{h=1}^m$ such that

$$f_k g_h =_{a.s.} 0 \tag{59}$$

the identity

$$\langle a_{f_1}^{\varepsilon(1)} \cdots a_{f_n}^{\varepsilon(N)} a_{g_1}^{\varepsilon'(1)} \cdots a_{g_m}^{\varepsilon'(m)} \rangle = \langle a_{f_1}^{\varepsilon(1)} \cdots a_{f_n}^{\varepsilon(N)} \rangle \langle a_{g_1}^{\varepsilon'(1)} \cdots a_{g_m}^{\varepsilon'(m)} \rangle \tag{60}$$

holds on the monotone Fock space over $L^2([0,T])$ for any $\varepsilon \in \{-1,1\}^n$, $\varepsilon' \in \{-1,1\}^m$. Notice that, if either $\varepsilon'(m) = -1$ or $\varepsilon(1) = +1$, then (60) holds trivially because both sides are zero. If $\varepsilon'(h) = +1$ for all $h$, then the same happens because, in this case, Lemma 3 implies that the left-hand side of (60) is zero and the right-hand side is zero because

$$\langle a^+_{g_1} \cdots a^+_{g_m} \rangle = 0$$

Therefore it is sufficient to prove (60) in the case when

$$\{h \, : \, \varepsilon'(h) = -1\} \neq \emptyset \quad ; \quad \varepsilon'(m) = +1 \quad ; \quad \varepsilon(1) = -1 \tag{61}$$

For $m = 1$, no $\varepsilon'$ can satisfy (61). For $m = 2$, (61) implies that,

$$(\varepsilon'(1), \varepsilon'(2)) = (-1, +1)$$

In this case (60) follows from

$$a^{\varepsilon'(1)}_{g_1} a^{\varepsilon'(2)}_{g_2} \Phi = a_{g_1} a^+_{g_2} \Phi = \langle g_1, g_2 \rangle \Phi = \langle a_{g_1} a^+_{g_2} \rangle \Phi = \langle a^{\varepsilon'(1)}_{g_1} a^{\varepsilon'(2)}_{g_2} \rangle \Phi$$

Suppose by induction that the thesis holds for $m \leq N$ and let $\varepsilon' \in \{-1,1\}^{N+1}$ satisfy (61). Denote

$$m_0 := \max \{h : \varepsilon'(h) = -1\}$$

Then (61) implies that $m_0 < N+1$. In this case

$$\begin{aligned} &a^{\varepsilon'(m_0)}_{g_{m_0}} a^{\varepsilon'(m_0+1)}_{g_{m_0+1}} \cdots a^{\varepsilon'(N+1)}_{g_{N+1}} \Phi = a_{g_{m_0}} a^+_{g_{m_0+1}} \cdots a^+_{g_{N+1}} \Phi \\ &= \int \left( \bar{g}_{m_0} g_{m_0+1} \right)(t) a^+_{g_{m_0+2}\chi_{[0,t]}} \cdots a^+_{g_{N+1}\chi_{[0,t]}} \Phi \end{aligned} \tag{62}$$

Since, for all $t \in \mathbb{R}$, condition (59) remains true if all the $g_h$ are multiplied by $\chi_{[0,t]}$, one can use the induction assumption and (62) to obtain

$$\begin{aligned} &\langle a^{\varepsilon(1)}_{f_1} \cdots a^{\varepsilon(n)}_{f_n} a^{\varepsilon'(1)}_{g_1} \cdots a^{\varepsilon'(N+1)}_{g_{N+1}} \rangle \\ &= \int \left( \bar{g}_{m_0} g_{m_0+1} \right)(t) \langle a^{\varepsilon(1)}_{f_1} \cdots a^{\varepsilon(n)}_{f_n} a^{\varepsilon'(1)}_{g_1} \cdots a^{\varepsilon'(m_0-1)}_{g_{m_0-1}} a^+_{g_{m_0+2}\chi_{[0,t]}} \cdots a^+_{g_{N+1}\chi_{[0,t]}} \rangle \\ &= \int \left( \bar{g}_{m_0} g_{m_0+1} \right)(t) \langle a^{\varepsilon(1)}_{f_1} \cdots a^{\varepsilon(n)}_{f_n} \rangle \langle a^{\varepsilon'(1)}_{g_1} \cdots a^{\varepsilon'(m_0-1)}_{g_{m_0-1}} a^+_{g_{m_0+2}\chi_{[0,t]}} \cdots a^+_{g_{N+1}\chi_{[0,t]}} \rangle \\ &= \langle a^{\varepsilon(1)}_{f_1} \cdots a^{\varepsilon(n)}_{f_n} \rangle \cdot \int \left( \bar{g}_{m_0} g_{m_0+1} \right)(t) \langle a^{\varepsilon'(1)}_{g_1} \cdots a^{\varepsilon'(m_0-1)}_{g_{m_0-1}} a^+_{g_{m_0+2}\chi_{[0,t]}} \cdots a^+_{g_{N+1}\chi_{[0,t]}} \rangle \\ &= \langle a^{\varepsilon(1)}_{f_1} \cdots a^{\varepsilon(n)}_{f_n} \rangle \langle a^{\varepsilon'(1)}_{g_1} \cdots a^{\varepsilon'(m_0-1)}_{g_{m_0-1}} a_{g_{m_0}} a^+_{g_{m_0+1}} \cdots a^+_{g_{N+1}} \rangle \end{aligned}$$

and this proves the statement. □

## References

1. Accardi, L.: Some trends and problems in quantum probability. In: Accardi, L., Frigerio, A., Gorini, V. (eds.) Quantum Probability and Applications to the Quantum Theory of Irreversible Processes, Proceedings of 2-d Conference: Quantum Probability and Applications to the Quantum Theory of Irreversible Processes Villa Mondragone (Rome), vol. 9, pp. 6–11 (1982), Springer LNM, vol. 1055, pp. 1–19. Springer, Berlin (1984)
2. Accardi, L., Bach, A.: Central limits of squeezing operator. In: Accardi, L., von Waldenfels, W. (eds.) Quantum Probability and Applications IV. Springer LNM, vol. 1396, pp. 7–19. Springer, Berlin (1987)
3. Accardi, L., Barhoumi, A., Dhahri, A.: Identification of the theory of orthogonal polynomials in $d$-indeterminates with the theory of 3-diagonal symmetric interacting Fock spaces on $\mathbb{C}^d$. Infin. Dimens. Anal. Quantum Probab. Relat. Top. (IDA-QP) **20**(1), 1–55 (2017) August (2015), revised 25 Jan 2016
4. Accardi, L., Crismale, V., Lu, Y.G.: Constructive universal central limit theorems based on interacting Fock spaces. Infin. Dimens. Anal. Quantum Probab. Relat. Top. (IDA-QP) **8**(4), 631–650 (2005). Preprint Volterra vol. 591 (2005)
5. Accardi, L., Kuo, H.-H., Stan, A.: Characterization of probability measures through the canonically associated interacting Fock spaces. Infin. Dimens. Anal. Quantum Probab. Relat. Top. (IDA-QP) **7**, 485–505 (2004)
6. Accardi, L., Lu, Y.G.: The quantum moment problem for a classical random variable and a classification of interacting Fock spaces, to be submitted to (2017)
7. Accardi, L., Lu, Y.G., Mastropietro, V.: Stochastic bosonization for a $d \geq 3$ Fermi system. Annales de l'Inst. Henri Poincaré **66**(2), 185–213 (1997). Preprint Volterra vol. 205 (1995)
8. Accardi, L., Lu, Y.G., Volovich, I.V.: Quantum Theory and Its Stochastic Limit. Springer, Berlin (2002)
9. Accardi, L., Lu, Y.G., Volovich, I.: The QED Hilbert Module and Interacting Fock Spaces. Publications of IIAS (Kyoto), N1997–008 (1997)
10. Arecchi, F.T., Courtens, E., Gilmore, R., Thomas, H.: Atomic coherent states in quantum optics. Phys. Rev. A **6**, 2211–2237 (1972)
11. Björck, G.: Functions of modulus 1 on $Z_n$, whose Fourier transforms have constant modulus, and cyclic $n$–roots. Recent Advances in Fourier Analysis and Its Applications. NATO, Advance Science Institutes Series C, Mathematical and Physical Sciences, vol. 315, pp. 131–140. Kluwer Academic Publisher, Netherlands (1990)
12. Busch, P., Lahti, P.J.: To what extent do position and momentum commute? Phys. Lett. A **115**(6), 259–264 (1986)
13. Cassinelli, G., Varadarajan, V.S.: On Accardi's notion of complementary observables. Infin. Dimens. Anal. Quantum Probab. Relat. Top. (IDAQP) **5**(2), 135–144 (2002)
14. Conway, J.B.: A Course in Functional Analysis. Springer, Berlin (1985)
15. Haagerup, U.: Cyclic $p$–roots of prime length $p$ and related complex Hadamard matrices (2008). arXiv:0803.2629 [math.AC]. Revised version arXiv:0803.2629v1 [math.AC]
16. Ivonovic, I.D.: Geometrical description of quantal state determination. J. Phys. A Math. Gen **14**, 3241 (1981)
17. Jammer, M.: The Conceptual Development of Quantum Mechanics. McGraw-Hill, New York (1966)
18. Kato, T.: Perturbation Theory for Linear Operators. Springer, Berlin (1967)
19. Kraus, K.: Complementary observables and uncertainty relations. Phys. Rev. D **35**, 3070–3075 (1987)
20. Lu, Y.G.: Interacting Fock space related to the Anderson model. Infin. Dimens. Anal. Quantum Probab. Relat. Top. (IDAQP) **1**(2), 247–283 (1998)
21. Maassen, H., Uffink, J.B.M.: Generalized entropic uncertainty relations. Phys. Rev. Lett. **60**(12), 1103–1106 (1988)
22. Ohya, M., Petz, D.: Quantum Entropy and Its Use. Texts and Monographs in Physics. Springer, Berlin (1993)

23. Parthasarathy, K.R.: On estimating the state of a finite level quantum system. Infin. Dimens. Anal. Quantum. Probab. Relat. Top. (IDAQP) **7**(4), 607–617 (2004)
24. Petz, D., Ruppert, L.: Efficient quantum tomography needs complementary and symmetric measurements. Rep. Math. Phys. **69**(2), 161–177 (2012)
25. Popa, S.: Orthogonal pairs of $*$-sub-algebras in finite von Neumann algebras. J. Oper. Theory **9**, 253–268 (1983)
26. Renyi, A.: Foundations of Probability. Holden-Day, San Francisco (1970)
27. Schwinger, J.: Unitary operator bases. Proc. Natl. Acad. Sci. **46**, 570–579 (1960)
28. Tadej, W., Zyczkowski, K.: A concise guide to complex Hadamard matrices. Open Syst. Inf. Dyn. **13**(2), 133–177 (2006)
29. von Neumann, J.: Mathematical Foundations of Quantum Mechanics. Princeton University Press, Princeton (1955)
30. Wootters, W.K., Fields, B.D.: On the U.V. mutually unbiased bases. Ann. Phys. **191**, 363 (1989)

# Norm Conditions for Separability in $\mathbb{M}_m \otimes \mathbb{M}_n$

**Tsuyoshi Ando**

**Abstract** An element $\mathbf{S}$ of the tensor product $\mathbb{M}_m \otimes \mathbb{M}_n$ is said to be separable if it admits a (separable) decomposition

$$\mathbf{S} = \sum_p X_p \otimes Y_p \quad \exists\, 0 \le X_p \in \mathbb{M}_m,\ \exists\, 0 \le Y_p \in \mathbb{M}_n.$$

This decomposition is not unique. We present some conditions on suitable norms of $\mathbf{S}$ which guarantee its separability. Even when separability of $\mathbf{S}$ is guaranteed by some method, its separable decomposition itself is difficult to construct. We present a general condition which makes it possible to find a way of an explicit separable decomposition.

## 1 Introduction

Let $\mathbb{M}_n$ denote the space of $n \times n$ (complex) matrices for each $n = 1, 2, \ldots$. Each element of $\mathbb{M}_n$ is considered as a linear map from $\mathbb{C}^n$ to itself. Here an element $x$ of $\mathbb{C}^n$ is understood as a column $n$-vector, and correspondingly $x^*$ is a row $n$-vector. Then given $x, y \in \mathbb{C}^n$, according to the rule of matrix multiplication, $x^*y$ is the inner product of $x$ and $y$, that is, $x^*y = \langle x|y\rangle$ while $yx^*$ is a matrix of rank $\le 1$ in $\mathbb{M}_n$.

Notice here that the inner product $\langle x|y\rangle$ is linear in $y$ and anti-linear in $x$. With this inner product and norm $\|x\| = \sqrt{\langle x|x\rangle}$, $\mathbb{C}^n$ becomes a Hilbert space.

Correspondingly the space $\mathbb{M}_n$ becomes a Hilbert space with the inner product

$$\langle T|S\rangle := \mathrm{Tr}(T^*S) \quad \forall\, S, T \in \mathbb{M}_n$$

T. Ando (✉)
Hokkaido University (Emeritus), Hokkaido, Japan
e-mail: ando@es.hokudai.ac.jp

© Springer Nature Switzerland AG 2019

T. M. Rassias and V. A. Zagrebnov (eds.), *Analysis and Operator Theory*, Springer Optimization and Its Applications 146,
https://doi.org/10.1007/978-3-030-12661-2_2

where $\mathrm{Tr}(S)$ is the trace of $S$, and the Hilbert–Schmidt (or *Frobenius*) norm

$$\|S\|_2 := \sqrt{\langle S|S\rangle} \quad \forall\, S \in \mathbb{M}_n.$$

For self-adjoint $S, T \in \mathbb{M}_n$, the order relation $S \geq T$ or equivalently $T \leq S$ is defined as $S - T$ is *positive semi-definite*. Therefore $S \geq 0$ or $0 \leq S$ means simply that $S$ is positive semi-definite.

The operator (or *spectral*) norm $\|S\|$ of $S \in \mathbb{M}_n$ is defined by

$$\|S\| := \sup\{\|Sx\|;\ \|x\| = 1\}. \tag{1}$$

When $S$ is selfadjoint, the norm $\|S\|$ is calculated simply as

$$S = S^* \quad \Longrightarrow \quad \|S\| = \sup\{|\langle x|Sx\rangle|;\ \|x\| = 1\}. \tag{2}$$

As an immediate consequence of (2),

$$S = S^*,\ \ \|I - S\| \leq 1 \quad \Longrightarrow \quad S \geq 0. \tag{3}$$

In fact, the assumption guarantees that

$$\|x\|^2 - \langle x|Sx\rangle \leq \|x\|^2 \quad \forall\, x.$$

Next let us turn to the *tensor product* $\mathbb{M}_m \otimes \mathbb{M}_n$ of two matrix spaces $\mathbb{M}_m$ and $\mathbb{M}_n$. There are canonical identifications

$$\mathbb{M}_m \otimes \mathbb{M}_n \sim \mathbb{M}_m(\mathbb{M}_n) \sim \mathbb{M}_{mn}.$$

Here $\mathbb{M}_m(\mathbb{M}_n)$ denotes the space of $m \times m$ block matrices with entries in $\mathbb{M}_n$. The first identification is understood in the following way:

$$X \otimes Y \sim [\xi_{j,k} Y]_{j,k} \quad \forall\, X = [\xi_{j,k}]_{j,k} \in \mathbb{M}_m,\ \ \forall\, Y \in \mathbb{M}_n. \tag{4}$$

The second identification is understood by cutting each $mn \times mn$ matrix into $m \times m$ block matrices in a natural way. Therefore by (4)

$$\sum_{j,k} E_{j,k} \otimes S_{j,k} \sim \mathbf{S} = [S_{j,k}]_{j,k}$$

where $E_{j,k}$ $(j, k = 1, 2, \ldots, m)$ is the *matrix unit* in $\mathbb{M}_m$, that is,

$$E_{j,k} = e_j e_k^* \quad (j, k = 1, 2, \ldots, m)$$

with the canonical orthonormal basis $e_j$ $(j = 1, 2, \ldots, m)$ of $\mathbb{C}^m$.

In the following we abuse the notation:

$$\mathbf{S} = \sum_{j,k} E_{j,k} \otimes S_{j,k} \quad \forall\, \mathbf{S} = [S_{j,k}]_{j,k} \in \mathbb{M}_m(\mathbb{M}_n). \tag{5}$$

According to the general rule, the space $\mathbb{M}_m(\mathbb{M}_n)$ becomes a Hilbert space with inner product and norm;

$$\langle \mathbf{S}|\mathbf{T}\rangle = \mathrm{Tr}(\mathbf{S}^*\mathbf{T}) = \sum_{j,k} \mathrm{Tr}(S_{jk}^* T_{jk}) \quad \text{and} \quad \|\mathbf{S}\|_2 = \sqrt{\langle \mathbf{S}|\mathbf{S}\rangle}.$$

Further the spectral norm (1) for $\mathbf{S} = [S_{jk}]_{j,k}$ takes the form

$$\|\mathbf{S}\| = \sup\Big\{ \{\sum_j \| \sum_k S_{j,k} x_k \|^2\}^{1/2};\ \sum_k \|x_k\|^2 = 1\ \forall\, x_k \in \mathbb{C}^n\ (k = 1, 2, \ldots, n) \Big\}. \tag{6}$$

In the following, $\mathbb{M}_{(m,n)}$ will denote the real subspace of $\mathbb{M}_m(\mathbb{M}_n)$ consisting of *selfadjoint* elements, that is, the subspace of $\mathbf{S} = [S_{j,k}]_{j,k}$ with $S_{j,k} = S_{k,j}^*$ $(j, k = 1, 2, \ldots, m)$. Let $\mathfrak{P}$ denote the cone of positive semi-definite elements in $\mathbb{M}_{(m,n)}$. We shall use $\mathbf{S} \geq 0$ as usual when $\mathbf{S} \in \mathfrak{P}$, and $\mathbf{S} > 0$ when $\mathbf{S} \geq 0$ and is invertible.

In the tensor product theory a fact of key importance is the following [3, p.23];

$$0 \leq X \in \mathbb{M}_m,\ 0 \leq Y \in \mathbb{M}_n \quad \Longrightarrow \quad X \otimes Y \in \mathfrak{P}. \tag{7}$$

The cone, generated by $X \otimes Y$ with $0 \leq X \in \mathbb{M}_m$ and $0 \leq Y \in \mathbb{M}_n$, will be denoted by $\mathfrak{P}_+$. The cone $\mathfrak{P}_+$ is contained in the cone $\mathfrak{P}$ by (7). A block matrix in $\mathfrak{P}_+$ is called *separable.*

By definition, $\mathbf{S}$ is separable if and only if it admits a decomposition

$$\mathbf{S} = \sum_p X_p \otimes Y_p \quad \exists\, 0 \leq X_p \in \mathbb{M}_m,\ \exists\, 0 \leq Y_p \in \mathbb{M}_n. \tag{8}$$

Difficulty is in the fact that decomposition (8) is *not* unique.

A nontrivial fact is that the cone $\mathfrak{P}_+$ is topologically *closed.* This comes from the finite dimensionality of $\mathbb{M}_m \otimes \mathbb{M}_n$. (See [1, p. 8]).

## 2 Norm Conditions for Separability

Let us begin with an inner characterization of the cone $\mathfrak{P}_+$. A linear map $\varphi : \mathbb{M}_n \to \mathbb{M}_N$ is said to be *positive* if $\varphi(S) \geq 0$ whenever $S \geq 0$. It is called *unital* if $\varphi(I_n) = I_N$ where $I_n$ and $I_N$ are the identity matrices in $\mathbb{M}_n$ and $\mathbb{M}_N$ respectively. It is known as Russo-Dye theorem [4, Theorem 2.3.7] that a *unital, positive* linear map $\varphi$ is *contractive* in the sense

$$\|\varphi(S)\| \le \|S\| \quad \forall\, S \in \mathbb{M}_n. \tag{9}$$

A linear map $\varphi : \mathbb{M}_n \to \mathbb{M}_N$ gives rise to a linear map $\tilde{\varphi} : \mathbb{M}_m(\mathbb{M}_n) \to \mathbb{M}_m(\mathbb{M}_N)$ by

$$\tilde{\varphi}\Big([S_{j,k}]_{j,k}\Big) := \Big[\varphi(S_{j,k})\Big]_{j,k} \quad \forall\, \mathbf{S} = [S_{j,k}]_{j,k}. \tag{10}$$

The map $\tilde{\varphi}$ is often denoted by $id_m \otimes \varphi$. If $\varphi$ is unital, so is $\tilde{\varphi}$. But positivity of $\varphi$ *does not* imply positivity of $\tilde{\varphi}$ in general.

Horodecki's [7] established an inner characterization of separability of $\mathbf{S} \in \mathbb{M}_{(m,n)}$ in the following form.

**Theorem 1** (Horodecki's [7]) *Let* $\mathbf{S} \in \mathbb{M}_{(m,n)}$. *Then*

$$\mathbf{S} \in \mathfrak{P}_+ \iff \tilde{\varphi}(\mathbf{S}) \ge 0 \;\; \forall\, N, \;\; \forall\, \textit{unital positive linear map } \varphi : \mathbb{M}_n \to \mathbb{M}_N.$$

In view of (3), an actual form of Theorem 1, useful in establishing separability of S, is the following.

**Corollary 2** *Let* $\mathbf{S} \in \mathbb{M}_{(m,n)}$. *Then,* $\forall\, N$, *with* $\mathbf{I} = I_m \otimes I_N$,

$$\|\mathbf{I} - \tilde{\varphi}(\mathbf{S})\| \le 1 \;\; \forall\, \textit{unital positive linear } \varphi : \mathbb{M}_n \to \mathbb{M}_N \implies \mathbf{S} \in \mathfrak{P}_+.$$

A norm $|||\cdot|||$ on $\mathbb{M}_m(\mathbb{M}_n)$ will be said to have *property* ($\sharp$) if $\forall\, N$

$$\|\tilde{\varphi}(\mathbf{S})\| \le |||\mathbf{S}||| \;\; \forall\, \textit{unital positive linear } \varphi : \mathbb{M}_n \to \mathbb{M}_N, \; \forall\, \mathbf{S} \in \mathbb{M}_m(\mathbb{M}_n). \tag{$\sharp$}$$

**Corollary 3** *If a norm* $|||\cdot|||$ *on* $\mathbb{M}_m(\mathbb{M}_n)$ *has property* ($\sharp$), *then for* $\mathbf{S} \in \mathbb{M}_{(m,n)}$

$$|||\mathbf{I} - \mathbf{S}||| \le 1 \implies \mathbf{S} \in \mathfrak{P}_+.$$

In fact, since for any unital positive linear map $\varphi : \mathbb{M}_n \to \mathbb{M}_N$,

$$\|\mathbf{I} - \tilde{\varphi}(\mathbf{S})\| = \|\tilde{\varphi}(\mathbf{I} - \mathbf{S})\| \le |||\mathbf{I} - \mathbf{S}||| \le 1,$$

where $\mathbf{I}$ on the left-hand term is the identity in $\mathbb{M}_m(\mathbb{M}_N)$ while $\mathbf{I}$ on the middle and the right hand terms are the identity in $\mathbb{M}_m(\mathbb{M}_n)$, appeal to (3) to see $\tilde{\varphi}(\mathbf{S}) \ge 0$ and then to Theorem 1 to see $\mathbf{S} \in \mathfrak{P}_+$.

Define a functional $\|\cdot\|_{(+)}$ for $\mathbf{S} = [S_{j,k}]_{j,k} \in \mathbb{M}_m(\mathbb{M}_n)$ by

$$\|\mathbf{S}\|_{(+)} := \Big\| \Big[\|S_{j,k}\|\Big]_{j,k} \Big\|, \tag{11}$$

where $\Big[\|S_{j,k}\|\Big]_{j,k}$ is the element of $\mathbb{M}_m$ whose $(j,k)$-entry is $\|S_{j,k}\|$ $(j,k = 1,2,\ldots,m)$.

**Lemma 4** *The functional* $\|\cdot\|_{(+)}$ *is a norm with property* (♮).

*Proof* That $\|\cdot\|_{(+)}$ becomes a norm on $\mathbb{M}_m(\mathbb{M}_n)$ is obvious. Let us show that

$$\|\mathbf{S}\| \le \|\mathbf{S}\|_{(+)} \quad \forall\, \mathbf{S} = [S_{jk}]_{j,k} \in \mathbb{M}_m(\mathbb{M}_n).$$

This results from (6) and definition (11) by the inequalities;

$$\|\sum_k S_{j,k}x_k\| \le \sum_k \|S_{j,k}\| \cdot \|x_k\| \quad \forall\, x_k \in \mathbb{C}^n \ (j = 1, 2, \ldots, n).$$

Take a unital positive linear map $\varphi : \mathbb{M}_n \to \mathbb{M}_N$. Since $\varphi$ is contractive by (9)

$$\|\varphi(S_{j,k})\| \le \|S_{j,k}\| \quad \forall\, j, k$$

we can see again from (6)

$$\|\tilde{\varphi}(\mathbf{S})\| \le \|\tilde{\varphi}(\mathbf{S})\|_{(+)} \le \|\mathbf{S}\|_{(+)}.$$

This completes the proof.

**Corollary 5** *Every norm* $|||\cdot|||$ *on* $\mathbb{M}_m(\mathbb{M}_n)$ *for which*

$$\|\mathbf{S}\|_{(+)} \le |||\mathbf{S}||| \quad \forall\, \mathbf{S} \in \mathbb{M}_m(\mathbb{M}_n)$$

*has property* (♮).

The following are among simple examples of norms with property (♮):

(A) $|||\mathbf{S}||| := m \cdot \|\mathbf{S}\|$.

(B) $|||\mathbf{S}||| := \|\mathbf{S}\|_2 = \sqrt{\sum_{j,k} \|S_{j,k}\|_2^2}$ (Hilbert–Schmidt norm).

(A) Since $\|S_{j,k}\| \le \|\mathbf{S}\| \ \forall\, j,\ k$, by (6)

$$\|\mathbf{S}\|_{(+)} \le \|\mathbf{S}\| \cdot \left\| \begin{bmatrix} 1 \\ \cdots \\ 1 \end{bmatrix} \cdot [1 \ldots 1] \right\| = m \cdot \|\mathbf{S}\|.$$

(B) Since $\|S_{j,k}\| \le \|S_{j,k}\|_2 \ \forall\, j, k$, and the operator norm $\le$ the Hilbert–Schmidt norm in $\mathbb{M}_m$,

$$\|\mathbf{S}\|_{(+)} \le \sqrt{\sum_{j,k} \|S_{j,k}\|^2} \le \sqrt{\sum_{j,k} \|S_{j,k}\|_2^2} = \|\mathbf{S}\|_2.$$

**Corollary 6** *Let* $\mathbf{S} \in \mathbb{M}_{(m,n)}$.

$$\|\mathbf{I} - \mathbf{S}\| \le 1/m \quad or \quad \|\mathbf{I} - \mathbf{S}\|_2 \le 1 \implies \mathbf{S} \in \mathfrak{P}_+.$$

The first assertion can be found in [1] while the second in [6].

## 3 Actual Conditions for Separability

Notice that $\mathbf{S}$ is separable if and only if $t \cdot \mathbf{S}$ is separable $\forall\, t > 0$ or $\exists\, t > 0$. It is easy to see that for $0 < S \in \mathbb{M}_n$

$$\|S\| = \lambda_{max}(S) = \text{the maximum eigenvalue of } S$$

and

$$\|S^{-1}\|^{-1} = \lambda_{min}(S) = \text{the minimum eigenvalue of } S.$$

**Theorem 7** *Let* $0 < \mathbf{S} \in \mathbb{M}_{(m,n)}$. *Then*

$$\|\mathbf{S}\| \cdot \|\mathbf{S}^{-1}\| \le \frac{m+1}{m-1} = 1 + \frac{2}{m-1} \implies \mathbf{S} \in \mathfrak{P}_+.$$

*Proof* The assumption means that there is $0 < t$ such that

$$\frac{m-1}{m} \cdot \|\mathbf{S}^{-1}\| \le t \le \frac{m+1}{m} \cdot \|\mathbf{S}\|^{-1}.$$

Then for such $t > 0$

$$\begin{aligned} t\mathbf{S} - \mathbf{I} &\le \{t \cdot \lambda_{max}(\mathbf{S}) - 1\} \cdot \mathbf{I} \\ &\le \{\frac{m+1}{m} - 1\} \cdot \mathbf{I} \le \frac{1}{m} \cdot \mathbf{I}, \end{aligned}$$

and

$$\mathbf{I} - t\mathbf{S} \le \{1 - t\lambda_{min}(\mathbf{S})\} \cdot \mathbf{I} \le \frac{1}{m} \cdot \mathbf{I}.$$

Therefore

$$\|\mathbf{I} - t\mathbf{S}\| \le \frac{1}{m}.$$

Now appeal to Corollary 6 to see that $\mathbf{S} \in \mathfrak{P}_+$.

Notice here the obvious relation;

$$1 \le \|\mathbf{S}\| \cdot \|\mathbf{S}^{-1}\|.$$

**Theorem 8** *Let* $0 \le \mathbf{S} \in \mathbb{M}_{(m,n)}$. *Then*

$$\|\mathbf{S}\|_2 \leq \frac{1}{\sqrt{mn-1}} \cdot \mathrm{Tr}(\mathbf{S}) \implies \mathbf{S} \in \mathfrak{P}_+.$$

*Proof* With $t := \frac{\mathrm{Tr}(\mathbf{S})}{\|\mathbf{S}\|_2^2} > 0$,

$$\|\mathbf{I} - t \cdot \mathbf{S}\|_2^2 - 1 = t^2 \cdot \|\mathbf{S}\|_2^2 - 2t \cdot \mathrm{Tr}(\mathbf{S}) + mn - 1$$
$$= mn - 1 - \frac{\mathrm{Tr}(\mathbf{S})^2}{\|\mathbf{S}\|_2^2} = \left\{\sqrt{mn-1} + \frac{\mathrm{Tr}(\mathbf{S})}{\|\mathbf{S}\|_2}\right\} \cdot \left\{\sqrt{mn-1} - \frac{\mathrm{Tr}(\mathbf{S})}{\|\mathbf{S}\|_2}\right\} \leq 0.$$

Now appeal to Corollary 6 to see that $\mathbf{S} \in \mathfrak{P}_+$.

Notice here the obvious Schwartz inequality;

$$\|\mathbf{S}\|_2 \geq \frac{1}{\sqrt{mn}} \cdot \mathrm{Tr}(\mathbf{S}) \quad \forall\ 0 \leq \mathbf{S} \in \mathbb{M}_{(m,n)}.$$

## 4 Explicit Separable Decomposition

Let us begin with some examples of $\mathbf{S}$'s which admit explicit separable decompositions.

**Lemma 9** *For* $0 \leq S_j \in \mathbb{M}_n$ $(j = 1, 2, \dots, m)$ *the block matrix*

$$\mathbf{S} := \mathrm{Diag}(S_1, S_2, \dots, S_m) = \begin{bmatrix} S_1 & 0 & 0 & \dots & 0 \\ 0 & S_2 & 0 & \dots & 0 \\ \vdots & \vdots & \vdots & \vdots & \vdots \\ 0 & 0 & 0 & \dots & S_m \end{bmatrix}$$

*admits an explicit separable decomposition.*

This is trivial because $\mathbf{S} = \sum_j E_{j,j} \otimes S_j$ by (5).

**Lemma 10** *If* $S \in \mathbb{M}_n$ *and*

$$\alpha \cdot \beta \geq \|S\|^2 \quad \exists\, \alpha, \beta > 0$$

*then for any* $1 \leq j < k \leq m$

$$E_{j,j} \otimes \alpha \cdot I + E_{j,k} \otimes S + E_{k,j} \otimes S^* + E_{k,k} \otimes \beta \cdot I$$

*admits an explicit separable decomposition in* $\mathbb{M}_m \otimes \mathbb{M}_n$ $\forall\, m \geq 2$.

*Proof* It suffices to prove that when $1 \geq \|S\|^2$ the block matrix $\mathbf{S} = \begin{bmatrix} I & S \\ S^* & I \end{bmatrix}$ admits an explicit separable decomposition in $\mathbb{M}_2 \otimes \mathbb{M}_n$.

It is immediate to see that

$$\mathbf{S} \geq 0 \iff 1 \geq \|S\|^2.$$

Take a unital positive linear map $\varphi : \mathbb{M}_n \to \mathbb{M}_N$. Since $\varphi$ is contractive by (9), definition (10) shows

$$\tilde{\varphi}\left(\begin{bmatrix} I & S \\ S^* & I \end{bmatrix}\right) = \begin{bmatrix} \varphi(I) & \varphi(S) \\ \varphi(S^*) & \varphi(I) \end{bmatrix} = \begin{bmatrix} I & \varphi(S) \\ \varphi(S)^* & I \end{bmatrix} \geq 0,$$

so that $\mathbf{S}$ is *separable* in $\mathbb{M}_2 \otimes \mathbb{M}_n$ by Theorem 1.

Let us present an explicit separable decomposition of $\mathbf{S}$. It is an exercise that the contraction $S$ is written as

$$S = \tfrac{1}{2}\{U + V\} \quad \exists \text{ unitary } U, V.$$

By the spectral theorem for unitary matrices in $\mathbb{M}_n$

$$U = \sum_{j=1}^{n} e^{i\theta_j} P_j \quad \text{and} \quad V = \sum_{j=1}^{n} e^{i\varphi_j} Q_j \quad \exists\, 0 \leq \theta_j,\ \varphi_j < 2\pi \ (j = 1, 2, \ldots, n)$$

where $P_j,\ Q_j$ are orthoprojections with

$$\sum_{j=1}^{n} P_j = \sum_{j=1}^{n} Q_j = I.$$

Then

$$\mathbf{S} = \tfrac{1}{2}\left\{\sum_{j=1}^{n} \begin{bmatrix} 1 & e^{i\theta_j} \\ e^{-i\theta_j} & 1 \end{bmatrix} \otimes P_j + \sum_{j=1}^{n} \begin{bmatrix} 1 & e^{\varphi_j} \\ e^{-\varphi_j} & 1 \end{bmatrix} \otimes Q_j\right\}.$$

Since $0 \leq \begin{bmatrix} 1 & e^{i\theta} \\ e^{-i\theta} & 1 \end{bmatrix}$ in $\mathbb{M}_2$ $\forall\, 0 \leq \theta < 2\pi$, each summand belongs to $\mathfrak{P}_+$, so does $\mathbf{S}$.

Now let us return to the case of the norm $\|\cdot\|_{(+)}$.

**Lemma 11** *Let* $0 \leq \mathbf{S} = [S_{j,k}]_{j,k} \leq \mathbf{I}$. *Then*

$$\|\mathbf{I} - \mathbf{S}\|_{(+)} \leq 1 \iff \begin{bmatrix} \lambda_{min}(S_{1,1}) & -\|S_{1,2}\| & -\|S_{1,3}\| & \ldots & -\|S_{1,m}\| \\ -\|S_{2,1}\| & \lambda_{min}(S_{2,2}) & -\|S_{2,3}\| & \ldots & -\|S_{2,m}\| \\ \vdots & \vdots & \vdots & \vdots & \vdots \\ -\|S_{m,1}\| & -\|S_{m,2}\| & -\|S_{m,3}\| & \ldots & \lambda_{min}(S_{m,m}) \end{bmatrix} \geq 0.$$

*Proof* Since the matrix

$$A := \begin{bmatrix} \|I - S_{1,1}\| & \|S_{1,2}\| & \|S_{1,3}\| & \dots & \|S_{1,m}\| \\ \|S_{2,1}\| & \|I - S_{2,2}\| & \|S_{2,3}\| & \dots & \|S_{2,m}\| \\ \vdots & \vdots & \vdots & \vdots & \vdots \\ \|S_{m,1}\| & \|S_{m,2}\| & \|S_{m,3}\| & \dots & \|I - S_{m,m}\| \end{bmatrix}$$

is entrywise nonnegative and symmetric, $\|A\| \leq 1$ simply means that $I - A \geq 0$, that is

$$\begin{bmatrix} 1 - \|I - S_{1,1}\| & -\|S_{1,2}\| & -\|S_{1,3}\| & \dots & -\|S_{1,m}\| \\ -\|S_{2,1}\| & 1 - \|I - S_{2,2}\| & -\|S_{2,3}\| & \dots & -\|S_{2,m}\| \\ \vdots & \vdots & \vdots & \vdots & \vdots \\ -\|S_{m,1}\| & -\|S_{m,2}\| & -\|S_{m,3}\| & \dots & 1 - \|I - S_{m,m}\| \end{bmatrix} \geq 0.$$

Further when $0 \leq \mathbf{S} \leq \mathbf{I}$,

$$1 - \|I - S_{j,j}\| = \lambda_{min}(S_{j,j}) \quad \forall\, j = 1, 2, \dots, m.$$

This completes the proof.

There are many interesting results, based on the Perron–Frobenius theory on entrywise nonnegative matrices. (See, for instance, [2].) The most basic is the following assertion.

If $A \in \mathbb{M}_m$ is entrywise nonnegative, then the eigenvalue with maximum modulus is real nonnegative and a corresponding eigenvector is entrywise nonnegative.

When applied to the case when $A$ is symmetric in addition, this produces the following result.

**Theorem 12** (Drew, Johnson and Loewy [5]) *Let* $A = [\alpha_{j,k}]_{j,k} \in \mathbb{M}_m$ *be entrywise nonnegative and symmetric. If its comparison matrix*

$$M(A) := \begin{bmatrix} \alpha_{1,1} & -\alpha_{1,2} & -\alpha_{1,3} & \dots & -\alpha_{1,m} \\ -\alpha_{2,1} & \alpha_{2,2} & -\alpha_{2,3} & \dots & -\alpha_{2,m} \\ \vdots & \vdots & \vdots & \dots & \vdots \\ -\alpha_{m,1} & -\alpha_{m,2} & -\alpha_{m,3} & \dots & \alpha_{m,m} \end{bmatrix}$$

*is positive semi-definite, then there is a decomposition of each diagonal entry*

$$\alpha_{j,j} = \sum_{k \neq j} \lambda_{j,k} \quad with\ \lambda_{j,k} \geq 0\ \forall\, k \neq j$$

*such that*

$$\lambda_{j,k}\lambda_{k,j} \geq \alpha_{j,k}\alpha_{k,j} = |\alpha_{j,k}|^2 \quad \forall\, j \neq k.$$

**Theorem 13** *If*

$$0 \leq \mathbf{S} \leq \mathbf{I} \quad and \quad \|\mathbf{I} - \mathbf{S}\|_{(+)} \leq 1$$

*then* **S** *admits an explicit separable decomposition.*

*Proof* Since by Lemma 11

$$\begin{bmatrix} \lambda_{min}(S_{1,1}) & -\|S_{1,2}\| & -\|S_{1,3}\| & \dots & -\|S_{1,m}\| \\ -\|S_{2,1}\| & \lambda_{min}(S_{2,2})\| & -\|S_{2,3}\| & \dots & -\|S_{2,m}\| \\ \vdots & \vdots & \vdots & \vdots & \vdots \\ -\|S_{m,1}\| & -\|S_{m,2}\| & -\|S_{m,3}\| & \dots & \lambda_{min}(S_{m,m}) \end{bmatrix} \geq 0,$$

by Theorem 12 each $\lambda_{min}(S_{j,j})$ admits a decomposition

$$\lambda_{min}(S_{j,j}) = \sum_{k \neq j} \lambda_{j,k} \quad \exists\, \lambda_{j,k} \geq 0 \quad \forall\, k \neq j$$

such that

$$\lambda_{j,k}\lambda_{k,j} \;\geq\; \|S_{j,k}\|^2 \quad \forall\, k \neq j.$$

Then by Lemma 10

$$E_{j,j} \otimes \lambda_{j,k} \cdot I + E_{j,k} \otimes S_{j,k} + E_{k,j} \otimes S_{k,j} + E_{k,k} \otimes \lambda_{k,j} \cdot I$$

admits an explicit separable decomposition in $\mathbb{M}_m \otimes \mathbb{M}_n$.

Finally by Lemma 9

$$\mathbf{S} \equiv \sum_j E_{j,j} \otimes \Big(S_{j,j} - \lambda_{min}(S_{j,j}) \cdot I\Big)$$

$$+ \sum_{j<k} \Big\{E_{j,j} \otimes \lambda_{j,k} \cdot I + E_{j,k} \otimes S_{j,k} + E_{k,j} \otimes S_{k,j} + E_{k,k} \otimes \lambda_{k,j} \cdot I\Big\}$$

shows that **S** admits an explicit separable decomposition.

**Corollary 14** *An ortho-projection* **S** *of co-rank* 1 *admits an explicit separable decomposition.*

The content of Theorem 13 can be reformulated in the following form.

**Theorem 15** *If* $\mathbf{S} \geq 0$ *and* $\lim_{t\downarrow 0} \frac{1 - \|\mathbf{I} - t \cdot \mathbf{S}\|_{(+)}}{t} \geq 0$, *then* **S** *admits an explicit separable decomposition.*

**Acknowledgements** Supported by KAKENHI Grant Number JP17K05285.

## References

1. Ando, T.: Cones and norms in the tensor product of matrix spaces. Linear Algebra Appl. **379**, 3–41 (2004)
2. Berman, A., Shaked-Monderer, N.: Completely Positive Matrices. World Scientific, New Jersey (2003)
3. Bhatia, R.: Matrix Analysis. Springer, New York (1997)
4. Bhatia, R.: Positive Definite Matrices. Princeton University Press, Princeton (2007)
5. Drew, J.H., Johnson, C.R., Loewy, R.: Completely positive matrices associated with M-matrices. Linear Multilinear Algebra **37**, 303–310 (1994)
6. Gurvits, L., Barnum, H.: Largest Separable Balls Around the Maximally Mixed Bipartite Quantum State. Phys. Rev. A **66**, 062311 (2002)
7. Horodecki, M., Horodecki, P., Horodecki, R.: Separability of mixed states, necessary and sufficient conditions. Phys. Lett. A **233**(1–2), 1–8 (1996)

# Kato's Inequality

**W. Arendt and A. F. M. ter Elst**

**Abstract** The aim of this article is to show some interesting consequences of Kato's inequality. First we show three striking properties of Schrödinger semigroups on $L_1(\mathbb{R}^d)$ (holomorphy and closedness of $-\Delta + V$, the test functions are a core) with the same elegant argument Kato gave, but extending the results to possibly non-symmetric elliptic operators. In the second part, we consider the Dirichlet problem

$$(-\Delta + V)u = 0, \quad u|_{\partial\Omega} = \varphi, \quad u \in C(\overline{\Omega}),$$

where $V \in L_\infty(\Omega, \mathbb{R})$ and $\Omega$ is a bounded Wiener regular set. Well-posedness has been studied in a recent paper (Arendt and ter Elst, Annales de l'Institut Fourier, 2019, [9]). Here we investigate when the maximum principle holds.

## 1 Introduction

In 1972 Kato proved the distributional inequality

$$\mathrm{Re}(\overline{\mathrm{sgn} u}\, \Delta u) \leq \Delta |u|$$

which holds for all $u \in L_{1,\mathrm{loc}}(\mathbb{R}^d)$ such that $\Delta u \in L_{1,\mathrm{loc}}(\mathbb{R}^d)$ and he surprised people proving self-adjointness of Schrödinger operators $-\Delta + V$ with the help of this inequality, and this in a very elegant way. In 1986, in the proceedings of the conference 'Aspects of Positivity in Functional Analysis' he considered the Schrödinger operator $-\Delta + V$ on $L_1(\mathbb{R}^d)$. In this framework Kato's inequality is even more effi-

W. Arendt
Institute of Applied Analysis, University of Ulm, Helmholtzstr. 18, 89069 Ulm, Germany
e-mail: wolfgang.arendt@uni-ulm.de

A. F. M. ter Elst (✉)
Department of Mathematics, University of Auckland,
Private Bag 92019, Auckland 1142, New Zealand
e-mail: terelst@math.auckland.ac.nz

© Springer Nature Switzerland AG 2019

T. M. Rassias and V. A. Zagrebnov (eds.), *Analysis and Operator Theory*,
Springer Optimization and Its Applications 146,
https://doi.org/10.1007/978-3-030-12661-2_3

cient and the results even more surprising. Let $0 \le V \in L_{1,\text{loc}}(\mathbb{R}^d)$. Then the operator $-(-\Delta_1 + V_1)$ with domain $D(-\Delta_1 + V_1) = \{u \in L_{1,\text{loc}}(\mathbb{R}^d) : \Delta u, V u \in L_1(\mathbb{R}^d)\}$ on $L_1(\mathbb{R}^d)$ generates a holomorphic positive $C_0$-semigroup on $L_1(\mathbb{R}^d)$. It is remarkable that the operator $-\Delta_1 + V_1$ is closed, which is a result on the maximal regularity of the equation

$$\lambda u - \Delta u + V u = f$$

in $L_1(\mathbb{R}^d)$, which is in the centre of interest still today. The third surprise is that the test functions are a core for the operator $-\Delta_1 + V_1$.

Our aim in this paper is twofold. First we present these beautiful and elegant arguments of Kato. But we choose a more general framework: the Laplacian is replaced by a possibly non-symmetric elliptic operator. For that choice some additional arguments are needed. The second subject concerns the Dirichlet problem

$$(-\Delta + V)u = 0, \quad u|_{\partial\Omega} = \varphi, \quad u \in C(\overline{\Omega}).$$

Here are given a Wiener regular bounded connected set $\Omega \subset \mathbb{R}^d$ and a potential $V \in L_\infty(\Omega, \mathbb{R})$. Let $-\Delta^D + V$ denote the operator $-\Delta + V$ on $L_2(\Omega)$ subject to Dirichlet boundary conditions. If $0 \notin \sigma(-\Delta^D + V)$, then the Dirichlet problem is well-posed, that is for every $\varphi \in C(\partial\Omega)$ there exists a unique solution $u \in C(\overline{\Omega})$. This result has been proved in [9] in a more general situation of complex elliptic operators. In the context of the Laplacian the arguments are much simpler and will be given here. The main result in [9] is that the Dirichlet problem is well-posed whenever $0 \notin \sigma(-\Delta^D + V)$ even though the maximum principle may fail. More precisely, we investigate two properties: positivity and the maximum principle. For the Dirichlet problem positivity means that the solution $u \ge 0$ whenever the boundary data $\varphi \ge 0$. We show in Theorem 5.1 that this holds if the first eigenvalue $\lambda_1^D$ of $\sigma(-\Delta^D + V)$ is positive. Moreover, it fails if $V \le -\lambda_1^D$ (see Theorem 5.3). In Theorem 5.8 we show that the maximum principle in the form 'if $c \ge 0$ is such that $\varphi \le c\,\mathbb{1}_{\partial\Omega}$, then $u \le c\,\mathbb{1}_{\overline{\Omega}}$' holds if $V \ge 0$. If $V = \alpha\,\mathbb{1}_\Omega$ is constant, then it holds if and only if $\alpha \ge 0$.

When in 1980 the first named author arrived as a postdoc in Berkeley, he was surprised that Kato announced a course entitled 'Positive operators'. The subject were positive operators on ordered spaces in the sense of the Dutch, the Russian and the Tübingen school (cf. [16] Theorem II). Kato used a paper of M. Krein for basic properties, but also the Kakutani representation theorem was in the table of content. He studied positive semigroups (which led to the paper [7]), quadratic forms and the Beurling–Deny criterion. The course culminated in a study of Schrödinger semigroups. Impressive was Kato's beautiful minimalistic style and his simple, elegant arguments. His influence has been and will be long-lasting, far beyond today, 100 years after his birthday.

## 2 Kato's Inequality and Schrödinger Semigroups

In [16] Kato considered the closure $T$ in $L_1(\mathbb{R}^d)$ of the operator $-\Delta + V$ on the domain $\mathscr{D}(\mathbb{R}^d)$, where $0 \leq V \in L_{1,\mathrm{loc}}(\mathbb{R}^d)$. Kato gave a beautiful argument that $-T$ is the generator of a holomorphic $C_0$-semigroup in $L_1(\mathbb{R}^d)$ and he gave a characterisation of the domain of $T$. These results are based on the ingenious Kato inequality

$$\mathrm{Re}(\overline{\mathrm{sgn} u}\, \Delta u) \leq \Delta|u| \tag{1}$$

in $\mathscr{D}'(\mathbb{R}^d)$, whenever $u \in L_{1,\mathrm{loc}}(\Omega)$ is such that $\Delta u \in L_{1,\mathrm{loc}}(\Omega)$, which Kato proved in [14] Lemma A. It means that

$$\int_{\mathbb{R}^d} \varphi\, \mathrm{Re}(\overline{\mathrm{sgn} u}\, \Delta u) \leq \int_{\mathbb{R}^d} |u|\, \Delta\varphi$$

for all $0 \leq \varphi \in \mathscr{D}(\mathbb{R}^d)$. Here

$$(\mathrm{sgn} u)(x) = \begin{cases} \dfrac{u(x)}{|u(x)|} & \text{if } u(x) \neq 0, \\ 0 & \text{if } u(x) = 0. \end{cases}$$

In this section we wish to extend his method to non-symmetric elliptic operators in divergence form with real $W^{1,\infty}$-coefficients.

Throughout this section we fix real-valued $c_{kl} \in W^{1,\infty}(\mathbb{R}^d)$ for all $k, l \in \{1, \ldots, d\}$ and $0 \leq V \in L_{1,\mathrm{loc}}(\mathbb{R}^d)$. We assume that there exists a $\mu > 0$ such that

$$\mathrm{Re} \sum_{k,l=1}^{d} c_{kl}(x)\, \xi_k\, \overline{\xi_k} \geq \mu\, |\xi|^2$$

for all $x \in \mathbb{R}^d$ and $\xi \in \mathbb{C}^d$. We emphasise that we do not assume that $c_{kl} = c_{lk}$ for all $k, l \in \{1, \ldots, d\}$. Consider the operator

$$A = -\sum_{k,l=1}^{d} \partial_k\, c_{kl}\, \partial_l \colon L_{1,\mathrm{loc}}(\mathbb{R}^d) \to \mathscr{D}'(\mathbb{R}^d).$$

Define the operator $A_1$ on $L_1(\mathbb{R}^d)$ with maximal domain

$$D(A_1) = \{u \in L_1(\mathbb{R}^d) : Au \in L_1(\mathbb{R}^d)\}$$

given by $A_1 u = Au$. Then $-A_1$ is the generator of a contractive $C_0$-semigroup, see for example [8] Theorem 3.1 and Lemma 3.7. The domain of $A_1$ is larger than the Sobolev space $W^{2,1}(\mathbb{R}^d)$. Next, define the operator $V_1$ in $L_1(\mathbb{R}^d)$ by

$$D(V_1) = \{u \in L_1(\mathbb{R}^d) : V\,u \in L_1(\mathbb{R}^d)\}$$

and $V_1 u = V\,u$. Then $-V_1$ generates a positive contraction semigroup $T$ on $L_1(\mathbb{R}^d)$ given by $(T_t u)(x) = e^{-tV(x)}\,u(x)$. It is clear that both $A_1$ and $V_1$ are m-accretive and densely defined. Surprising is that $A_1 + V_1$ is m-accretive, with the natural domain $D(A_1) \cap D(V_1)$. Moreover, it is also surprising that the semigroup generated by $-(A_1 + V_1)$ is holomorphic on $L_1$. These two regularity results are still of big actuality. For the Laplacian they were proved by Kato with very elegant arguments involving Kato's inequality. We now extend these results to (possibly non-symmetric) elliptic operators.

**Theorem 2.1** *Consider the operator $A_1 + V_1$ with natural domain*

$$\begin{aligned} D(A_1 + V_1) &= D(A_1) \cap D(V_1) \\ &= \{u \in L_1(\mathbb{R}^d) : Au \in L_1(\mathbb{R}^d) \text{ and } V\,u \in L_1(\mathbb{R}^d)\} \end{aligned}$$

*given by $(A_1 + V_1)u = Au + V\,u$. Then $-(A_1 + V_1)$ generates a positive contractive $C_0$-semigroup on $L_1(\mathbb{R}^d)$, which is a holomorphic semigroup. Moreover, $\mathcal{D}(\mathbb{R}^d)$ is a core for $A_1 + V_1$.*

We need two lemmas. The first is an extension of Kato's inequality. Kato proved this for real symmetric $c_{kl}$ in [14] Lemma A. We allow non-symmetric coefficients and need an additional argument for that.

**Lemma 2.2** *If $u \in L_{1,\mathrm{loc}}(\mathbb{R}^d)$ and $Au \in L_{1,\mathrm{loc}}(\mathbb{R}^d)$, then*

$$\mathrm{Re}(\overline{\mathrm{sgn} u}\; Au) \ge A|u|$$

*in $\mathcal{D}'(\mathbb{R}^d)$.*

*Proof* First suppose in addition that $u \in C^\infty(\mathbb{R}^d)$. Let $\varepsilon > 0$. Define $u_\varepsilon = \sqrt{|u|^2 + \varepsilon}$. If $l \in \{1, \ldots, d\}$ then $u_\varepsilon\, \partial_l u_\varepsilon = \mathrm{Re}(\overline{u}\, \partial_l u)$. Multiplication with the real-valued $c_{kl}$ and taking a partial derivative $\partial_k$ gives

$$\sum_{k,l=1}^d (\partial_k u_\varepsilon)\, c_{kl}\, \partial_l u_\varepsilon + \sum_{k,l=1}^d u_\varepsilon\, \partial_k\, c_{kl}\, \partial_l u_\varepsilon = \mathrm{Re} \sum_{k,l=1}^d \overline{\partial_k u}\, c_{kl}\, \partial_l u + \mathrm{Re} \sum_{k,l=1}^d \overline{u}\, \partial_k\, c_{kl}\, \partial_l u. \qquad (2)$$

We next need an elementary matrix inequality.

If $C$ is a real-valued positive definite $d \times d$ matrix, $C_s = \frac{1}{2}(C + C^*)$ and $\xi, \eta \in \mathbb{R}^d$, then

$$\begin{aligned} \mathrm{Re}\langle C(\xi + i\,\eta), \xi + i\,\eta\rangle &= \langle C_s(\xi + i\,\eta), \xi + i\,\eta\rangle \\ &= \langle C_s\xi, \xi\rangle + \langle C_s\eta, \eta\rangle \ge \langle C_s\xi, \xi\rangle = \langle C\xi, \xi\rangle, \end{aligned}$$

where $\langle \cdot, \cdot\rangle$ is the inner product on $\mathbb{C}^d$.

If $x \in \mathbb{R}^d$, choose $\xi = \mathrm{Re}(\overline{u}\,\nabla u)(x) = (u_\varepsilon \nabla u_\varepsilon)(x)$ and $\eta = \mathrm{Im}(\overline{u}\,\nabla u)(x)$. Then

$$|u_\varepsilon|^2 \sum_{k,l=1}^d (\partial_k u_\varepsilon)\, c_{kl}\, \partial_l u_\varepsilon \le |u|^2 \mathrm{Re} \sum_{k,l=1}^d \overline{\partial_k u}\, c_{kl}\, \partial_l u$$
$$\le |u_\varepsilon|^2 \mathrm{Re} \sum_{k,l=1}^d \overline{\partial_k u}\, c_{kl}\, \partial_l u.$$

Dividing by $|u_\varepsilon|^2$ and using (2) gives

$$\sum_{k,l=1}^d u_\varepsilon\, \partial_k\, c_{kl}\, \partial_l u_\varepsilon \ge \mathrm{Re} \sum_{k,l=1}^d \overline{u}\, \partial_k\, c_{kl}\, \partial_l u.$$

Dividing by $u_\varepsilon$ and taking the limit $\varepsilon \downarrow 0$ gives $A|u| \le \mathrm{Re}(\overline{\mathrm{sgn} u}\, Au)$ in $\mathscr{D}'(\mathbb{R}^d)$.

The condition $u \in C^\infty(\mathbb{R}^d)$ can be dropped by regularising $u$ and is left to the reader.

Occasionally we need duality arguments. We denote by $A^t$ the operator with coefficients $c_{kl}$ replaced by the transposed coefficients $c_{lk}$.

**Lemma 2.3** *Let* $\lambda \in \mathbb{C}$ *with* $\mathrm{Re}\,\lambda \ge 0$. *Moreover, let* $u \in D(A_1 + V_1)$. *Then* $\|V_1 u\|_{L_1(\mathbb{R}^d)} \le \|(\lambda I + A_1 + V_1)u\|_{L_1(\mathbb{R}^d)}$.

*Proof* Let $u \in D(A_1 + V_1)$. Write $f = (\lambda I + A_1 + V_1)u$. Then $\lambda\,|u| + \overline{\mathrm{sgn} u}\, Au + V\,|u| = \overline{\mathrm{sgn} u}\, f$. Taking the real part, it follows from Kato's inequality (1) that

$$\int_{\mathbb{R}^d} \varphi\, V\,|u| = \int_{\mathbb{R}^d} \varphi\, \mathrm{Re}(\overline{\mathrm{sgn} u}\, f) - \int_{\mathbb{R}^d} \varphi\, \mathrm{Re}(\overline{\mathrm{sgn} u}\, Au) - \int_{\mathbb{R}^d} \varphi\,|u|\,\mathrm{Re}\lambda$$
$$\le \int_{\mathbb{R}^d} \varphi\,|f| - \int_{\mathbb{R}^d} |u|\, A^t\varphi \tag{3}$$

for all $0 \le \varphi \in \mathscr{D}(\mathbb{R}^d)$. Fix $\tau \in \mathscr{D}(\mathbb{R}^d)$ such that $0 \le \tau \le 1$ and $\tau|_{B_1(0)} = \mathbb{1}$. For all $n \in \mathbb{N}$ define $\tau_n \in \mathscr{D}(\mathbb{R}^d)$ by $\tau_n(x) = \tau(n^{-1}x)$. Then $\lim_{n\to\infty} \tau_n(x) = 1$ for all $x \in \mathbb{R}^d$ and $\lim_{n\to\infty}(A\tau_n)(x) = 0$ uniformly for all $x \in \mathbb{R}^d$. Moreover, choosing $\varphi = \tau_n$ in (3) and taking the limit $n \to \infty$ gives $\int_{\mathbb{R}^d} V\,|u| \le \int_{\mathbb{R}^d} |f|$, that is $\|V_1 u\|_{L_1(\mathbb{R}^d)} \le \|(\lambda I + A_1 + V_1)u\|_{L_1(\mathbb{R}^d)}$.

*Proof* (*of Theorem* 2.1) Since $\mathscr{D}(\mathbb{R}^d) \subset D(A_1) \cap D(V_1) = D(A_1 + V_1)$, the operator $A_1 + V_1$ is densely defined and accretive. Let $A_{\min} = A|_{\mathscr{D}(\mathbb{R}^d)}$. We shall show that $\mathrm{ran}(I + A_{\min} + V)$ is dense in $L_1(\mathbb{R}^d)$ and $A_1 + V_1$ is closed. Then it follows from the Lumer–Phillips theorem [17] Theorem 3.1 that $-(A_1 + V_1)$ generates a contraction $C_0$-semigroup on $L_1(\mathbb{R}^d)$ and $\mathscr{D}(\mathbb{R}^d)$ is a core.

We first prove that $\mathrm{ran}(I + A_{\min} + V)$ is dense in $L_1(\mathbb{R}^d)$. Let $f \in L_\infty(\mathbb{R}^d)$ and suppose that $\int (\varphi + A\varphi + V\,\varphi)\, f = 0$ for all $\varphi \in \mathscr{D}(\mathbb{R}^d)$. Then $f + A^t f + V f = 0$ in $\mathscr{D}'(\mathbb{R}^d)$. Note that this implies that $-A^t f = f + V f \in L_{1,\mathrm{loc}}(\mathbb{R}^d)$. Moreover,

$$|f| + \overline{\operatorname{sgn} f}\, A^t f + V\,|f| = \overline{\operatorname{sgn} f}\,(f + A^t f + V f) = 0.$$

Therefore $|f| + V\,|f| = -\operatorname{Re}(\overline{\operatorname{sgn} f}\, A^t f) \leq -A^t|f|$ by Kato's inequality in Lemma 2.2 and $(I + A^t)|f| \leq -V\,|f| \leq 0$. Hence

$$\int_{\mathbb{R}^d} |f|\,(I + A_1)\varphi \leq 0 \tag{4}$$

for all $0 \leq \varphi \in \mathscr{D}(\mathbb{R}^d)$. It is well known that $\mathscr{D}(\mathbb{R}^d)$ is a core for $A_1$ (see for example [10] Theorem 9.8). We will improve this by showing that the positive test functions are dense in the positive cone of the domain. Fix $\psi, \tau \in \mathscr{D}(\mathbb{R}^d)$ with $\psi \geq 0$, $\int \psi = 1$, $0 \leq \tau \leq \mathbb{1}$ and $\tau|_{B_1(0)} = \mathbb{1}$. For all $n \in \mathbb{N}$ define $\psi_n, \tau_n \in \mathscr{D}(\mathbb{R}^d)$ by $\psi_n(x) = n^d\, \psi(n\,x)$ and $\tau_n(x) = \tau(n^{-1}\,x)$. Let $0 \leq u \in D(A_1)$. Then $\psi_n * u \in W^{2,1}(\mathbb{R}^d) \subset D(A_1)$ and $\lim_{n\to\infty} \psi_n * u = u$ in $D(A_1)$. Moreover, $\lim_{m\to\infty} \tau_m\,(\psi_n * u) = \psi_n * u$ in $D(A_1)$ for all $n \in \mathbb{N}$. Now $0 \leq \tau_m\,(\psi_n * u) \in \mathscr{D}(\mathbb{R}^d)$ for all $n, m \in \mathbb{N}$. Using (4) one deduces that $\int |f|\,(I + A_1)u \leq 0$. This is for all $0 \leq u \in D(A_1)$. Next let $0 \leq u \in L_1(\mathbb{R}^d)$. Since the semigroup generated by $-A_1$ is a positive semigroup, one deduces that $(I + A_1)^{-1}u \in D(A_1)^+$. So $\int |f|\,u \leq 0$. Therefore $|f| = 0$ and $f = 0$. Consequently $\operatorname{ran}(I + A_{\min} + V)$ is dense in $L_1(\mathbb{R}^d)$.

Next we show that the operator $A_1 + V_1$ is closed. It follows from Lemma 2.3 that $\|V_1 u\|_{L_1(\mathbb{R}^d)} \leq \|(A_1 + V_1)u\|_{L_1(\mathbb{R}^d)}$ for all $u \in D(A_1 + V_1)$. Since both $A_1$ and $V_1$ are closed, also $A_1 + V_1$ is closed.

We proved that $-(A_1 + V_1)$ generates a contraction $C_0$-semigroup on $L_1(\mathbb{R}^d)$ and that $\mathscr{D}(\mathbb{R}^d)$ is a core.

Next we show that the semigroup generated by $-(A_1 + V_1)$ is positive. Let $0 \leq f \in L_1(\mathbb{R}^d)$, $\lambda > 0$ and define $u = (\lambda\, I + A_1 + V_1)^{-1} f$. Then both $(\lambda\, I + A_1 + V_1)u = f$ and $(\lambda\, I + A_1 + V_1)\overline{u} = \overline{f} = f$, so $u = \overline{u}$ and $u$ is real-valued. Moreover, Lemma 2.2 implies that

$$(\lambda\, I + A + V)|u| \leq \lambda\,|u| + (\operatorname{sgn} u)\, Au + V\,|u| = (\operatorname{sgn} u)\, f \leq f = (\lambda\, I + A + V)u.$$

Therefore $(\lambda\, I + A + V)(|u| - u) \leq 0$ and hence $(\lambda\, I + A)(|u| - u) \leq 0$ since $V \geq 0$. So

$$\int_{\mathbb{R}^d} (|u| - u)(\lambda I + A^t)\varphi \leq 0$$

for all $0 \leq \varphi \in \mathscr{D}(\mathbb{R}^d)$. Let $B_\infty$ be the adjoint of the operator $A_1$ on the space $L_\infty(\mathbb{R}^d)$ with weak*-topology. Then

$$\int_{\mathbb{R}^d} (|u| - u)(\lambda I + B_\infty)\varphi \leq 0 \tag{5}$$

for all $0 \leq \varphi \in \mathscr{D}(\mathbb{R}^d)$. Note that $|u| - u \in L_1(\mathbb{R}^d)$. Using a same approximation as above, this time in the weak*-topology, one deduces that (5) is valid for all $0 \leq \varphi \in D(B_\infty)$. Now $-B_\infty$ is the generator of a positive semigroup on $L_\infty(\mathbb{R}^d)$ with the weak*-topology. Therefore if $0 \le \psi \in L_\infty(\mathbb{R}^d)$, then $(\lambda I + B_\infty)^{-1}\psi \in D(B_\infty)^+$ and $\int(|u| - u)\,\psi \le 0$. Hence $|u| - u = 0$ and $u = |u| \ge 0$. So $(\lambda I + A_1 + V_1)^{-1}$ is a positive operator for all $\lambda > 0$. Therefore the semigroup generated by $-(A_1 + V_1)$ is positive.

Finally we show that $-(A_1 + V_1)$ generates a holomorphic semigroup on $L_1(\mathbb{R}^d)$. Since $-(I + A_1)$ generates a bounded holomorphic semigroup on $L_1(\mathbb{R}^d)$ by [8] Theorem 5.4, there exists a $c > 0$ such that $\|(\lambda I + I + A_1)u\|_{L_1(\mathbb{R}^d)} \ge c\,|\lambda|\,\|u\|_{L_1(\mathbb{R}^d)}$ for all $u \in D(\Delta_1)$ and $\lambda \in \mathbb{C}$ with $\mathrm{Re}\lambda > 0$. Then it follows from Lemma 2.3 that

$$c\,|\lambda|\,\|u\|_{L_1(\mathbb{R}^d)} \le \|(\lambda I + I + A_1)u\|_{L_1(\mathbb{R}^d)} \le 2\|(\lambda I + I + A_1 + V_1)u\|_{L_1(\mathbb{R}^d)}$$

for all $u \in D(A_1 + V)$ and $\lambda \in \mathbb{C}$ with $\mathrm{Re}\lambda > 0$. Hence $-(I + A_1 + V_1)$ generates a bounded holomorphic semigroup on $L_1(\mathbb{R}^d)$. This completes the proof of Theorem 2.1.

*Remark 2.4* Holomorphy on $L_1$ is always difficult to prove. Kato's trick, which we used in the above proof and which is based on Kato's inequality, allows one to deduce holomorphy of the perturbed semigroup generated by $-(A_1 + V_1)$ from the holomorphy of the semigroup generated by $-A_1$. In the case of the Laplacian, the latter is the Gaussian semigroup and holomorphy on $L_1$ is obvious. In our more general case, holomorphy on $L_1(\mathbb{R}^d)$ is not so easy to prove. It follows from Gaussian estimates of the semigroup generated by $-A_1$, as we say in the proof. In this case, however, we could have used Gaussian estimates directly. Since $0 \le e^{-t(A_1+V_1)} \le e^{-tA_1}$ for all $t > 0$ (remember that $V \ge 0$), also the perturbed semigroup satisfies Gaussian estimates. The Gaussian kernel bounds imply that there exists a $C_0$-semigroup on $L_2(\mathbb{R}^d)$ which is consistent with $(e^{-t(A_1+V_1)})_{t>0}$. But it is unclear whether this semigroup on $L_2(\mathbb{R}^d)$ is holomorphic since merely $V \in L_{1,\mathrm{loc}}(\mathbb{R}^d)$. Kato's beautiful trick avoids this detour and problem.

For further applications of Kato's inequality we refer to Kato's papers [14, 16], and [19] Section X.4. In [3] the Kato inequality has been used to show that the Dirichlet Laplacian generates a holomorphic semigroup on $L_1(\Omega)$, where $\Omega \subset \mathbb{R}^d$ is open. One may use Kato's inequality in order to give a characterisation of the positivity of the semigroup in terms of its generator, see [1, 18].

*Remark 2.5* Most of Theorem 2.1 can be extended to complex valued $V \in L_{1,\mathrm{loc}}(\mathbb{R}^d)$ for which there exists an $M > 0$ such that $|\mathrm{Im}\,V| \le M\,\mathrm{Re}\,V$. One can define similarly the operator $A_1 + V_1$. It follows that $-(A_1 + V_1)$ is an m-accretive operator, that is, it is the generator of a contraction $C_0$-semigroup on $L_1(\mathbb{R}^d)$. Moreover, this semigroup is holomorphic. The main difference is in Lemma 2.3. If $u \in D(A_1 + V_1)$, then

$$\|(\mathrm{Re}\,V_1)u\|_{L_1(\mathbb{R}^d)} \le \|(\lambda I + A_1 + V_1)u\|_{L_1(\mathbb{R}^d)}$$

for all $\lambda \in \mathbb{C}$ with $\mathrm{Re}\lambda \ge 0$. Since $\|V_1 u\|_{L_1(\mathbb{R}^d)} \le (1 + M)\,\|(\mathrm{Re}\,V_1)u\|_{L_1(\mathbb{R}^d)}$, all other estimates follow with obvious modifications. Of course, the semigroup is no longer positive if $V$ is not real-valued.

A similar extension on $L_2(\mathbb{R}^d)$ for the Laplacian was done by Kato in [15].

Without the ellipticity condition, but merely with a sectoriality condition on the coefficients, it is an open problem whether $\mathscr{D}(\mathbb{R}^d)$ is a core for the generator, even if $V = 0$. In one dimension this is the case on $L_1(\mathbb{R})$ by [12] Corollary 1.6.

## 3 Kato's Inequality and the Maximum Principle

There are other versions of Kato's inequality. For simplicity we restrict ourselves to the Laplacian in the rest of this paper and we restrict to real-valued functions and merely consider real potentials.

Instead of $j(r) = |r|$ one may consider an arbitrary convex, lower semicontinuous function $j : \mathbb{R} \to (-\infty, \infty]$ satisfying $j(0) = 0$. Denote by

$$\partial_j(r) = \{w \in \mathbb{R} : j(s) - j(r) \geq w\,(s - r) \text{ for all } s \in \mathbb{R}\}$$

the **subdifferential** of $j$ at $r$.

**Theorem 3.1** *Let $\Omega \subset \mathbb{R}^d$ be open. Let $u \in L_{1,\mathrm{loc}}(\Omega, \mathbb{R})$ and suppose that $\Delta u \in L_{1,\mathrm{loc}}(\Omega)$. Let $w : \Omega \to \mathbb{R}$ be a measurable function such that $w(x) \in \partial_j(u(x))$ for almost all $x \in \Omega$ and $w\,\Delta u \in L_{1,\mathrm{loc}}(\Omega)$. Then $j \circ u \in L_{1,\mathrm{loc}}(\Omega)$ and*

$$w\,\Delta u \leq \Delta(j \circ u) \tag{6}$$

*in $\mathscr{D}'(\Omega)$.*

*Proof* See [5] Proposition 5.

The inequality (6) means that

$$\int_\Omega w\,(\Delta u)\,\varphi \leq \int_\Omega (j \circ u)\,\Delta\varphi$$

for all $0 \leq \varphi \in \mathscr{D}(\Omega)$. A concrete example is $j(r) = r^+$, which gives the next special case of Theorem 3.1.

**Corollary 3.2** *Let $u \in L_{1,\mathrm{loc}}(\Omega, \mathbb{R})$ be such that $\Delta u \in L_{1,\mathrm{loc}}(\Omega)$. Then*

$$\mathbb{1}_{[u>0]}\,\Delta u \leq \Delta(u^+) \tag{7}$$

*in $\mathscr{D}'(\Omega)$.*

This inequality implies Kato's inequality (1), but the converse implication is not clear since $\Delta$ is not known to be local on $L_{1,\mathrm{loc}}(\Omega)$.

An interesting version of the maximum principle which can be proved with the help of Kato's inequality is the following.

**Theorem 3.3** *Let $\Omega \subset \mathbb{R}^d$ be open. Further let $u \in L_{1,\mathrm{loc}}(\Omega, \mathbb{R})$ be such that $\Delta u \in L_{1,\mathrm{loc}}(\Omega)$ and $-\Delta u \leq 0$ in $\mathscr{D}'(\Omega)$. Let $c \in \mathbb{R}$ be such that $(u-c)^+ \in H_0^1(\Omega)$. Then $u \leq c$.*

Note that $(u-c)^+ \in H_0^1(\Omega)$ is a weak formulation to express that $u \leq c$ at the boundary. If $u \in H^1(\Omega)$, then the result is a special case of [13] Theorem 8.1. But here we merely suppose that $u, \Delta u \in L_{1,\mathrm{loc}}(\Omega)$.

*Proof* (*of Theorem* 3.3) It follows from Kato's inequality (7) that

$$-\Delta((u-c)^+) \leq -\mathbb{1}_{[u>c]}\, \Delta(u-c) = -\mathbb{1}_{[u>c]}\, \Delta u \leq 0.$$

Hence

$$\int_\Omega (\nabla((u-c)^+))\, \nabla v = -\int_\Omega (\Delta((u-c)^+))\, v \leq 0$$

for all $0 \leq v \in \mathscr{D}(\Omega)$. Therefore $\int_\Omega (\nabla((u-c)^+))\, \nabla v \leq 0$ for all $0 \leq v \in H_0^1(\Omega)$ by density. Since $(u-c)^+ \in H_0^1(\Omega)$ one can choose $v = (u-c)^+$ to deduce that $\int_\Omega |\nabla((u-c)^+)|^2 \leq 0$. Using once again that $(u-c)^+ \in H_0^1(\Omega)$ one concludes that $(u-c)^+ = 0$, that is $u \leq c$ almost everywhere on $\Omega$.

## 4 The Dirichlet Problem for $-\Delta + V$

Let $\Omega \subset \mathbb{R}^d$ be open bounded and connected with boundary $\Gamma = \partial\Omega$. We assume that $\Omega$ is **Wiener regular**, that is for all $\varphi \in C(\Gamma)$ there exists a unique $u \in C(\overline{\Omega})$ such that $\Delta u = 0$ and $u|_\Gamma = \varphi$. So the Dirichlet problem is well-posed. Here $\Delta u \in \mathscr{D}'(\Omega)$ is understood as a distribution.

Now let $V \in L_\infty(\Omega, \mathbb{R})$. We consider the **Dirichlet problem**

$$\begin{cases} u \in C(\overline{\Omega}), \\ -\Delta u + V\, u = 0 \text{ in } \mathscr{D}'(\Omega), \\ u|_\Gamma = \varphi. \end{cases}$$

We denote by $\Delta^D$ the **Dirichlet Laplacian**, that is

$$D(\Delta^D) = \{u \in H_0^1(\Omega) : \Delta u \in L_2(\Omega)\}$$

and $\Delta^D u = \Delta u$. Then $-\Delta^D$ is a self-adjoint operator with compact resolvent. Consequently also $-\Delta^D + V$ is self-adjoint with compact resolvent. We denote by $\sigma(-\Delta^D + V)$ its spectrum. Now we can describe well-posedness. The following results are special cases of [9], where elliptic second-order operators in divergence form with complex first-order terms and potentials are considered instead of the Laplacian. The proofs in our case are much simpler.

**Theorem 4.1** *The following are equivalent.*

(i) *For all $\varphi \in C(\Gamma)$ there exists a unique $u \in C(\overline{\Omega})$ such that $-\Delta u + V u = 0$ and $u|_\Gamma = \varphi$.*
(ii) $0 \notin \sigma(-\Delta^D + V)$.

For the proof we need two lemmas.

**Lemma 4.2** *Let $f \in L_\infty(\Omega)$.*

(a) *If $u \in H_0^1(\Omega)$ and $-\Delta u + V u = f$, then $u \in C_0(\Omega)$.*
(b) *If $u \in C_0(\Omega)$ and $-\Delta u + V u = f$, then $u \in H_0^1(\Omega)$.*

*Proof* '(a)' Since $f \in L_\infty(\Omega) \subset L_2(\Omega)$, it follows that $u = (-\Delta^D + V)^{-1} f$. Then the Gaussian kernel bounds of [8] Theorem 3.1 of the semigroup generated by $-(-\Delta^D + V)$ imply that $u \in L_\infty(\Omega)$. Denoting by $\Delta_\infty^D$ the part of $\Delta^D$ in $L_\infty(\Omega)$, the assumption implies that $u \in D(-\Delta_\infty^D + V) = D(-\Delta_\infty^D)$. Now the statement follows from [6] Theorem 2.4(i)⇒(iii).

'(b)' This follows from [6] Lemma 2.2a.

**Lemma 4.3** *Let $V_1, V_2 \in L_\infty(\Omega)$ and suppose that $0 \notin \sigma(-\Delta^D + V_1)$. Let $\varphi \in C(\Gamma)$. Let $u_2 \in C(\overline{\Omega})$ and assume that $-\Delta u_2 + V_2 u_2 = 0$ and $u_2|_\Gamma = \varphi$. Define*

$$u_1 = u_2 + (-\Delta^D + V_1)^{-1}(V_2 - V_1)u_2.$$

*Then $u_1 \in C(\overline{\Omega})$. Moreover, $-\Delta u_1 + V_1 u_1 = 0$ and $u_1|_\Gamma = \varphi$.*

*Proof* It follows from Lemma 4.2(a) that $(-\Delta^D + V_1)^{-1}(V_2 - V_1)u_2 \in C_0(\Omega)$. Therefore $u_1 \in C(\overline{\Omega})$ and $u_1|_\Gamma = \varphi$. Moreover,

$$-\Delta u_1 + V_1 u_1 = -\Delta u_2 + V_1 u_2 + (V_2 - V_1)\, u_2 = -\Delta u_2 + V_2 u_2 = 0$$

as required.

*Proof* *(of Theorem 4.1)* '(i)⇒(ii)' Suppose that $0 \in \sigma(-\Delta^D + V)$. Since $-\Delta^D + V$ has compact resolvent, there exists a $u \in \ker(-\Delta^D + V) \setminus \{0\}$. Then $u \in H_0^1(\Omega)$ and $-\Delta u + V u = 0$. It follows from Lemma 4.2(a) that $u \in C_0(\Omega)$. Hence the uniqueness in Condition (i) of Theorem 4.1 fails if $\varphi = 0$.

'(ii)⇒(i)' Let $\varphi \in C(\Gamma)$. Since $\Omega$ is Wiener regular there exists a unique $u_0 \in C(\overline{\Omega})$ such that $-\Delta u_0 = 0$ and $u_0|_\Gamma = \varphi$. By assumption $0 \notin \sigma(-\Delta^D + V)$. Then one can apply Lemma 4.3 with $V_1 = V$ and $V_2 = 0$ to obtain the existence of a solution of the Dirichlet problem. For the uniqueness part, suppose that $u \in C_0(\Omega)$ and $-\Delta u + V u = 0$. Then $u \in H_0^1(\Omega)$ by Lemma 4.2(b) and $\Delta u = V u \in L_2(\Omega)$. Consequently $u \in D(-\Delta^D + V)$ and $-\Delta^D u + V u = 0$. Since $0 \notin \sigma(-\Delta^D + V)$, one deduces that $u = 0$.

If $0 \notin \sigma(-\Delta^D + V)$, then by Theorem 4.1 one can define the map $B_V : C(\Gamma) \to C(\overline{\Omega})$ by $B_V \varphi = u$, where $u \in C(\overline{\Omega})$ is the unique element such that $-\Delta u + V u =$

0 and $u|_\Gamma = \varphi$. Then $B_V$ is linear and bounded by the closed graph theorem. If $V$ is constant, say $V = \lambda \mathbb{1}_\Omega$ with $\lambda \in \mathbb{R}$, then for simplicity we write $B_\lambda = B_{\lambda \mathbb{1}_\Omega}$.

We obtain from Lemma 4.3 the following useful formula:

**Lemma 4.4** *Let $V_1, V_2 \in L_\infty(\Omega)$. Assume that $0 \notin \sigma(-\Delta^D + V_1)$ and also $0 \notin \sigma(-\Delta^D + V_2)$. Then*

$$B_{V_1} - B_{V_2} = (-\Delta^D + V_1)^{-1}(V_2 - V_1)\, B_{V_2}.$$

## 5 Kato's Inequality and the Dirichlet Problem

Let $\Omega \subset \mathbb{R}^d$ be open bounded connected and Wiener regular, with boundary $\Gamma = \partial\Omega$. Let $V \in L_\infty(\Omega)$. Then $-\Delta^D + V$ is a lower bounded self-adjoint operator with a discrete spectrum. Define

$$\lambda_1(-\Delta^D + V) = \min \sigma(-\Delta^D + V).$$

Then $\lambda_1(-\Delta^D + V)$ is the smallest eigenvalue of the operator $-\Delta^D + V$. Note that $\lambda_1(-\Delta^D + V) \geq 0$ if and only if $-\Delta^D + V$ is a positive self-adjoint operator.

**Theorem 5.1** *Let $V \in L_\infty(\Omega)$. Suppose that $\lambda_1(-\Delta^D + V) > 0$. Then the operator $B_V$ is positive, that is $\varphi \geq 0$ implies $B_V \varphi \geq 0$.*

Theorem 5.1 follows from the following more general assertion.

**Proposition 5.2** *Let $V \in L_\infty(\Omega)$ and assume that $\lambda_1(-\Delta^D + V) > 0$. Let $u \in C(\overline{\Omega})$ and assume that $f = -\Delta u + V\, u \in L_{2,\text{loc}}(\Omega)$. Further assume that $f \leq 0$ and $u|_\Gamma \leq 0$. Then $u \leq 0$.*

*Proof* Since $\Delta u \in L_{2,\text{loc}}(\Omega)$ it follows from elliptic regularity that $u \in H^2_{\text{loc}}(\Omega)$, see [11] Proposition II.3.8. Let $\varepsilon > 0$. Then $(u - \varepsilon)^+ \in C_c(\Omega) \cap H^1_{\text{loc}}(\Omega) \subset H^1_0(\Omega)$. Moreover,

$$-\Delta(u - \varepsilon) + V\,(u - \varepsilon) = f - \varepsilon\, V \leq -\varepsilon\, V \leq \varepsilon\, V^-.$$

Hence $-\mathbb{1}_{[u>\varepsilon]}\, \Delta(u - \varepsilon) + V\,(u - \varepsilon)^+ \leq \varepsilon\, V^-$. By Kato's inequality (7) it follows that $-\Delta((u - \varepsilon)^+) + V\,(u - \varepsilon)^+ \leq \varepsilon\, V^-$. Therefore

$$\int_\Omega (\nabla((u - \varepsilon)^+))\, \nabla v + V\,(u - \varepsilon)^+\, v \leq \varepsilon \int_\Omega V^-\, v$$

first for all $0 \leq v \in \mathscr{D}(\Omega)$ and then by density for all $0 \leq v \in H^1_0(\Omega)$. Choosing $v = (u - \varepsilon)^+$ gives $\int_\Omega |\nabla((u - \varepsilon)^+)|^2 + \int_\Omega V\,|(u - \varepsilon)^+|^2 \leq \varepsilon \int_\Omega V^-\,(u - \varepsilon)^+$. Since

$$\lambda_1(-\Delta^D + V) = \min\{\int_\Omega |\nabla w|^2 + \int_\Omega V\,|w|^2 : w \in H^1_0(\Omega) \text{ and } \|w\|_2 = 1\}$$

it follows that

$$\begin{aligned}\|(u-\varepsilon)^+\|_2\,\lambda_1(-\Delta^D+V) &\le \int_\Omega |\nabla((u-\varepsilon)^+)|^2 + \int_\Omega V\,|(u-\varepsilon)^+|^2 \\ &\le \varepsilon \int_\Omega V^-\,(u-\varepsilon)^+.\end{aligned}$$

Taking the limit $\varepsilon \downarrow 0$ one deduces that $\|u^+\|_2\,\lambda_1(-\Delta^D+V) \le 0$. Hence $u^+ = 0$.

Next we prove a partial converse of Theorem 5.1. Define $\lambda_1^D = \lambda_1^D(-\Delta^D)$.

**Theorem 5.3** *Let $V \in L_\infty(\Omega)$ and suppose that $0 \notin \sigma(-\Delta^D+V)$. Assume that $V \le -\lambda_1^D$. Then the operator $B_V$ is not positive.*

For the proof we use the following facts.

**Lemma 5.4** *Let $\alpha \in \mathbb{R} \cap \rho(\Delta^D)$.*

(a) *The operator $(\alpha I - \Delta^D)^{-1}$ is positive if and only if $\alpha > -\lambda_1^D$.*
(b) $\lim_{\alpha\downarrow -\lambda_1^D} \|(\alpha I - \Delta^D)^{-1}\mathbb{1}_\Omega\|_{L_\infty(\Omega)} = \infty$.

*Proof* Statement (a) follows from [4] Proposition 3.11.2.

'(b)' Let $\Delta_\infty^D$ denote the part of the operator $\Delta^D$ in $L_\infty(\Omega)$. Then $\sigma(-\Delta_\infty^D) = \sigma(-\Delta^D)$ by [6] Proposition 1.4. Hence $\lim_{\alpha\downarrow -\lambda_1^D} \|(\alpha I - \Delta_\infty^D)^{-1}\|_{\mathcal{L}(L_\infty(\Omega))} = \infty$. If $\alpha \in (-\lambda_1^D, \infty)$, then the operator $(\alpha I - \Delta_\infty^D)^{-1}$ is a positive operator, because the kernel of the semigroup generated by $\Delta_\infty^D$ is positive. Finally, if $S \in \mathcal{L}(L_\infty(\Omega))$ is a positive operator, then $\|S\|_{\mathcal{L}(L_\infty(\Omega))} = \|S\mathbb{1}_\Omega\|_{L_\infty(\Omega)}$ and Statement (b) follows.

*Proof (of Theorem 5.3)* Assume that $B_V$ is positive. Let $\alpha \in (-\lambda_1^D, 0)$. Then $\alpha \in \rho(\Delta^D)$ and the operator $(\alpha I - \Delta^D)^{-1}$ is positive by Lemma 5.4(a). Since $V \le -\lambda_1^D \le \alpha$ one has $(V-\alpha) \le 0$. Hence by Lemma 4.4 one deduces

$$B_\alpha = B_V + (\alpha I - \Delta^D)^{-1}\,(V-\alpha)\,B_V \le B_V.$$

In particular, $0 \le B_\alpha \mathbb{1}_\Gamma \le B_V \mathbb{1}_\Gamma$. On the other hand, using again Lemma 4.4 one deduces

$$B_\alpha \mathbb{1}_\Gamma = B_0 \mathbb{1}_\Gamma + (\alpha I - \Delta^D)^{-1}(-\alpha)\,B_0 \mathbb{1}_\Gamma \ge (-\alpha)(\alpha I - \Delta^D)^{-1}\mathbb{1}_\Omega.$$

Since $\lim_{\alpha\downarrow -\lambda_1^D} \|(\alpha I - \Delta^D)^{-1}\mathbb{1}_\Omega\|_{L_\infty(\Omega)} = \infty$ by Lemma 5.4(b), this leads to a contradiction.

Next we discuss comparison. Let $V_1, V_2 \in L_\infty(\Omega)$ be such that $V_1 \le V_2$. Then obviously $\lambda_1(-\Delta^D+V_1) \le \lambda_1(-\Delta^D+V_2)$. If $\Omega$ is connected and $V_1 \ne V_2$, then $\lambda_1(-\Delta^D+V_1) < \lambda_1(-\Delta^D+V_2)$ by [2] Theorem 1.3.

**Proposition 5.5** *Let $V_1, V_2 \in L_\infty(\Omega)$ be such that $V_1 \le V_2$. Moreover, suppose $\lambda_1(-\Delta^D+V_1) > 0$ and the operator $(-\Delta^D+V_1)^{-1}$ is positive. Then $B_{V_2} \le B_{V_1}$.*

*Proof* Since $\lambda_1(-\Delta^D + V_2) \geq \lambda_1(-\Delta^D + V_1) > 0$ the operator $B_{V_2}$ is positive by Theorem 5.1. Then Lemma 4.4 gives that $B_{V_1} - B_{V_2} = (-\Delta^D + V_1)^{-1}(V_2 - V_1)\, B_{V_2}$ is positive.

We next determine the asymptotic behaviour of the solution of the Dirichlet problem when the potential tends to infinity.

**Proposition 5.6** *If $\varphi \in C(\Gamma)$, then $\lim_{\alpha\to\infty} B_\alpha\varphi = 0$.*

*Proof* Using again Lemma 4.4 one deduces that

$$\lim_{\alpha\to\infty} B_\alpha\varphi = \lim_{\alpha\to\infty} B_0\varphi - \alpha\,(\alpha\, I - \Delta^D)^{-1}\, B_0\varphi = 0$$

as required.

Finally we discuss the maximum principle. The next lemma is easy to prove.

**Lemma 5.7** *Let $V \in L_\infty(\Omega)$ and assume that $0 \notin \sigma(-\Delta^D + V)$. Then the following conditions are equivalent.*

(i) *If $c > 0$, $\varphi \in C(\Gamma)$ and $\varphi \leq c\,\mathbb{1}_\Gamma$, then $B_V\varphi \leq c\,\mathbb{1}_{\overline{\Omega}}$.*
(ii) *The operator $B_V$ is positive and $B_V\mathbb{1}_\Gamma \leq \mathbb{1}_{\overline{\Omega}}$.*

Condition (ii) in Lemma 5.7 means that the operator $B_V$ is submarkovian.

**Theorem 5.8**

(a) *Let $0 \leq V \in L_\infty(\Omega)$. Then $B_V$ is submarkovian. Moreover, $B_V\mathbb{1}_\Gamma = \mathbb{1}_{\overline{\Omega}}$ if and only if $V = 0$.*
(b) *Let $\alpha \in \mathbb{R} \setminus \sigma(\Delta^D)$. Then $B_\alpha$ is submarkovian if and only if $\alpha \geq 0$.*

*Proof* '(a)' We already know from Theorem 5.1 that the operator $B_V$ is positive. Lemma 4.4 gives

$$B_V\mathbb{1}_\Gamma = B_0\mathbb{1}_\Gamma + (-\Delta^D + V)^{-1}\Big((0 - V)\, B_0\mathbb{1}_\Gamma\Big) = \mathbb{1}_{\overline{\Omega}} - (-\Delta^D + V)^{-1}V.$$

Hence $B_V\mathbb{1}_\Gamma \leq \mathbb{1}_{\overline{\Omega}}$ and $B_V$ is submarkovian. Moreover, $B_V\mathbb{1}_\Gamma = \mathbb{1}_{\overline{\Omega}}$ if and only if the function $(-\Delta^D + V)^{-1}V = 0$, which is equivalent with $V = 0$.

'(b)' The implication '$\Leftarrow$' follows from Statement (a). So it remains to show '$\Rightarrow$'. Suppose $B_\alpha$ is submarkovian. Then $B_\alpha$ is positive and it follows from Theorem 5.3 that $\alpha > -\lambda_1^D$. If $\alpha < 0$, then Lemma 4.4 gives

$$B_\alpha\mathbb{1}_\Gamma = B_0\mathbb{1}_\Gamma + (\alpha\, I - \Delta^D)^{-1}\,(-\alpha)\, B_0\mathbb{1}_\Gamma = \mathbb{1}_{\overline{\Omega}} + (-\alpha)\,(\alpha\, I - \Delta^D)^{-1}\mathbb{1}_{\overline{\Omega}} > \mathbb{1}_{\overline{\Omega}},$$

where the last inequality follows from Lemma 5.4(a). This gives a contradiction.

**Acknowledgements** Part of this work is supported by the Marsden Fund Council from Government funding, administered by the Royal Society of New Zealand.

## References


1. Arendt, W.: Kato's inequality: a characterisation of generators of positive semigroups. Proc. R. Irish Acad. Sect. A **84**, 155–175 (1984)
2. Arendt, W., Batty, C.J.K.: Domination and ergodicity for positive semigroups. Proc. Am. Math. Soc. **114**, 743–747 (1992)
3. Arendt, W., Batty, C.J.K.: L'holomorphie du semi-groupe engendré par le laplacian Dirichlet sur $L^1(\Omega)$. C. R. Acad. Sci. Paris, Série I **315**, 31–35 (1992)
4. Arendt, W., Batty, C.J.K., Hieber, M., Neubrander, F.: Vector-Valued Laplace Transforms and Cauchy Problems. Monographs in Mathematics, vol. 96. Birkhäuser, Basel (2001)
5. Arendt, W., Bénilan, P.: Inegalités de Kato et semi-groups sous-Markoviens. Revista Matematica Univ. Complutense de Madrid **5**, 279–308 (1992)
6. Arendt, W., Bénilan, P.: Wiener regularity and heat semigroups on spaces of continuous functions. In: Topics in Nonlinear Analysis. The Herbert Amann Anniversary Volume. Progress in Nonlinear Differential Equations, vol. 35, pp. 29–49. Birkhäuser Verlag, Basel (1999)
7. Arendt, W., Chernoff, P.R., Kato, T.: A generalization of dissipativity and positive semigroups. J. Oper. Theory **8**, 167–180 (1982)
8. Arendt, W., ter Elst, A.F.M.: Gaussian estimates for second order elliptic operators with boundary conditions. J. Oper. Theory **38**, 87–130 (1997)
9. Arendt, W., ter Elst, A.F.M.: The Dirichlet problem without the maximum principle. Annales de l'Institut Fourier (2019). In Press
10. Agmon, S.: Lectures on Elliptic Boundary Value Problems. AMS Chelsea Publishing, Providence (2010)
11. Dautray, R., Lions, J.L.: Mathematical Analysis and Numerical Methods for Science and Technology, vol. 1. Springer, Berlin (1990)
12. Do, T.D., ter Elst, A.F.M.: One-dimensional degenerate elliptic operators on $L_p$-spaces with complex coefficients. Semigr. Forum **92**, 559–586 (2016)
13. Gilbarg, D., Trudinger, N.S.: Elliptic Partial Differential Equations of Second Order. Grundlehren der mathematischen Wissenschaften, vol. 224, 2nd edn. Springer, Berlin (1983)
14. Kato, T.: Schrödinger operators with singular potentials. Israel J. Math. **13**, 135–148 (1972)
15. Kato, T.: Remarks on Schrödinger operators with vector potentials. Integral Equ. Oper. Theory **1**, 103–113 (1978)
16. Kato, T.: $L^p$-theory of Schrödinger operators with a singular potential. In: Nagel, R., Schlotterbeck, U., Wolf, M.P.H. (eds.) Aspects of Positivity in Functional Analysis. North-Holland Mathematics Studies, vol. 122, pp. 63–78. Elsevier Science Publishers B.V. (North-Holland), Amsterdam (1986)
17. Lumer, G., Phillips, R.S.: Dissipative operators in a Banach space. Pac. J. Math. **11**, 679–698 (1961)
18. Nagel, R., Uhlig, H.: An abstract Kato inequality for generators of positive operators semigroups on Banach lattices. J. Oper. Theory **6**, 113–123 (1981)
19. Reed, M., Simon, B.: Methods of Modern Mathematical Physics II. Fourier Analysis, Self-Adjointness. Academic Press, New York (1975)

# Tosio Kato's Unpublished Paper

**Claude Bardos and Hisashi Okamoto**

**Abstract** An unpublished manuscript of Tosio Kato was recently found. His paper is published here, after nearly 30 years, together with some comments by the authors.

When Tosio Kato died in Oakland, California, in 1999, he left very many research notes, which were later sent to Japan. According to the will of Mizue Kato (Tosio's widow) who died in 2011, Hiroshi Fujita, Shige Toshi Kuroda and Makio Ishiguro were endowed with these materials and they subsequently asked the second author of the present paper to help them by ordering and inspecting these materials for proper uses in future.

It is only in 2017 that the second author found that one of his notes remained unpublished. It was submitted to Themistocles M. Rassias in 1990 and was later returned to Tosio Kato in the same year. Since then it remained unpublished until now. With the permission of H. Fujita and S.T. Kuroda, we now let Kato's paper be open to public. We are very happy with this publication, since this paper of Kato seems to have historical merit, as will be seen from our comments right after his paper.

The original manuscript was typeset by a word processor by Kato. It is the second author who typeset it by LaTeX and any typo, if there is any, is responsible to him.

On this occasion the authors would like to direct the reader's attention to a volume which was authored by Tosio Kato and edited by Shige Toshi Kuroda entitled "Mathematical Theory of Quantum Mechanics" in Japanese. This volume was finished in 1945 but published only in 2017.[1] The reader of this volume will surely note that

[1] By Kindaikagakusya: ISBN 978-4-7649-0545-0.

C. Bardos
Laboratoire J.-L. Lions, Paris Cedex 05, France

H. Okamoto (✉)
Department of Mathematics, Gakushuin University, Tokyo 171-8588, Japan
e-mail: hisashi.okamoto@gakushuin.ac.jp

© Springer Nature Switzerland AG 2019

T. M. Rassias and V. A. Zagrebnov (eds.), *Analysis and Operator Theory*,
Springer Optimization and Its Applications 146,
https://doi.org/10.1007/978-3-030-12661-2_4

some of Kato's innovative ideas are already there as early as in 1945.

* * *

## A remark on a theorem of C. Bardos on the 2D–Euler equation

Tosio Kato

Department of Mathematics, University of California, Berkeley, CA 94720, USA

The object of this note is to complete the proof of a uniqueness theorem due to Bardos [1]

Consider the Euler equation for an incompressible fluid in a bounded domain $\Omega \subset \mathbb{R}^2$ with $C^2$-boundary $\Gamma$:

$$\begin{aligned} \partial_t u + (u \cdot \partial)u = f - \partial\pi, \qquad \partial \cdot u = 0 \qquad \text{in} \quad \Omega \times (0, T), \\ u \cdot n = 0 \qquad \text{on} \quad \Gamma \times (0, T), \qquad u(x, 0) = a(x); \end{aligned} \tag{1}$$

here $u$ is the velocity field, $\pi$ the pressure, $f$ the external force, and $n$ is the unit normal to $\Gamma$. Denote by $H$ the space of $u \in L^2(\Omega)^2$ such that $\partial \cdot u = 0$ in $\Omega$ and $u \cdot n = 0$ on $\Gamma$, and define the space $V$ similarly with $L^2$ replaced by $H^1$ (the Sobolev space).

We denote by $\| \ \|_p$ the $L^p(\Omega)$-norm, and write $\| \ \|$ for $\| \ \|_2$.

In [1] it is proved, among other things, that if $a \in V$, $\partial \wedge a \in L^\infty(\Omega)$, $f \in L^2((0, T); H)$ and $\partial \wedge f \in L^\infty(\Omega \times (0, T))$, then (1) has a solution $u$ such that

$$u \in L^\infty((0, T); V) \cap C([0, T); H), \tag{2}$$

and

$$\|\partial u(t)\|_p \leq cp, \qquad t \in (0, T), \quad \text{for} \quad 2 \leq p < \infty, \tag{3}$$

where $c$ is a constant depending on $a$ and $f$ but not on $p$. (In what follows we denote similar constants indiscriminately by $c$.) Moreover, $u$ is the *unique solution in the class* (2).

The uniqueness proof in [1] aims at showing that $w \equiv v - u = 0$ if $v$ is another solution of (1) in the class (2). Using (1) and (3), one first deduces the estimate

$$\partial_t \|w\|^2 \leq cp\|w\|_{2p'}^2 \qquad \text{for} \quad 2 \leq p < \infty, \tag{4}$$

where $1/p + 1/p' = 1$. Then one wants to show that $\partial_t \|w\|^2 \leq cp(M(t))^{2/p}\|w\|^{2-2/p}$, where $M(t) = \|w(t)\|_\infty$. Here is a difficulty, however, since it is not known that $w(t) \in L^\infty$. In fact (3) implies that $\|u(t)\|_\infty \leq c$ (because $\Omega$ is bounded), but $v$ need not have this property.

In what follows we shall show how one may overcome this difficulty. For this we need a lemma.

**Lemma 1** $\|\varphi\|_p \leq cp^{1/2}\|\varphi\|_{H^1}$ *for* $2 \leq p < \infty$, $\varphi \in H^1(\Omega)$.

*Proof* By the extension theorem we may assume that $\Omega = \mathbb{R}^2$. If $\hat{\varphi}$ is the Fourier transform of $\varphi$, it follows from the Hölder inequality that

$$\|\hat{\varphi}\|_{p'} \leq \|\langle\cdot\rangle^{-1}\|_q \left\|\langle\cdot\rangle\hat{\varphi}\right\|, \qquad 2 \leq p < \infty, \qquad \frac{1}{q} = \frac{1}{2} - \frac{1}{p} = \frac{1}{p'} - \frac{1}{2},$$

where $\langle\xi\rangle = \left(1 + |\xi|^2\right)^{1/2}$. But it is easy to see that $\|\langle\cdot\rangle^{-1}\|_q \leq c(q-2)^{-1/q} = c(p/2q)^{1/q} \leq cp^{1/2}$ because $q > 2$. Since $\|\langle\cdot\rangle\hat{\varphi}\| = \|\varphi\|_{H^1}$, the desired result follows by the Hausdorff–Young theorem.

To complete the uniqueness proof, we use the Hölder inequality to obtain

$$\|w\|_{2p'} \leq \|w\|^{1-\lambda}\|w\|_p^{\lambda}, \qquad \lambda = 1/(p-2) \leq 1, \qquad 3 \leq p < \infty.$$

Hence (4) gives

$$\partial_t\|w\|^{2\lambda} = \lambda\|w\|^{2\lambda-2}\partial_t\|w\|^2 \leq \lambda\|w\|^{2\lambda-2}cp\|w\|_{2p'}^2 \leq \lambda cp\|w\|_p^{2\lambda}. \tag{5}$$

But $\|w\|_p \leq (cp)^{1/2}$ by Lemma 1, since the $H^1$-norm of $w = v - u$ is bounded due to $u, v, \in L^\infty((0, T); V)$. Since $\lambda = 1/(p-2)$ and $w(0) = 0$, integration of (5) leads to the estimate

$$\|w(t)\|^{2/(p-2)} \leq ct(p/(p-2))(cp)^{1/(p-2)},$$

or

$$\|w(t)\| \leq (ct)^{(p-2)/2}(p/(p-2))^{(p-2)/2}(cp)^{1/2}.$$

Suppose now that $t$ is so small that $ct < 1$. If we let $p \to \infty$, then $(ct)^{(p-2)/2}$ $(cp)^{1/2} \to 0$ while $(p/(p-2))^{(p-2)/2}$ stays bounded. Hence $w(t) = 0$ if $ct < 1$. Since the argument can be repeated, we have proved the required uniqueness.

*Remarks* (a) The theorem of Bardos is a refinement of a similar one due to Judovič [2], in which the stream function is used. In [2] the uniqueness is proved, roughly, in the class $\partial \wedge u \in L^\infty(\Omega \times (0, T))$, which is (essentially) more restrictive than (2).

(b) The same proof applies without change to the case that $\Omega = \mathbb{R}^2$. Also, it is not difficult to extend the existence theorem of [1] to this case.

1. Bardos, C.: Existence et unicité de la solution de l'équation d'Euler en dimension deux. J. Math. Anal. Appl. **40**, 769–790 (1972)
2. Judovič, V.I.: Non-stationary flows of an ideal incompressible fluid (Russian). Ž. Vyčisl. Mat. i Mat. Fiz. **3**, 1032–1066 (1963)

***

The first author is very well aware of this short note of Tosio Kato. In the summer of 1970, he had a fantastic chance to spend 2 weeks with Kato in a workshop organized by Aronszajn at the University of Kansas devoted to general properties of evolution equations in functional analysis. We had a lot of time to discuss the

Euler equations and zero viscosity limit of the Navier–Stokes equations in presence of different types of boundary conditions. Back in France he prepared the paper of 1972 which was sent to Kato and then he later wrote the present unpublished short note.

The first result on the subject goes back to Judovič and several refinements were made (first by J.-L. Lions, described in his book "Quelques Méthodes de Résolution des Problèmes aux Limites Non Linéaires", p. 88) then the first author's contribution and the above note of Kato. The improvement of Kato was in some sense a forerunner of the modern approach which one could describe as weak–strong uniqueness.

The first author takes this opportunity to add a few lines about his scientific debt to Kato: He wrote his thesis in 1969 and by that time the best and the most user-friendly book available for young researchers was Kato's book "Perturbation Theory for Linear Operators" published by Springer. It was bread and butter of the authors' generation.

The first scientific workshop that the first author was attendant was 1969 meeting on Scattering Theory organized by C. H. Wilcox in Flagstaff. Kato was (with Lax and Phillips) among the main lecturers. Not only science but also the super-friendly atmosphere of this meeting marked a corner stone in the research activities of his generation.

The authors guess that most of the contributions of Kato to the mathematical theory of fluid mechanics is quite well documented. However we would like to underline the importance of the Berkeley Note: "Remarks on zero viscosity limit for nonstationary Navier–Stokes flows with boundary.[2] His paper written in Japanese, "On well-posedness",[3] is also related to the present note and highly recommended.

The present note was for many years ignored by our community, probably since it was not published in a genuine journal with good circulation. However in view of recent developments carried out by many authors including the first author, we may say that the present note is a visionary forerunner for the understanding of problems unavoidable both in theory and also in engineering science.

Although Kato did not take action to submit this note after 1990 to any journal, we hope that publishing it in this volume is quite appropriate. In particular with the above comments it gives us an opportunity to emphasize the importance of his contributions in diverse fields and up to the present time. Hence we believe that he would pardon our use of his note in this form.

**Acknowledgements** With great pleasure the second author acknowledges that he owes H. Fujita and S. T. Kuroda a debt of gratitude for their very useful comments. He is supported by JSPS Kakenhi 18H01137.

[2]Seminar on nonlinear partial differential equations, ed. S. S. Chern, Springer (1984), pages 85–98.
[3]Sugaku, *48* (1996), 298–300.

# On the Border Lines Between the Regions of Distinct Solution Type for Solutions of the Friedmann Equation

**Hellmut Baumgärtel**

**Abstract** It is well known that there are four distinct basic types (two Big Bang types, Lemaitre and Big Crunch type) for solutions of the general Friedmann equation with positive cosmological constant, where radiation and matter do not couple (see e.g. Baumgärtel in Journal of Mathematical Physics 122505, 2012, [2]). In that paper, the system of case distinction parameters contains a "critical radiation parameter" $\sigma_{cr}$. The present note contains the constructive description of the so-called *border lines* between Big Bang/Big Crunch type and Big Bang/Lemaitre type for the so-called Hubble solutions of the Friedmann equation by two smooth function branches, expressing the cosmological constant as unique functions of the matter and radiation density (which is considered as a parameter). These functions satisfy simple asymptotic relations with respect to the matter density. They are constructed as the solutions of the equation $\sigma = \sigma_{cr}$.

## 1 Introduction

Einstein's famous field equation of the general relativity theory, linking together the gravitational field with the stress-energy tensor, is also a basic tool in cosmology to solve problems of the large-scale structure of the cosmos. The idea of the existence of a definite large-scale structure of the cosmos is expressed by the "cosmological principle" (E.A. Milne 1933). This principle appears as a boundary condition of Einstein's field equation. This boundary condition consists of a special ansatz of the gravitational field given by a so-called Robertson–Walker(RW-)metric, and of the special form of of the stress-energy tensor as that of a perfect fluid ([1], p. 69). Moreover, it is assumed that matter and radiation do not couple. Then Einstein's field equation reduces to the so-called Friedmann equation

$$\left(\frac{dR}{dt}\right)^2 = \frac{\alpha}{R} + \frac{\sigma}{R^2} + \frac{1}{3}\Lambda c^2 R^2 - \varepsilon c^2 \tag{1}$$

H. Baumgärtel (✉)
Department of Mathematics, University of Potsdam, Potsdam, Germany
e-mail: baumg@uni-potsdam.de

© Springer Nature Switzerland AG 2019

T. M. Rassias and V. A. Zagrebnov (eds.), *Analysis and Operator Theory*,
Springer Optimization and Its Applications 146,
https://doi.org/10.1007/978-3-030-12661-2_5

for the so-called scale-factor $R(t) > 0$, a dimensionless number. The constant parameters are: velocity of light $c$, cosmological constant $\Lambda$, curvature constant $\varepsilon = 0, \pm 1$, matter invariant

$$\alpha := \frac{8\pi G}{3}\rho_{mat}(\tau)R(\tau)^3, \tag{2}$$

and radiation invariant

$$\sigma := \frac{8\pi G}{3}\rho_{rad}(\tau)R(\tau)^4, \tag{3}$$

where $\rho_{mat}$ and $\rho_{rad}$ denote mass and radiation density. Note that the time invariance of the right-hand sides of (2) and (3) is an implication of the assumption that matter and radiation do not couple.

Admissible initial conditions $t_0$, $R_0$ for a solution of Eq. (1) have the property that the right- hand side of (1) is positive for $R := R_0$. For an admissible solution $R(\cdot)$, the so-called Hubble parameter $t \to H(t)$ is defined by

$$H(t) := \left(\frac{1}{R}\frac{dR}{dt}\right)(t). \tag{4}$$

Solutions of (1) whose Hubble parameter at the present time T coincides with the famous *Hubble constant* $H_0$, $H(T) = H_0$, are called *Hubble solutions* in this note.

Using the equations (2), (3), the Hubble solutions satisfy the equation

$$H_0^2 = \frac{8\pi G}{3}\rho_{mat}(T) + \frac{8\pi G}{3}\rho_{rad}(T) + \frac{1}{3}\Lambda c^2 - \varepsilon\frac{c^2}{R(T)^2}. \tag{5}$$

In the special case, $\Lambda = 0, \sigma = 0, \varepsilon = 0$ it reads

$$H_0^2 = \frac{8\pi G}{3}\rho_{mat}(T). \tag{6}$$

This case is the Euclidean–Einstein–de Sitter model of 1932. Since that time the term

$$\rho_{cr} := \frac{3}{8\pi G}H_0^2$$

is called the *critical density*. Introducing instead of $\rho_{mat}(T)$, $\rho_{rad}(T)$, $\Lambda$ the variables

$$x := \frac{\rho_{mat}(T)}{\rho_{cr}}, \quad y := \frac{\frac{1}{3}\Lambda c^2}{H_0^2}, \quad z := \frac{\rho_{rad}(T)}{\rho_{cr}}, \tag{7}$$

then Eq. (5) can be written as

$$1 = x + y + z - \varepsilon\frac{c^2}{H_0^2 R(T)^2} \tag{8}$$

and the constants $\alpha, \sigma, \Lambda$ are given by

$$\alpha = xH_0^2 R(T)^3,\ \sigma = zH_0^2 R(T)^4,\ \Lambda = y\frac{3H_0^2}{c^2}. \tag{9}$$

If $x + y + z - 1 \neq 0$ then

$$R(T)^2 = \left(\frac{c}{H_0}\right)^2 \frac{\varepsilon}{x + y + z - 1}, \tag{10}$$

i.e., in this case the scale-factor $R(T)$ in the present time is uniquely determined. Moreover, one obtains

$$\text{If } x + y + z > 1 \text{ then } \varepsilon = +1, \quad \text{if } x + y + z < 1 \text{ then } \varepsilon = -1.$$

That is,

$$x + y + z = 1 \text{ iff } \varepsilon = 0. \tag{11}$$

This means: the Euclidean case is quite singular, realized only on the plane (11) and in this case the scale-factor is *not* uniquely determined. In any case, the geometric structure of the cosmos is uniquely determined by the expression $x + y + z - 1$. Sometimes the term $\frac{c}{H_0}$ is called the Hubble radius.

## 2 Discriminant and Critical Radiation Parameter

The right-hand side of equation (1) can be written in the form

$$\frac{\Lambda c^2}{3R^2} p(R),$$

where

$$p(R) := Rq(R) + \frac{3}{\Lambda c^2}\sigma,$$

and

$$q(R) := R^3 - \frac{3\varepsilon}{\Lambda} R + \frac{3\alpha}{\Lambda c^2}.$$

is a polynomial of the third degree with discriminant

$$\Delta := \frac{1}{4\Lambda^3 c^4}\left(9\alpha^2 \Lambda - 4\varepsilon c^4\right).$$

Obviously, the type of a solution of (1) depends on the mutual position of $R(T)$ and the positive zeros of $p$. The shape of $p$ is independent of $\sigma$. This fact suggests to use the polynomial $q$ and its discriminant as a first parameter to distinguish different types of solutions.

If $R(\cdot)$ is a Hubble solution, then $\Delta$ can be written in the form

$$\Delta = \frac{1}{4\Lambda^3}\left(\frac{R(T)H_0}{c}\right)^6 D,$$

where

$$D = D(x, y; z) := 27x^2y - 4(x + y + z - 1)^3, \quad x \geq 0, y \geq 0, z \geq 0. \tag{12}$$

Note that for $z = 1$, one has

$$D(x, y; 1) = 27x^2y - 4(x + y)^3 = -(x - 2y)^2(4x + y) \leq 0,$$

hence $D(x, y; 1) = 0$ if and only if $x = 2y$.

For $z > 1$, one has

$$-D(x, y; z) = 4(x + y + z - 1)^3 - 27x^2y > 4(x + y)^3 - 27x^2y = -D(x, y; 1) \geq 0,$$

and $D(x, y; z) < D(x, y; 1) \leq 0$, i.e. $D(x, y; z) < 0$ for all $z > 1$.

This means: only in the case $0 \leq z < 1$ there are regions $D > 0$ in the first quadrant $x \geq 0$, $y \geq 0$.

If $D(x, y; z) < 0$ then we introduce the angle $\phi$, $\frac{\pi}{2} < \phi < \pi$ by

$$\cos\phi = -\frac{\sqrt{27}}{2}\frac{xy^{\frac{1}{2}}}{(x + y + z - 1)^{\frac{3}{2}}}, \quad \frac{\pi}{2} < \phi < \pi, \tag{13}$$

which is associated to the triple $(x, y; z)$. The boundary value $\phi = \pi$ characterizes the triples $(x, y; z)$, where $D = 0$, i.e., the boundary of the region $D < 0$, for example, in the case $z = 1$ the line $2x = y$. That is: fixing a parameter $z = z_0$, then by equation (13) to each $(x, y)$ of the first quadrant $x > 0$, $y > 0$ with $D < 0$ is associated a uniquely determined angle $\phi$, $\frac{\pi}{2} < \phi < \pi$ and the algebraic curves $\phi = const$ exhaust the region $D < 0$ within the first quadrant $x > 0$, $y > 0$. In the following these curves are called the $\phi$-curves.

The second parameter to distinguish different types of solutions of (1) is the critical radiation parameter $\sigma_{cr}$, introduced in [2], defined for triples, where $D < 0$, by

$$\sigma_{cr} = \frac{c^4}{H_0^2}\frac{2}{3}y^{-1}\left(\cos\frac{\psi}{3}\right)^2\left(2\left(\cos\frac{\psi}{3}\right)^2 - 1\right), \quad \cos\psi = \frac{1}{\sqrt{2}}\cos\phi, \quad \frac{3\pi}{4} > \psi > \frac{\pi}{2} \tag{14}$$

a slight modification compared with the expression (12) in [2]. It should be emphasized that $\sigma_{cr}$ is—according to (12) and (13)—a function of $x, y; z$, just as $\sigma$ is according to (9) and (10).

$D$ and $\sigma_{cr}$ are sufficient to describe the two essential statements on the zeros of the polynomial $p$ ([2], Sect. III)

(i) If $D < 0$ and $0 \leq \sigma < \sigma_{cr}$, then there are two positive zeros $R_1 < R_2$ (if $x = 0$ then $R_1 = 0$).
(ii) If $D \geq 0$ and $\sigma > 0$ or $D < 0$ and $\sigma > \sigma_{cr}$ then there is no positive zero.

Hence, in the case $D < 0$ the $\sigma$-value $\sigma := \sigma_{cr}$ separates the cases (i) and (ii), i.e., the solutions of the equation $\sigma = \sigma_{cr}$ are the corresponding border lines within the region $D < 0$.

In the case $D < 0$ and $\sigma = 0$, i.e., $z = 0$, one obtains the explicit expressions

$$R_1(x, y; 0) := \frac{c}{H_0}\frac{2}{\sqrt{3}}y^{-\frac{1}{2}}\cos\left(\frac{1}{3}(2\pi - \phi)\right), \quad R_2(x, y; 0) := \frac{c}{H_0}\frac{2}{\sqrt{3}}y^{-\frac{1}{2}}\cos\frac{1}{3}\phi. \tag{15}$$

Since in the case (i), one has $p(R) < 0$ for $R \in (R_1, R_2)$, this is an forbidden interval for initial conditions $t, R$. However, for Hubble solutions, $R(T) = R(x, y; z)$ given by equation (10) is admissible, i.e., one obtains that either

$$R(x, y; z) > R_2(x, y; z) \tag{16}$$

or

$$R(x, y; z) < R_1(x, y; z). \tag{17}$$

## 3 The Equation $\sigma = \sigma_{cr}$

Also, in the following $z$ is considered as a parameter and the equation is considered in the first quadrant $x \geq 0$, $y \geq 0$. As already mentioned, the terms $\sigma$ and $\sigma_{cr}$ are functions of $(x, y)$ and of the parameter $z$. The function $\sigma_{cr}$ is given by (13). According to equations (9) and (10), one obtains for $\sigma$

$$\sigma = \frac{c^4}{H_0^2}\frac{z}{(x + y + z - 1)^2}. \tag{18}$$

Then, the equation $\sigma = \sigma_{cr}$ reads

$$\frac{4zy}{(x + y + z - 1)^2} = \frac{8}{3}\left(\cos\frac{\psi}{3}\right)^2\left(2\left(\cos\frac{\psi}{3}\right)^2 - 1\right), \tag{19}$$

where $\psi$ is a function of $x$, $y$, and $z$, according to Eqs. (13) and (14). Solution of equation (19) means the construction of functions $x \to Y(x; z_0)$, $x \geq 0$, for every parameter $z_0 > 0$, which satisfy this equation. As a function of $\phi$, the right-hand side of Eq. (19) has a simple structure. In the following, we put

$$\frac{8}{3}\left(\cos\frac{\psi}{3}\right)^2\left(2\left(\cos\frac{\psi}{3}\right)^2 - 1\right) =: F(\phi), \quad \frac{\pi}{2} \leq \phi \leq \pi. \tag{20}$$

One obtains $F(\frac{\pi}{2}) = 1$, $F(\pi) = 0$ and $F(\cdot)$ is strongly monotonically decreasing.

The proofs for the solution of Eq. (19) are considered separately for the parameter regions $0 \leq z < 1, z = 1$ and $z > 1$. The case $z = 1$ is considered first. On the one hand, it is a rather singular case, but on the other hand it is explicitly solvable.

# 4 Results

## *4.1 Theorem*

(i) Let $z = 1$. Then the solution of Eq. (19) consists of two functions (branches)

$$(0, \infty) \ni x \to Y_\pm(x),$$

given by the parameter representation

$$x := x_\pm(\phi), \quad Y_\pm(x) := \mu_\pm(\phi)x_\pm(\phi), \quad \frac{\pi}{2} < \phi < \pi,$$

where

$$\mu_\pm(\phi) := 3\nu_\pm(\phi) - 1, \quad \nu_+(\phi) := \frac{\cos\frac{1}{3}\phi}{|\cos\phi|}, \quad \nu_-(\phi) := \frac{\cos\frac{1}{3}(2\pi - \phi)}{|\cos\phi|} \tag{21}$$

and

$$x_\pm(\phi) := \frac{4}{9}\frac{1}{F(\phi)}\left(\frac{3}{\nu_\pm(\phi)} - \frac{1}{\nu_\pm(\phi)^2}\right). \tag{22}$$

The solutions $Y_\pm(\cdot)$ have the properties

$$0 < Y_-(x) < \frac{1}{2}x < Y_+(x), \quad x > 0, \tag{23}$$

such that

$$Y_\pm(x) = \frac{1}{2}x \pm u_\pm(x), \quad u_\pm(x) > 0,$$

$$\lim_{x\to 0} Y_+(x) = 4, \quad \lim_{x\to 0} Y_-(x) = 0, \tag{24}$$

and

$$\lim_{x \to \infty} \frac{u_{\pm}(x)}{x} = 0. \tag{25}$$

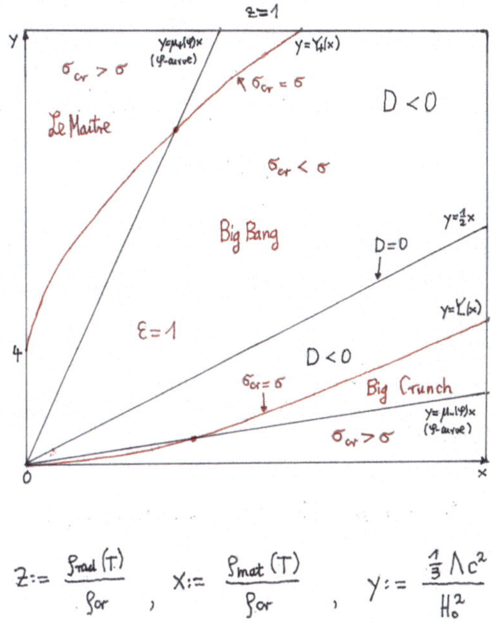


(ii) Let $z > 1$. Then, the solution of Eq. (19) consists of two functions (branches)

$$(0, \infty) \ni x \to Y_{\pm}(x)$$

given by the parameter representation

$$x = x(\phi, \rho_{\pm}(\phi)), \quad Y_{\pm}(x) = \rho_{\pm}(\phi)x(\phi, \rho_{\pm}(\phi)), \quad \frac{\pi}{2} < \phi < \pi, \tag{26}$$

where

$$x(\phi, \mu) := (z-1)\frac{2^{2/3}(\cos\phi)^{2/3}}{3\mu^{1/3} - 2^{2/3}(\cos\phi)^{2/3}(\mu+1)}, \quad \mu_{-}(\phi) < \mu < \mu_{+}(\phi), \tag{27}$$

and the corresponding $\phi$-curve is given by $y_{\pm}(x, \phi) := \mu x(\phi, \mu)$. The terms $\mu := \rho_{\pm}(\phi)$ are uniquely determined solutions of the equation

$$(\cos\phi)^{2/3}\mu^{1/3}(3\mu^{1/3} - 2^{2/3}(\cos\phi)^{2/3}(\mu+1)) = \frac{3^2}{2^{8/3}}\frac{z-1}{z}F(\phi), \tag{28}$$

where

$$\mu_-(\phi) < \rho_-(\phi) < \rho_+(\phi) < \mu_+(\phi). \tag{29}$$

The solutions $Y_\pm(\cdot)$ have the properties $0 < Y_-(x) < Y_+(x),\ x > 0,$

$$0 < y_-(x;\phi) < Y_-(x) < \frac{1}{2}x < Y_+(x) < y_+(x;\phi), \quad x \text{ sufficiently large}, \quad \frac{\pi}{2} < \phi < \pi, \tag{30}$$

such that

$$Y_\pm(x) = \frac{1}{2}x \pm u_\pm(x), \quad u_\pm(x) > 0, \quad x \text{ sufficiently large},$$

$$\lim_{x\to 0} Y_\pm(x) = (1 \pm \sqrt{z})^2, \tag{31}$$

$$\lim_{x\to\infty} \frac{u_\pm(x)}{x} = 0. \tag{32}$$

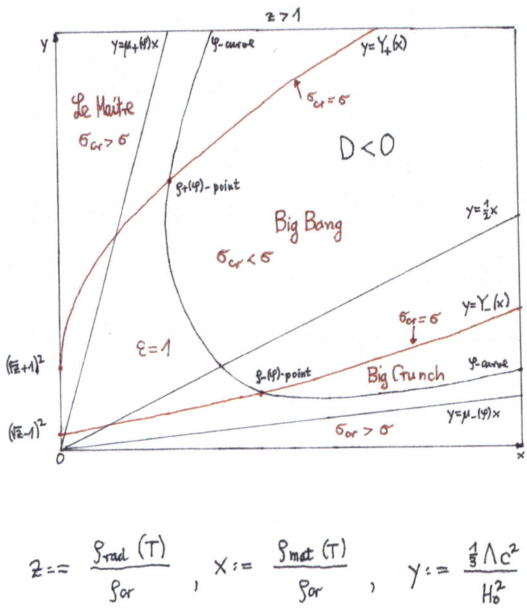


(iii) Let $0 < z < 1$. In this case every $\phi$-curve, $\frac{\pi}{2} < \phi < \pi$ has two branches, a lower one $x \to y_-(x;\phi)$ for $x \geq a$, where $a := 1 - z$, starting at $(a, 0)$, and an upper one $x \to y_+(x;\phi)$ for $x \geq 0$, starting at $(0, a)$. The branches $y_\pm(\cdot;\pi)$ of the $\pi$-curve form the boundary of the region defined by $D < 0$. It corresponds to $D = 0$ (cf. Sect. 2). The solution of Eq. (19) consists of two functions (branches)

$$(0,\infty) \ni x \to Y_+(x), \quad (a,\infty) \ni x \to Y_-(x).$$

The branch $Y_+(\cdot)$ is given by the parameter representation

$$x = x(\phi, \rho_+(\phi)), \quad Y_+(x) = \rho_+(\phi)x(\phi, \rho_+(\phi)), \quad \frac{\pi}{2} < \phi < \pi, \tag{33}$$

where

$$x(\phi, \mu) := (1 - z)\frac{2^{2/3}(\cos\phi)^{2/3}}{2^{2/3}(\cos\phi)^{2/3}(\mu + 1) - 3\mu^{1/3}}, \quad \mu > \mu_+(\phi), \tag{34}$$

and the term $\mu := \rho_+(\phi)$ is the uniquely determined solution of the equation

$$(\cos\phi)^{2/3}\mu^{1/3}(2^{2/3}(\cos\phi)^{2/3}(\mu + 1) - 3\mu^{1/3}) = \frac{3^2}{2^{8/3}}\frac{1-z}{z}F(\phi), \tag{35}$$

where $\rho_+(\phi) > \mu_+(\phi)$. This means: every $\phi$-curve has exactly one intersection with the branch $Y_+(\cdot)$, realized by the upper branch $y_+$ of the $\phi$-curve.

In contrast to this property of $Y_+(\cdot)$, the branch $Y_-(\cdot)$ has either exactly two intersections with a $\phi$-curve or there is no intersection. More precisely:

To every $a, 0 < a < 1$, there is an angle $\phi(a), \frac{\pi}{2} < \phi(a) < \pi$, such that the parameter representation of $Y_-(\cdot)$ is given by

$$x = x(\phi, \rho_-^{\pm}(\phi)), \quad Y_-(x) = \rho_-^{\pm}(\phi)x(\phi, \rho_-^{\pm}(\phi)), \quad \phi(a) < \phi < \pi, \tag{36}$$

where

$$0 < \rho_-^{-}(\phi) < \mu_{max} < \rho_-^{+}(\phi) < \mu_-(\phi), \tag{37}$$

and the left-hand side of Eq. (35) takes its maximum at $\mu_{max}$. Note that for $\phi := \phi(a)$ the terms $\rho_-^{-}(\phi(a))$ and $\rho_-^{+}(\phi(a))$ coincide, $\rho_-^{+}(\phi(a)) = \rho_-^{-}(\phi(a))$. If $\phi < \phi(a)$ then there is no intersection with this $\phi$-curve.

The solutions $Y_\pm(\cdot)$ have the properties

$$0 < Y_-(x) < \frac{1}{2}x < Y_+(x), \; x > a, \quad 0 < Y_+(x), \; x > 0$$

$$0 < y_-(x, \phi) < Y_-(x) < \frac{1}{2}x < Y_+(x;) < y_+(x, \phi), \quad x \text{ sufficiently large}, \tag{38}$$

$$Y_\pm(x) = \frac{1}{2}x \pm u_\pm(x), \quad u_\pm(x) > 0,$$

$$\lim_{x\to a} Y_-(x) = 0, \tag{39}$$

$$\lim_{x\to 0} Y_+(x) = (1 + \sqrt{z})^2, \tag{40}$$

$$\lim_{x\to\infty} \frac{u_{\pm}(x)}{x} = 0. \tag{41}$$

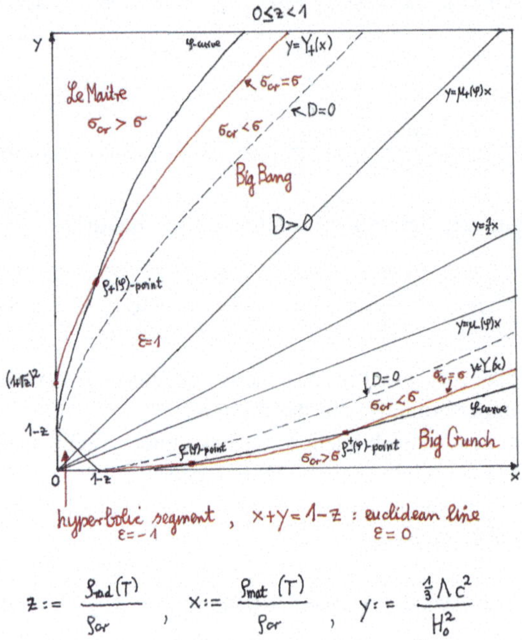


(iv) Let $z = 0$. In this case, one has $\sigma = 0$, i.e., the solution branches of the equation $\sigma = \sigma_{cr}$ coincide with the branches $y_{\pm}(\cdot; \pi)$ of the $\pi$-curve

$$4(x + y - 1)^3 = 27x^2 y,$$

i.e., they coincide with the curve where $D = 0$.

## 4.2 The Regions $G_{\pm}(z)$

In the case $D(x, y; z) < 0$, these regions are defined for $z \geq 1$ by

$$G_+(z) := \{(x, y) : x > 0, y > Y_+(x; z)\}, \quad G_-(z) := \{(x, y) : x > 0, y < Y_-(x; z)\},$$

and for $0 \leq z < 1$ by

$$G_+(z) := \{(x, y) : x > 0, y > Y_+(x : z)\}, \quad G_-(z) := \{(x, y) : x > 1 - z, y < Y_-(x; z)\},$$

such that the region $D(x, y; z) < 0$ is the union $G_+(z) \cup G_-(z)$.

These regions can be also characterized by the mutual position of $R(T) = R(x, y; z)$ and the roots $R_1(x, y; z) < R_2(x, y; z)$ of the polynomial $p(\cdot)$. Actually, one obtains.

### 4.3 Lemma

The regions $G_\pm$ can be characterized as follows:

$$G_+(z) = \{(x, y) : R(x, y; z) > R_2(x, y; z)\} \quad G_-(z) := \{(x, y) : R(x, y; z) < R_1(x, y; z)\}. \tag{42}$$

## 5 Proofs

### 5.1 Proof of the Theorem

(i) Let $z = 1$. First note that

$$\mu_+(\phi) \to \infty \quad \text{and} \quad \mu_-(\phi) \to 0 \quad \text{for} \quad \phi \to \frac{\pi}{2},$$

further

$$\mu_-(\phi) < \frac{1}{2} < \mu_+(\phi), \ \frac{\pi}{2} < \phi < \pi; \quad \mu_\pm(\phi) \to \mu_\pm(\pi) = \frac{1}{2} \ \text{ for } \phi \to \pi. \tag{43}$$

The equation for the $\phi$-curves reads

$$4(x + y)^3(\cos\phi)^2 = 27x^2y, \quad \frac{\pi}{2} < \phi < \pi.$$

The terms $\nu_\pm(\phi)$ satisfy the relation

$$4\nu_\pm(\phi)^3(\cos\phi)^2 + 1 = 3\nu_\pm(\phi)$$

and for the terms $\mu_\pm(\phi)$ one obtains

$$27\mu_\pm(\phi) - 4(1 + \mu_\pm(\phi))^3(\cos\phi)^2 = 0. \tag{44}$$

In the case, $\phi = \pi$ for the left-hand side of (44) one has $4(1 + \mu)^3 - 27\mu = (2\mu - 1)^2(\mu + 4)$. Moreover, one obtains

$$27\mu - 4(1 + \mu)^3(\cos\phi)^2 > 0, \quad \mu_-(\phi) < \mu < \mu_+(\phi). \tag{45}$$

Therefore, the $\phi$-curve consists of the two rays

$$y = \mu_{\pm}(\phi)x, \ x > 0, \quad \frac{\pi}{2} < \phi < \pi. \tag{46}$$

In the case $\phi = \pi$, there is only one ray and for $\phi = \frac{\pi}{2}$ the limit rays are the half axes $x = 0$ and $y = 0$.

The equation $\sigma = \sigma_{cr}$ reads

$$4y = (x + y)^2 F(\phi).$$

For a point $(x, y)$ on a $\phi$-ray (46), this means

$$4\mu_{\pm}(\phi)x = F(\phi)(1 + \mu_{\pm}(\phi))^2 x^2.$$

The solution of this equation is given by (22), which implies $x_-(\phi) < x_+(\phi)$ and

$$x_{\pm}(\phi) \to 0 \quad \text{if } \phi \to \frac{\pi}{2}, \quad x_{\pm}(\phi) \to \infty \quad \text{if } \phi \to \pi.$$

Then $\mu_-(\phi)x_-(\phi) \to 0$ and $\mu_+(\phi)x_+(\phi) \to 4$ for $\phi \to \frac{\pi}{2}$. This proves (24). If $x = x_+(\phi_+) = x_-(\phi_-)$ then $x \to \infty$ iff $\phi_{\pm} \to \pi$. In this case one has $Y_{\pm}(x) = \mu_{\pm}(\phi_{\pm})x$. Because of (43), this proves (23). Finally, if $x \to \infty$ then $\mu_{\pm}(\phi_{\pm}) \to \mu_{\pm}(\pi) = \frac{1}{2}$. This proves (25).

(ii) Let $z > 1$. For convenience put $a := z - 1, a > 0$. The equation for the $\phi$-curves reads

$$4(x + y + a)^3(\cos\phi)^2 = 27x^2 y, \quad \frac{\pi}{2} < \phi < \pi. \tag{47}$$

Note that in this case the limit case $\phi = \pi$ is excluded because $D < 0$ everywhere. The limit case $\phi = \frac{\pi}{2}$ corresponds to the half axes $x = 0, y = 0$. The parameter representation for the $\phi$-curve using $\mu$-rays $y = \mu x, \mu > 0$ yields the term $x = x(\phi, \mu)$ given by Eq. (27). According to equations (44) and (45), one has $x \to \infty$ for $\mu \to \mu_{\pm}(\phi)$. That is, in this case the $\phi$-curve (47) has only a single branch, where $x \geq a\frac{|\cos\phi|}{1-|\cos\phi|}$ for fixed $\phi$. Moreover, the expression $3\mu^{1/3} - 2^{2/3}(1 + \mu)(\cos\phi)^{2/3}$ takes its maximum at $\mu := \frac{1}{2(\cos\phi)^{2/3}}$, i.e., for any $x > a\frac{|\cos\phi|}{1-|\cos\phi|}$ there are parameters $\mu_+, \mu_-$ such that

$$\mu_- < \frac{1}{2(\cos\phi)^{2/3}} < \mu_+$$

and $x = x(\phi, \mu_-) = x(\phi, \mu_+)$. The corresponding values for $y$ are $y_{\pm}(x, \phi) = \mu_{\pm}x$, i.e., one obtains

$$\lim_{x\to\infty} \frac{y_{\pm}(x, \phi)}{x} = \mu_{\pm}(\phi). \tag{48}$$

The equation (19) for a point $(x, y) = (x, \mu x)$ on a fixed $\phi$-curve leads the equation (28). The left-hand side of this equation is positive for parameters $\mu$ satisfying $\mu_-(\phi) < \mu < \mu_+(\phi)$ and vanishes for $\mu = \mu_\pm(\phi)$. It takes its maximal value in this interval at

$$\mu_{max}^{1/3} := \frac{2^{1/6}}{|\cos\phi|^{1/3}} \cos\frac{\psi}{3},$$

where the angle $\psi$ is given by Eq. (14). The corresponding maximum is given by $2^{1/3}\cos\frac{\psi}{3}(2(\cos\frac{\psi}{3})^3 - 3\cos\psi)$. Now the inequality

$$2^{1/3}\cos\frac{\psi}{3}\left(2\left(\cos\frac{\psi}{3}\right)^3 - 3\cos\psi\right) > \frac{3^2}{2^{8/3}} F(\phi)$$

is true, it is equivalent with the inequality $\frac{1}{2}\sqrt{3} > \cos\frac{\psi}{3}$ and $\frac{1}{2}\sqrt{3}$ is the maximal value of $\cos\frac{\psi}{3}$ in the admissible interval for $\psi$ (see Eq. (14)), which is taken at the limit case $\phi = \frac{\pi}{2}$. The consequence is that for every $\phi \in (\frac{\pi}{2}, \pi)$ there are exactly two solutions $\mu := \rho_\pm(\phi)$ of Eq. (28), where the inequality (29) is satisfied. This proves (26). The relation (30) follows from (29). The relations (31) can be obtained by solving Eq. (19) directly for $x = 0$ which implies $\phi = \frac{\pi}{2}$. Concerning relation (32) note that $x \to \infty$ on a fixed $\phi$-curve corresponds to $\mu \to \mu_\pm(\phi)$ and for each $\phi < \pi$ there is a solution $x = x(\phi, \mu)$ with $\mu := \rho_\pm(\phi)$, according to Eq. (26). Taking the limit $\phi \to \pi$ then with Eqs. (48) and (30) one obtains (32).

(iii) Let $0 < z < 1$. The equation for the $\phi$-curve reads

$$4(x + y - a)^3(\cos\phi)^2 = 27x^2 y, \quad \frac{\pi}{2} < \phi < \pi. \tag{49}$$

The limit case $\phi = \frac{\pi}{2}$ corresponds to the half axes $x = 0,\ y \geq a$ and $y = 0,\ x \geq a$. The case $\phi = \pi$ corresponds to $D = 0$. The parameter representation using $\mu$-rays $y = \mu x,\ \mu > 0$, yields the term

$$x = x(\phi, \mu) = a\frac{2^{2/3}(\cos\phi)^{2/3}}{2^{2/3}(\cos\phi)^{2/3}(\mu + 1) - 3\mu^{1/3}}, \quad \mu > \mu_+(\phi),\ \mu < \mu_-(\phi) \tag{50}$$

for the $x$-coordinate of the point of the $\phi$-curve. The parameter values $\mu > \mu_+(\phi)$, $\mu < \mu_-(\phi)$ describe the upper and lower branch $y_\pm(\cdot; \phi)$ of the $\phi$-curve, respectively. Inserting (50) with $\mu > \mu_+(\phi)$ and $y = \mu x$ into Eq. (19) then the resulting Eq. (35) is the condition for those $\mu$, where the intersection point of the $\mu$-ray with the $\phi$-curve is simultaneously a solution point of Eq. (19). The left-hand side of Eq. (35) vanishes for $\mu := \mu_+(\phi)$, it tends to infinity for $\mu \to \infty$ and it is strongly monotonically increasing for $\mu > \mu_+(\phi)$. If $\frac{\pi}{2} < \phi < \pi$ then $0 < F(\phi) < 1$ and $0 < \frac{1-z}{z}F(\phi) < \infty$ because of $0 < z < 1$. That is, for every pair $\{\phi, z\}$ there is exactly one solution $\rho_+(\phi)$ of Eq. (35) and one has $\mu_+(\phi) < \rho_+(\phi)$.

For the investigation of solutions of Eq. (35) in the interval $(0, \mu_-(\phi))$, we write this equation in the form

$$(1-a)(\cos\phi)^{2/3}\mu^{1/3}(2^{2/3}(\cos\phi)^{2/3}(\mu+1)-3\mu^{1/3}) = \frac{3^2}{2^{8/3}}aF(\phi). \tag{51}$$

The extrema of the left-hand side of Eq. (51) as a function of $\mu$ are

$$\mu_{min}^{1/3} := \frac{2^{1/6}}{|\cos\phi|^{1/3}}\cos\frac{\psi}{3}, \quad \mu_{max}^{1/3} := \frac{2^{1/6}}{|\cos\phi|^{1/3}}\cos\frac{1}{3}(2\pi-\psi).$$

First one obtains $\mu_-(\phi) < \mu_{min} < \mu_+(\phi)$. This term has been used for the solution of Eq. (19) in the case $z > 1$. Further, $0 < \mu_{max} < \mu_-(\phi)$. At this term, the left-hand side takes its maximum. Inserting this value into the left-hand side one gets $(1-a)G(\phi)$, where

$$G(\phi) := 3\cdot 2^{1/3}\left(\cos\frac{1}{3}(2\pi-\psi)\right)^2\left(1-2\left(\cos\frac{1}{3}(2\pi-\psi)\right)^2\right). \tag{52}$$

Recall Eq. (14) which implies $\frac{\pi}{4} > \frac{\psi}{3} > \frac{\pi}{6}$ and $\frac{5}{12}\pi < \frac{1}{3}(2\pi-\psi) < \frac{\pi}{2}$. The function $\phi \to G(\phi)$, $\frac{\pi}{2} < \phi < \pi$, is monotonically increasing and

$$G(\frac{\pi}{2}) = 0, \quad G(\pi) = 3\cdot 2^{1/3}\left(\cos\frac{5}{12}\pi\right)^2\left(1-2\left(\cos\frac{5}{12}\pi\right)^2\right) > 0. \tag{53}$$

Comparing the maximum value $G(\phi)$ of the left-hand side of Eq. (51) with the right-hand side $a\cdot 3^2 2^{-8/3}F(\phi)$ and taking into account the monotony properties of $G(\cdot)$ and $F(\cdot)$, further (51) and $F(\frac{\pi}{2}) = 1$, $F(\pi) = 0$ then one obtains: There is exactly one angle $\phi(a)$, $\frac{\pi}{2} < \phi(a) < \pi$, such that

$$(1-a)G(\phi(a)) = a\cdot\frac{3^2}{2^{8/3}}F(\phi(a)).$$

Further, if $\frac{\pi}{2} < \phi < \phi(a)$ then the left-hand side of Eq. (51) at the maximum point $\mu_{max}$ is smaller than the right-hand side at $\phi$; however, if $\phi(a) < \phi < \pi$ then the left-hand side of Eq. (51) at the maximum point is larger than the right-hand side at $\phi$. This implies: In the case $\frac{\pi}{2} < \phi < \phi(a)$, there is no solution $\mu$ of Eq. (51). If $\phi(a) < \phi < \pi$ then there are exactly two solutions $\rho_-^{\pm}(\phi)$ of Eq. (51) such that Eq. (36) is true.

From Eq. (50), it follows that $\lim_{\mu\to 0} x(\phi,\mu) = a$ for all $\phi$, $\frac{\pi}{2} < \phi < \pi$, i.e., the lower branch of the $\phi$-curve starts at $(a, 0)$. This implies (39), according to (37).

Relation (40) can be obtained by solving Eq. (19) directly for $x = 0$, which implies $\phi = \frac{\pi}{2}$. Note that in this case the second (formal) solution $(1-\sqrt{z})^2$ is excluded because of $(1-\sqrt{z})^2 < 1-z$ which implies $D(0,(1-\sqrt{z})^2, z) > 0$, i.e. the pair $(0,(1-\sqrt{z})^2)$ belongs to the region $D > 0$.

Concerning relation (41) note that it follows from (50) that for $\mu > \mu_+(\phi),\ \mu \to \mu_+(\phi)$ or $\mu < \mu_-(\phi),\ \mu \to \mu_-(\phi)$ one has $x(\phi, \mu) \to \infty$. This means that for the two branches $y_\pm(\cdot;\phi)$ of the $\phi$-curves one has

$$\lim_{x\to\infty} \frac{y_\pm(x;\phi)}{x} = \mu_\pm(\phi). \tag{54}$$

However, for each $\phi$, $\phi(a) < \phi < \pi$, there is a solution $x := x(\phi, \mu)$ of equation (19), either with $\mu := \rho_+(\phi)$ according to (35) or with $\mu := \rho_-^+(\phi)$ according to (36) and (37). Taking the limit $\phi \to \pi$ then with Eq. (52) one obtains (41).

## 5.2 *Proof of the Lemma*

First note that, according to equations (10) and (15), in the case $z_0 = 0$ one has to show that

$$(x+y-1)^{-1} > \frac{4}{3}y^{-1}\left(\cos\frac{\phi}{3}\right)^2, \quad (x, y) \in G_+,$$

and

$$(x+y-1)^{-1} < \frac{4}{3}y^{-1}\left(\cos\frac{1}{3}(2\pi-\phi)\right)^2, \quad (x, y) \in G_-.$$

These inequalities can be verified by a direct estimation.

The corresponding property is also valid for $z > 0$. First there are special points satisfying this property, for example, $(0, y)$, where $y > (1+\sqrt{z})^2$, for $G_+(z)$, and, on the other hand, $(x, y)$, where $x > 1$ large and $0 < y < 1$, for $G_-(z)$. Further both regions are simply connected, i.e., an arbitrary point of a region can be smoothly joined within the region with such a special point. Choose a smooth path joining the special point $(x_0, y_0)$, say, of $G_+(z)$ with an arbitrary point $(x_1, y_1)$ of this region. Consider the function $R(x, y; z) - R_2(x, y; z)$ for points of the path. At $(x_0, y_0)$ it is positive. Assume that $R(x_1, y_1; z) < R_1(x_1, y_1; z)$, then the value of the function at this point is smaller than $R_1(x_1, y_1; z) - R_2(x_1, y_1; z) < 0$, i.e. then there is a point on the path such that value of the function at this point is zero which is a contradiction.

# 6 Remark

The asymptotic relation for the functions $u_\pm$ points to a logarithm-like behavior. The asymptotics of these functions could be investigated more precisely. The ray $y = \frac{1}{2}x$ in the description of the border lines characterizes in the case of vanishing radiation the the transition points from deceleration to acceleration.

## References

1. Wald, R.M.: General Relativity. University of Chicago Press, Chicago (1984)
2. Baumgärtel, H.: On a critical radiation density in the Friedmann equation. JMP **53**, 122505 (2012)

# Scattering on Leaky Wires in Dimension Three

**Pavel Exner and Sylwia Kondej**

**Abstract** We consider the scattering problem for a class of strongly singular Schrödinger operators in $L^2(\mathbb{R}^3)$ which can be formally written as $H_{\alpha,\Gamma} = -\Delta + \delta_\alpha(x - \Gamma)$, where $\alpha \in \mathbb{R}$ is the coupling parameter and $\Gamma$ is an infinite curve which is a local smooth deformation of a straight line $\Sigma \subset \mathbb{R}^3$. Using Kato–Birman method, we prove that the wave operators $\Omega_\pm(H_{\alpha,\Gamma}, H_{\alpha,\Sigma})$ exist and are complete.

## 1 Introduction

It is often said that when a great scientist says that something can be done, it can be done, while if the claim is that it cannot be done, he or she is usually wrong; sooner or later a younger one will come and do the impossible work earning a deserved fame. A nice illustration of this effect can be found in the biography of *Tosio Kato* to the centenary of whom the present volume is dedicated. There are various testimonies [23] that John von Neumann who otherwise did so much for the mathematical foundations of quantum mechanics discouraged people from attempts to prove the self-adjointness of atomic Hamiltonians because he considered the task hopelessly beyond reach. Kato's elegant and, in the matter of fact, simple proof [14] was a starting point of the rigorous theory of Schrödinger operators which in the subsequent decades brought a plethora of results and managed to address fundamental questions such as those concerning the stability of matter [18].

P. Exner (✉)
Department of Theoretical Physics, Nuclear Physics Institute, Czech Academy of Sciences, 25068 Řež near Prague, Czechia
e-mail: exner@ujf.cas.cz

Doppler Institute for Mathematical Physics and Applied Mathematics, Czech Technical University, Břehová 7, 11519 Prague, Czechia

S. Kondej
Institute of Physics, University of Zielona Góra, ul. Szafrana 4a, 65246 Zielona Góra, Poland
e-mail: s.kondej@if.uz.zgora.pl

© Springer Nature Switzerland AG 2019

T. M. Rassias and V. A. Zagrebnov (eds.), *Analysis and Operator Theory*, Springer Optimization and Its Applications 146,
https://doi.org/10.1007/978-3-030-12661-2_6

While this may be arguably the most important result of Tosio Kato, from the other point of view, it is just one item in the broad spectrum of his achievements which extended also to some less well-known directions [12]. We believe that other contributions to this volume, in combination with recent reviews [23, 24] will provide a full picture showing how much mathematical physics owes to him. Different people may have different preferences but his *opus magnum*, the monograph [13], will probably come to everybody's mind first. As anybody in the field, the present authors use it regularly and employed also other Kato's results, for instance, his contribution to the theory of product formulæ [17] that inspired us in the discussion of quantum Zeno dynamics [6–8].

The result we are going to present in this note is based on a method at the origin of which Kato left his footprint and which bears his name together with that of Mikhail Birman. The starting point here was two of his 1957 papers [15, 16], which together with the paper by Rosenblum [22] were the origin of the trace-class perturbation theory. Later substantial contributions were made by the others, the said Birman, Kuroda, Putnam, and Pearson, to name just the main ones—for a description of the history we refer to the review [24] or the notes to Sect. XI.3 in [21]—but the starting point was here.

In this note, we are interested in Schrödinger operators with singular "potentials" supported by zero measure sets. In recent years, they were studied as models of "leaky" quantum wires and networks made of them, cf. [11, Chap. 10] for an introduction to the subject and a bibliography. From the mathematical point of view, the parameter which matters is the codimension of the interaction support. If the latter is one, the operators can be treated naturally using the associated quadratic forms in the spirit of [3], for codimension two, the problem is more subtle. The scattering problem in the codimension one case was discussed in [10] where we considered the situation where the singular interaction support is a curve $\Gamma$ in the plane, or more generally a family of curves, which can be regarded as a local deformation of a single straight line $\Sigma$. Under suitable regularity assumptions, we proved there the existence and completeness of the wave operators.

In the present note, we address a similar question in the codimension two case, and for simplicity, we restrict ourselves to the situation when $\Gamma$ is a single curve in $\mathbb{R}^3$ being a smooth local deformation of a straight line $\Sigma$. Note that the scattering problem with singular interactions supported by curves in $\mathbb{R}^3$ has been considered recently[1] in [2], however, our task here is different. The curves in the said paper were supposed to be finite and the Hamiltonian was compared to the one describing the free motion in the three-dimensional space, $-\Delta$ with the usual mathematical-physics license concerning the units. In our case, the comparison operator can be formally written as $H_{\alpha,\Sigma} = -\Delta + \delta_\alpha(x - \Sigma)$, where $\alpha \in \mathbb{R}$ is the parameter characterizing the interaction strength. Using separation of variables and the well-known result about two-dimensional point interactions [1], we find easily that the spectrum is $[\xi_\alpha, \infty)$ where $\xi_\alpha < 0$ is given by (3) below; in addition to motion at positive energies, the system has now a guided mode in which the particle can move remaining localized

[1]See also recent related results in [5, 19].

in the vicinity of $\Sigma$. The scattering will now mean a comparison between $H_{\alpha,\Sigma}$ and the "full" Hamiltonian formally written as

$$H_{\alpha,\Gamma} = -\Delta + \delta_\alpha(x - \Gamma)\,; \tag{1}$$

and a rigorous definition of these operators will be given in the next section. Our aim is to show that the scattering is well defined in this setting, in other words, that the wave operators for the given pair exist and are complete. In fact, the wave operators are also asymptotically complete as one has $\sigma_{sc}(H_{\alpha,\Gamma}) = \emptyset$ under our assumptions, but we leave the proof of this property together with extensions of the result to a wider class of the interaction supports $\Gamma$ to a subsequent paper.

## 2 Preliminaries: The Operator

First, we have to introduce the main notions. Let $\Gamma \subset \mathbb{R}^3$ be an infinite curve of class $C^1$ and piecewise $C^2$ which coincides asymptotically with a straight line $\Sigma$ in a sense to be made precise below. With the usual abuse of notation we regard $\Gamma$ both as a map $\mathbb{R} \to \mathbb{R}^3$ and its image. Furthermore, without loss of generality, we may fix $\Sigma := \{(x_1, 0, 0)\; :\; x_1 \in \mathbb{R}\}$ and to parameterize $\Gamma$ by its arc length $s$ so that we have $|\dot{\Gamma}| = 1$ and $|\Gamma(s) - \Gamma(s')| \le |s - s'|$. We will also suppose that

$$\exists c \in (0, 1) \quad \text{such that} \quad |\Gamma(s) - \Gamma(s')| \ge c|s - s'| \quad \text{for} \quad \forall s, s' \in \mathbb{R}\,, \tag{1}$$

which means, in particular, that the curve $\Gamma$ has no self-intersections and that it cannot be of a U-shaped form.

Our next task is to introduce the Hamiltonian, which will be a singular Schrödinger operator with an interaction supported by the curve $\Gamma$, in other words, a singular perturbation of the "free" operator $H_0$ which is the Laplacian in $L^2(\mathbb{R}^3)$ with the natural domain. There are various ways to do that using, for instance, quadratic forms or a Krein-type formula [2, 4, 20, 25]. For the purpose of the present paper, we recall the method employed in [9] inspired by the classical definition of the two-dimensional $\delta$ interaction [1]; its advantage is that it has a local character. The curve regularity allows us to associate with $\Gamma$, apart from a discrete set, the Frenet's frame, i.e., the triple $(t(s), b(s), n(s))$ of the tangent, binormal, and normal vectors, which are by assumption piecewise continuous functions of $s$. Moreover, at the discontinuity points of $\ddot{\Gamma}$, the Frenet frame limits from the two sides differ by a rotation around the tangent vector, and hence one can construct a globally smooth coordinate system and, with an abuse of notation, employ the symbols $b(s), n(s)$ for the rotated binormal and normal, respectively.

Using this system, we may further introduce a family of "shifted" curve: given $\xi, \eta \in \mathbb{R}$ we denote $r = (\xi^2 + \eta^2)^{1/2}$ and set

$$\Gamma_r^{\xi\eta} := \{\,\Gamma_r^{\xi\eta}(s) := \Gamma(s) + \xi b(s) + \eta n(s)\,,\ s \in \mathbb{R}\}\,,$$

in view of (1) and the smoothness of $\Gamma$ there is an $r_0 > 0$ such that $\Gamma_r^{\xi\eta} \cap \Gamma = \emptyset$ holds for all $r < r_0$. This allows us to define generalized boundary values of a function $f \in H^2_{\mathrm{loc}}(\mathbb{R}^3 \setminus \Gamma)$ using its restriction $f \restriction_{\Gamma_r^{\xi\eta}}(s)$ to $\Gamma_r^{\xi\eta}$ which is by assumption well defined for $0 < r < r_0$. We shall say that a function $f \in H^2_{\mathrm{loc}}(\mathbb{R}^3 \setminus \Gamma) \cap L^2(\mathbb{R}^3)$ belongs to $\Upsilon$ if the limits

$$\begin{aligned} \Xi(f)(s) &:= -\lim_{r\to 0} \frac{1}{\ln r} f \restriction_{\Gamma_r^{\xi\eta}}(s), \\ \Omega(f)(s) &:= \lim_{r\to 0} \left[ f \restriction_{\Gamma_r^{\xi\eta}}(s) + \Xi(f)(s) \ln r \right] \end{aligned}$$

exist a.e. in $\mathbb{R}$ independently of the direction $\frac{1}{r}(\xi, \eta)$ in which they are taken and belong to $L^2(\mathbb{R})$ as functions of $s$. This makes it possible to define the sought singular Schrödinger operator as the restriction of the Laplacian acting on $\mathbb{R}^3 \setminus \Gamma$ to a suitable subset of $\Upsilon$.

To be specific, we fix an $\alpha \in \mathbb{R}$ and define the operator $H_{\alpha,\Gamma}$ as follows:

$$D(H_{\alpha,\Gamma}) = \Upsilon_\alpha := \{ g \in \Upsilon : \; 2\pi\alpha\Xi(g)(s) = \Omega(g)(s) \}, \tag{2a}$$

$$H_{\alpha,\Gamma} f = -\Delta f \quad \text{for} \quad x \in \mathbb{R}^3 \setminus \Gamma. \tag{2b}$$

It was shown in [9] that such an operator is self-adjoint. Note that the absence of a singular interaction means that the singular boundary value $\Xi(f)$ vanishes identically, in other words, the free operator $H_0$ corresponds to $\alpha = \infty$. This fact leads some authors to write the operator in question as $-\Delta - \frac{1}{\alpha}\delta(\cdot - \Gamma)$, see e.g. [2]. This, however, does not fit well with the fact that the two-dimensional $\delta$ interaction is "always attractive", and hence we avoid such formal expressions showing the interaction "strength" and restrict ourselves to the definition (2) in the spirit of [1, Sect. I.5].

Before proceeding further, let us say a few words about the spectrum of $H_{\alpha,\Gamma}$. If the interaction support is a straight line, $\Gamma = \Sigma$, one finds it easily by separation of variables: it is absolutely continuous and equal to

$$\sigma(H_{\alpha,\Gamma}) = [\xi_\alpha, \infty),$$

where

$$\xi_\alpha = -4\, \mathrm{e}^{2(-2\pi\alpha + \psi(1))} \tag{3}$$

is the eigenvalue of the corresponding one-center two-dimensional $\delta$ interaction, with $-\psi(1) \approx 0.57721$ being the Euler–Mascheroni constant. For a non-straight $\Gamma$, the spectrum of $H_{\alpha,\Gamma}$ may be different and depends, in general, on the geometry of $\Gamma$. One of the interesting situations concerns curves that are asymptotically straight. In [9], for instance, we assumed that there are $\omega \in (0, 1)$, $\mu \geq 0$ and $\varepsilon, d > 0$ such that for all $(s, s') \in S_{\omega,\varepsilon}$, we have

$$1 - \frac{|\Gamma(s) - \Gamma(s')|}{|s - s'|} \leq \frac{d|s - s'|}{(|s - s'|+1)(1 + (s^2+s'^2)^{\mu})^{1/2}}, \tag{4}$$

where $S_{\omega,\varepsilon}$ is the subset of $\mathbb{R}^2$ consisting of points $(s, s')$ satisfying $\omega < \frac{s}{s'} < \omega^{-1}$ if $|s + s'| > \varepsilon \, \frac{1+\omega}{1-\omega}$ and $|s - s'| < \varepsilon$ if $|s + s'| < \varepsilon \, \frac{1+\omega}{1-\omega}$. If this assumption is satisfied with some $\mu > \frac{1}{2}$, together with (1), then the essential spectrum is preserved,

$$\sigma_{\text{ess}}(H_{\alpha,\Gamma}) = [\xi_\alpha, \infty),$$

and, in addition, the operator $H_{\alpha,\Gamma}$ has a non-void discrete spectrum whenever the deformation is nontrivial, $\Gamma \neq \Sigma$.

## 3 Preliminaries: The Resolvent

In what follows, we adopt a more restrictive assumption about the curve, namely, we suppose that there exists a compact set $M \subset \mathbb{R}^3$, such that

$$\Gamma \setminus \Sigma \subset M\,. \tag{1}$$

To analyze the scattering problem for the pair $(H_{\alpha,\Gamma}, H_{\alpha,\Sigma})$, we need to know more about the resolvent of singular Schrödinger operator (2). In analogy with the considerations of [9], we begin from the embedding of the free resolvent $R^z := (-\Delta - z)^{-1} : L^2(\mathbb{R}^3) \to W^{2,2}(\mathbb{R}^3)$ to $L^2(\Gamma)$. It is sufficient to restrict the spectral parameter $z$ to negative real values, and hence we consider $z = -\kappa^2$ with $\kappa > 0$ and denote $\mathbf{R}^\kappa = R^{-\kappa^2}$. It is well known that $\mathbf{R}^\kappa$ is integral operator with the kernel determined by the function

$$G^\kappa(x) := \frac{e^{-\kappa|x|}}{4\pi|x|}\,.$$

Specifically, $\breve{\mathbf{R}}^\kappa_\Gamma : L^2(\Gamma) \to L^2(\mathbb{R}^3)$ acts as $\breve{\mathbf{R}}^\kappa_\Gamma f := \int_{\mathbb{R}^3} G^\kappa(\cdot - x) f(x) \delta(x - \Gamma) \mathrm{d}x$, and furthermore, we define $\mathbf{R}^\kappa_\Gamma : L^2(\mathbb{R}^3) \to L^2(\Gamma)$ as the adjoint of $\breve{\mathbf{R}}^\kappa_\Gamma$. To find out the resolvent of $H_{\alpha,\Gamma}$, we define the operator $\check{\mathbf{T}}^\kappa : L^2(\Gamma) \to L^2(\Gamma)$ by

$$(\check{\mathbf{T}}^\kappa f)(s) = -\frac{1}{(2\pi)^2} \int_{\mathbb{R}} \ln(p^2 + \kappa^2)^{1/2} \, e^{ips} \hat{f}(p) \, \mathrm{d}p, \tag{2}$$

where $\hat{f}$ stands for the Fourier transform of $f$ and the maximal domain of this operator is $D(\check{\mathbf{T}}^\kappa) = \{f : \check{\mathbf{T}}^\kappa f \in L^2(\Gamma)\}$. Furthermore, we set

$$\mathbf{T}^\kappa = \check{\mathbf{T}}^\kappa + \frac{1}{2\pi}(\ln 2 + \psi(1)), \tag{3}$$

where $\psi(1)$ is, up to the sign, the Euler–Mascheroni constant mentioned above. Finally, we define the integral operator $\mathbf{B}^\kappa : L^2(\Gamma) \to L^2(\Gamma)$ with the kernel of the form

$$\mathbf{B}^\kappa(s, s') := G^\kappa(\Gamma(s) - \Gamma(s')) - G^\kappa(s - s'),$$

and the operator

$$\mathbf{Q}^\kappa := \mathbf{T}^\kappa + \mathbf{B}^\kappa : D := D(\check{\mathbf{T}}^\kappa) \to L^2(\Gamma). \tag{4}$$

Note that the operator $\mathbf{B}^\kappa$ is positive because the function $G^\kappa$ is monotonous and by assumption, we have $|\Gamma(s) - \Gamma(s')| \leq |s - s'|$; this fact was crucial in [9] to prove that a curve deformation gives rise to the existence of a discrete spectrum of $H_{\alpha,\Gamma}$. By [9, Thm. 2.1] the operator $\alpha - \mathbf{Q}^\kappa : L^2(\Gamma) \to L^2(\Gamma)$ is invertible for all $\kappa$ large enough and

$$\mathbf{R}^\kappa_{\alpha,\Gamma} = \mathbf{R}^\kappa + \check{\mathbf{R}}^\kappa_\Gamma(\alpha - \mathbf{Q}^\kappa)^{-1}\mathbf{R}^\kappa_\Gamma \tag{5}$$

is the resolvent of $H_{\alpha,\Gamma}$. It is not by a chance that this expression has a Krein-like form because $H_{\alpha,\Gamma}$ is a self-adjoint extension of the symmetric operator $-\Delta : C_0^\infty(\mathbb{R}^3 \setminus \Gamma) \to L^2(\mathbb{R}^3)$. Note also that the geometric perturbation is encoded in the part $\mathbf{B}^\kappa$ of (4), we have $\mathbf{B}^\kappa = 0$ if $\Gamma = \Sigma$, and consequently, $\mathbf{Q}^\kappa = \mathbf{T}^\kappa$ holds in this case. The resolvent expression (5) is a tool to prove the spectral properties of $H_{\alpha,\Gamma}$ mentioned at the end of the preceding section.

**Lemma 1** *The operator* $(\alpha - \mathbf{Q}^\kappa)^{-1}$ *is bounded for all* $\kappa$ *large enough.*

*Proof* It follows from (2) that

$$\|\check{\mathbf{T}}^\kappa f\|^2 = \frac{1}{4(2\pi)^3}\int_{\mathbb{R}} \left(\ln(p^2 + \kappa^2)\right)^2 |\hat{f}(p)|^2\, \mathrm{d}p$$

and therefore for all $\kappa$ large enough, we have

$$\|\mathbf{T}^\kappa f\|^2 \geq C(\ln\kappa)^2\|f\|^2$$

with a suitable constant $C$. For the sake of simplicity, we use the symbol $C$ as a generic positive constant which may vary case from case. Furthermore, by [9, Lemma 5.3], the operator $\mathbf{B}^\kappa$ belongs to the Hilbert–Schmidt class under assumption (4), and therefore, *a fortiori*, if we assume (1). This allows to conclude that

$$\|(\alpha - \mathbf{Q}^\kappa)f\|^2 = ((\alpha - \mathbf{T}^\kappa - \mathbf{B}^\kappa)f, (\alpha - \mathbf{T}^\kappa - \mathbf{B}^\kappa)f) \geq C(\ln\kappa)^2\|f\|^2$$

with another constant $C$. On the other hand, we know that operator $(\alpha - \mathbf{Q}^\kappa)^{-1}$ exists and from the above inequality we can conclude that it is bounded. □

## 4 Existence and Completeness of Wave Operators

Now, we are able to pass to our main task in this note, namely, the existence and completeness of the wave operators given by

$$\Omega_{\pm}(H_{\alpha,\Gamma}, H_{\alpha,\Sigma}) := \operatorname{s-lim}_{t\to\pm\infty} \mathrm{e}^{iH_{\alpha,\Gamma}t}\, \mathrm{e}^{-iH_{\alpha,\Sigma}t},$$

where we can skip the projection $P_{\mathrm{ac}}(H_{\alpha,\Sigma})$ usually appearing in the definition because the spectrum of $H_{\alpha,\Sigma}$ is purely absolutely continuous as we have recalled above. For notational convenience, we decompose the line $\Sigma$ into three parts,

$$\Sigma = \Sigma_M \cup \Sigma_+ \cup \Sigma_-,$$

where $\Sigma_M := M \cap \Sigma$ and $\Sigma_\pm$ are the straight components of $\Sigma \setminus M$ which correspond, respectively, to $x_1 \to \pm\infty$ in the chosen coordinate system. Without loss of generality, we may assume that $(0,0,0) \in M$ and

$$\Sigma_\pm = \{x : x = (x_1, 0, 0),\ x_1 \in (-\infty, x_-) \cup (x_+, \infty) \text{ and } x_\pm \gtrless 0\}.$$

In a similar way, one can dissect the curve $\Gamma$ into three parts, $\Gamma_M$ and $\Gamma_\pm = \Sigma_\pm$. The inverses of the "full" and "free" Birman–Schwinger operators, $(\alpha - \mathbf{Q}^\kappa)^{-1}$ and $(\alpha - \mathbf{T}^\kappa)^{-1}$, act, respectively, in $L^2(\Gamma)$ and $L^2(\Sigma)$. To compare the resolvents of $H_{\alpha,\Gamma}$ and $H_{\alpha,\Sigma}$, we introduce the following embeddings:

$$(\alpha - \mathbf{Q}^\kappa)^{-1}_{\Gamma_i\Gamma_j} := \chi_{\Gamma_i}(\alpha - \mathbf{Q}^\kappa)^{-1}\chi_{\Gamma_j} : L^2(\Gamma_j) \to L^2(\Gamma_i),$$

where $i, j = \pm, M$, $\chi_{\Gamma_i}$ is the characteristic functions of $\Gamma_i$, and in the analogous way we define $(\alpha - \mathbf{T}^\kappa)^{-1}_{\Sigma_i\Sigma_j}$. Let us now consider the resolvent difference

$$\mathbf{R}^\kappa_{\alpha,\Gamma} - \mathbf{R}^\kappa_{\alpha,\Sigma} = \breve{\mathbf{R}}^\kappa_\Gamma(\alpha - \mathbf{Q}^\kappa)^{-1}\mathbf{R}^\kappa_\Gamma - \breve{\mathbf{R}}^\kappa_\Sigma(\alpha - \mathbf{T}^\kappa)^{-1}\mathbf{R}^\kappa_\Sigma.$$

Using the obvious fact

$$(\alpha - \mathbf{Q}^\kappa)^{-1}_{\Gamma_\pm\Gamma_\pm} = (\alpha - \mathbf{T}^\kappa)^{-1}_{\Sigma_\pm\Gamma_\pm},$$

we are coming to the conclusion that

$$\mathbf{R}^\kappa_{\alpha,\Gamma} - \mathbf{R}^\kappa_{\alpha,\Sigma} = \sum_{i,j\in\mathscr{X}} \breve{\mathbf{R}}^\kappa_{\Gamma_i}(\alpha - \mathbf{Q}^\kappa)^{-1}_{\Gamma_i\Gamma_j}\mathbf{R}^\kappa_{\Gamma_j} - \sum_{i,j\in\mathscr{X}} \breve{\mathbf{R}}^\kappa_{\Sigma_i}(\alpha - \mathbf{T}^\kappa)^{-1}_{\Sigma_i\Sigma_j}\mathbf{R}^\kappa_{\Sigma_j}, \quad (1)$$

where $\mathscr{X} := \{(i,j) : i, j = +, -, M \wedge (i,j) \neq (+,+), (-,-)\}$. This will be used to prove the following result:

**Theorem 1** *The operator $\mathbf{R}^{\kappa}_{\alpha,\Gamma} - \mathbf{R}^{\kappa}_{\alpha,\Sigma}$ belongs to the trace class for all $\kappa$ large enough.*

Let us start from an auxiliary claim:

**Lemma 2** *We have*

$$\int_{\mathbb{R}^3} G^{\kappa}(y-x)G^{\kappa}(x-z)\,\mathrm{d}x = \frac{1}{8\pi\kappa}\,\mathrm{e}^{-\kappa|y-z|}\,. \tag{2}$$

*Proof* We use the Fourier representation of the Green function

$$\int_{\mathbb{R}^3} G^{\kappa}(y-x)\, f(x)\,\mathrm{d}x = \frac{1}{(2\pi)^{3/2}}\int_{\mathbb{R}^3}\frac{\mathrm{e}^{ipy}}{p^2+\kappa^2}\,\hat{f}(p)\,\mathrm{d}p$$

and apply it to the Green function itself,

$$\int_{\mathbb{R}^3} G^{\kappa}(y-x)G^{\kappa}(x-z)\,\mathrm{d}x = \frac{1}{(2\pi)^3}\int_{\mathbb{R}^3}\frac{\mathrm{e}^{ip(y-z)}}{(p^2+\kappa^2)^2}\,\mathrm{d}p\,.$$

Performing the integration over angles in the integral on the right-hand side, we get

$$I = \int_{\mathbb{R}^3}\frac{\mathrm{e}^{ipy}}{(p^2+\kappa^2)^2}\,\mathrm{d}p = \frac{4\pi}{|y|}\int_0^{\infty}\frac{p\,\sin p|y|}{(p^2+\kappa^2)^2}\,\mathrm{d}p = \frac{2\pi}{i|y|}\int_{\mathbb{R}}\frac{p\,\mathrm{e}^{ip|y|}}{(p^2+\kappa^2)^2}\,\mathrm{d}p\,.$$

Next, we extend in a standard way the integration forming the contour by adding upper semicircle and using the Jordan's lemma which implies that the integral over the semicircle vanishes in the limit of infinite radius. This gives

$$I = \frac{2\pi}{i|y|}\oint\frac{z\,\mathrm{e}^{iz|y|}}{(z^2+\kappa^2)^2}\,\mathrm{d}z = \frac{4\pi^2}{|y|}\sum_{\textit{upper halfplane}}\mathrm{Res}\,\frac{z\mathrm{e}^{iz|y|}}{(z^2+\kappa^2)^2}\,.$$

Using now the generalized Cauchy integral formulae, one gets

$$I = \frac{4\pi^2}{|y|}\left(\frac{d}{dz}\frac{z\mathrm{e}^{iz|y|}}{(z+i\kappa)^2}\right)\Big|_{z=i\kappa} = \frac{\pi^2}{\kappa}\,\mathrm{e}^{-\kappa|y|}\,.$$

On putting these results together, we arrive at the formula (2). □

*Proof of Theorem* 1. Let us pick one of the components of $\mathbf{R}^{\kappa}_{\alpha,\Gamma} - \mathbf{R}^{\kappa}_{\alpha,\Sigma}$, for instance, $\breve{\mathbf{R}}^{\kappa}_{\Gamma_+}(\alpha-\mathbf{Q}^{\kappa})^{-1}_{\Gamma_+\Gamma_-}\mathbf{R}^{\kappa}_{\Gamma_-}$. The symbol $\mathscr{B}_\delta$ will conventionally denote the ball of radius $\delta$ centered at the origin and $\chi_{\mathscr{B}_\delta}$ stands for the characteristic function of the ball $\mathscr{B}_\delta$. We define the "cut-off" operator family

$$S^{+-}_{\delta} \equiv S_\delta := \chi_{\mathscr{B}_\delta}\left(\breve{\mathbf{R}}^{\kappa}_{\Gamma_+}(\alpha-\mathbf{Q}^{\kappa})^{-1}_{\Gamma_+\Gamma_-}\mathbf{R}^{\kappa}_{\Gamma_-}\right)\chi_{\mathscr{B}_\delta}$$

and ask for its weak limit as $\delta \to \infty$. We have, in particular,

$$\int_{\mathbb{R}^3} S_\delta(x,x)\mathrm{d}x = \int_{\mathbb{R}^3} \Big( G^\kappa(\cdot - x)\chi_{\mathscr{B}_\delta}(x), (\alpha - \mathbf{Q}^\kappa)^{-1}_{\Gamma_+\Gamma_-} G^\kappa(\cdot - x)\chi_{\mathscr{B}_\delta}(x) \Big)_{L^2(\Gamma_+)} \mathrm{d}x\,. \tag{3}$$

Using now the Lebesgue's dominated convergence theorem in combination with Lemma 2 and the equivalence that $\mathrm{e}^{-\kappa|x-y|} = \mathrm{e}^{-\kappa(|x|+|y|)}$ holds, in view of that fact that $\Gamma_\pm = \Sigma_\pm = \{x\,:\, x = (x_1, 0, 0),\ x_1 \in (-\infty, x_-) \cup (x_+, \infty)\ \text{ and }\ x_\pm \gtrless 0\}$, we obtain

$$\lim_{\delta\to\infty} \int_{\mathbb{R}^3} S_\delta(x,x)\,\mathrm{d}x = \frac{\pi^4}{\kappa^2} \int_{\Gamma_+} e^{-\kappa|s|} \left( (\alpha - \mathbf{Q}^\kappa)^{-1}_{\Gamma_+\Gamma_-} e^{-\kappa|\cdot|} \right)(s)\,\mathrm{d}s\,. \tag{4}$$

Using further the boundedness of $(\alpha - \mathbf{Q}^\kappa)^{-1}_{\Gamma_+\Gamma_-}$ in combination with Schwarz inequality, we get from (4) the following estimate:

$$\lim_{\delta\to\infty} \int_{\mathbb{R}^3} S_\delta(x,x)\mathrm{d}x \le C\frac{\pi^4}{\kappa^2} \|e^{-\kappa|\cdot|}\|_{L^2(\Gamma_+)} \|e^{-\kappa|\cdot|}\|_{L^2(\Gamma_-)}\,,$$

where the constant $C$ is the norm of $(\alpha - \mathbf{Q}^\kappa)^{-1}_{\Gamma_+\Gamma_-}$. We want to conclude that

$$\int_{\mathbb{R}^3} S_\delta(x,x)\mathrm{d}x \to \operatorname{Tr}\, \breve{\mathbf{R}}^\kappa_{\Gamma_+} (\alpha - \mathbf{Q}^\kappa)^{-1}_{\Gamma_+\Gamma_-} \mathbf{R}^\kappa_{\Gamma_-}$$

as $\delta \to \infty$ and to show in this way that the operator $\breve{\mathbf{R}}^\kappa_{\Gamma_+} (\alpha - \mathbf{Q}^\kappa)^{-1}_{\Gamma_+\Gamma_-} \mathbf{R}^\kappa_{\Gamma_-}$ belongs to the trace class. According to the lemma following Theorem XI.31 in [21], the trace can be expressed through the integral of the kernel diagonal provided the latter is continuous in both arguments and the operator is positive. The continuity was mentioned already, the positivity follows from the fact that the operator $\alpha - \mathbf{Q}^\kappa$ is positive from all $\kappa$ large enough, cf. [9, Lemma 5.5].

Let us next consider the component of (1) referring to the operator acting between the spaces $L^2(\Gamma_+)$ and $L^2(\Gamma_M)$. We put

$$S_\delta^{M+} \equiv S_\delta := \chi_{\mathscr{B}_\delta} \left( \breve{\mathbf{R}}^\kappa_{\Gamma_M} (\alpha - \mathbf{Q}^\kappa)^{-1}_{\Gamma_M\Gamma_+} \mathbf{R}^\kappa_{\Gamma_+} \right) \chi_{\mathscr{B}_\delta}\,.$$

Applying again Lemma 2 in combination with the Lebesgue's dominated convergence theorem, one obtains

$$\lim_{\delta\to\infty} \int_{\mathbb{R}^3} S_\delta(x,x)\,\mathrm{d}x = \frac{\pi^4}{\kappa^2} \int_{\Gamma_M} \left( (\alpha - \mathbf{Q}^\kappa)^{-1}_{\Gamma_M\Gamma_+} \mathrm{e}^{-\kappa|\Gamma(\cdot)-\Gamma(s)|} \right) \mathrm{d}s\,. \tag{5}$$

Using further the boundedness of $(\alpha - \mathbf{Q}^\kappa)^{-1}_{\Gamma_M\Gamma_+}$ and the continuous imbedding of spaces $L^2(\Gamma_M) \hookrightarrow L^1(\Gamma_M)$ together with the Fubini's theorem and Schwarz inequality, we infer that

$$\left|\lim_{\delta\to\infty}\int_{\mathbb{R}^3} S_\delta(x,x)\,\mathrm{d}x\right| \le \frac{\pi^4}{\kappa^2}\int_{\Gamma_M}\left|\left((\alpha-\mathbf{Q}^\kappa)^{-1}_{\Gamma_M\Gamma_+}\mathrm{e}^{-\kappa|\Gamma(\cdot)-\Gamma(s)|}\right)\right|\mathrm{d}s$$
$$\le \frac{\pi^4}{\kappa^2}|\Gamma_M|\left(\int_{\Gamma_M}\left|(\alpha-\mathbf{Q}^\kappa)^{-1}_{\Gamma_M\Gamma_+}\mathrm{e}^{-\kappa|\Gamma(\cdot)-\Gamma(s)|}\right|^2\mathrm{d}s\right)^{1/2}. \qquad (6)$$

The integral on the right-hand side of (6) is finite because the integrated function belongs to $L^2(\Gamma_M)$. This implies that $\check{\mathbf{R}}^\kappa_{\Gamma_M}(\alpha-\mathbf{Q}^\kappa)^{-1}_{\Gamma_M\Gamma_+}\mathbf{R}^\kappa_{\Gamma_+}$ belongs to the trace class in the same way as above.

The remaining components of $\mathbf{R}^\kappa_{\alpha,\Gamma}-\mathbf{R}^\kappa_{\alpha,\Sigma}$ contributing to formula (1) can be dealt with in the analogous way. The only terms which do not allow for such a treatment are those containing $(\alpha-\mathbf{Q}^\kappa)^{-1}_{\Gamma_\pm\Gamma_\pm}$, however, they cancel when we subtract $\mathbf{R}^\kappa_{\alpha,\Gamma}$ from $\mathbf{R}^\kappa_{\alpha,\Sigma}$. Concluding the above discussion, we thus find that the difference $\mathbf{R}^\kappa_{\alpha,\Gamma}-\mathbf{R}^\kappa_{\alpha,\Sigma}$ is a trace-class operator for all $\kappa$ large enough what we have set out to prove. □

Now, we are in a position to present the result indicated in the introduction:

**Corollary 1** *In the stated assumptions, the wave operators $\Omega_\pm(H_{\alpha,\Gamma},H_{\alpha,\Sigma})$ exist and are complete.*

*Proof* In view of Theorem 1, the claim follows immediately from Kuroda–Birman theorem, cf. [21, Thm. XI.9]. □

**Acknowledgements** The research was supported by the Czech Science Foundation (GAČR) within the project 17-01706S and the EU project CZ.02.1.01/0.0/0.0/16_019/0000778.

## References

1. Albeverio, S., Gesztesy, F., Høegh-Krohn, R., Holden, H.: Solvable Models in Quantum Mechanics, 2nd edn. AMS Chelsea Publishing, Providence (2005)
2. Behrndt, J., Frank, R.L., Kühn, Ch., Lotoreichik, V., Rohleder, J.: Spectral theory for Schrödinger operators with $\delta$-interactions supported on curves in $\mathbb{R}^3$. Ann. H. Poincaré **18**, 1305–1347 (2017)
3. Brasche, J.F., Exner, P., Kuperin, YuA, Šeba, P.: Schrödinger operators with singular interactions. J. Math. Anal. Appl. **184**, 112–139 (1994)
4. Brasche, J.F., Teta, A.: Spectral analysis and scattering for Schrödinger operators with an interaction supported by a regular curve. In: Albeverio, S., Fenstadt, J.E., Holden, H., Lindstrøm, T. (eds.) Ideas and Methods in Quantum and Statistical Physics, pp. 197–211. Cambridge University Press, Cambridge (1992)
5. Cacciapuoti, A., Fermi, D., Posilicano, A.: Scattering from local deformations of a semitransparent plane. J. Math. Anal. Appl. **473**, 215–257 (2019)
6. Exner, P., Ichinose, T.: A product formula related to quantum Zeno dynamics. Ann. H. Poincaré **6**, 195–215 (2005)
7. Exner, P., Ichinose, T., Kondej, S.: On relations between stable and Zeno dynamics in a leaky graph decay model. In: Proceedings of the Conference "Operator Theory and Mathematical Physics" (Będlewo 2004); Operator Theory: Advances and Applications, vol. 174, pp. 21–34. Basel, Birkhäuser (2007)

8. Exner, P., Ichinose, T., Neidhardt, H., Zagrebnov, V.A.: Zeno product formula revisited. Int. Eq. Oper. Theory **57**, 67–81 (2007)
9. Exner, P., Kondej, S.: Curvature-induced bound states for a delta interaction supported by a curve in $R^3$. Ann. H. Poincaré **3**, 967–981 (2002)
10. Exner, P., Kondej, S.: Scattering by local deformations of a straight leaky wire. J. Phys. A: Math. Gen. **38**, 4865–4874 (2005)
11. Exner, P., Kovařík, H.: Quantum Waveguides, p. xxii+382. Springer International, Heidelberg (2015)
12. Fujita, H., Okamoto, H.: Tosio Kato as an applied mathematician: a historical study of a Japanese mathematician, pp. 15–16. ICIAM Newsletter, October (2018)
13. Kato, T.: Perturbation Theory for Linear Operators, 2nd edn. Springer, Berlin (1976)
14. Kato, T.: Fundamental properties of Hamiltonian of the Schrödinger type. Trans. Am. Math. Soc. **70**, 195–211 (1951)
15. Kato, T.: On finite dimensional perturbations of self-adjoint operators. J. Math. Soc. Jpn. **9**, 239–249 (1957)
16. Kato, T.: Perturbations of continuous spectra by trace class operators. Proc. Jpn. Acad. **33**, 260–264 (1957)
17. Kato, T.: Trotter's product formula for an arbitrary pair of self-adjoint contraction semigroups. In: Topics in Functional Analysis (Essays Dedicated to M.G. Krein on the Occasion of his 70th Birthday). Advances in Mathematics: Supplementary Studies, vol. 3, pp. 185–195. Academic, New York (1978)
18. Lieb, E.H., Seiringer, R.: The Stability of Matter in Quantum Mechanics. Cambridge University Press, Cambridge (2010)
19. Mantile, A., Posilicano, A.: Asymptotic completeness and S-matrix for singular perturbations. J. Math. Pures Appl., to appear; arXiv:1711.07556
20. Posilicano, A.: A Krein-like formula for singular perturbations of self-adjoint operators and applications. J. Funct. Anal. **183**, 109–147 (2001)
21. Reed, M., Simon, B.: Methods of Modern Mathematical Physics, III. Scattering Theory IV. Analysis of Operators. Academic Press, New York (1979)
22. Rosenblum, M.: Perturbation of the continuous spectrum and unitary equivalence. Pac. J. Math. **7**, 997–1010 (1957)
23. Simon, B.: Tosio Kato's work on non-relativistic quantum mechanics: part 1. Bull. Math. Sci. **8**, 121–232 (2018)
24. Simon, B.: Tosio Kato's work on non-relativistic quantum mechanics: part 2. Bull. Math. Sci., to appear
25. Teta, A.: Quadratic forms for singular perturbations of the Laplacian. Publ. RIMS **26**, 803–817 (1990)

# Computing Traces, Determinants, and $\zeta$-Functions for Sturm–Liouville Operators: A Survey

**Fritz Gesztesy and Klaus Kirsten**

*Dedicated, with deep admiration, to the memory of Tosio Kato (1917–1999)*

**Abstract** The principal aim of this contribution is to survey an effective and unified approach to the computation of traces of resolvents (and resolvent differences), (modified) Fredholm determinants, $\zeta$-functions, and $\zeta$-function regularized determinants associated with linear operators in a Hilbert space. In particular, we detail the connection between Fredholm and $\zeta$-function regularized determinants. Concrete applications of our formalism to general (i.e., three-coefficient) regular Sturm–Liouville operators on bounded intervals with various (separated and coupled) boundary conditions, and Schrödinger operators on a half-line, are provided and further illustrated with an array of examples. In addition, we consider a class of half-line Schrödinger operators $(-d^2/dx^2) + q$ on $(0, \infty)$ with purely discrete spectra. Roughly speaking, the class considered is generated by potentials $q$ that, for some fixed $C_0 > 0$, $\varepsilon > 0$, $x_0 \in (0, \infty)$, diverge at infinity of the type $q(x) \geq C_0 x^{(2/3)+\varepsilon_0}$ for all $x \geq x_0$. We treat all self-adjoint boundary conditions at the left endpoint 0. This manuscript surveys our recent two papers [19, 20].

K.K. was supported by the Baylor University Summer Sabbatical Program.

F. Gesztesy (✉)
Department of Mathematics, Baylor University,
One Bear Place #97328, Waco, TX 76798-7328, USA
e-mail: Fritz_Gesztesy@baylor.edu
URL: http://www.baylor.edu/math/index.php?id=935340

K. Kirsten
GCAP-CASPER, Department of Mathematics, Baylor University,
One Bear Place #97328, Waco, TX 76798-7328, USA
e-mail: Klaus_Kirsten@baylor.edu
URL: http://www.baylor.edu/math/index.php?id=54012

© Springer Nature Switzerland AG 2019 
T. M. Rassias and V. A. Zagrebnov (eds.), *Analysis and Operator Theory*,
Springer Optimization and Its Applications 146,
https://doi.org/10.1007/978-3-030-12661-2_7

**2010 Mathematics Subject Classiffication** Primary: 47A10 · 47B10 · 47G10 · Secondary: 34B27 · 34L40

## 1 Introduction

This survey of our recent papers [19, 20] details the relationship between Fredholm and $\zeta$-function regularized determinants of Hilbert space operators and provides a unified approach to these determinants as well as traces of resolvents.

The case of traces of resolvents (and resolvent differences), (modified) Fredholm determinants, $\zeta$-functions, and $\zeta$-function regularized determinants associated with linear operators in a Hilbert space is described in detail in Section 2. In particular, under an appropriate trace class hypothesis on the resolvent of a self-adjoint operator $S$ in $\mathcal{H}$ we define its $\zeta$-function and derive a formula for it in terms of the self-adjoint functional calculus and the resolvent of $S$. In the case of trace class resolvent differences for a pair of self-adjoint operators $(S_1, S_2)$ in $\mathcal{H}$, we also describe the underlying relative spectral $\zeta$-function for such a pair and relate the corresponding $\zeta$-function regularized relative determinant and a symmetrized Fredholm perturbation determinant.

In Section 3, we provide an exhaustive discussion of regular (three-coefficient) self-adjoint Sturm–Liouville operators, that is, operators in $L^2((a, b); rdx)$ associated with self-adjoint realizations of differential expressions of the type $\tau = -r^{-1}(d/dx)p(d/dx) + q$, with arbitrary (separated and coupled) self-adjoint boundary conditions on compact intervals $[a, b]$. Their traces of resolvents and associated perturbation determinants are calculated for all self-adjoint boundary conditions in terms of concrete expressions involving a canonical system of fundamental solutions $\phi(z, \cdot, a)$ and $\theta(z, \cdot, a)$ of $\tau\psi = z\psi$, and special examples such as Floquet boundary conditions and the Krein–von Neumann extension are highlighted. The $\zeta$-function regularized determinants are determined for all self-adjoint boundary conditions and a variety of concrete examples complete this section.

Section 4 then illustrates some of the abstract notions in Section 2 with the help of self-adjoint Schrödinger operators associated with differential expressions of the type $-(d^2/dx^2) + q$ on the half-line $\mathbb{R}_+ = (0, \infty)$ with short-range potentials $q$ (i.e., we treat the scattering theory situation). Again, we study all self-adjoint boundary conditions at $x = 0$. The assumed short-range nature of $q$ then necessitates a comparison with the case $q = 0$ and invoking the concept of Jost functions and Jost solutions, illustrates the case of relative perturbation determinants, relative $\zeta$-functions, and relative $\zeta$-function regularized determinants abstractly discussed in Section 2.

Our final Section 5 again treats the case of half-line Schrödinger operators on $\mathbb{R}_+$, but this time with potentials diverging at infinity, and hence with purely discrete spectra (i.e., empty essential spectra). In particular, we assume that $q$ satisfies $q \in$

$L^1_{loc}(\mathbb{R}_+; dx)$, $q$ real-valued a.e. on $\mathbb{R}_+$, and that for some $\varepsilon_0 > 0$, $C_0 > 0$, and sufficiently large $x_0 > 0$,

$$q(x) \geq C_0 \, x^{(2/3)+\varepsilon_0}, \quad x \in (x_0, \infty). \tag{1.1}$$

Applying modified Fredholm determinants $\det_2(\,\cdot\,)$ associated to Hilbert–Schmidt operators, we compute traces of resolvent differences and modified Fredholm determinants for half-line Schrödinger operators in terms of a natural analog of the Jost function and the Jost solution when compared with the short-range case in Section 4. Again our discussion involves all self-adjoint boundary conditions at $x = 0$. The theory is illustrated with the help of the explicitly solvable example $q(x) = x$, $x \in \mathbb{R}_+$, recently also studied in [48].

Finally, we summarize some of the basic notation used in this paper (especially, in Section 2): Let $\mathcal{H}$ and $\mathcal{K}$ be separable complex Hilbert spaces, $(\,\cdot\,,\,\cdot\,)_{\mathcal{H}}$ and $(\,\cdot\,,\,\cdot\,)_{\mathcal{K}}$ the scalar products in $\mathcal{H}$ and $\mathcal{K}$ (linear in the second factor), and $I_{\mathcal{H}}$ and $I_{\mathcal{K}}$ the identity operators in $\mathcal{H}$ and $\mathcal{K}$, respectively.

Next, let $T$ be a closed linear operator from $\operatorname{dom}(T) \subseteq \mathcal{H}$ to $\operatorname{ran}(T) \subseteq \mathcal{K}$, with $\operatorname{dom}(T)$ and $\operatorname{ran}(T)$ denoting the domain and range of $T$. The closure of a closable operator $S$ is denoted by $\overline{S}$. The kernel (null space) of $T$ is denoted by $\ker(T)$.

The spectrum, point spectrum, and resolvent set of a closed linear operator in $\mathcal{H}$ will be denoted by $\sigma(\cdot)$, $\sigma_p(\cdot)$, and $\rho(\cdot)$; the discrete spectrum of $T$ (i.e., points in $\sigma_p(T)$ which are isolated from the rest of $\sigma(T)$, and which are eigenvalues of $T$ of finite algebraic multiplicity) is abbreviated by $\sigma_d(T)$. The *algebraic multiplicity* $m_a(z_0; T)$ of an eigenvalue $z_0 \in \sigma_d(T)$ is the dimension of the range of the corresponding *Riesz projection* $P(z_0; T)$,

$$m_a(z_0; T) = \dim(\operatorname{ran}(P(z_0; T))) = \operatorname{tr}_{\mathcal{H}}(P(z_0; T)), \tag{1.2}$$

where (with the symbol $\oint$ denoting counterclockwise-oriented contour integrals)

$$P(z_0; T) = \frac{-1}{2\pi i} \oint_{C(z_0;\varepsilon)} d\zeta \, (T - \zeta I_{\mathcal{H}})^{-1}, \tag{1.3}$$

for $0 < \varepsilon < \varepsilon_0$ and $D(z_0; \varepsilon_0)\backslash\{z_0\} \subset \rho(T)$; here $D(z_0; r_0) \subset \mathbb{C}$ is the open disk with center $z_0$ and radius $r_0 > 0$, and $C(z_0; r_0) = \partial D(z_0; r_0)$ the corresponding circle. The *geometric multiplicity* $m_g(z_0; T)$ of an eigenvalue $z_0 \in \sigma_p(T)$ is defined by

$$m_g(z_0; T) = \dim(\ker((T - z_0 I_{\mathcal{H}}))). \tag{1.4}$$

The essential spectrum of $T$ is defined by $\sigma_{ess}(T) = \sigma(T)\backslash\sigma_d(T)$.

The Banach spaces of bounded and compact linear operators in $\mathcal{H}$ are denoted by $\mathcal{B}(\mathcal{H})$ and $\mathcal{B}_\infty(\mathcal{H})$, respectively. Similarly, the Schatten–von Neumann (trace) ideals will subsequently be denoted by $\mathcal{B}_p(\mathcal{H})$, $p \in [1, \infty)$, and the subspace of all finite rank operators in $\mathcal{B}_1(\mathcal{H})$ will be abbreviated by $\mathcal{F}(\mathcal{H})$. Analogous notation $\mathcal{B}(\mathcal{H}_1, \mathcal{H}_2)$, $\mathcal{B}_\infty(\mathcal{H}_1, \mathcal{H}_2)$, etc., will be used for bounded, compact, etc., operators between two Hilbert spaces $\mathcal{H}_1$ and $\mathcal{H}_2$. In addition, $\operatorname{tr}_{\mathcal{H}}(T)$ denotes the trace of a trace class operator $T \in \mathcal{B}_1(\mathcal{H})$, $\det_{\mathcal{H}}(I_{\mathcal{H}} - T)$ the Fredholm determinant of $I_{\mathcal{H}} - T$, and for $p \in \mathbb{N}$, $p \geq 2$, $\det_{\mathcal{H},p}(I_{\mathcal{H}} - T)$ abbreviates the $p$th modified Fredholm determinant of $I_{\mathcal{H}} - T$.

Finally, we find it convenient to abbreviate $\mathbb{N}_0 = \mathbb{N} \cup \{0\}$.

## 2 Traces, (Modified) Fredholm Determinants, and Zeta Functions of Operators

In this section, we recall some well-known formulas relating traces and (modified) Fredholm determinants and also discuss the notion of $\zeta$-functions of self-adjoint operators. For background on the material used in this section see, for instance, [30, 31], [32, Chap. XIII], [29, Chap. IV], [42], [55, Chap. 17], [58], [59, Chap. 3].

To set the stage, we start with densely defined, closed, linear operators $A$ in $\mathcal{H}$ having a trace class resolvent, and hence introduce the following assumption:

**Hypothesis 2.1** Suppose that $A$ is densely defined and closed in $\mathcal{H}$ with $\rho(A) \neq \emptyset$, and $(A - zI_{\mathcal{H}})^{-1} \in \mathcal{B}_1(\mathcal{H})$ for some (and hence for all) $z \in \rho(A)$.

Given Hypothesis 2.1 and $z_0 \in \rho(A)$, consider the bounded, entire family $A(\cdot)$ defined by

$$A(z) := I_{\mathcal{H}} - (A - zI_{\mathcal{H}})(A - z_0 I_{\mathcal{H}})^{-1} = (z - z_0)(A - z_0 I_{\mathcal{H}})^{-1}, \quad z \in \mathbb{C}. \tag{2.1}$$

Employing the formula (cf. [29, Sect. IV.1], see also [42], [65, Sect. I.7]),

$$\operatorname{tr}_{\mathcal{H}}\big((I_{\mathcal{H}} - T(z))^{-1}T'(z)\big) = -(d/dz)\ln(\det_{\mathcal{H}}(I_{\mathcal{H}} - T(z))), \tag{2.2}$$

valid for a trace class-valued analytic family $T(\cdot)$ on an open set $\Omega \subset \mathbb{C}$ such that $(I_{\mathcal{H}} - T(\cdot))^{-1} \in \mathcal{B}(\mathcal{H})$, and applying it to the entire family $A(\cdot)$ then results in

$$\begin{aligned} \operatorname{tr}_{\mathcal{H}}\big((A - zI_{\mathcal{H}})^{-1}\big) &= -(d/dz)\ln\big(\det_{\mathcal{H}}\big(I_{\mathcal{H}} - (z - z_0)(A - z_0 I_{\mathcal{H}})^{-1}\big)\big) \\ &= -(d/dz)\ln\big(\det_{\mathcal{H}}\big((A - zI_{\mathcal{H}})(A - z_0 I_{\mathcal{H}})^{-1}\big)\big), \\ &\qquad z \in \rho(A). \end{aligned} \tag{2.3}$$

One notes that the left- and hence the right-hand side of (2.3) is independent of the choice of $z_0 \in \rho(A)$.

Next, following the proof of [59, Theorem 3.5 (c)] step by step, and employing a Weierstrass-type product formula (see, e.g., [59, Theorem 3.7]), yields the subsequent result (see also [25]).

**Lemma 2.2** *Assume Hypothesis 2.1 and let* $\lambda_k \in \sigma(A)$ *then*

$$\det_{\mathcal{H}}\big(I_{\mathcal{H}} - (z-z_0)(A - z_0 I_{\mathcal{H}})^{-1}\big) \underset{z\to\lambda_k}{=} (\lambda_k - z)^{m_a(\lambda_k)}[C_k + O(\lambda_k - z)], \quad C_k \neq 0, \tag{2.4}$$

*that is, the multiplicity of the zero of the Fredholm determinant* $\det_{\mathcal{H}}\big(I_{\mathcal{H}} - (z - z_0)(A - z_0 I_{\mathcal{H}})^{-1}\big)$ *at* $z = \lambda_k$ *equals the algebraic multiplicity of the eigenvalue* $\lambda_k$ *of* $A$.

*In addition, denote the spectrum of* $A$ *by* $\sigma(A) = \{\lambda_k\}_{k\in\mathbb{N}}$, $\lambda_k \neq \lambda_{k'}$ *for* $k \neq k'$. *Then*

$$\begin{aligned} \det_{\mathcal{H}}(I_{\mathcal{H}} - (z-z_0)(A - z_0 I_{\mathcal{H}})^{-1}) &= \prod_{k\in\mathbb{N}} \big[1 - (z-z_0)(\lambda_k - z_0)^{-1}\big]^{m_a(\lambda_k)} \\ &= \prod_{k\in\mathbb{N}} \bigg(\frac{\lambda_k - z}{\lambda_k - z_0}\bigg)^{m_a(\lambda_k)}, \end{aligned} \tag{2.5}$$

*with absolutely convergent products in* (2.5).

The case of trace class resolvent operators is tailor-made for a number of one-dimensional Sturm–Liouville operators (e.g., finite interval problems). But for applications to half-line problems with potentials behaving like $x$, or increasing slower than $x$ at $+\infty$, and similarly for partial differential operators, traces of higher order powers of resolvents need to be involved which naturally lead to modified Fredholm determinants as follows.

**Hypothesis 2.3** Let $p \in \mathbb{N}$, $p \geq 2$, and suppose that $A$ is densely defined and closed in $\mathcal{H}$ with $\rho(A) \neq \emptyset$, and $(A - zI_{\mathcal{H}})^{-1} \in \mathcal{B}_p(\mathcal{H})$ for some (and hence for all) $z \in \rho(A)$.

Applying the formula

$$\operatorname{tr}_{\mathcal{H}}\big((I_{\mathcal{H}} - T(z))^{-1} T(z)^{p-1} T'(z)\big) = -(d/dz)\ln(\det_{\mathcal{H},p}(I_{\mathcal{H}} - T(z))), \tag{2.6}$$

valid for a $\mathcal{B}_p(\mathcal{H})$-valued analytic family $T(\cdot)$ on an open set $\Omega \subset \mathbb{C}$ such that $(I_{\mathcal{H}} - T(\cdot))^{-1} \in \mathcal{B}(\mathcal{H})$, [29, Sect. IV.2] (see also [65, Sect. I.7]) to the entire family $A(\cdot)$ in (2.1), assuming Hypothesis 2.3, then yields

$$\begin{aligned} &(z-z_0)^{p-1}\operatorname{tr}_{\mathcal{H}}\big((A - zI_{\mathcal{H}})^{-1}(A - z_0 I_{\mathcal{H}})^{1-p}\big) \\ &\quad = -(d/dz)\ln\big(\det_{\mathcal{H},p}\big(I_{\mathcal{H}} - (z-z_0)(A - z_0 I_{\mathcal{H}})^{-1}\big)\big) \\ &\quad = -(d/dz)\ln\big(\det_{\mathcal{H},p}\big((A - zI_{\mathcal{H}})(A - z_0 I_{\mathcal{H}})^{-1}\big)\big), \quad z \in \rho(A). \end{aligned} \tag{2.7}$$

In the special case, $p = 2$ this yields

$$\begin{aligned} &\operatorname{tr}_{\mathcal{H}}\big((A - zI_{\mathcal{H}})^{-1} - (A - z_0 I_{\mathcal{H}})^{-1}\big) \\ &\quad = -(d/dz)\ln\big(\det_{\mathcal{H},2}\big(I_{\mathcal{H}} - (z-z_0)(A - z_0 I_{\mathcal{H}})^{-1}\big)\big). \end{aligned} \tag{2.8}$$

We refer to Section 5 for an application of (2.8) to half-line Schrödinger operators with potentials diverging at infinity.

In many cases of interest not a single resolvent, but a difference of two resolvents is trace class and hence one is naturally led to the following generalization discussed in detail in [23, 26, 27] (see also [22]). To avoid technicalities, we will now consider the case of self-adjoint operators $A$, $B$ below, but note that [27] considers the general case of densely defined, closed linear operators with nonempty resolvent sets:

**Hypothesis 2.4** Suppose $A$ and $B$ are self-adjoint operators in $\mathcal{H}$ with $A$ bounded from below.
$(i)$ Assume that $B$ can be written as the form sum (denoted by $+_q$) of $A$ and a self-adjoint operator $W$ in $\mathcal{H}$

$$B = A +_q W, \tag{2.9}$$

where $W$ can be factorized into

$$W = W_1^* W_2, \tag{2.10}$$

such that

$$\operatorname{dom}(W_j) \supseteq \operatorname{dom}\big(|A|^{1/2}\big), \quad j = 1, 2. \tag{2.11}$$

$(ii)$ Suppose that for some (and hence for all) $z \in \rho(A)$,

$$W_2(A - zI_{\mathcal{H}})^{-1/2}, \ \overline{(A - zI_{\mathcal{H}})^{-1/2} W_1^*} \in \mathcal{B}_2(\mathcal{H}). \tag{2.12}$$

Given Hypothesis 2.4, one observes that

$$\operatorname{dom}\big(|B|^{1/2}\big) = \operatorname{dom}\big(|A|^{1/2}\big), \tag{2.13}$$

and that the resolvent of $B$ can be written as (cf., e.g., the detailed discussion in [27] and the references therein)

$$\begin{aligned} (B - zI_{\mathcal{H}})^{-1} &= (A - zI_{\mathcal{H}})^{-1} \\ &\quad - \overline{(A - zI_{\mathcal{H}})^{-1} W_1^*}\big[I_{\mathcal{H}} + \overline{W_2(A - zI_{\mathcal{H}})^{-1} W_1^*}\big]^{-1} W_2(A - zI_{\mathcal{H}})^{-1}, \\ &\qquad z \in \rho(B) \cap \rho(A). \end{aligned} \tag{2.14}$$

We also note the analog of Tiktopoulos' formula (cf. [57, p. 45]),

$$\begin{aligned} (B - zI_{\mathcal{H}})^{-1} &= (A - zI_{\mathcal{H}})^{-1/2}\big[I_{\mathcal{H}} + \overline{(A - zI_{\mathcal{H}})^{-1/2} W (A - zI_{\mathcal{H}})^{-1/2}}\big]^{-1} \\ &\quad \times (A - zI_{\mathcal{H}})^{-1/2}, \quad z \in \rho(B) \cap \rho(A). \end{aligned} \tag{2.15}$$

Here the closures in (2.14) and (2.15) are well-defined employing (2.12). In addition, one observes that the resolvent formulas (2.14) and (2.15) are symmetric with respect to $A$ and $B$ employing $A = B -_q W$.

As a consequence, $B$ is bounded from below in $\mathcal{H}$ and one concludes that for some (and hence for all ) $z \in \rho(B) \cap \rho(A)$,

$$\big[(B - zI_{\mathcal{H}})^{-1} - (A - zI_{\mathcal{H}})^{-1}\big] \in \mathcal{B}_1(\mathcal{H}). \tag{2.16}$$

Moreover, one infers that (cf. [26])

$$\begin{aligned} &\operatorname{tr}_{\mathcal{H}}\big((B - zI_{\mathcal{H}})^{-1} - (A - zI_{\mathcal{H}})^{-1}\big) \\ &\quad = -\frac{d}{dz}\ln\Big(\det_{\mathcal{H}}\Big(\overline{(B - zI_{\mathcal{H}})^{1/2}(A - zI_{\mathcal{H}})^{-1}(B - zI_{\mathcal{H}})^{1/2}}\Big)\Big) && (2.17) \\ &\quad = -\frac{d}{dz}\ln\Big(\det_{\mathcal{H}}\Big(I_{\mathcal{H}} + \overline{W_2(A - zI_{\mathcal{H}})^{-1}W_1^*}\Big)\Big), \quad z \in \rho(B) \cap \rho(A). && (2.18) \end{aligned}$$

Here any choice of branch cut of the normal operator $(B - zI_{\mathcal{H}})^{1/2}$ (employing the spectral theorem) is permissible. The first equality in (2.18) follows as in the proof of [26, Theorem 2.8]. (The details are actually a bit simpler now since $A$, $B$ are self-adjoint and bounded from below, and hence of positive type after some translation, which is the case considered in [26].) For completeness, we mention that the second equality in (2.18) can be arrived at as follows: Employing the commutation formula (cf., [11]),

$$C[I_{\mathcal{H}} - DC]^{-1}D = -I_{\mathcal{H}} + [I_{\mathcal{H}} - CD]^{-1} \tag{2.19}$$

for $C, D \in \mathcal{B}(\mathcal{H})$ with $1 \in \rho(DC)$ (and hence $1 \in \rho(CD)$ since $\sigma(CD)\backslash\{0\} = \sigma(DC)\backslash\{0\}$), the resolvent identity (2.15) with $A$ and $B$ interchanged yields

$$\begin{aligned} &\overline{(B - zI_{\mathcal{H}})^{1/2}(A - zI_{\mathcal{H}})^{-1}(B - zI_{\mathcal{H}})^{1/2}} \\ &\quad = \big[I_{\mathcal{H}} - \overline{(B - zI_{\mathcal{H}})^{-1/2}W(B - zI_{\mathcal{H}})^{-1/2}}\big]^{-1} \\ &\quad - \Big[I_{\mathcal{H}} - \overline{\big[(B - zI_{\mathcal{H}})^{-1/2}W_1^*\big]}\big[W_2(B - zI_{\mathcal{H}})^{-1/2}\big]\Big]^{-1}. \end{aligned} \tag{2.20}$$

On the other hand, it is well-known (cf. [27, 37]) that

$$\begin{aligned} &I_{\mathcal{H}} + \overline{W_2(A - zI_{\mathcal{H}})^{-1}W_1^*} \\ &\quad = \big[I_{\mathcal{H}} - \overline{W_2(B - zI_{\mathcal{H}})^{-1}W_1^*}\big]^{-1} \\ &\quad = \Big[I_{\mathcal{H}} - \big[W_2(B - zI_{\mathcal{H}})^{-1/2}\big]\overline{\big[(B - zI_{\mathcal{H}})^{-1/2}W_1^*\big]}\Big]^{-1}, \end{aligned} \tag{2.21}$$

and hence using the fact

$$\det_{\mathcal{H}}(I_{\mathcal{H}} - ST) = \det_{\mathcal{H}}(I_{\mathcal{H}} - TS) \tag{2.22}$$

for $S, T \in \mathcal{B}(\mathcal{H})$ with $ST, TS \in \mathcal{B}_1(\mathcal{H})$ (again, since $\sigma(ST)\backslash\{0\} = \sigma(TS)\backslash\{0\}$) then proves the second equality in (2.18).

While this approach based on Hypothesis 2.4 is tailor-made for a discussion of perturbations of the potential coefficient in the context of Schrödinger and, more generally, Sturm–Liouville operators, we also mention the following variant that is best suited for changes in the boundary conditions:

**Hypothesis 2.5** Suppose that $A$ and $B$ are self-adjoint operators in $\mathcal{H}$ bounded from below. In addition, assume that

$$\operatorname{dom}\big(|A|^{1/2}\big) \subseteq \operatorname{dom}\big(|B|^{1/2}\big), \tag{2.23}$$

$$\overline{(B - tI_{\mathcal{H}})^{1/2}\big[(B - tI_{\mathcal{H}})^{-1} - (A - tI_{\mathcal{H}})^{-1}\big](B - tI_{\mathcal{H}})^{1/2}} \in \mathcal{B}_1(\mathcal{H}) \tag{2.24}$$

for some $t < \inf(\sigma(A) \cup \sigma(B))$.

Given Hypothesis 2.5 it has been proven in [26] (actually, in a more general context) that for $z \in \mathbb{C}\backslash[\inf(\sigma(A) \cup \sigma(B)), \infty)$,

$$\begin{aligned}
&\overline{(B - zI_{\mathcal{H}})^{1/2}\big[(B - zI_{\mathcal{H}})^{-1} - (A - zI_{\mathcal{H}})^{-1}\big](B - zI_{\mathcal{H}})^{1/2}} \\
&\quad = I_{\mathcal{H}} - \overline{(B - zI_{\mathcal{H}})^{1/2}(A - zI_{\mathcal{H}})^{-1}(B - zI_{\mathcal{H}})^{1/2}} \qquad (2.25)\\
&\quad = I_{\mathcal{H}} - \big[(B - zI_{\mathcal{H}})^{1/2}(A - zI_{\mathcal{H}})^{-1/2}\big]\big[(B - \overline{z}I_{\mathcal{H}})^{1/2}(A - \overline{z}I_{\mathcal{H}})^{-1/2}\big]^*
\end{aligned}$$

implying that $\det_{\mathcal{H}}\big(\overline{(B - zI_{\mathcal{H}})^{1/2}(A - zI_{\mathcal{H}})^{-1}(B - zI_{\mathcal{H}})^{1/2}}\big)$ is well-defined. Moreover,

$$\begin{aligned}
&\operatorname{tr}_{\mathcal{H}}\big((B - zI_{\mathcal{H}})^{-1} - (A - zI_{\mathcal{H}})^{-1}\big) \\
&\quad = -\frac{d}{dz}\ln\Big(\det_{\mathcal{H}}\Big(\overline{(B - zI_{\mathcal{H}})^{1/2}(A - zI_{\mathcal{H}})^{-1}(B - zI_{\mathcal{H}})^{1/2}}\Big)\Big), \qquad (2.26)\\
&\qquad\qquad z \in \mathbb{C}\backslash[\inf(\sigma(A) \cup \sigma(B)), \infty).
\end{aligned}$$

Next, we briefly turn to spectral $\zeta$-functions of self-adjoint operators $S$ with a trace class resolvent (and hence purely discrete spectrum).

**Hypothesis 2.6** Suppose $S$ is a self-adjoint operator in $\mathcal{H}$, bounded from below, satisfying

$$(S - zI_{\mathcal{H}})^{-1} \in \mathcal{B}_1(\mathcal{H}) \tag{2.27}$$

for some (and hence for all) $z \in \rho(S)$. We denote the spectrum of $S$ by $\sigma(S) = \{\lambda_j\}_{j \in J}$ (with $J \subset \mathbb{Z}$ an appropriate index set), with every eigenvalue repeated according to its multiplicity.

Given Hypothesis 2.6, the spectral zeta function of $S$ is then defined by

$$\zeta(s; S) := \sum_{\substack{j \in J \\ \lambda_j \neq 0}} \lambda_j^{-s} \tag{2.28}$$

for $\operatorname{Re}(s) > 0$ sufficiently large such that (2.28) converges absolutely.

**Fig. 1** Contour $\gamma$ in the complex $z$-plane

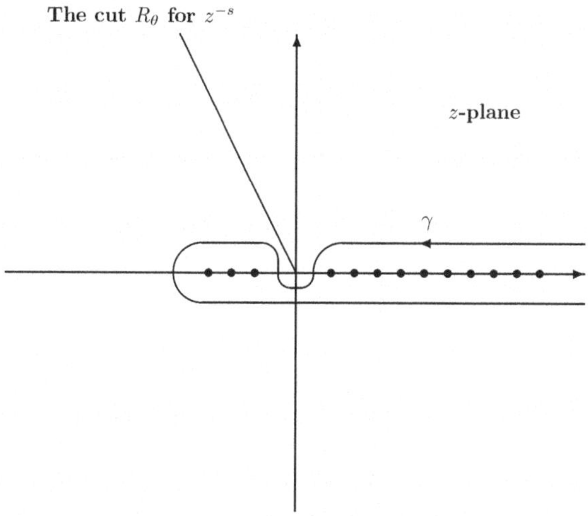


Next, let $P(0; S)$ be the spectral projection of $S$ corresponding to the eigenvalue 0 and denote by $m(\lambda_0; S)$ the multiplicity of the eigenvalue $\lambda_0$ of $S$, in particular,

$$m(0; S) = \dim(\ker(S)). \tag{2.29}$$

(One recalls that since $S$ is self-adjoint, the geometric and algebraic multiplicity of each eigenvalue of $S$ coincide and hence the subscript "$g$" or "$a$" is simply omitted from $m(\,\cdot\,; S)$.) In addition, we introduce the simple contour $\gamma$ encircling $\sigma(S)\backslash\{0\}$ in a counterclockwise manner so as to dip under (and hence avoid) the point 0 (cf. Figure 1). In fact, following [40] (see also [39]), we will henceforth choose as the branch cut of $z^{-s}$ the ray

$$R_\theta = \big\{z = te^{i\theta} \big| t \in [0, \infty)\big\} \quad \theta \in (\pi/2, \pi), \tag{2.30}$$

and note that the contour $\gamma$ avoids any contact with $R_\theta$ (cf. Figure 1).

We note in passing that one could also use a semigroup approach via

$$\begin{aligned} \zeta(s; S) &= \Gamma(s)^{-1} \int_0^\infty dt\, t^{s-1} \operatorname{tr}_{\mathcal{H}}\big(e^{-tS}[I_{\mathcal{H}} - P(0; S)]\big) \\ &= \Gamma(s)^{-1} \int_0^\infty dt\, t^{s-1} \big[\operatorname{tr}_{\mathcal{H}}\big(e^{-tS}\big) - m(0; S)\big], \end{aligned} \tag{2.31}$$

for $\operatorname{Re}(s) > 0$ sufficiently large, but we prefer to work with resolvents in this paper.

**Lemma 2.7** *In addition to Hypothesis 2.6 and the counterclockwise oriented contour $\gamma$ just described (cf. Figure 1), suppose that $\big|\operatorname{tr}_{\mathcal{H}}\big((S - zI_{\mathcal{H}})^{-1}[I_{\mathcal{H}} - P(0; S)]\big)\big|$ is polynomially bounded on $\gamma$. Then*

$$\zeta(s;S) = -(2\pi i)^{-1} \oint_\gamma dz\, z^{-s}\big[\mathrm{tr}_{\mathcal{H}}\big((S - zI_{\mathcal{H}})^{-1}\big) + z^{-1} m(0;S)\big] \tag{2.32}$$

*for* $\mathrm{Re}(s) > 0$ *sufficiently large.*

*Proof* Assuming $\mathrm{Re}(s) > 0$ sufficiently large, a contour integration argument yields

$$\begin{aligned}\zeta(s;S) &= -(2\pi i)^{-1} \oint_\gamma dz\, z^{-s} \mathrm{tr}_{\mathcal{H}}\big((S - zI_{\mathcal{H}})^{-1}[I_{\mathcal{H}} - P(0;S)]\big) \\ &= -(2\pi i)^{-1} \oint_\gamma dz\, z^{-s} \mathrm{tr}_{\mathcal{H}}\big((S - zI_{\mathcal{H}})^{-1} + z^{-1} P(0;S)\big) \\ &= -(2\pi i)^{-1} \oint_\gamma dz\, z^{-s}\big[\mathrm{tr}_{\mathcal{H}}\big((S - zI_{\mathcal{H}})^{-1}\big) + z^{-1} m(0;S)\big],\end{aligned} \tag{2.33}$$

taking into account that $\mathrm{tr}_{\mathcal{H}}\big((S - zI_{\mathcal{H}})^{-1}[I_{\mathcal{H}} - P(0;S)]\big)$ is meromorphic with poles precisely at the nonzero eigenvalues $\lambda_j \neq 0$ of $S$, with residues given by (self-adjoint) spectral projections of $S$ of rank equal to $m(\lambda_j;S)$. □

It is very tempting to continue the computation leading to (2.32) and now deform the contour $\gamma$ so as to "hug" the branch cut $R_\theta$, but this requires the right asymptotic behavior of $\mathrm{tr}_{\mathcal{H}}\big((S - zI_{\mathcal{H}})^{-1}[I_{\mathcal{H}} - P(0;S)]\big)$ as $|z| \to \infty$ as well as $|z| \to 0$, and we will investigate this in the context of relative $\zeta$-functions next.

In cases where $(S - zI_{\mathcal{H}})^{-1}$ is not trace class, but one is dealing with a pair of operators $(S_1, S_2)$ such that the difference of their resolvents lies in the trace class, (2.33) naturally leads to the notion of a relative $\zeta$-function as follows. For pertinent background information on this circle of ideas we refer, for instance, to Forman [16, 17], Müller [49].

**Hypothesis 2.8** Suppose $S_j$, $j = 1, 2$, are self-adjoint operators in $\mathcal{H}$, bounded from below, satisfying

$$\big[(S_2 - zI_{\mathcal{H}})^{-1} - (S_1 - zI_{\mathcal{H}})^{-1}\big] \in \mathcal{B}_1(\mathcal{H}) \tag{2.34}$$

for some (and hence for all) $z \in \rho(S_1) \cap \rho(S_2)$. In addition, assume that $S_j$, $j = 1, 2$, have essential spectrum contained in $(0, \infty)$, that is, for some $\lambda_1 > 0$,

$$\sigma_{ess}(S_j) \subseteq [\lambda_1, \infty), \quad j = 1, 2. \tag{2.35}$$

We note, in particular, the essential spectrum hypothesis (2.35) includes the case of purely discrete spectra of $S_j$ (i.e., $\sigma_{ess}(S_j) = \emptyset$, $j = 1, 2$.) Since $S_j$ were assumed to be bounded from below, adding a suitable constant to $S_j$, $j = 1, 2$, will shift their (essential) spectra accordingly.

Given Hypothesis 2.8, and again choosing a counterclockwise-oriented simple contour $\gamma$ that encircles $\sigma(S_1) \cup \sigma(S_2)$, however, with the stipulation that 0 does not lie in the interior of $\gamma$, and $0 \notin \gamma$ (cf. Figure 1), the relative spectral $\zeta$-function for the pair $(S_1, S_2)$ is defined by

$$\zeta(s; S_1, S_2) = -(2\pi i)^{-1} \oint_\gamma dz\, z^{-s} \operatorname{tr}_{\mathcal{H}}\big((S_2 - zI_{\mathcal{H}})^{-1}[I_{\mathcal{H}} - P(0; S_2)] \\ - (S_1 - zI_{\mathcal{H}})^{-1}[I_{\mathcal{H}} - P(0; S_1)]\big) \tag{2.36}$$

for $\operatorname{Re}(s) > 0$ sufficiently large, ensuring convergence of (2.36).

Employing the contour $\gamma$ and branch cut $R_\theta$ as in Figure 1, and deforming $\gamma$ so it eventually surrounds $R_\theta$, one arrives at the following result.

**Theorem 2.9** *Suppose $S_j$, $j = 1, 2$, are self-adjoint operators in $\mathcal{H}$ satisfying Hypothesis 2.8 and that (cf.* (2.17) *and* (2.26))

$$\begin{aligned} &\operatorname{tr}_{\mathcal{H}}\big((S_2 - zI_{\mathcal{H}})^{-1} - (S_1 - zI_{\mathcal{H}})^{-1}\big) \\ &\quad = -\frac{d}{dz} \ln\Big(\operatorname{det}_{\mathcal{H}}\Big(\overline{(S_2 - zI_{\mathcal{H}})^{1/2}(S_1 - zI_{\mathcal{H}})^{-1}(S_2 - zI_{\mathcal{H}})^{1/2}}\Big)\Big), \\ &\qquad\qquad z \in \mathbb{C}\backslash[\inf(\sigma(S_1) \cap \sigma(S_2)), \infty). \end{aligned} \tag{2.37}$$

*In addition, assume that for some $\varepsilon > 0$,*

$$\begin{aligned} &\big|\operatorname{tr}_{\mathcal{H}}\big((S_2 - zI_{\mathcal{H}})^{-1}[I_{\mathcal{H}} - P(0; S_2)] - (S_1 - zI_{\mathcal{H}})^{-1}[I_{\mathcal{H}} - P(0; S_1)]\big)\big| \\ &\quad = \begin{cases} O\big(|z|^{-1-\varepsilon}\big), & as\ |z| \to \infty, \\ O(1), & as\ |z| \to 0. \end{cases} \end{aligned} \tag{2.38}$$

*Then, for* $\operatorname{Re}(s) \in (-\varepsilon, 1)$,

$$\begin{aligned} \zeta(s; S_1, S_2) &= e^{is(\pi-\theta)} \pi^{-1} \sin(\pi s) \int_0^\infty dt\, t^{-s} \\ &\quad \times \frac{d}{dt} \ln\Big((e^{i\theta} t)^{[m(0;S_1) - m(0;S_2)]} \\ &\quad \times \operatorname{det}_{\mathcal{H}}\big((S_2 - e^{i\theta} t I_{\mathcal{H}})^{1/2}(S_1 - e^{i\theta} t I_{\mathcal{H}})^{-1}(S_2 - e^{i\theta} t I_{\mathcal{H}})^{1/2}\big)\Big). \end{aligned} \tag{2.39}$$

*Proof* Due to hypothesis (2.38) one can deform the contour $\gamma$ so that it wraps around the branch cut $R_\theta$,

$$\begin{aligned} \zeta(s; S_1, S_2) &= -(2\pi i)^{-1} \oint_\gamma dz\, z^{-s} \operatorname{tr}_{\mathcal{H}}\big((S_2 - zI_{\mathcal{H}})^{-1}[I_{\mathcal{H}} - P(0; S_2)] \\ &\qquad - (S_1 - zI_{\mathcal{H}})^{-1}[I_{\mathcal{H}} - P(0; S_1)]\big) \\ &= -(2\pi i)^{-1} \oint_\gamma dz\, z^{-s} \big\{\operatorname{tr}_{\mathcal{H}}\big((S_2 - zI_{\mathcal{H}})^{-1} - (S_1 - zI_{\mathcal{H}})^{-1}\big) \\ &\qquad + z^{-1}[m(0; S_2) - m(0; S_1)]\big\} \\ &= (2\pi i)^{-1} \oint_\gamma dz\, z^{-s} \bigg\{\frac{d}{dz} \ln\Big(\operatorname{det}_{\mathcal{H}}\big((S_2 - zI_{\mathcal{H}})^{1/2}(S_1 - zI_{\mathcal{H}})^{-1}(S_2 - zI_{\mathcal{H}})^{1/2}\big)\Big) \\ &\qquad - z^{-1}[m(0; S_2) - m(0; S_1)]\bigg\} \end{aligned}$$

$$= (2\pi i)^{-1} \oint_{\gamma} dz\, z^{-s} \frac{d}{dz} \ln\Big( z^{[m(0;S_1)-m(0;S_2)]} \\ \times \det_{\mathcal{H}}\Big( (S_2 - zI_{\mathcal{H}})^{1/2} (S_1 - zI_{\mathcal{H}})^{-1} (S_2 - zI_{\mathcal{H}})^{1/2} \Big)\Big)$$

$$= e^{is(\pi-\theta)} \pi^{-1} \sin(\pi s) \int_0^{\infty} dt\, t^{-s} \frac{d}{dt} \ln\Big( \big(te^{i\theta}\big)^{[m(0;S_1)-m(0;S_2)]} \\ \times \det_{\mathcal{H}}\Big( \big(S_2 - e^{i\theta} t I_{\mathcal{H}}\big)^{1/2} \big(S_1 - e^{i\theta} t I_{\mathcal{H}}\big)^{-1} \big(S_2 - e^{i\theta} t I_{\mathcal{H}}\big)^{1/2} \Big)\Big). \tag{2.40}$$

Here we first applied $\operatorname{tr}_{\mathcal{H}}(P(0; S_j)) = m(0; S_j)$ (cf. also (2.36)) and then (2.18). Carefully paying attention to the phases when shrinking the contour to the branch cut $R_\theta$, one obtains (2.39). □

**Theorem 2.10** *Suppose $S_j$, $j = 1, 2$, are self-adjoint operators in $\mathcal{H}$ satisfying Hypothesis 2.8 and* (2.38). *Then*

$$\zeta'(0; S_1, S_2) = i\pi(n_2 - n_1) - \lim_{t\downarrow 0} \Big| \ln\Big( (e^{i\theta} t)^{[m(0;S_1)-m(0;S_2)]} \\ \times \det_{\mathcal{H}}\big((S_2 - e^{i\theta} t I_{\mathcal{H}})^{1/2} (S_1 - e^{i\theta} t I_{\mathcal{H}})^{-1} (S_2 - e^{i\theta} t I_{\mathcal{H}})^{1/2}\big)\Big)\Big|, \tag{2.41}$$

*where $n_j$ is the number of negative eigenvalues of $S_j$, $j = 1, 2$. If $n_j = 0$, $j = 1, 2$, then*

$$\zeta'(0; S_1, S_2) = -\lim_{t\downarrow 0} \ln\Big( (e^{i\theta} t)^{[m(0;S_1)-m(0;S_2)]} \\ \times \det_{\mathcal{H}}\big((S_2 - e^{i\theta} t I_{\mathcal{H}})^{1/2} (S_1 - e^{i\theta} t I_{\mathcal{H}})^{-1} (S_2 - e^{i\theta} t I_{\mathcal{H}})^{1/2}\big)\Big) \\ = -\ln(C_0), \tag{2.42}$$

*where*

$$\det_{\mathcal{H}}\big((S_2 - zI_{\mathcal{H}})^{1/2} (S_1 - zI_{\mathcal{H}})^{-1} (S_2 - zI_{\mathcal{H}})^{1/2}\big) \\ \underset{z\to 0}{=} z^{[m(0;S_2)-m(0;S_1)]} [C_0 + O(z)], \quad C_0 > 0. \tag{2.43}$$

*Proof* First we note that (2.39) implies

$$\zeta'(0; S_1, S_2) = \int_0^{\infty} dt\, \frac{d}{dt} \ln\Big\{ \big(te^{i\theta}\big)^{[m(0;S_1)-m(0;S_2)]} \\ \times \det_{\mathcal{H}}\Big( \big(S_2 - e^{i\theta} t I_{\mathcal{H}}\big)^{1/2} \big(S_1 - e^{i\theta} t I_{\mathcal{H}}\big)^{-1} \big(S_2 - e^{i\theta} t I_{\mathcal{H}}\big)^{1/2} \Big)\Big\}. \tag{2.44}$$

In computing this quantity, one notes that for $t \in [0, \infty)$, the graph of the function

$$G(t) = \big(te^{i\theta}\big)^{[m(0;S_1)-m(0;S_2)]} \\ \times \det_{\mathcal{H}}\Big( \big(S_2 - e^{i\theta} t I_{\mathcal{H}}\big)^{1/2} \big(S_1 - e^{i\theta} t I_{\mathcal{H}}\big)^{-1} \big(S_2 - e^{i\theta} t I_{\mathcal{H}}\big)^{1/2} \Big) \tag{2.45}$$

can cross the branch cut $R_\theta$ at several $t$-values. Therefore, the integral has to be split at these $t$-values and pursuant real and imaginary parts have to be summed. The real part between consecutive segments cancels except for the contributions from zero. This explains the real part of (2.41). The resulting imaginary part is found as follows. The sum defining the $\zeta$-function can be split into contributions from negative and positive eigenvalues, namely,

$$\zeta(s; S_j) = \sum_{\substack{\ell=1 \\ \lambda_{-\ell}<0}}^{n_j} \lambda_{-\ell}^{-s} + \sum_{\substack{k\in J \\ \lambda_k>0}} \lambda_k^{-s}. \tag{2.46}$$

For each negative eigenvalue, one computes

$$\frac{d}{ds}\bigg|_{s=0} \lambda_{-\ell}^{-s} = -\ln(\lambda_{-\ell}) = i\pi - \ln(|\lambda_{-\ell}|). \tag{2.47}$$

This yields the imaginary part in (2.41).

Since 0 lies outside the essential spectra of $S_j$, $j = 1, 2$, (2.43) follows, for instance, from [65, pp. 271–272]. Given the relation (2.43), the fact (2.42) follows from (2.44) and the Lebesgue dominated convergence theorem. □

In the special case, where $0 \in \rho(S_1) \cap \rho(S_2)$ (i.e., $m(0; S_1) = m(0; S_2) = 0$), one thus obtains

$$e^{-\zeta'(0; S_1, S_2)} = \det_{\mathcal{H}}\big(S_2^{1/2} S_1^{-1} S_2^{1/2}\big) \tag{2.48}$$

and hence the $\zeta$-function regularized relative determinant now equals the symmetrized (Fredholm) perturbation determinant for the pair $(S_1, S_2)$. Here any choice of branch cut of the self-adjoint operator $S_2^{1/2}$ (employing the spectral theorem) is permissible.

For additional background and applications of (modified) Fredholm determinants to ordinary differential operators, we also refer to [5, 9, 13, 19, 21, 26, 38–46, 53], and the extensive literature cited therein.

## 3 Sturm–Liouville Operators on Bounded Intervals

To illustrate the material of Section 2, we now apply it to the case of self-adjoint Sturm–Liouville operators on bounded intervals.

We start by recalling a convenient parametrization of all self-adjoint extensions associated with a regular, symmetric, second-order differential expression as discussed in detail, for instance, in [64, Theorem 13.15] and [66, Theorem 10.4.3], and recorded in [9].

Throughout this section, we make the following set of assumptions:

**Hypothesis 3.1** Suppose $p, q, r$ satisfy the following conditions:
$(i)$ $r > 0$ a.e. on $(a, b)$, $r \in L^1((a, b); dx)$.
$(ii)$ $p > 0$ a.e. on $(a, b)$, $1/p \in L^1((a, b); dx)$.
$(iii)$ $q \in L^1((a, b); dx)$, $q$ is real-valued a.e. on $(a, b)$.

Given Hypothesis 3.1, we take $\tau$ to be the Sturm–Liouville-type differential expression defined by

$$\tau = \frac{1}{r(x)}\left[-\frac{d}{dx}p(x)\frac{d}{dx} + q(x)\right] \text{ for a.e. } x \in (a, b), \quad -\infty < a < b < \infty. \tag{3.1}$$

By definition, $\tau$ is called *regular on* $[a, b]$ if $p, q, r$ satisfy Hypothesis 3.1. In addition, the following convenient notation for the *first quasi-derivative* is introduced,

$$y^{[1]}(x) = p(x)y'(x) \text{ for a.e. } x \in (a, b),\ y \in AC([a, b]). \tag{3.2}$$

Here $AC([a, b])$ denotes the set of absolutely continuous functions on $[a, b]$ and $' = d/dx$. For notational convenience, we will occasionally abbreviate $L^2_r((a, b)) := L^2((a, b); rdx)$.

Given that $\tau$ is regular on $[a, b]$, the *maximal operator* $H_{max}$ in $L^2((a, b); rdx)$ associated with $\tau$ is defined by

$$\begin{aligned} &H_{max} f = \tau f, \\ &f \in \text{dom}(H_{max}) = \big\{g \in L^2((a, b); rdx) \,\big|\, g, g^{[1]} \in AC([a, b]);\ \tau g \in L^2((a, b); rdx)\big\}, \end{aligned} \tag{3.3}$$

while the *minimal operator* $H_{min}$ in $L^2((a, b); rdx)$ associated with $\tau$ is given by

$$\begin{aligned} &H_{min} f = \tau f, \\ &f \in \text{dom}(H_{min}) = \big\{g \in L^2((a, b); rdx) \,\big|\, g, g^{[1]} \in AC([a, b]); \\ &\qquad g(a) = g^{[1]}(a) = g(b) = g^{[1]}(b) = 0;\ \tau g \in L^2((a, b); rdx)\big\}. \end{aligned} \tag{3.4}$$

One notes that the operator $H_{min}$ is symmetric and that

$$H^*_{min} = H_{max}, \quad H^*_{max} = H_{min}, \tag{3.5}$$

(cf. Weidmann [63, Theorem 13.8]). Next, we summarize material found, for instance, in [64, Chap. 13] and [66, Sects. 10.3, 10.4] in which self-adjoint extensions of the minimal operator $H_{min}$ are characterized.

**Theorem 3.2** *(See, [62, Satz 2.6], [64, Theorem 13.14], [66, Theorem 10.4.2].) Assume Hypothesis 3.1 and suppose that $\widetilde{H}$ is an extension of the minimal operator $H_{min}$ defined in (3.4). Then the following hold:*
*(i) $\widetilde{H}$ is a self-adjoint extension of $H_{min}$ if and only if there exist $2 \times 2$ matrices A and B with complex-valued entries satisfying*[1]

$$\operatorname{rank}(A\ B) = 2, \quad AJA^* = BJB^*, \quad J = \begin{pmatrix} 0 & -1 \\ 1 & 0 \end{pmatrix}, \tag{3.6}$$

*with*

$$\begin{aligned} &\widetilde{H} f = \tau f \\ &\operatorname{dom}\big(\widetilde{H}\big) = \left\{ g \in \operatorname{dom}(H_{max}) \,\middle|\, A\begin{pmatrix} g(a) \\ g^{[1]}(a) \end{pmatrix} = B\begin{pmatrix} g(b) \\ g^{[1]}(b) \end{pmatrix} \right\}. \end{aligned} \tag{3.7}$$

*Henceforth, the self-adjoint extension $\widetilde{H}$ corresponding to the matrices A and B will be denoted by $H_{A,B}$.*
*(ii) For $z \in \rho(H_{A,B})$, the resolvent $H_{A,B}$ is of the form*

$$\begin{aligned} \big((H_{A,B} - zI_{L^2_r((a,b))})^{-1} f\big)(x) = \int_a^b r(x')dx'\, G_{A,B}(z,x,x') f(x'), \\ f \in L^2((a,b); rdx), \end{aligned} \tag{3.8}$$

*where the Green's function $G_{A,B}(z,x,x')$ is of the form given by*

$$G_{A,B}(z,x,x') = \begin{cases} \sum_{j,k=1}^2 m_{j,k}^+(x) u_j(z,x) u_k(z,x'), & a \le x' \le x \le b, \\ \sum_{j,k=1}^2 m_{j,k}^-(x) u_j(z,x) u_k(z,x'), & a \le x < x' \le b. \end{cases} \tag{3.9}$$

*Here $\{u_1(z,\cdot), u_2(z,\cdot)\}$ represents a fundamental set of solutions for $(\tau - z)u = 0$ and $m_{j,k}^{\pm}(z)$, $1 \le j,k \le 2$, are appropriate constants.*
*(iii) $H_{A,B}$ has purely discrete spectrum with eigenvalues of multiplicity at most 2. Moreover, if $\sigma(H_{A,B}) = \{\lambda_{A,B,j}\}_{j \in \mathbb{N}}$, then*

$$\sum_{\substack{j \in \mathbb{N} \\ \lambda_{A,B,j} \neq 0}} |\lambda_{A,B,j}|^{-1} < \infty. \tag{3.10}$$

*In particular,*

$$(H_{A,B} - zI_{L^2_r((a,b))})^{-1} \in \mathcal{B}_1\big(L^2((a,b); rdx)\big), \quad z \in \rho(H_{A,B}). \tag{3.11}$$

[1] The expression $(A\ B)$ represents a $2 \times 4$ rectangular matrix.

The characterization of self-adjoint extensions of $H_{\min}$ in terms of pairs of matrices $(A, B) \in \mathbb{C}^{2\times 2} \times \mathbb{C}^{2\times 2}$ satisfying (3.6) is not unique in the sense that different pairs may lead to the same self-adjoint extension. The next result recalls a unique characterization for all self-adjoint extensions of $H_{min}$ and hence can be viewed as a refinement of Theorem 3.2.

**Theorem 3.3** (*See, e.g., [64, Theorem 13.15], [66, Theorem 10.4.3]*) *Assume Hypothesis 3.1. Let $H_{min}$ be the minimal operator associated with $\tau$ and defined in* (3.4) *and $H_{A,B}$ a self-adjoint extension of the minimal operator as characterized in Theorem 3.2; then, the following hold:*
*(i) $H_{A,B}$ is a self-adjoint extension of $H_{min}$, with* $\operatorname{rank}(A) = \operatorname{rank}(B) = 1$ *if and only if A and B can be put in the form*

$$A = \begin{pmatrix} \cos(\alpha) & \sin(\alpha) \\ 0 & 0 \end{pmatrix}, \quad B = \begin{pmatrix} 0 & 0 \\ -\cos(\beta) & \sin(\beta) \end{pmatrix}, \tag{3.12}$$

*for a unique pair $\alpha, \beta \in [0, \pi)$. Hence, upon identifying $H_{A,B}$ with $H_{\alpha,\beta}$, all self-adjoint extensions of $H_{min}$ with separated boundary conditions are of the form*

$$\begin{aligned} &H_{\alpha,\beta} f = \tau f, \quad \alpha, \beta \in [0, \pi), \\ &f \in \operatorname{dom}(H_{\alpha,\beta}) = \{g \in \operatorname{dom}(H_{max}) \mid \sin(\alpha) g^{[1]}(a) + \cos(\alpha) g(a) = 0, \\ &\qquad\qquad\qquad\qquad -\sin(\beta) g^{[1]}(b) + \cos(\beta) g(b) = 0\}. \end{aligned} \tag{3.13}$$

*(ii) $H_{A,B}$ is a self-adjoint extension of $H_{min}$ with* $\operatorname{rank}(A) = \operatorname{rank}(B) = 2$ *if and only if A and B can be put in the form*

$$A = e^{i\varphi} R, \quad B = I_2, \tag{3.14}$$

*for a unique $\varphi \in [0, 2\pi)$, and unique $R \in \mathrm{SL}_2(\mathbb{R})$. Hence, upon identifying $H_{A,B}$ with $H_{\varphi,R}$, all self-adjoint extensions of $H_{min}$ with coupled boundary conditions are of the form*

$$\begin{aligned} &H_{\varphi,R} f = \tau f, \quad \varphi \in [0, 2\pi), \ R \in \mathrm{SL}_2(\mathbb{R}), \\ &f \in \operatorname{dom}(H_{\varphi,R}) = \left\{ g \in \operatorname{dom}(H_{max}) \,\middle|\, \begin{pmatrix} g(b) \\ g^{[1]}(b) \end{pmatrix} = e^{i\varphi} R \begin{pmatrix} g(a) \\ g^{[1]}(a) \end{pmatrix} \right\}. \end{aligned} \tag{3.15}$$

*(iii) All self-adjoint extensions of $H_{min}$ are either of type (i) (i.e., separated) or of type (ii) (i.e., coupled).*

Here $\mathrm{SL}_2(\mathbb{R})$ denotes the group of $2 \times 2$ matrices with real-valued entries and determinant one.

For notational convenience, we will adhere to the notation $H_{\alpha,\beta}$ and $H_{\varphi,R}$ in the following.

Next, we recall some results of [25]. For this purpose, we introduce the fundamental system of solutions $\theta(z, x, a)$, $\phi(z, x, a)$ of $\tau y = zy$ defined by

$$\theta(z, a, a) = \phi^{[1]}(z, a, a) = 1, \quad \theta^{[1]}(z, a, a) = \phi(z, a, a) = 0, \tag{3.16}$$

such that

$$W(\theta(z, \cdot, a), \phi(z, \cdot, a)) = 1, \tag{3.17}$$

where, for $f, g$ (locally) absolutely continuous,

$$W(f, g)(\cdot) = f(\cdot)g^{[1]}(\cdot) - f^{[1]}(\cdot)g(\cdot). \tag{3.18}$$

Furthermore, we introduce the boundary values for $g, g^{[1]} \in AC([a, b])$, see [51, Sect. 1.2], [66, Sect. 3.2],

$$\begin{aligned} U_{\alpha,\beta,1}(g) &= \sin(\alpha)g^{[1]}(a) + \cos(\alpha)g(a), \\ U_{\alpha,\beta,2}(g) &= -\sin(\beta)g^{[1]}(b) + \cos(\beta)g(b), \end{aligned} \tag{3.19}$$

in the case $(i)$ of separated boundary conditions in Theorem 3.3, and

$$\begin{aligned} V_{\varphi,R,1}(g) &= g(b) - e^{i\varphi}R_{1,1}g(a) - e^{i\varphi}R_{1,2}g^{[1]}(a), \\ V_{\varphi,R,2}(g) &= g^{[1]}(b) - e^{i\varphi}R_{2,1}g(a) - e^{i\varphi}R_{2,2}g^{[1]}(a), \end{aligned} \tag{3.20}$$

in the case $(ii)$ of coupled boundary conditions in Theorem 3.3. Moreover, we define

$$F_{\alpha,\beta}(z) = \det\begin{pmatrix} U_{\alpha,\beta,1}(\theta(z, \cdot, a)) & U_{\alpha,\beta,1}(\phi(z, \cdot, a)) \\ U_{\alpha,\beta,2}(\theta(z, \cdot, a)) & U_{\alpha,\beta,2}(\phi(z, \cdot, a)) \end{pmatrix}, \tag{3.21}$$

and

$$F_{\varphi,R}(z) = \det\begin{pmatrix} V_{\varphi,R,1}(\theta(z, \cdot, a)) & V_{\varphi,R,1}(\phi(z, \cdot, a)) \\ V_{\varphi,R,2}(\theta(z, \cdot, a)) & V_{\varphi,R,2}(\phi(z, \cdot, a)) \end{pmatrix}, \tag{3.22}$$

and note, in particular, that

$$\begin{aligned} &F_{\alpha,\beta}(z) \\ &= \begin{cases} \phi(z, b, a), & \alpha = \beta = 0, \\ -\sin(\beta)\phi^{[1]}(z, b, a) + \cos(\beta)\phi(z, b, a), & \alpha = 0,\ \beta \in (0, \pi), \\ \cos(\alpha)\phi(z, b, a) - \sin(\alpha)\theta(z, b, a), & \alpha \in (0, \pi),\ \beta = 0, \\ \cos(\alpha)[-\sin(\beta)\phi^{[1]}(z, b, a) + \cos(\beta)\phi(z, b, a)] & \\ -\sin(\alpha)[-\sin(\beta)\theta^{[1]}(z, b, a) + \cos(\beta)\theta(z, b, a)], & \alpha, \beta \in (0, \pi). \end{cases} \end{aligned} \tag{3.23}$$

Given these preparations, we can state our first result concerning the computation of traces and determinants.

**Theorem 3.4** *Assume Hypothesis 3.1 and denote by $H_{\alpha,\beta}$ and $H_{\varphi,R}$ the self-adjoint extensions of $H_{min}$ as described in cases $(i)$ and $(ii)$ of Theorem 3.3, respectively.*

(*i*) *Suppose* $z_0 \in \rho(H_{\alpha,\beta})$, *then*

$$\begin{aligned}&\det_{L^2_r((a,b))}\big(I_{L^2_r((a,b))} - (z - z_0)(H_{\alpha,\beta} - z_0 I_{L^2_r((a,b))})^{-1}\big)\\&\quad = F_{\alpha,\beta}(z)/F_{\alpha,\beta}(z_0), \quad z \in \mathbb{C}.\end{aligned} \tag{3.24}$$

*In particular,*

$$\operatorname{tr}_{L^2_r((a,b))}\big((H_{\alpha,\beta} - zI_{L^2_r((a,b))})^{-1}\big) = -(d/dz)\ln(F_{\alpha,\beta}(z)), \quad z \in \rho(H_{\alpha,\beta}). \tag{3.25}$$

(*ii*) *Suppose* $z_0 \in \rho(H_{\varphi,R})$, *then*

$$\begin{aligned}&\det_{L^2_r((a,b))}\big(I_{L^2_r((a,b))} - (z - z_0)(H_{\varphi,R} - z_0 I_{L^2_r((a,b))})^{-1}\big)\\&\quad = F_{\varphi,R}(z)/F_{\varphi,R}(z_0), \quad z \in \mathbb{C}.\end{aligned} \tag{3.26}$$

*In particular,*

$$\operatorname{tr}_{L^2_r((a,b))}\big((H_{\varphi,R} - zI_{L^2_r((a,b))})^{-1}\big) = -(d/dz)\ln(F_{\varphi,R}(z)), \quad z \in \rho(H_{\varphi,R}). \tag{3.27}$$

*Proof* In the special case $p = r = 1$ and given separated (but generally, nonself-adjoint) boundary conditions, the fact (3.24) was proved in [25] upon combining the eigenvalue results in [51, Sect. 1.2] and Lemma 2.2. The proof in [25] extends to the present situation with $p, r$ satisfying Hypothesis 3.1 in the cases (*i*) and (*ii*) since actually the eigenvalues $\lambda_{A,B,j}$ of $H_{A,B}$ all have the universal leading Weyl-type asymptotics [33, Sect. VI.7] (see also [3, 66, Theorem 4.3.1])

$$\lim_{j\to\infty} j^{-2}\lambda_{A,B,j} = \pi^2\bigg(\int_a^b dx\,[r(x)/p(x)]^{1/2}\bigg)^{-2}, \tag{3.28}$$

independently of the chosen boundary conditions, improving upon relation (3.10). More precisely, [33, Sect. VI.7] determines the leading asymptotic behavior of the eigenvalue counting function (independently of the underlying choice of boundary conditions) which is known to yield (3.28).

Relations (3.25) and (3.27) are then clear from combining formula (2.3) with (3.24) and (3.26), respectively. □

*Remark 3.5* Considering traces of resolvent differences for various boundary conditions (separated and/or coupled) permits one to make a direct connection with the boundary data maps studied in [8, 9, 26]. More precisely, suppose the pairs $A, B \in \mathbb{C}^{2\times 2}$ and $A', B' \in \mathbb{C}^{2\times 2}$ satisfy (3.6) and define $H_{A,B}$ and $H_{A',B'}$ according to (3.7). Then

$$\begin{aligned}&\operatorname{tr}_{L^2((a,b);rdx)}\big((H_{A',B'} - zI_{L^2_r((a,b))})^{-1} - (H_{A,B} - zI_{L^2_r((a,b))})^{-1}\big)\\&\quad = -\frac{d}{dz}\ln\Big(\det_{\mathbb{C}^2}\Big(\Lambda_{A,B}^{A',B'}(z)\Big)\Big), \quad z \in \rho(H_{A,B}) \cap \rho(H_{A',B'}),\end{aligned} \tag{3.29}$$

where the $2 \times 2$ matrix $\Lambda_{A,B}^{A',B'}(z)$ represents the boundary data map associated to the pair $(H_{A,B}, H_{A',B'})$. A comparison of (3.29) with (3.25) and (3.27) yields

$$\det_{\mathbb{C}^2}\Big(\Lambda_{A,B}^{A',B'}(z)\Big) = C_0\, F_{A',B'}(z)/F_{A,B}(z), \quad z \in \rho(H_{A,B}) \cap \rho(H_{A',B'}), \tag{3.30}$$

where, in obvious notation, $F_{A',B'}(z)$, $F_{A,B}(z)$ represent either $F_{\alpha,\beta}(z)$ or $F_{\varphi,R}(z)$, depending on which boundary conditions (separated or coupled) are represented by the pairs $(A', B')$, $(A, B)$, and $C_0 = C_0(A', B', A, B)$ is a $z$-independent constant. Indeed, $C_0 = 1$ for separated as well as coupled boundary conditions. The separated case was shown in [26], Lemma 3.4, using the explicit matrix representation of $\Lambda_{AB}^{A'B'}(z)$ for that case, and the result for coupled boundary conditions follows along the same lines. ◇

For the case of determinants of general higher order differential operators (with matrix-valued coefficients) and general boundary conditions on bounded intervals we refer to Burghelea, Friedlander, and Kappeler [5], Dreyfus and Dym [13], Falco, Fedorenko, and Gruzberg [15], and Forman [16, Sect. 3]. We also refer to [46] for a closed form of an infinite product of ratios of eigenvalues of Sturm–Liouville operators on bounded intervals.

We briefly look at two prominent examples associated with coupled boundary conditions next:

*Example 3.6* (*Floquet boundary conditions*) Consider the family of operators $H_{\varphi,I_2}$, familiar from Floquet theory, defined by taking $R = B = I_2$ in (3.15). In this case

$$F_{\varphi,I_2}(z) = -2e^{i\varphi}[\Delta(z) - \cos(\varphi)], \quad z \in \mathbb{C}, \tag{3.31}$$

where

$$\Delta(z) = \big[\theta(z,b,a) + \phi^{[1]}(z,b,a)\big]/2, \quad z \in \mathbb{C}, \tag{3.32}$$

represents the well-known Floquet discriminant and hence

$$\begin{aligned} &\det_{L^2_r((a,b))}\big(I_{L^2_r((a,b))} - (z-z_0)(H_{\varphi,I_2} - z_0 I_{L^2_r((a,b))})^{-1}\big) \\ &\quad = [\Delta(z) - \cos(\varphi)]/[\Delta(z_0) - \cos(\varphi)], \quad z \in \mathbb{C}, \end{aligned} \tag{3.33}$$

$$\begin{aligned} &\operatorname{tr}_{L^2_r((a,b))}\big((H_{\varphi,I_2} - zI_{L^2_r((a,b))})^{-1}\big) \\ &\quad = -(d/dz)\ln([\Delta(z) - \cos(\varphi)]), \quad z \in \rho(H_{\varphi,I_2}). \end{aligned} \tag{3.34}$$

*Example 3.7* (*The Krein–von Neumann extension*) Consider the case $\varphi = 0$, $A = R_K$, with

$$R_K = \begin{pmatrix} \theta(0,b,a) & \phi(0,b,a) \\ \theta^{[1]}(0,b,a) & \phi^{[1]}(0,b,a) \end{pmatrix}. \tag{3.35}$$

As shown in [9, Example 3.3], the resulting operator $H_{0,R_K}$ represents the Krein–von Neumann extension of $H_{min}$. Thus,

$$F_{0,R_K}(z) = -2[D_K(z) - 1], \quad z \in \mathbb{C}, \tag{3.36}$$

where

$$D_K(z) = \big[\phi^{[1]}(0,b,a)\theta(z,b,a) + \theta(0,b,a)\phi^{[1]}(z,b,a) - \phi(0,b,a)\theta^{[1]}(z,b,a) \\ - \theta^{[1]}(0,b,a)\phi(z,b,a)\big]/2, \quad z \in \mathbb{C}. \tag{3.37}$$

Hence,

$$\det_{L^2_r((a,b))}\big(I_{L^2_r((a,b))} - (z-z_0)(H_{0,R_K} - z_0 I_{L^2_r((a,b))})^{-1}\big) \\ = [D_K(z) - 1]/[D_K(z_0) - 1], \quad z \in \mathbb{C}, \tag{3.38}$$

$$\operatorname{tr}_{L^2_r((a,b))}\big((H_{0,R_K} - z I_{L^2_r((a,b))})^{-1}\big) \\ = -(d/dz)\ln(D_K(z) - 1), \quad z \in \rho(H_{0,R_K}). \tag{3.39}$$

Because of the Wronskian relation (3.17), $D_K(0) = 1$, furthermore

$$D_K(z) - 1 \underset{z\to 0}{=} z^2[c + O(z)], \tag{3.40}$$

where

$$c = \frac{1}{2}\Big(\overset{\bullet}{\phi}(0,b,a)\,[\overset{\bullet}{\theta}]^{[1]}(0,b,a) - [\overset{\bullet}{\phi}]^{[1]}(0,b,a)\,\overset{\bullet}{\theta}(0,b,a)\Big) \\ = \frac{1}{2}W\Big(\overset{\bullet}{\phi}(0,\cdot\,,a), \overset{\bullet}{\theta}(0,\cdot\,,a)\Big)(b), \tag{3.41}$$

abbreviating $\bullet = d/dz$.

The small-$z$ behavior (3.40) is clear from general results in [51, Sect. 1.2] or [66, Sect. 3.2] and is in accordance with a two-dimensional nullspace of the Krein–von Neumann extension on a bounded interval. To actually show that $c \neq 0$ (in fact, $c < 0$) in (3.40) one recalls that $\overset{\bullet}{\phi}(z,x,a)$ and $\overset{\bullet}{\theta}(z,x,a)$ satisfy an inhomogeneous Sturm–Liouville equation and as a result one obtains (for $z \in \mathbb{C}$, $x \in [a,b]$)

$$\overset{\bullet}{\theta}(z,x,a) = \theta(z,x,a)\int_a^x r(x')dx'\,\phi(z,x',a)\theta(z,x',a) \\ - \phi(z,x,a)\int_a^x r(x')dx'\,\theta(z,x',a)^2, \\ [\overset{\bullet}{\theta}]^{[1]}(z,x,a) = \theta^{[1]}(z,x,a)\int_a^x r(x')dx'\,\phi(z,x',a)\theta(z,x',a) \\ - \phi^{[1]}(z,x,a)\int_a^x r(x')dx'\,\theta(z,x',a)^2, \tag{3.42}$$

$$\dot{\phi}(z,x,a) = \theta(z,x,a)\int_a^x r(x')dx'\,\phi(z,x',a)^2 \\ - \phi(z,x,a)\int_a^x r(x')dx'\,\phi(z,x',a)\theta(z,x',a),$$

$$[\dot{\phi}]^{[1]}(z,x,a) = \theta^{[1]}(z,x,a)\int_a^x r(x')dx'\,\phi(z,x',a)^2 \\ - \phi^{[1]}(z,x,a)\int_a^x r(x')dx'\,\phi(z,x',a)\theta(z,x',a).$$

Equations (3.42) and Cauchy's inequality then imply

$$W\big(\dot{\phi}(\lambda,\cdot,a),\dot{\theta}(\lambda,\cdot,a)\big)(x) = \bigg[\int_a^x r(x')dx'\,\phi(\lambda,x',a)\theta(\lambda,x',a)\bigg]^2 \\ - \bigg[\int_a^x r(x')dx'\,\theta(\lambda,x',a)^2\bigg]\bigg[\int_a^x r(x')dx'\,\phi(\lambda,x',a)^2\bigg] \leq 0, \quad \lambda\in\mathbb{R},\ x\in[a,b], \tag{3.43}$$

using the fact that $\phi(\lambda,x,a)$ and $\theta(\lambda,x,a)$ are real-valued for $\lambda\in\mathbb{R}$, $x\in[a,b]$. Equality in Cauchy's inequality for $x>a$ would imply that for some $\alpha,\beta\in[0,\infty)$, $(\alpha,\beta)\neq(0,0)$,

$$\alpha\,\phi(\lambda,x',a) = \beta\,\theta(\lambda,x',a), \quad x'\in(a,x], \tag{3.44}$$

a contradiction. Thus, $W\big(\dot{\phi}(\lambda,\cdot,a),\dot{\theta}(\lambda,\cdot,a)\big)(x) < 0$ for $\lambda\in\mathbb{R}$, $x\in(a,b]$ and hence $c<0$ in (3.40), (3.41).

In the case of separated boundary conditions, that is, case $(i)$ in Theorem 3.3, one can shed more light on $F_{\alpha,\beta}(z)$ in terms of appropriate Weyl solutions $\psi_-(z,\cdot,a,\alpha)$ and $\psi_+(z,\cdot,a,\beta)$ that satisfy the boundary conditions in $\mathrm{dom}(H_{\alpha,\beta})$ in (3.13) at $a$ and $b$, respectively. Up to normalizations, $\psi_\pm(z,\cdot,a,\alpha)$ are given by

$$\psi_-(z,\cdot,a,\alpha) = c_-\begin{cases}\cos(\alpha)\phi(z,\cdot,a)-\sin(\alpha)\theta(z,\cdot,a), & \alpha\in(0,\pi),\\ \phi(z,\cdot,a), & \alpha=0,\end{cases} \tag{3.45}$$

$$\psi_+(z,\cdot,a,\beta) = c_+\begin{cases}\big\{\big[-\sin(\beta)\theta^{[1]}(z,b,a)+\cos(\beta)\theta(z,b,a)\big]\phi(z,\cdot,a) \\ +\big[\sin(\beta)\phi^{[1]}(z,b,a)-\cos(\beta)\phi(z,b,a)\big]\theta(z,\cdot,a)\big\}, \\ \hfill \beta\in(0,\pi),\\ \theta(z,b,a)\phi(z,\cdot,a)-\phi(z,b,a)\theta(z,\cdot,a), \quad \beta=0,\end{cases} \tag{3.46}$$

and hence the Green's function of $H_{\alpha,\beta}$ is of the semi-separable form,

$$G_{\alpha,\beta}(z,x,x') = \big(H_{\alpha,\beta}-zI_{L^2_r((a,b))}\big)^{-1}(x,x') \tag{3.47}$$

$$= \frac{1}{W(\psi_+(z,\,\cdot\,,a,\beta),\psi_-(z,\,\cdot\,,a,\alpha))} \times \begin{cases} \psi_-(z,x,a,\alpha)\psi_+(z,x',a,\beta), & a \le x \le x' \le b, \\ \psi_-(z,x',a,\alpha)\psi_+(z,x,a,\beta), & a \le x' \le x \le b. \end{cases}$$

A direct computation then reveals the following connection between $F_{\alpha,\beta}(z)$, $\psi_-(z,\,\cdot\,,a,\alpha)$, and $\psi_+(z,\,\cdot\,,a,\beta)$,

$$F_{\alpha,\beta}(z) = \frac{1}{W(\psi_+(z,\,\cdot\,,a,\beta),\psi_-(z,\,\cdot\,,a,\alpha))} \times \begin{cases} \psi_-(z,b,a,0)\psi_+(z,a,a,0), \quad \alpha=\beta=0, \\ \big[-\sin(\beta)\psi_-^{[1]}(z,b,a,0)+\cos(\beta)\psi_-(z,b,a,0)\big]\psi_+(z,a,a,\beta), \\ \qquad\qquad \alpha=0,\ \beta\in(0,\pi), \\ \psi_-(z,b,a,\alpha)\big[\sin(\alpha)\psi_+^{[1]}(z,a,a,0)+\cos(\alpha)\psi_+(z,a,a,0)\big], \\ \qquad\qquad \alpha\in(0,\pi),\ \beta=0, \\ \big[-\sin(\beta)\psi_-^{[1]}(z,b,a,\alpha)+\cos(\beta)\psi_-(z,b,a,\alpha)\big] \\ \times\big[\sin(\alpha)\psi_+^{[1]}(z,a,a,\beta)+\cos(\alpha)\psi_+(z,a,a,\beta)\big], \quad \alpha,\beta\in(0,\pi). \end{cases} \tag{3.48}$$

Combining (3.48) and (3.25) thus yields

$$\operatorname{tr}_{L^2_r((a,b))}\big((H_{\alpha,\beta}-zI_{L^2_r((a,b))})^{-1}\big) = \begin{cases} -(d/dz)\ln\big(\psi_+(z,a,a,0)/\psi_+^{[1]}(z,b,a,0)\big) \\ \quad = -(d/dz)\ln\big(\psi_-(z,b,a,0)/\psi_-^{[1]}(z,a,a,0)\big), & \alpha=\beta=0, \\ -(d/dz)\ln(\psi_+(z,a,a,\beta)/\psi_+(z,b,a,\beta)), & \alpha=0,\ \beta\in(0,\pi), \\ -(d/dz)\ln(\psi_-(z,b,a,\alpha)/\psi_-(z,a,a,\alpha)), & \alpha\in(0,\pi),\ \beta=0, \\ -(d/dz)\ln\Big(\frac{W(\psi_+(z,\,\cdot\,,a,\beta),\psi_-(z,\,\cdot\,,a,\alpha))}{\psi_+(z,b,a,\beta)\psi_-(z,a,a,\alpha)}\Big), & \alpha,\beta\in(0,\pi), \end{cases} \quad z\in\rho(H_{\alpha,\beta}). \tag{3.49}$$

Next, applying Theorem 3.4 in the context of Theorem 2.10 immediately yields results about $\zeta$-regularized determinants (see also [18, 38, Chaps. 2, 3], [44]).

*Remark 3.8* In Example 3.19, we will consider a simple case with negative eigenvalues present. Otherwise, in the examples of this section, we will always assume that eigenvalues are nonnegative. If that is not the case, an appropriate imaginary part according to Theorem 2.10 has to be included. ◇

To deal with $\zeta$-regularized determinants we now strengthen Hypothesis 3.1 and introduce the following assumptions on $p, q, r$:

**Hypothesis 3.9** Suppose $p, q, r$ satisfy the following conditions:
(*i*) $r > 0$ a.e. on $(a,b)$, $r \in L^1((a,b);dx)$, $1/r \in L^\infty((a,b);dx)$.

(*ii*) $p > 0$ a.e. on $(a, b)$, $1/p \in L^1((a, b); dx)$.
(*iii*) $q \in L^1((a, b); dx)$, $q$ is real-valued a.e. on $(a, b)$.
(*iv*) $pr$ and $(pr)'/r$ are absolutely continuous on $[a, b]$.

The substitutions (cf. [47, p. 2])

$$v(x) = \frac{1}{c}\int_a^x dt\,[r(t)/p(t)]^{1/2}, \quad u(x) = [r(x)p(x)]^{1/4}y(x), \tag{3.50}$$

where $c$ is given by

$$c = \int_a^b dt\,[r(t)/p(t)]^{1/2}, \tag{3.51}$$

such that $v \in [0, 1]$, transforms the Sturm–Liouville problem $(\tau y)(x) = zy(x)$ into

$$-u''(v) + V(v)u(v) = c^2 z u(v), \tag{3.52}$$

with ($\mu(v) = [r(x)p(x)]^{1/4}$)

$$\begin{aligned} V(v) &= \frac{\mu''(v)}{\mu(v)} + c^2\frac{q(x)}{r(x)} \\ &= -\frac{c^2}{16}\frac{1}{r(x)p(x)}\frac{(r(x)p(x))'}{r(x)} + \frac{c^2}{4}\frac{1}{r(x)}\frac{d}{dx}\left[\frac{(r(x)p(x))'}{r(x)}\right] + c^2\frac{q(x)}{r(x)}. \end{aligned} \tag{3.53}$$

Hypothesis 3.9 guarantees[2] that $V \in L^1((0, 1); dx)$, and as a consequence one has asymptotically for $x \in (a, b]$,

$$\begin{aligned} \phi(z, x, a) &\underset{|z|\to\infty}{=} 2^{-1}z^{-1/2}[p(x)p(a)r(x)r(a)]^{-1/4} \\ &\qquad \times \exp\left(z^{1/2}\int_a^x dt\,[r(t)/p(t)]^{1/2}\right)\left(1 + O\left(|z|^{-1/2}\right)\right), \\ \phi^{[1]}(z, x, a) &\underset{|z|\to\infty}{=} 2^{-1}[p(a)r(a)]^{-1/4}[p(x)r(x)]^{1/4} \\ &\qquad \times \exp\left(z^{1/2}\int_a^x dt\,[r(t)/p(t)]^{1/2}\right)\left(1 + O\left(|z|^{-1/2}\right)\right), \\ \theta(z, x, a) &\underset{|z|\to\infty}{=} 2^{-1}[r(a)/p(a)]^{1/2}[p(x)p(a)r(x)r(a)]^{-1/4} \\ &\qquad \times \exp\left(z^{1/2}\int_a^x dt\,[r(t)/p(t)]^{1/2}\right)\left(1 + O\left(|z|^{-1/2}\right)\right), \end{aligned} \tag{3.54}$$

[2]We are indebted to Bennewitz [4] for a very helpful discussion of this issue.

$$\theta^{[1]}(z,x,a) \underset{|z|\to\infty}{=} 2^{-1}z^{1/2}p(a)^{-1}[p(x)p(a)r(x)r(a)]^{1/4}$$
$$\times \exp\left(z^{1/2}\int_a^x dt\,[r(t)/p(t)]^{1/2}\right)\left(1+O\left(|z|^{-1/2}\right)\right).$$

This asymptotic behavior is used to guarantee that assumption (2.38) is satisfied in several of the following examples.

**Theorem 3.10** *Assume Hypothesis 3.9 and denote by $H_{\alpha,\beta,j}$ and $H_{\varphi,R,j}$, $j=1,2$, the self-adjoint extensions of $H_{min}$ as described in Theorem 3.3 $(i)$ and $(ii)$, respectively. Here the index $j$ refers to a potential $q_j$ in (3.1), $j=1,2$. Then the following items $(i)$ and $(ii)$ hold:*
$(i)$

$$\zeta'\left(0; H_{\alpha,\beta,1}, H_{\alpha,\beta,2}\right) = -\lim_{t\downarrow 0}\ln\left(\left|\frac{F_{\alpha,\beta,2}(te^{i\theta})}{F_{\alpha,\beta,1}(te^{i\theta})}\,(te^{i\theta})^{[m(0,H_{\alpha,\beta,1})-m(0,H_{\alpha,\beta,2})]}\right|\right). \tag{3.55}$$

$(ii)$

$$\zeta'\left(0; H_{\varphi,R,1}, H_{\varphi,R,2}\right) = -\lim_{t\downarrow 0}\ln\left(\left|\frac{F_{\varphi,R,2}(te^{i\theta})}{F_{\varphi,R,1}(te^{i\theta})}\,(te^{i\theta})^{[m(0,H_{\varphi,R,1})-m(0,H_{\varphi,R,2})]}\right|\right). \tag{3.56}$$

*Proof* This follows immediately from Theorems 2.9 and 2.10 applied to $S_j = H_{\alpha,\beta,j}$, respectively, $S_j = H_{\varphi,R,j}$, $j=1,2$. □

*Remark 3.11* In the absence of zero eigenvalues of $H_{\alpha,\beta,j}$, respectively, $H_{\varphi,R,j}$, $j=1,2$, these results simplify to

$$\zeta'\left(0; H_{\alpha,\beta,1}, H_{\alpha,\beta,2}\right) = \ln\left(\left|\frac{F_{\alpha,\beta,1}(0)}{F_{\alpha,\beta,2}(0)}\right|\right), \tag{3.57}$$

respectively,

$$\zeta'\left(0; H_{\varphi,R,1}, H_{\varphi,R,2}\right) = \ln\left(\left|\frac{F_{\varphi,R,1}(0)}{F_{\varphi,R,2}(0)}\right|\right). \tag{3.58}$$

In case there are zero eigenvalues, a suitable energy shift will again lead to this case. ◇

*Example 3.12* If one of the potentials vanishes, say $q_1 = 0$, more explicit results can be obtained. We consider separated boundary conditions with no zero eigenvalues present for $j=1,2$. Then

$$\begin{aligned} F_{\alpha,\beta,j}(0) &= \cos(\alpha)\left[-\sin(\beta)\phi_j^{[1]}(0,b,a)+\cos(\beta)\phi_j(0,b,a)\right] \\ &\quad -\sin(\alpha)\left[-\sin(\beta)\theta_j^{[1]}(0,b,a)+\cos(\beta)\theta_j(0,b,a)\right], \quad j=1,2. \end{aligned} \tag{3.59}$$

For $q_1 = 0$, $\theta_1(0, x, a)$ and $\phi_1(0, x, a)$ satisfy the initial value problems

$$-\left(\frac{d}{dx}p(x)\frac{d}{dx}\right)\theta_1(0, x, a) = 0, \qquad \theta_1(0, a, a) = 1, \quad \theta_1^{[1]}(0, a, a) = 0,$$
$$-\left(\frac{d}{dx}p(x)\frac{d}{dx}\right)\phi_1(0, x, a) = 0, \qquad \phi_1(0, a, a) = 0, \quad \phi_1^{[1]}(0, a, a) = 1, \tag{3.60}$$

solutions of which are

$$\phi_1(0, x, a) = \int_a^x dt\, p(t)^{-1}, \quad \theta_1(0, x, a) = 1. \tag{3.61}$$

Then one can show that

$$F_{\alpha,\beta,1}(0) = -\sin(\alpha+\beta) + \cos(\alpha)\cos(\beta)\int_a^b dt\, p(t)^{-1}. \tag{3.62}$$

The relative ζ-regularized determinant in this case then reads

$$\zeta'\left(0; H_{\alpha,\beta,1}, H_{\alpha,\beta,2}\right) = \ln\left(\left|\frac{\cos(\alpha)\cos(\beta)\int_a^b dt\, p(t)^{-1} - \sin(\alpha+\beta)}{F_{\alpha,\beta,2}(0)}\right|\right). \tag{3.63}$$

*Example 3.13* Restricting Example 3.12 to the case $p(x) = r(x) = 1$, the ζ-determinant $\zeta'(0; H_{\alpha,\beta,2})$ can be computed explicitly. First, one notes that under Hypothesis 3.9, the ζ-function can be analytically continued to a neighborhood of $s = 0$ and $\zeta'(0; H_{\alpha,\beta,2})$ is a well-defined quantity; similarly, $\zeta'(0; H_{\alpha,\beta,1})$ is well-defined for $q_1 = 0$. Computing $\zeta'(0; H_{\alpha,\beta,1})$, employing (3.63), one obtains $\zeta'(0; H_{\alpha,\beta,2})$.

First we assume there are no zero eigenvalues for $j = 1$, which is the case if

$$(b-a)\cos(\alpha)\cos(\beta) - \sin(\alpha+\beta) \neq 0. \tag{3.64}$$

Next, one notes that

$$\phi_1(z, x, a) = z^{-1/2}\sin\left(z^{1/2}(x-a)\right), \quad \theta_1(z, x, a) = \cos\left(z^{1/2}(x-a)\right), \tag{3.65}$$

and thus from (3.23),

$$\begin{aligned} F_{\alpha,\beta,1}(z) = {} & \cos(\alpha)\left[-\sin(\beta)\cos\left(z^{1/2}(b-a)\right) + z^{-1/2}\cos(\beta)\sin\left(z^{1/2}(b-a)\right)\right] \\ & - \sin(\alpha)\left[z^{1/2}\sin(\beta)\sin\left(z^{1/2}(b-a)\right) + \cos(\beta)\cos\left(z^{1/2}(b-a)\right)\right]. \end{aligned} \tag{3.66}$$

The details of what follows depend on the boundary conditions imposed. In this example, we consider $\alpha, \beta \in (0, \pi)$. Along the relevant contour, as $|z| \to \infty$, one has $\mathrm{Im}\big(z^{1/2}\big) > 0$ and the asymptotics for the boundary conditions considered reads

$$F_{\alpha,\beta,1}(z) = F^{asym}_{\alpha,\beta,1}(z)\big[1 + O\big(z^{-1/2}\big)\big], \tag{3.67}$$

where

$$F^{asym}_{\alpha,\beta,1}(z) = -(i/2)z^{1/2}\sin(\alpha)\sin(\beta)e^{-iz^{1/2}(b-a)}. \tag{3.68}$$

Adding and subtracting this asymptotics where applicable, we rewrite the $\zeta$-function for $H_{\alpha,\beta,1}$ in the form

$$\begin{aligned}\zeta(s; H_{\alpha,\beta,1}) &= e^{is(\pi-\theta)}\frac{\sin(\pi s)}{\pi}\int_0^1 dt\, t^{-s}\frac{d}{dt}\ln\big(F_{\alpha,\beta,1}\big(te^{i\theta}\big)\big)\\ &\quad + e^{is(\pi-\theta)}\frac{\sin(\pi s)}{\pi}\int_1^\infty dt\, t^{-s}\frac{d}{dt}\ln\left(\frac{F_{\alpha,\beta,1}\big(te^{i\theta}\big)}{F^{asym}_{\alpha,\beta,1}\big(te^{i\theta}\big)}\right)\\ &\quad + \zeta^{asym}(s; H_{\alpha,\beta,1}),\end{aligned} \tag{3.69}$$

where

$$\zeta^{asym}(s; H_{\alpha,\beta,1}) = e^{is(\pi-\theta)}\frac{\sin(\pi s)}{\pi}\int_1^\infty dt\, t^{-s}\frac{d}{dt}\ln\big(F^{asym}_{\alpha,\beta,1}\big(te^{i\theta}\big)\big). \tag{3.70}$$

This representation is valid for $\frac{1}{2} < \mathrm{Re}(s) < 1$. The term (3.70) is easily computed and yields

$$\zeta^{asym}(s; H_{\alpha,\beta,1}) = e^{is(\pi-\theta)}\frac{\sin(\pi s)}{\pi}\left[\frac{1}{2s} - \frac{i}{2}e^{i\theta/2}\frac{(b-a)}{s-(1/2)}\right], \tag{3.71}$$

yielding its analytic continuation to $-\frac{1}{2} < \mathrm{Re}(s) < 1$. The $\zeta$-determinant for $H_{\alpha,\beta,1}$ then follows from

$$\begin{aligned}\zeta'(0; H_{\alpha,\beta,1}) &= \mathrm{Re}\Big(\ln\big(F^{asym}_{\alpha,\beta,1}\big(e^{i\theta}\big)\big) - \ln(F_{\alpha,\beta,1}(0)) + \zeta^{asym\,\prime}(0; H_{\alpha,\beta,1})\Big)\\ &= -\ln\left(\left|\frac{2\,[\cos(\alpha)\cos(\beta)\,(b-a) - \sin(\alpha+\beta)]}{\sin(\alpha)\sin(\beta)}\right|\right).\end{aligned} \tag{3.72}$$

Using expression (3.72) in (3.63) yields

$$\zeta'(0; H_{\alpha,\beta,2}) = \ln\left(\left|\frac{\sin(\alpha)\sin(\beta)}{2F_{\alpha,\beta,2}(0)}\right|\right). \tag{3.73}$$

If there is a zero eigenvalue for $j = 1$, namely, if

$$(b-a)\cos(\alpha)\cos(\beta) - \sin(\alpha+\beta) = 0, \tag{3.74}$$

the relevant formula for the relative ζ-determinant is

$$\zeta'(0; H_{\alpha,\beta,1}, H_{\alpha,\beta,2}) = \ln\left( \left| \frac{F'_{\alpha,\beta,1}(0)}{F_{\alpha,\beta,2}(0)} \right| \right). \tag{3.75}$$

Employing (3.66), this can be cast in the form

$$\zeta'(0; H_{\alpha,\beta,1}, H_{\alpha,\beta,2}) = \ln\left( \left| \frac{(b-a)\,[(b-a)\sin(\alpha+\beta) - 3\sin(\alpha)\sin\beta]}{3F_{\alpha,\beta,2}(0)} \right| \right). \tag{3.76}$$

In order to compute $\zeta'(0; H_{\alpha,\beta,2})$, the part $\zeta'(0; H_{\alpha,\beta,1})$ can be computed as before, the only difference being that $F_{\alpha,\beta,1}(z)$ is replaced by $F_{\alpha,\beta,1}(z)/z$. The final answer then reads

$$\zeta'(0; H_{\alpha,\beta,1}) = -\ln\left( \left| 2(b-a)\left[1 - \frac{(b-a)\sin(\alpha+\beta)}{3\sin(\alpha)\sin\beta}\right] \right| \right). \tag{3.77}$$

(This confirms the known result in the case of Neumann boundary conditions at $x = a$ and $x = b$, that is, for $\alpha = \beta = \pi/2$.) From (3.77) it is immediate that

$$\zeta'(0; H_{\alpha,\beta,2}) = \ln\left( \left| \frac{\sin(\alpha)\sin(\beta)}{2F_{\alpha,\beta,2}(0)} \right| \right). \tag{3.78}$$

*Example 3.14* Next, consider the Dirichlet boundary condition at $x = a$ and a Robin boundary condition at $x = b$, that is, $\alpha = 0$ and $\beta \in (0, \pi)$. In the absence of zero eigenvalues one then has

$$\zeta'(0; H_{0,\beta,1}, H_{0,\beta,2}) = \ln\left( \left| \frac{F_{0,\beta,1}(0)}{F_{0,\beta,2}(0)} \right| \right). \tag{3.79}$$

In order to find the ζ-determinant $\zeta'(0; H_{0,\beta,2})$ we consider as before $q_1 = 0$. Then,

$$F_{0,\beta,1}(z) = -\sin(\beta)\cos\left(z^{1/2}(b-a)\right) + z^{-1/2}\cos\left(\beta\right)\sin(z^{1/2}(b-a)). \tag{3.80}$$

The relevant asymptotic large-$|z|$ behavior is

$$F^{asym}_{0,\beta,1}(z) = -\sin(\beta)e^{-iz^{1/2}(b-a)}/2, \tag{3.81}$$

and proceeding along the lines of previous computations, one finds

$$\zeta'(0; H_{0,\beta,1}) = \ln\left( \left| \frac{\sin(\beta)}{2\,[\sin(\beta) - (b-a)\cos(\beta)]} \right| \right). \tag{3.82}$$

(Again, this confirms the known case where $\beta = \pi/2$.) From (3.82) it is immediate that

$$\zeta'(0; H_{0,\beta,2}) = \ln\left( \left| \frac{\sin(\beta)}{2F_{0,\beta,2}(0)} \right| \right). \tag{3.83}$$

*Example 3.15* For Dirichlet boundary conditions at both endpoints, that is, for $\alpha = \beta = 0$, the relative zeta-determinant follows from

$$\zeta'(0; H_{0,0,1}, H_{0,0,2}) = \ln\left( \left| \frac{F_{0,0,1}(0)}{F_{0,0,2}(0)} \right| \right). \tag{3.84}$$

For $q_1 = 0$, from

$$F_{0,0,1}(z) = z^{-1/2} \sin\left(z^{1/2}(b-a)\right), \tag{3.85}$$

the relevant asymptotics is

$$F_{0,0,1}^{asym}(z) = (i/2)z^{-1/2} e^{-iz^{1/2}(b-a)}. \tag{3.86}$$

Thus, one finds

$$\zeta'(0; H_{0,0,1}) = -\ln(2(b-a)), \tag{3.87}$$

and hence,

$$\zeta'(0; H_{0,0,2}) = -\ln(|2F_{0,0,2}(0)|). \tag{3.88}$$

*Remark 3.16* Under the assumptions of Example 3.13 no additional computations are needed when considering certain relative determinants for different boundary conditions. Indeed, for $\alpha_j, \beta_j \in (0, \pi)$, $j = 1, 2$, Eq. (3.73) is valid replacing $\alpha, \beta \to \alpha_1, \beta_1$ and $F_{\alpha,\beta,2} \to F_{\alpha_2,\beta_2,2}$, and Eq. (3.83) is valid replacing $\beta \to \beta_1$, and $F_{0,\beta,2} \to F_{0,\beta_2,2}$. ◇

*Example 3.17* As an example for coupled boundary conditions we reconsider the Krein–von Neumann extension, Example 3.7. We first note that different potentials $q_1 \neq q_2$ lead to different Krein–von Neumann extensions $R_{K_1} \neq R_{K_2}$; see, for instance, (3.35). Nevertheless, under the assumptions made, Theorem 3.10 $(ii)$ remains valid and

$$\zeta'(0; H_{0,R_{K_1},1}, H_{0,R_{K_2},2}) = \lim_{z\to 0} \ln\left( \left| \frac{F_{0,R_{K_1},1}(z)}{F_{0,R_{K_2},2}(z)} \right| \right), \tag{3.89}$$

and from (3.40) one finds

$$\zeta'(0; H_{0,R_{K_1},1}, H_{0,R_{K_2},2}) = \ln(|c_1/c_2|), \tag{3.90}$$

with

$$c_j = \frac{1}{2}\left(\overset{\bullet}{\theta}{}_j^{[1]}(0,b,a)\,\overset{\bullet}{\phi}_j(0,b,a) - \overset{\bullet}{\theta}_j(0,b,a)\overset{\bullet}{\phi}{}_j^{[1]}(0,b,a)\right), \quad j = 1, 2. \quad (3.91)$$

For the case of vanishing potential, $q_1 = 0$, the constant $c_1$ can be determined explicitly. To this end, we need the small-$z$ expansion of the solutions of

$$-\left(\frac{1}{r(x)}\frac{d}{dx}p(x)\frac{d}{dx}\right)\theta_1(z,x,a) = z\theta_1(z,x,a), \quad \theta_1(z,a,a) = 1, \ \theta_1^{[1]}(z,a,a) = 0, \quad (3.92)$$

and

$$-\left(\frac{1}{r(x)}\frac{d}{dx}p(x)\frac{d}{dx}\right)\phi_1(z,x,a) = z\phi_1(z,x,a), \quad \phi_1(z,a,a) = 0, \ \phi_1^{[1]}(z,a,a) = 1. \quad (3.93)$$

Expanding

$$\theta_1(z,x,a) = \theta_1(0,x,a) + z\overset{\bullet}{\theta}_1(0,x,a) + O\big(z^2\big), \quad (3.94)$$

one compares $O(z)$-terms in (3.92) to find with (3.61)

$$\left(\frac{d}{dx}p(x)\frac{d}{dx}\right)\overset{\bullet}{\theta}_1(0,x,a) = -r(x). \quad (3.95)$$

Integrating, this yields

$$\overset{\bullet}{\theta}{}_1^{[1]}(0,x,a) - \overset{\bullet}{\theta}{}_1^{[1]}(0,a,a) = -\int_a^x du\, r(u), \quad (3.96)$$

but given $\theta_1^{[1]}(z,a,a) = 0$, one concludes that

$$\overset{\bullet}{\theta}{}_1^{[1]}(0,x,a) = -\int_a^x du\, r(u). \quad (3.97)$$

Similarly, integrating (3.97),

$$\overset{\bullet}{\theta}_1(0,b,a) = -\int_a^b dv\, p(v)^{-1}\int_a^v du\, r(u). \quad (3.98)$$

Proceeding in the same way for $\phi_1(z,x,a)$, one first shows

$$\left(\frac{d}{dx}p(x)\frac{d}{dx}\right)\overset{\bullet}{\phi}_1(0,x,a) = -r(x)\int_a^x du\, p(u)^{-1}, \quad (3.99)$$

and thus

$$\overset{\bullet}{\phi}_1^{[1]}(0,x,a) = -\int_a^x dv\, r(v) \int_a^v du\, p(u)^{-1}. \tag{3.100}$$

Furthermore,

$$\overset{\bullet}{\phi}_1(0,x,a) = -\int_a^x dw\, p(w)^{-1} \int_a^w dv\, r(v) \int_a^v du\, p(u)^{-1}. \tag{3.101}$$

Altogether this yields

$$\begin{aligned} c_1 = \frac{1}{2}\bigg[&\bigg(\int_a^b dt\, r(t)\bigg)\bigg(\int_a^b dw \int_a^w dv \int_a^v du\, \frac{r(v)}{p(w)p(u)}\bigg) \\ &-\bigg(\int_a^b dv \int_a^v du\, \frac{r(u)}{p(v)}\bigg)\bigg(\int_a^b dw \int_a^w dt\, \frac{r(w)}{p(t)}\bigg)\bigg]. \end{aligned} \tag{3.102}$$

*Example 3.18* For the particular case $r(x) = p(x) = 1$, we now recompute the $\zeta$-determinant $\zeta'(0; H_{0,R_{K_2},2})$ by choosing $q_1 = 0$. First, from (3.102) one finds that

$$c_1 = -(b-a)^4/24. \tag{3.103}$$

From Example 3.7, one determines

$$F_{0,R_{K_1},1}(z) = 2\big[1 - \cos\big(z^{1/2}(b-a)\big)\big] - (b-a)z^{1/2}\sin\big(z^{1/2}(b-a)\big). \tag{3.104}$$

The relevant leading asymptotics reads

$$F^{asym}_{0,R_{K_1},1}(z) = -i(b-a)z^{1/2}e^{-iz^{1/2}(b-a)}/2. \tag{3.105}$$

The *zeta*-function for the Krein–von Neumann extension is therefore analyzed using

$$\begin{aligned} \zeta(s; H_{0,R_{K_1},1}) &= e^{is(\pi-\theta)}\frac{\sin(\pi s)}{\pi}\int_0^\infty dt\, t^{-s}\frac{d}{dt}\ln\left(\frac{F_{0,R_{K_1},1}(te^{i\theta})}{t^2e^{2i\theta}}\right) \\ &= e^{is(\pi-\theta)}\frac{\sin(\pi s)}{\pi}\int_0^1 dt\, t^{-s}\frac{d}{dt}\ln\left(\frac{F_{0,R_{K_1},1}(te^{i\theta})}{t^2e^{2i\theta}}\right) \\ &\quad + e^{is(\pi-\theta)}\frac{\sin(\pi s)}{\pi}\int_1^\infty dt\, t^{-s}\frac{d}{dt}\ln\left(\frac{F_{0,R_{K_1},1}(te^{i\theta})}{F^{asym}_{0,R_{K_1},1}(te^{i\theta})}\right) \\ &\quad + \zeta^{asym}(s; H_{0,R_{K_1},1}), \end{aligned} \tag{3.106}$$

where

$$\zeta^{asym}(s; H_{0,R_{K_1},1}) = e^{is(\pi-\theta)}\frac{\sin(\pi s)}{\pi}\int_1^\infty dt\, t^{-s}\frac{d}{dt}\ln\left(\frac{F^{asym}_{0,R_{K_1},1}(te^{i\theta})}{t^2e^{2i\theta}}\right)$$

$$= e^{is(\pi-\theta)}\frac{\sin(\pi s)}{\pi}\left(-\frac{3}{2s}-\frac{i}{2}(b-a)e^{i\theta/2}\frac{1}{s-(1/2)}\right), \tag{3.107}$$

such that

$$\zeta^{asym\prime}(0; H_{0,R_{K_1},1}) = i(b-a)e^{i\theta/2} + [3i(\theta-\pi)/2]. \tag{3.108}$$

From (3.106) we then find

$$\zeta'(0; H_{0,R_{K_1},1}) = -\ln\big((b-a)^3/6\big), \tag{3.109}$$

in agreement with [50]. Finally, this proves

$$\zeta'(0; H_{0,R_{K_2},2}) = \zeta'(0; H_{0,R_{K_1},1}) + \ln(|c_1/c_2|) = \ln\left(\left|\frac{b-a}{4c_2}\right|\right). \tag{3.110}$$

*Example 3.19* As our final example, we consider a case where negative eigenvalues are present. Let $x \in (0,\pi)$ and $\tau_j = -(d^2/dx^2) - m_j^2$ with $m_j \in (n_j, n_j+1)$, $n_j \in \mathbb{N}$, $j = 1, 2$. Imposing Dirichlet boundary conditions at both endpoints, the eigenvalues are $\lambda_\ell^{(j)} = \ell^2 - m_j^2$, $\ell \in \mathbb{N}$, such that there are $n_j$ negative eigenvalues for $H_{0,0,j}$. The ζ-function representation for each $j = 1, 2$, can be found following Examples 3.13 and 3.15. We note that

$$\begin{aligned} F_{0,0,j}(z) &= \big(z+m_j^2\big)^{-1/2}\sin\Big(\big(z+m_j^2\big)^{1/2}\pi\Big) \\ &= \big(z+m_j^2\big)^{-1/2}(i/2)e^{-i(z+m_j^2)^{1/2}\pi}\Big(1-e^{2\pi i(z+m_j^2)^{1/2}}\Big) \\ &=: F_{0,0,j}^{asym}(z)\Big(1-e^{2\pi i(z+m_j^2)^{1/2}}\Big), \end{aligned} \tag{3.111}$$

and hence,

$$\zeta(s; H_{0,0,j}) = e^{is(\pi-\theta)}\frac{\sin(\pi s)}{\pi}\int_0^\infty dt\, t^{-s}\frac{d}{dt}\ln\left(\frac{F_{0,0,j}\left(te^{i\theta}\right)}{F_{0,0,j}^{asym}\left(te^{i\theta}\right)}\right) + \zeta^{asym}(s; H_{0,0,j}), \tag{3.112}$$

where

$$\zeta^{asym}(s; H_{0,0,j}) = e^{is(\pi-\theta)}\frac{\sin(\pi s)}{\pi}\int_0^\infty dt\, t^{-s}\frac{d}{dt}\ln\big(F_{0,0,j}^{asym}\left(te^{i\theta}\right)\big), \tag{3.113}$$

this representation being valid for $(1/2) < \mathrm{Re}(s) < 1$. From [34, 3.193, 3.194] one infers

$$\int_0^\infty dt\,\frac{t^{-s}}{(t+M)^{1/2}} = \frac{M^{(1/2)-s}\Gamma(1-s)\Gamma\left(s-\frac{1}{2}\right)}{\pi^{1/2}}, \tag{3.114}$$

$$\int_0^\infty dt\, \frac{t^{-s}}{t+M} = \frac{\pi}{M^s \sin(\pi s)}, \tag{3.115}$$

and hence one obtains

$$\zeta^{asym}(s; H_{0,0,j}) = -\frac{e^{is\pi}}{2m_j^{2s}} \left[ \frac{i\pi^{1/2} m_j \Gamma(s-(1/2))}{\Gamma(s)} + 1 \right], \tag{3.116}$$

implying

$$\zeta^{asym\prime}(0; H_{0,0,j}) = -i(\pi/2) + i m_j \pi + \ln(m_j). \tag{3.117}$$

It then follows that

$$\zeta'(0; H_{0,0,j}) = -\ln\left(1 - e^{2\pi i m_j}\right) - i(\pi/2) + i m_j \pi + \ln(m_j). \tag{3.118}$$

In order to obtain the final answer explicitly, showing the relation between the imaginary part and the number of negative eigenvalues, we first note that

$$1 - e^{2\pi i m_j} = -2i e^{\pi i m_j} \sin(\pi m_j). \tag{3.119}$$

A careful analysis of the argument of $1 - e^{2\pi i m_j}$ then shows that it equals $\pi[m_j - n_j - (1/2)]$, such that

$$\zeta'(0; H_{0,0,j}) = i\pi n_j - \ln\left( \left| \frac{2\sin(\pi m_j)}{m_j} \right| \right). \tag{3.120}$$

Considering instead

$$\zeta(s; H_{0,0,1}, H_{0,0,2}) = e^{is(\pi-\theta)} \frac{\sin(\pi s)}{\pi} \int_0^\infty dt\, t^{-s} \frac{d}{dt} \ln\left( \frac{F_{0,0,2}(te^{i\theta})}{F_{0,0,1}(te^{i\theta})} \right), \tag{3.121}$$

in (3.120), one obtains

$$\zeta'(0; H_{0,0,1}, H_{0,0,2}) = i\pi(n_2 - n_1) + \ln\left( \left| \frac{\sin(\pi m_1)}{\sin(\pi m_2)} \, \frac{m_2}{m_1} \right| \right). \tag{3.122}$$

The real part of this answer is readily reproduced from (2.41) in Theorem 2.10. However, even for this simple example, the behavior of $F_{0,0,1}(te^{i\theta})/F_{0,0,2}(te^{i\theta})$ along the integration range $t \in [0, \infty)$ is quite intricate so that finding the correct imaginary part from (3.121), namely, from

$$\zeta'(0; H_{0,0,1}, H_{0,0,2}) = \int_0^\infty dt\, \frac{d}{dt} \ln\left( \frac{F_{0,0,1}(te^{i\theta})}{F_{0,0,2}(te^{i\theta})} \right), \tag{3.123}$$

is rather involved, and appears to be next to impossible for more general cases.

For the case of Schrödinger operators with strongly singular potentials at one or both endpoints of a bounded interval, see [41, 43, 45, 61].

## 4 Schrödinger Operators on a Half-Line: The Short-Range Case

In this section, we illustrate some of the abstract notions in Section 2 with the help of self-adjoint Schrödinger operators on the half-line $\mathbb{R}_+ = (0, \infty)$. We will focus on the case of short-range potentials $q$ (cf. (4.1)) and hence the scattering theory situation which necessitates a comparison with the case $q = 0$ and thus illustrates the case of relative perturbation determinants, relative $\zeta$-functions, and relative $\zeta$-function regularized determinants.

We assume that the potential coefficient $q$ satisfies the following conditions.

**Hypothesis 4.1** Suppose $q$ satisfies the short-range assumption

$$q \in L^1(\mathbb{R}_+; (1 + |x|)dx), \ q \text{ is real-valued a.e. on } \mathbb{R}_+. \tag{4.1}$$

Given Hypothesis 4.1, we take $\tau_+$ to be the Schrödinger differential expression

$$\tau_+ = -\frac{d^2}{dx^2} + q(x) \ \text{ for a.e. } x \in \mathbb{R}_+, \tag{4.2}$$

and note that $\tau_+$ is regular at 0 and in the limit point case at $+\infty$. The *maximal operator* $H_{+,max}$ in $L^2(\mathbb{R}_+; dx)$ associated with $\tau_+$ is defined by

$$\begin{aligned} &H_{+,max} f = \tau_+ f, \\ &f \in \operatorname{dom}(H_{+,max}) = \big\{g \in L^2(\mathbb{R}_+; dx) \,\big|\, g, g' \in AC([0, b]) \text{ for all } b > 0; \\ &\qquad\qquad\qquad\qquad\qquad\qquad\qquad\qquad\qquad \tau_+ g \in L^2(\mathbb{R}_+; dx)\big\}, \end{aligned} \tag{4.3}$$

while the *minimal operator* $H_{+,min}$ in $L^2(\mathbb{R}_+; dx)$ associated with $\tau_+$ is given by

$$\begin{aligned} &H_{+,min} f = \tau_+ f, \\ &f \in \operatorname{dom}(H_{+,min}) = \big\{g \in L^2(\mathbb{R}_+; dx) \,\big|\, g, g' \in AC([0, b]) \text{ for all } b > 0; \\ &\qquad\qquad\qquad\qquad\qquad g(0) = g'(0) = 0; \ \tau_+ g \in L^2(\mathbb{R}_+; dx)\big\}. \end{aligned} \tag{4.4}$$

Again, one notes that the operator $H_{+,min}$ is symmetric and that

$$H_{+,min}^* = H_{+,max}, \quad H_{+,max}^* = H_{+,min}. \tag{4.5}$$

Moreover, all self-adjoint extensions of $H_{+,min}$ are given by the one-parameter family $H_{+,\alpha}$ in $L^2(\mathbb{R}_+; dx)$,

$$\begin{aligned}
&H_{+,\alpha} f = \tau_+ f, \\
&f \in \operatorname{dom}(H_{+,\alpha}) = \big\{ g \in L^2(\mathbb{R}_+; dx) \,\big|\, g, g' \in AC([0,b]) \text{ for all } b > 0; \\
&\qquad\qquad \sin(\alpha) g'(0) + \cos(\alpha) g(0) = 0;\ \tau_+ g \in L^2(\mathbb{R}_+; dx) \big\}, \\
&\qquad\qquad\qquad \alpha \in [0, \pi).
\end{aligned} \tag{4.6}$$

The corresponding comparison operator with vanishing potential coefficient $q \equiv 0$ will be denoted by $H_{+,\alpha}^{(0)}$, $\alpha \in [0, \pi)$.

Next, introducing the Jost solutions

$$f_+(z,x) = f_+^{(0)}(z,x) - \int_x^\infty dx'\, z^{-1/2} \sin(z^{1/2}(x - x')) q(x') f_+(z,x'), \tag{4.7}$$

$$f_+^{(0)}(z,x) = e^{iz^{1/2}x}, \quad z \in \mathbb{C},\ \operatorname{Im}\big(z^{1/2}\big) \geq 0,\ x \geq 0, \tag{4.8}$$

satisfying $\tau_+ y = zy$, $z \in \mathbb{C}$, on $\mathbb{R}_+$, and abbreviating $I_{L^2(\mathbb{R}_+;dx)} = I_+$, and

$$v = |q|^{1/2}, \quad u = v \operatorname{sign}(q), \quad \text{such that } q = uv = vu, \tag{4.9}$$

one infers the following facts:

$$\begin{aligned}
&\operatorname{det}_{L^2(\mathbb{R}_+;dx)}\Big(\overline{(H_{+,\alpha} - zI_+)^{1/2}\big(H_{+,\alpha}^{(0)} - zI_+\big)^{-1}(H_{+,\alpha} - zI_+)^{1/2}}\Big) \\
&\quad = \operatorname{det}_{L^2(\mathbb{R}_+;dx)}\Big(I_+ + \overline{u\big(H_{+,\alpha}^{(0)} - zI_+\big)^{-1} v}\Big) \\
&\quad = \frac{\sin(\alpha) f_+'(z,0) + \cos(\alpha) f_+(z,0)}{\sin(\alpha) i z^{1/2} + \cos(\alpha)}, \quad \alpha \in [0,\pi),\ z \in \rho(H_{+,\alpha}^{(0)}) \cap \rho(H_{+,\alpha}).
\end{aligned} \tag{4.10}$$

Here the first equality in (4.10) is shown as in the abstract context (2.18)–(2.22), and the second equality in (4.10) for the Dirichlet and Neumann cases $\alpha = 0$, $\alpha = \pi/2$ has been discussed in [21, 27, 28]; the general case $\alpha \in [0, \pi)$ is proved in [22, Theorem 2.6].

Since

$$\sigma_{ess}(H_{+,\alpha}) = [0, \infty), \quad \alpha \in [0, \pi), \tag{4.11}$$

we now shift all operators $H_{+,\alpha}$ by $\lambda_1 I_+$, with a fixed $\lambda_1 > 0$, and consider

$$H_{+,\alpha}(\lambda_1) = H_{+,\alpha} + \lambda_1 I_+, \quad \alpha \in [0, \pi), \tag{4.12}$$

from this point on and hence obtain

$$\begin{aligned}
&\operatorname{det}_{L^2(\mathbb{R}_+;dx)}\Big(\overline{(H_{+,\alpha}(\lambda_1) - zI_+)^{1/2}\big(H_{+,\alpha}^{(0)}(\lambda_1) - zI_+\big)^{-1}(H_{+,\alpha}(\lambda_1) - zI_+)^{1/2}}\Big) \\
&\quad = \operatorname{det}_{L^2(\mathbb{R}_+;dx)}\Big(I_+ + \overline{u\big(H_{+,\alpha}^{(0)}(\lambda_1) - zI_+\big)^{-1} v}\Big)
\end{aligned}$$

$$= \frac{\sin(\alpha) f_+'(z-\lambda_1,0) + \cos(\alpha) f_+(z-\lambda_1,0)}{\sin(\alpha) i (z-\lambda_1)^{1/2} + \cos(\alpha)}, \tag{4.13}$$
$$\alpha \in [0,\pi), \ z \in \rho\big(H_{+,\alpha}^{(0)}(\lambda_1)\big) \cap \rho(H_{+,\alpha}(\lambda_1)).$$

In this half-line context all discrete eigenvalues $H_{+,\alpha}(\lambda_1)$ (i.e., all eigenvalues of $H_{+,\alpha}(\lambda_1)$ below $\lambda_1$) are simple and hence

$$m(0; H_{+,\alpha}(\lambda_1)) \in \{0,1\}, \quad \alpha \in [0,\pi). \tag{4.14}$$

In addition, it is known that under Hypothesis 4.1, the threshold of the essential spectrum of $H_{+,\alpha}(\lambda_1)$, $\lambda_1$, is never an eigenvalue of $H_{+,\alpha}(\lambda_1)$.

Iterating the Volterra integral equation (4.7) for $f_+(z-\lambda_1,0)$, and analogously for its $x$-derivative, yields uniform asymptotic expansions near $z=0$ and as $z\to\infty$ (in terms of powers of $|z|^{-1/2}$). The same applies to their $z$-derivatives (cf., e.g., [7, Chap. I]) and explicit computations yield the following. For fixed $0<\varepsilon_0$ sufficiently small, and using the abbreviation $\mathbb{C}_{\varepsilon_0} = \mathbb{C}\backslash B(\lambda_1;\varepsilon_0)$, with $B(z_0;r_0)$ the open ball in $\mathbb{C}$ of radius $r_0>0$ centered at $z_0\in\mathbb{C}$, one obtains

$$\begin{aligned}
f_+(z-\lambda_1,0) &\underset{\substack{|z|\to\infty\\ z\in\mathbb{C}_{\varepsilon_0}}}{=} 1 - \frac{i}{2z^{1/2}} \int_0^\infty dx_1 \, \big[e^{2i(z-\lambda_1)^{1/2}x_1} - 1\big] q(x_1) + O\big(|z|^{-1}\big),\\
f_+'(z-\lambda_1,0) &\underset{\substack{|z|\to\infty\\ z\in\mathbb{C}_{\varepsilon_0}}}{=} iz^{1/2} - \frac{1}{2}\int_0^\infty dx_1 \, \big[e^{2i(z-\lambda_1)^{1/2}x_1} + 1\big] q(x_1) + O\big(|z|^{-1/2}\big),\\
\overset{\bullet}{f}_+(z-\lambda_1,0) &\underset{\substack{|z|\to\infty\\ z\in\mathbb{C}_{\varepsilon_0}}}{=} \frac{1}{2z}\int_0^\infty dx_1 \, e^{2i(z-\lambda_1)^{1/2}x_1} x_1 q(x_1) + O\big(|z|^{-3/2}\big),\\
\overset{\bullet}{f}{}_+'(z-\lambda_1,0) &\underset{\substack{|z|\to\infty\\ z\in\mathbb{C}_{\varepsilon_0}}}{=} \frac{i}{2z^{1/2}} - \frac{i}{2z^{1/2}}\int_0^\infty dx_1 \, e^{2i(z-\lambda_1)^{1/2}x_1} x_1 q(x_1) + O\big(|z|^{-1}\big)
\end{aligned} \tag{4.15}$$

(abbreviating again $\bullet = d/dz$).

Given the asymptotic expansions (4.15) as $|z|\to\infty$, and employing the fact that $f_+(\cdot-\lambda_1,0)$, $f_+'(\cdot-\lambda_1,0)$, $\overset{\bullet}{f}_+(\cdot-\lambda_1,0)$, $\overset{\bullet}{f}{}_+'(\cdot-\lambda_1,0)$ are all analytic with respect to $z$ around $z=0$, investigating the case distinctions $\alpha\in[0,\pi)\backslash\{0,\pi/2\}$, $\alpha=0, \alpha=\pi/2$, $f_+(z-\lambda_1,0)\neq 0$, $f_+(z-\lambda_1,0)=0$, $f_+'(z-\lambda_1,0)\neq 0$, $f_+'(z-\lambda_1,0)=0$, etc., one verifies in each case that the logarithmic $z$-derivative of (4.10) satisfies the hypotheses in Theorem 2.9, hence the latter applies with $\varepsilon=1/2$ to the pairs $\big(H_{+,\alpha}^{(0)}(\lambda_1), H_{+,\alpha}(\lambda_1)\big)$, $\alpha\in[0,\pi)$.

More generally, we now replace the pair $(0,q)$ by $(q_1,q_2)$, where $q_j$, $j=1,2$, satisfy Hypothesis 4.1, and denote the corresponding Schrödinger operators in $L^2(\mathbb{R}_+;dx)$ with $q$ (resp., $u$, $v$) replaced by $q_j$ (resp., $u_j$, $v_j$) by $H_{+,\alpha,j}$ and similarly, after the shift with $\lambda_1$, by $H_{+,\alpha,j}(\lambda_1)$, $j=1,2$. Analogously, we denote the corresponding Jost solutions by $f_{+,j}(z,\cdot)$, $j=1,2$. This then yields the following results:

**Theorem 4.2** *Suppose $q_j$, $j=1,2$, satisfy Hypothesis 4.1. Then,*

$$\begin{aligned}
&\det_{L^2(\mathbb{R}_+;dx)}\Big(\overline{(H_{+,\alpha,2}(\lambda_1)-zI_+)^{1/2}\big(H_{+,\alpha,1}(\lambda_1)-zI_+\big)^{-1}(H_{+,\alpha,2}(\lambda_1)-zI_+)^{1/2}}\Big)\\
&\quad=\det_{L^2(\mathbb{R}_+;dx)}\Big(I_+ + \overline{u_{1,2}\big(H_{+,\alpha,1}(\lambda_1)-zI_+\big)^{-1}v_{1,2}}\Big)\\
&\quad=\frac{\sin(\alpha)f'_{+,2}(z-\lambda_1,0)+\cos(\alpha)f_{+,2}(z-\lambda_1,0)}{\sin(\alpha)f'_{+,1}(z-\lambda_1,0)+\cos(\alpha)f_{+,1}(z-\lambda_1,0)},\\
&\qquad\alpha\in[0,\pi),\ z\in\rho\big(H_{+,\alpha,1}(\lambda_1)\big)\cap\rho(H_{+,\alpha,2}(\lambda_1)),
\end{aligned}\tag{4.16}$$

*where*

$$\begin{aligned}
&v_{1,2}=|q_2-q_1|^{1/2},\ u_{1,2}=v_{1,2}\operatorname{sign}(q_2-q_1),\\
&\text{such that } q_2-q_1=u_{1,2}v_{1,2}=v_{1,2}u_{1,2}.
\end{aligned}\tag{4.17}$$

*In addition,*

$$\begin{aligned}
&\zeta(s;H_{+,\alpha,1}(\lambda_1),H_{+,\alpha,2}(\lambda_1))=e^{is(\pi-\theta)}\pi^{-1}\sin(\pi s)\\
&\quad\times\int_0^\infty dt\,t^{-s}\frac{d}{dt}\ln\bigg((e^{i\theta}t)^{[m(0;H_{+,\alpha,1}(\lambda_1))-m(0;H_{+,\alpha,2}(\lambda_1))]}\\
&\qquad\times\frac{\sin(\alpha)f'_{+,2}(e^{i\theta}t-\lambda_1,0)+\cos(\alpha)f_{+,2}(e^{i\theta}t-\lambda_1,0)}{\sin(\alpha)f'_{+,1}(e^{i\theta}t-\lambda_1,0)+\cos(\alpha)f_{+,1}(e^{i\theta}t-\lambda_1,0)}\bigg),\\
&\qquad\alpha\in[0,\pi),\ \operatorname{Re}(s)\in(-1/2,1).
\end{aligned}\tag{4.18}$$

*Proof* Relations (4.16), (4.17) follow as summarized in (4.7)–(4.15) and the two paragraphs preceding Theorem 4.2. Thus, Theorem 2.9 applies to the pairs of self-adjoint operators $(H_{+,\alpha,1}(\lambda_1),H_{+,\alpha,2}(\lambda_1))$, $\alpha\in[0,\pi)$. □

The relative $\zeta$-function regularized determinant now follows immediately from Theorem 2.10.

Special cases of (4.16) (pertaining to the Dirichlet boundary conditions $\alpha_j=0$, $j=1,2$) appeared in the celebrated work by Jost and Pais [36] and Buslaev and Faddeev [6] (see also [12, 21, 52, 53, 56]).

Up to this point we kept the boundary condition, that is, $\alpha$, fixed and varied the potential coefficient $q$. Next, we keep $q$ fixed, but vary $\alpha$. Returning to the operator $H_{+,\alpha}$, $\alpha\in[0,\pi)$, we turn to its underlying quadratic form $Q_{H_{+,\alpha}}$ next,

$$\begin{aligned}
&Q_{H_{+,\alpha}}(f,g)=\int_0^\infty dx\big[\overline{f'(x)}g'(x)+q(x)\overline{f(x)}g(x)\big]-\cot(\alpha)\overline{f(0)}g(0),\\
&f,g\in\operatorname{dom}(Q_{H_{+,\alpha}})=\operatorname{dom}\big(|H_{+,\alpha}|^{1/2}\big)=H^1(\mathbb{R}_+)
\end{aligned}\tag{4.19}$$

$$= \{h \in L^2(\mathbb{R}_+; dx) \,|\, h \in AC([0,b]) \text{ for all } b > 0;\ h' \in L^2(\mathbb{R}_+; dx)\}, \quad \alpha \in (0,\pi),$$

$$Q_{H_{+,0}}(f,g) = \int_0^\infty dx \big[\overline{f'(x)}g'(x) + q(x)\overline{f(x)}g(x)\big],$$

$$f,g \in \operatorname{dom}(Q_{H_{+,0}}) = \operatorname{dom}\big(|H_{+,0}|^{1/2}\big) = H_0^1(\mathbb{R}_+) \tag{4.20}$$
$$= \{h \in L^2(\mathbb{R}_+; dx) \,|\, h \in AC([0,b]) \text{ for all } b > 0;\ h(0) = 0;\ h' \in L^2(\mathbb{R}_+; dx)\}.$$

Moreover, introducing the regular solution $\phi_\alpha(z,\,\cdot\,)$ associated with $H_{+,\alpha}$ satisfying $\tau_+ y = zy$, $z \in \mathbb{C}$, on $\mathbb{R}_+$, and

$$\sin(\alpha)\phi'_\alpha(z,0) + \cos(\alpha)\phi_\alpha(z,0) = 0, \quad \alpha \in [0,\pi),\ z \in \mathbb{C}, \tag{4.21}$$

one infers

$$\phi_\alpha(z,x) = \phi_\alpha^{(0)}(z,x) + \int_0^x dx'\, z^{-1/2} \sin(z^{1/2}(x-x'))q(x')\phi_\alpha(z,x'), \quad z \in \mathbb{C},\ \operatorname{Im}(z^{1/2}) \geq 0,\ x \geq 0, \tag{4.22}$$

with

$$\phi_\alpha^{(0)}(z,x) = \cos(\alpha) z^{-1/2} \sin(z^{1/2}x) - \sin(\alpha)\cos(z^{1/2}x), \quad z \in \mathbb{C},\ \operatorname{Im}(z^{1/2}) \geq 0,\ x \geq 0. \tag{4.23}$$

Given the solutions $\phi_\alpha(z,\,\cdot\,)$, $f_+(z,\,\cdot\,)$ of $\tau_+ y = z$, $z \in \mathbb{C}$, the resolvent of $H_{+,\alpha}$ is given by

$$\big((H_{+,\alpha} - zI_+)^{-1} f\big)(x) = \int_0^\infty dx'\, G_{+,\alpha}(z,x,x') f(x'), \quad z \in \rho(H_{+,\alpha}),\ x \geq 0,\ f \in L^2(\mathbb{R}_+; dx), \tag{4.24}$$

with the Green's function $G_{+,\alpha}$ of $H_{+,\alpha}$ expressed in terms of $\phi_\alpha$ and $f_+$ by

$$G_{+,\alpha}(z,x,x') = (H_{+,\alpha} - zI_+)^{-1}(x,x')$$
$$= \frac{1}{W(f_+(z,\cdot),\phi_\alpha(z,\cdot))} \begin{cases} \phi_\alpha(z,x)\, f_+(z,x'), & 0 \leq x \leq x' < \infty, \\ \phi_\alpha(z,x')\, f_+(z,x), & 0 \leq x' \leq x < \infty, \end{cases} \quad z \in \rho(H_{+,\alpha}). \tag{4.25}$$

In the special case $\alpha = 0$ one verifies

$$W(f_+(z,\cdot),\phi_0(z,\cdot)) = f_+(z,0), \quad z \in \rho(H_{+,0}), \tag{4.26}$$

with $f_+(z,0)$ the well-known Jost function.

Next, we compare the half-line Green's functions $G_{+,\alpha_1}$ and $G_{+,\alpha_2}$, that is, we investigate the integral kernels associated with a special case of Krein's formula for resolvents (cf. [2, & 106]): Assume $\alpha_1, \alpha_2 \in [0,\pi)$, with $\alpha_1 \neq \alpha_2$. Then,

$$\begin{aligned} G_{+,\alpha_2}(z,x,x') = G_{+,\alpha_1}(z,x,x') - \frac{\psi_{+,\alpha_1}(z,x)\psi_{+,\alpha_1}(z,x')}{\cot(\alpha_2-\alpha_1)+m_{+,\alpha_1}(z)}, \\ z\in\rho(H_{+,\alpha_1})\cap\rho(H_{+,\alpha_2}),\ x,x'\in[0,\infty), \end{aligned} \tag{4.27}$$

and

$$\int_0^\infty dx\, \psi_{+,\alpha}(z,x)^2 = \dot{m}_{+,\alpha}(z), \tag{4.28}$$

implying

$$\begin{aligned} &\operatorname{tr}_{L^2(\mathbb{R}_+;dx)}\big((H_{+,\alpha_2}(\lambda_1)-zI_+)^{-1}-(H_{+,\alpha_1}(\lambda_1)-zI_+)^{-1}\big) \\ &\quad = -\frac{d}{dz}\ln[\cot(\alpha_2-\alpha_1)+m_{+,\alpha_1}(z)],\quad z\in\rho(H_{+,\alpha_1})\cap\rho(H_{+,\alpha_2}), \end{aligned} \tag{4.29}$$

according to Lemma 2.2 and (A.44) in [24]. Here $\psi_{+,\alpha}(z,\cdot)$ and $m_{+,\alpha}(z)$ are the Weyl–Titchmarsh solution and $m$-function corresponding to $H_{+,\alpha}$. More precisely,

$$\psi_{+,\alpha}(z,\cdot) = \theta_\alpha(z,\cdot) + m_{+,\alpha}(z)\phi_\alpha(z,\cdot),\quad z\in\rho(H_{+,\alpha}), \tag{4.30}$$

where

$$\begin{aligned} \theta_\alpha(z,x) = \theta_\alpha^{(0)}(z,x) + \int_0^x dx'\, z^{-1/2}\sin(z^{1/2}(x-x'))q(x')\theta_\alpha(z,x'), \\ z\in\mathbb{C},\ \operatorname{Im}(z^{1/2})\geq 0,\ x\geq 0, \end{aligned} \tag{4.31}$$

with

$$\begin{aligned} \theta_\alpha^{(0)}(z,x) = \cos(\alpha)\cos(z^{1/2}x) + \sin(\alpha)z^{-1/2}\sin(z^{1/2}x), \\ z\in\mathbb{C},\ \operatorname{Im}(z^{1/2})\geq 0,\ x\geq 0. \end{aligned} \tag{4.32}$$

Due to the limit point property of $\tau_+$ at $+\infty$, one actually has

$$\psi_{+,\alpha}(z,\cdot) = C_\alpha(z)f_+(z,\cdot),\quad z\in\rho(H_{+,\alpha}), \tag{4.33}$$

for some $z$-dependent complex-valued constant $C_\alpha(z)$. Actually, since $\psi_{+,\alpha}(z,0) = C_\alpha f_+(z,0)$, one can show (using (A.18) in [24]) that

$$C_\alpha(z) = \frac{\cos(\alpha)-\sin(\alpha)m_{+,\alpha}(z)}{f_+(z,0)} = \frac{1}{\cos(\alpha)f_+(z,0)+\sin(\alpha)f_+'(z,0)}. \tag{4.34}$$

Similarly,

$$m_{+,0}(z) = \psi'_{+,0}(z,0)/\psi_{+,0}(z,0) = f'_+(z,0)/f_+(z,0), \quad z \in \rho(H_{+,\alpha}), \tag{4.35}$$

and

$$m_{+,\alpha}(z) = \frac{-\sin(\alpha) + \cos(\alpha)m_{+,0}(z)}{\cos(\alpha) + \sin(\alpha)m_{+,0}(z)} = \frac{\cos(\alpha)f'_+(z,0) - \sin(\alpha)f_+(z,0)}{\sin(\alpha)f'_+(z,0) + \cos(\alpha)f_+(z,0)}, \\ z \in \rho(H_{+,\alpha}). \tag{4.36}$$

Moreover, one verifies that

$$\cos(\alpha_2 - \alpha_1) + \sin(\alpha_2 - \alpha_1)m_{+,\alpha_1}(z) = \frac{\sin(\alpha_2)f'_+(z,0) + \cos(\alpha_2)f_+(z,0)}{\sin(\alpha_1)f'_+(z,0) + \cos(\alpha_1)f_+(z,0)}. \tag{4.37}$$

Combining (4.29) and (4.37) yields

$$\begin{aligned} &\operatorname{tr}_{L^2(\mathbb{R}_+;dx)}\big((H_{+,\alpha_2}(\lambda_1) - zI_+)^{-1} - (H_{+,\alpha_1}(\lambda_1) - zI_+)^{-1}\big) \\ &\quad = -\frac{d}{dz}\ln\left(\frac{\sin(\alpha_2)f'_+(z-\lambda_1,0) + \cos(\alpha_2)f_+(z-\lambda_1,0)}{\sin(\alpha_1)f'_+(z-\lambda_1,0) + \cos(\alpha_1)f_+(z-\lambda_1,0)}\right), \\ &\qquad z \in \rho(H_{+,\alpha_1}(\lambda_1)) \cap \rho(H_{+,\alpha_2}(\lambda_1)), \ \alpha_1, \alpha_2 \in [0,\pi). \end{aligned} \tag{4.38}$$

In particular, $\big[(H_{+,\alpha_2} - zI_+)^{-1} - (H_{+,\alpha_1} - zI_+)^{-1}\big]$ is rank-one and hence trace class. Moreover, since by (4.19)

$$f_+(z,\,\cdot\,) \in \operatorname{dom}\big(|H_{+,\alpha}|^{1/2}\big) = H^1(\mathbb{R}_+), \quad \alpha \in (0,\pi), \tag{4.39}$$

Hypothesis 2.5 and hence relations (2.25), (2.26) are now satisfied for the pair $(H_{+,\alpha_1}(\lambda_1), H_{+,\alpha_2}(\lambda_1))$ for $\alpha_1 \in [0,\pi)$, $\alpha_2 \in (0,\pi)$, implying the following result:

**Theorem 4.3** *Suppose $q_j$, $j = 1, 2$, satisfy Hypothesis 4.1. Then,*

$$\begin{aligned} &\operatorname{tr}_{L^2(\mathbb{R}_+;dx)}\big((H_{+,\alpha_2}(\lambda_1) - zI_+)^{-1} - (H_{+,\alpha_1}(\lambda_1) - zI_+)^{-1}\big) \\ &\quad = -\frac{d}{dz}\ln\Big(\det{}_{L^2(\mathbb{R}_+;dx)}\Big(\big\{(H_{+,\alpha_2}(\lambda_1) - zI_+)^{1/2}\big(H_{+,\alpha_1}(\lambda_1) - zI_+\big)^{-1} \\ &\qquad\qquad \times (H_{+,\alpha_2}(\lambda_1) - zI_+)^{1/2}\big\}^c\Big)\Big) \qquad (4.40) \\ &\quad = -\frac{d}{dz}\ln\left(\frac{\sin(\alpha_2)f'_+(z-\lambda_1,0) + \cos(\alpha_2)f_+(z-\lambda_1,0)}{\sin(\alpha_1)f'_+(z-\lambda_1,0) + \cos(\alpha_1)f_+(z-\lambda_1,0)}\right), \qquad (4.41) \\ &z \in \rho(H_{+,\alpha_1}(\lambda_1)) \cap \rho(H_{+,\alpha_2}(\lambda_1)), \ \alpha_1 \in [0,\pi), \ \alpha_2 \in (0,\pi), \ \alpha_1 \neq \alpha_2, \end{aligned}$$

*temporarily abbreviating the operator closure symbol by $\{\cdots\}^c$ due to lack of space in* (4.40). *In addition,*

$$\begin{aligned}\zeta(s; H_{+,\alpha_1}(\lambda_1), H_{+,\alpha_2}(\lambda_1)) &= e^{is(\pi-\theta)}\pi^{-1}\sin(\pi s)\\ &\times \int_0^\infty dt\, t^{-s}\frac{d}{dt}\ln\bigg((te^{i\theta})^{[m(0,H_{+,\alpha_1}(\lambda_1))-m(0,H_{+,\alpha_2}(\lambda_1))]}\\ &\times \frac{\sin(\alpha_2)f_+'(te^{i\theta}-\lambda_1,0)+\cos(\alpha_2)f_+(te^{i\theta}-\lambda_1,0)}{\sin(\alpha_1)f_+'(te^{i\theta}-\lambda_1,0)+\cos(\alpha_1)f_+(te^{i\theta}-\lambda_1,0)}\bigg),\\ &\qquad \alpha_j\in(0,\pi),\ j=1,2,\ \operatorname{Re}(s)\in(-1/2,1).\end{aligned} \tag{4.42}$$

*Proof* Relations (4.40), (4.41) summarize the discussion in (4.21)–(4.39). Applying Theorem 2.9 in the case $\alpha_1, \alpha_2 \in (0,\pi)$ then yields (4.42). □

The relative $\zeta$-function regularized determinant again follows immediately from Theorem 2.10,

$$\begin{aligned}\zeta'(0; H_{+,\alpha_1}(\lambda_1), H_{+,\alpha_2}(\lambda_1)) &= -\lim_{t\downarrow 0}\frac{d}{dt}\ln\bigg(\left(te^{i\theta}\right)^{[m(0,H_{+,\alpha_1}(\lambda_1))-m(0,H_{+,\alpha_2}(\lambda_1))]}\\ &\times \frac{\sin(\alpha_2)f_+'(te^{i\theta}-\lambda_1,0)+\cos(\alpha_2)f_+(te^{i\theta}-\lambda_1,0)}{\sin(\alpha_1)f_+'(te^{i\theta}-\lambda_1,0)+\cos(\alpha_1)f_+(te^{i\theta}-\lambda_1,0)}\bigg).\end{aligned} \tag{4.43}$$

*Remark 4.4* The case $\alpha_1 = 0, \alpha_2 \in (0,\pi)$, is more involved in that the representation (4.42) is only valid for $\operatorname{Re}(s) \in (0,1)$. This is due to the fact that

$$\frac{\sin(\alpha_2)f_+'(z-\lambda_1,0)+\cos(\alpha_2)f_+(z-\lambda_1,0)}{f_+(z-\lambda_1,0)} \underset{|z|\to\infty}{=} O\big(|z|^{1/2}\big), \tag{4.44}$$

and hence assumption (2.38) is not satisfied. One then proceeds as follows. Let $H_{+,\alpha}^{(0)}(\lambda_1)$ denote the case with vanishing potential $q$. We rewrite

$$\begin{aligned}&\operatorname{tr}_{L^2(\mathbb{R}_+;dx)}\big((H_{+,\alpha_2}(\lambda_1)-zI_+)^{-1}-(H_{+,0}(\lambda_1)-zI_+)^{-1}\big)\\ &\quad= \operatorname{tr}_{L^2(\mathbb{R}_+;dx)}\big((H_{+,\alpha_2}(\lambda_1)-zI_+)^{-1}-(H_{+,\alpha_2}^{(0)}(\lambda_1)-zI_+)^{-1}\big)\\ &\qquad+ \operatorname{tr}_{L^2(\mathbb{R}_+;dx)}\big((H_{+,0}^{(0)}(\lambda_1)-zI_+)^{-1}-(H_{+,0}(\lambda_1)-zI_+)^{-1}\big)\\ &\qquad+ \operatorname{tr}_{L^2(\mathbb{R}_+;dx)}\big((H_{+,\alpha_2}^{(0)}(\lambda_1)-zI_+)^{-1}-(H_{+,0}^{(0)}(\lambda_1)-zI_+)^{-1}\big).\end{aligned} \tag{4.45}$$

For the first two of the three contributions on the right-hand side of (4.45) the relative $\zeta$-determinants can be computed from Theorem 4.2. For the third contribution more care is needed as the subleading large-$|z|$ behavior differs due to one boundary condition being Dirichlet and the other being Robin. The starting point is the representation

$$\begin{aligned}\zeta\Big(s; H_{+,0}^{(0)}(\lambda_1), H_{+,\alpha_2}^{(0)}(\lambda_1)\Big) &= e^{is(\pi-\theta)}\pi^{-1}\sin(\pi s)\\ &\times \int_0^\infty dt\, t^{-s}\frac{d}{dt}\ln\Big(\sin(\alpha_2)i\big(te^{i\theta}-\lambda_1\big)^{1/2}+\cos(\alpha_2)\Big).\end{aligned} \tag{4.46}$$

Along the lines of (3.69) one rewrites this as

$$\begin{aligned}\zeta\Big(s; H^{(0)}_{+,0}(\lambda_1), H^{(0)}_{+,\alpha_2}(\lambda_1)\Big) &= e^{is(\pi-\theta)}\pi^{-1}\sin(\pi s)\int_0^\infty dt\, t^{-s} \\ &\times \frac{d}{dt}\ln\left(\frac{\sin(\alpha_2)i\big(te^{i\theta}-\lambda_1\big)^{1/2}+\cos(\alpha_2)}{\sin(\alpha_2)i\big(te^{i\theta}-\lambda_1\big)^{1/2}}\right)+\zeta^{asym}(s),\end{aligned} \tag{4.47}$$

where, using (3.115),

$$\begin{aligned}\zeta^{asym}(s) &= e^{is(\pi-\theta)}\pi^{-1}\sin(\pi s)\int_0^\infty dt\, t^{-s}\frac{d}{dt}\ln\Big(i\sin(\alpha_2)\big(te^{i\theta}-\lambda_1\big)^{1/2}\Big) \\ &= \lambda_1^{-s}/2.\end{aligned} \tag{4.48}$$

The relative ζ-determinant for the last term in (4.45) then follows from

$$\begin{aligned}\zeta'\Big(0; H^{(0)}_{+,0}(\lambda_1), H^{(0)}_{+,\alpha_2}(\lambda_1)\Big) &= \mathrm{Re}\left(-\ln\left(\frac{\sin(\alpha_2)i(-\lambda_1)^{1/2}+\cos(\alpha_2)}{\sin(\alpha_2)i(-\lambda_1)^{1/2}}\right)-\ln\big(\lambda_1^{1/2}\big)\right) \\ &= -\ln\big(\big|\lambda_1^{1/2}-\cot(\alpha_2)\big|\big).\end{aligned} \tag{4.49}$$

◇

One notes that formally, (4.41) extends to the trivial case $\alpha_1 = \alpha_2$.

For the case of a strongly singular potential on the half-line with $x^{-2}$-type singularity at $x = 0$ we refer to [41].

## 5 Schrödinger Operators on a Half-Line: The Case of Purely Discrete Spectra

We now illustrate (2.8) with the help of self-adjoint Schrödinger operators $-\frac{d^2}{dx^2}+q$ on the half-line $\mathbb{R}_+ = (0,\infty)$ in the particular case where the potential $q$ diverges at $\infty$ and hence gives rise to a purely discrete spectrum (i.e, the absence of essential spectrum).

To this end we introduce the following set of assumptions on $q$:

**Hypothesis 5.1** Suppose $q$ satisfies

$$q \in L^1_{loc}(\mathbb{R}_+; dx),\ q \text{ is real-valued a.e. on } \mathbb{R}_+, \tag{5.1}$$

and for some $\varepsilon_0 > 0$, $C_0 > 0$, and sufficiently large $x_0 > 0$,

$$q, q' \in AC([x_0, R]) \text{ for all } R > x_0, \tag{5.2}$$

$$q(x) \geq C_0 x^{(2/3)+\varepsilon_0}, \quad x \in (x_0, \infty), \tag{5.3}$$

$$q'/q \underset{x\to\infty}{=} o\big(q^{1/2}\big), \tag{5.4}$$

$$\big(q^{-3/2} q'\big)' \in L^1((x_0, \infty); dx). \tag{5.5}$$

Condition (5.3) guarantees that traces and modified determinants in this paper (see, e.g., the ones in Theorem 5.3) are well-defined. Conditions (5.4) and (5.5) are imposed so that [14, Corollary 2.2.1] is applicable, implying the asymptotic behavior (5.21).

Given Hypothesis 5.1, we take $\tau_+$ to be the Schrödinger differential expression

$$\tau_+ = -\frac{d^2}{dx^2} + q(x) \text{ for a.e. } x \in \mathbb{R}_+, \tag{5.6}$$

and note that $\tau_+$ is regular at 0 and in the limit point case at $+\infty$. The *maximal operator* $H_{+,max}$ in $L^2(\mathbb{R}_+; dx)$ associated with $\tau_+$ is defined by

$$\begin{aligned} &H_{+,max} f = \tau_+ f, \\ &f \in \mathrm{dom}(H_{+,max}) = \big\{g \in L^2(\mathbb{R}_+; dx) \,\big|\, g, g' \in AC([0, b]) \text{ for all } b > 0; \\ &\qquad\qquad \tau_+ g \in L^2(\mathbb{R}_+; dx)\big\}, \end{aligned} \tag{5.7}$$

while the *minimal operator* $H_{+,min}$ in $L^2(\mathbb{R}_+; dx)$ associated with $\tau_+$ is given by

$$\begin{aligned} &H_{+,min} f = \tau_+ f, \\ &f \in \mathrm{dom}(H_{+,min}) = \big\{g \in L^2(\mathbb{R}_+; dx) \,\big|\, g, g' \in AC([0, b]) \text{ for all } b > 0; \\ &\qquad\qquad g(0) = g'(0) = 0;\ \tau_+ g \in L^2(\mathbb{R}_+; dx)\big\}. \end{aligned} \tag{5.8}$$

One notes that the operator $H_{+,min}$ is symmetric and that

$$H_{+,min}^* = H_{+,max}, \quad H_{+,max}^* = H_{+,min} \tag{5.9}$$

(cf., eg., [64, Theorem 13.8]). Moreover, all self-adjoint extensions of $H_{+,min}$ are given by the one-parameter family in $L^2(\mathbb{R}_+; dx)$

$$\begin{aligned} &H_{+,\alpha} f = \tau_+ f, \\ &f \in \mathrm{dom}(H_{+,\alpha}) = \big\{g \in L^2(\mathbb{R}_+; dx) \,\big|\, g, g' \in AC([0, b]) \text{ for all } b > 0; \\ &\qquad\qquad \sin(\alpha) g'(0) + \cos(\alpha) g(0) = 0;\ \tau_+ g \in L^2(\mathbb{R}_+; dx)\big\}, \\ &\qquad\qquad\qquad\qquad \alpha \in [0, \pi). \end{aligned} \tag{5.10}$$

Next, we introduce the fundamental system of solutions $\phi_\alpha(z, \cdot)$ and $\theta_\alpha(z, \cdot)$, $\alpha \in [0, \pi)$, $z \in \mathbb{C}$, associated with $H_{+,\alpha}$ satisfying

$$(\tau_+\psi(z,\,\cdot\,))(x) = z\psi(z,x), \quad z \in \mathbb{C}, \; x \in \mathbb{R}_+, \tag{5.11}$$

and the initial conditions

$$\begin{aligned} &\phi_\alpha(z,0) = -\sin(\alpha), \quad \phi'_\alpha(z,0) = \cos(\alpha), \\ &\theta_\alpha(z,0) = \cos(\alpha), \qquad \theta'_\alpha(z,0) = \sin(\alpha). \end{aligned} \tag{5.12}$$

Explicitly, one infers

$$\begin{aligned} \phi_\alpha(z,x) = \phi_\alpha^{(0)}(z,x) + \int_0^x dx' \, \frac{\sin(z^{1/2}(x-x'))}{z^{1/2}} q(x')\phi_\alpha(z,x'), \\ z \in \mathbb{C}, \; \mathrm{Im}(z^{1/2}) \geq 0, \; x \geq 0, \end{aligned} \tag{5.13}$$

with

$$\phi_\alpha^{(0)}(z,x) = \cos(\alpha)\frac{\sin(z^{1/2}x)}{z^{1/2}} - \sin(\alpha)\cos(z^{1/2}x), \quad z \in \mathbb{C}, \; \mathrm{Im}(z^{1/2}) \geq 0, \; x \geq 0, \tag{5.14}$$

and

$$\begin{aligned} \theta_\alpha(z,x) = \theta_\alpha^{(0)}(z,x) + \int_0^x dx' \, \frac{\sin(z^{1/2}(x-x'))}{z^{1/2}} q(x')\theta_\alpha(z,x'), \\ z \in \mathbb{C}, \; \mathrm{Im}(z^{1/2}) \geq 0, \; x \geq 0, \end{aligned} \tag{5.15}$$

with

$$\theta_\alpha^{(0)}(z,x) = \cos(\alpha)\cos(z^{1/2}x) + \sin(\alpha)\frac{\sin(z^{1/2}x)}{z^{1/2}}, \quad z \in \mathbb{C}, \; \mathrm{Im}(z^{1/2}) \geq 0, \; x \geq 0. \tag{5.16}$$

The Weyl–Titchmarsh solution, $\psi_{+,\alpha}(z,\,\cdot\,)$, and Weyl–Titchmarsh $m$-function, $m_{+,\alpha}(z)$, corresponding to $H_{+,\alpha}$, $\alpha \in [0,\pi)$, are then related via,

$$\psi_{+,\alpha}(z,\,\cdot\,) = \theta_\alpha(z,\,\cdot\,) + m_{+,\alpha}(z)\phi_\alpha(z,\,\cdot\,), \quad z \in \rho(H_{+,\alpha}), \; \alpha \in [0,\pi), \tag{5.17}$$

where

$$\psi_{+,\alpha}(z,\,\cdot\,) \in L^2(\mathbb{R}_+; dx), \quad z \in \rho(H_{+,\alpha}), \; \alpha \in [0,\pi). \tag{5.18}$$

One then obtains for the Green's function $G_{+,\alpha}$ of $H_{+,\alpha}$ expressed in terms of $\phi_\alpha$ and $\psi_{+,\alpha}$,

$$\begin{aligned} G_{+,\alpha}(z,x,x') &= (H_{+,\alpha} - zI_+)^{-1}(x,x') \\ &= \begin{cases} \phi_\alpha(z,x)\,\psi_{+,\alpha}(z,x'), & 0 \leq x \leq x' < \infty, \\ \phi_\alpha(z,x')\,\psi_{+,\alpha}(z,x), & 0 \leq x' \leq x < \infty, \end{cases} \quad z \in \rho(H_{+,\alpha}), \; \alpha \in [0,\pi), \end{aligned} \tag{5.19}$$

utilizing

$$W(\theta_\alpha(z,\cdot),\phi_\alpha(z,\cdot)) = 1, \quad z \in \mathbb{C},\ \alpha \in [0,\pi), \tag{5.20}$$

implying $W(\psi_{+,\alpha}(z,\cdot),\phi_\alpha(z,\cdot)) = 1$, $z \in \rho(H_{+,\alpha})$.

By [14, Corollary 2.2.1], Hypothesis 5.1 implies the existence of two solutions $f_{+,j}(\lambda,\,\cdot\,,x_0)$, $j = 1,2$, of $\tau_+\psi(\lambda,\,\cdot\,) = \lambda\psi(\lambda,\,\cdot\,)$, $\lambda < 0$ sufficiently negative (and below $\inf(\sigma(H_{+,\alpha}))$), satisfying

$$\begin{aligned} f_{+,j}(\lambda,x,x_0) &\underset{x\to\infty}{=} 2^{-1/2}[q(x)-\lambda]^{-1/4}\exp\bigg((-1)^j\int_{x_0}^x dx'[q(x')-\lambda]^{1/2}\bigg) \\ &\qquad\times[1+o(1)], \\ f'_{+,j}(\lambda,x,x_0) &\underset{x\to\infty}{=} (-1)^j 2^{-1/2}[q(x)-\lambda]^{1/4}\exp\bigg((-1)^j\int_{x_0}^x dx'[q(x')-\lambda]^{1/2}\bigg) \\ &\qquad\times[1+o(1)], \quad j=1,2, \end{aligned} \tag{5.21}$$

with

$$W\big(f_{+,1}(\lambda,\,\cdot\,,x_0), f_{+,2}(\lambda,\,\cdot\,,x_0)\big) = 1. \tag{5.22}$$

(Here we explicitly introduced the $x_0$ dependence of $f_{+,j}$, implied by the choice of normalization in (5.21), as keeping track of it later on will become a necessity.) In particular, $f_{+,1}(\lambda,\,\cdot\,,x_0)$ now plays the analog of the Jost solution in the case of a short-range potential $q$ (i.e., $q \in L^1(\mathbb{R}_+;(1+x)dx)$, $q$ real-valued a.e. on $\mathbb{R}_+$).

By the limit point property of $\tau_+$ at $+\infty$ and the asymptotic behavior of $f_{+,1}$ in (5.21) one infers, in addition,

$$\psi_{+,\alpha}(\lambda,\,\cdot\,) = f_{+,1}(\lambda,\,\cdot\,,x_0)\big/\big[\sin(\alpha)f'_{+,1}(\lambda,0,x_0)+\cos(\alpha)f_{+,1}(\lambda,0,x_0)\big], \tag{5.23}$$

$$\begin{aligned} \phi_\alpha(\lambda,\,\cdot\,) = \big[\cos(\alpha)f_{+,1}(\lambda,0,x_0)+\sin(\alpha)f'_{+,1}(\lambda,0,x_0)\big]f_{+,2}(\lambda,\,\cdot\,,x_0) \\ -\big[\cos(\alpha)f_{+,2}(\lambda,0,x_0)+\sin(\alpha)f'_{+,2}(\lambda,0,x_0)\big]f_{+,1}(\lambda,\,\cdot\,,x_0) \end{aligned} \tag{5.24}$$

for $\lambda < 0$ sufficiently negative. Analytic continuation with respect to $\lambda$ in (5.23) then yields the existence of a unique Jost-type solution $f_{+,1}(z,\,\cdot\,,x_0)$ satisfying

$$\tau_+ f_{+,1}(z,\,\cdot\,,x_0) = z f_{+,1}(z,\,\cdot\,,x_0), \quad z \in \mathbb{C}\backslash\mathbb{R}, \tag{5.25}$$

$$f_{+,1}(z,\,\cdot\,,x_0) \in L^2(\mathbb{R}_+;dx), \quad z \in \mathbb{C}\backslash\mathbb{R}, \tag{5.26}$$

coinciding with $f_{+,1}(\lambda,\,\cdot\,,x_0)$ for $z = \lambda < 0$ sufficiently negative. In addition one has

$$W\big(f_{+,1}(z,\,\cdot\,,x_0),\phi_\alpha(z,\,\cdot\,,x_0)\big) = \cos(\alpha)f_{+,1}(z,0,x_0)+\sin(\alpha)f'_{+,1}(z,0,x_0), \quad z \in \rho(H_{+,\alpha}), \tag{5.27}$$

which should be compared with the Jost function $f_+(z, 0)$ in the case where $q$ represents a short-range potential and $\alpha = 0$.

In the following, we want to illustrate how Hypothesis 2.3 and (2.7) apply to $H_{+,\alpha}$ in the case $p = 2$. For this purpose, we first recall the following standard convergence property for trace ideals in $\mathcal{H}$:

**Lemma 5.2** *Let $q \in [1, \infty)$ and assume that $R, R_n, T, T_n \in \mathcal{B}(\mathcal{H})$, $n \in \mathbb{N}$, satisfy $\operatorname{s-lim}_{n\to\infty} R_n = R$ and $\operatorname{s-lim}_{n\to\infty} T_n = T$ and that $S, S_n \in \mathcal{B}_q(\mathcal{H})$, $n \in \mathbb{N}$, satisfy $\lim_{n\to\infty} \|S_n - S\|_{\mathcal{B}_q(\mathcal{H})} = 0$. Then $\lim_{n\to\infty} \|R_n S_n T_n^* - RST^*\|_{\mathcal{B}_q(\mathcal{H})} = 0$.*

(Here the strong limit of a sequence of bounded operators $B_n$, $n \in \mathbb{N}$, as $n \to \infty$, was abbreviated by $\operatorname{s-lim}_{n\to\infty} B_n$.) Lemma 5.2 follows, for instance, from [35, Theorem 1], [59, pp. 28–29], or [65, Lemma 6.1.3] with a minor additional effort (taking adjoints, etc.).

Next, we recall a few facts that enable one to compute the trace of a nonnegative trace class integral operator in a straightforward manner:
$(i)$ Let $0 \leq A \in \mathcal{B}(\mathcal{H})$, $\{\phi_m\}_{m\in\mathbb{N}}$ an orthonormal basis in $\mathcal{H}$ (without loss of generality we assume $\dim(\mathcal{H}) = \infty$), then

$$\sum_{m\in\mathbb{N}} (\phi_m, A\phi_m)_{\mathcal{H}} \in [0, \infty) \cup \{\infty\} \tag{5.28}$$

is independent of the orthonormal basis $\{\phi_m\}_{m\in\mathbb{N}}$ in $\mathcal{H}$. Moreover,

$$\sum_{m\in\mathbb{N}} (\phi_m, A\phi_m)_{\mathcal{H}} < \infty \text{ if and only if } A \in \mathcal{B}_1(\mathcal{H}), \tag{5.29}$$

and if $A \in \mathcal{B}_1(\mathcal{H})$,

$$\operatorname{tr}_{\mathcal{H}}(A) = \sum_{m\in\mathbb{N}} (\phi_m, A\phi_m)_{\mathcal{H}} - \sum_{j\in J} \lambda_j(A) = \|A\|_{\mathcal{B}_1(\mathcal{H})}, \tag{5.30}$$

where $0 \leq \lambda_j(A)$, $j \in J$, $J \subseteq \mathbb{N}$, denote the eigenvalues of $A$ counting multiplicity. (For details, see, e.g., [59, Theorems 2.14 and 3.1].)
$(ii)$ Let $0 \leq K \in \mathcal{B}(L^2(\Omega; d^n x))$, $\Omega \subseteq \mathbb{R}^n$, and suppose that $K$ is an integral operator with continuous integral kernel $K(\cdot, \cdot)$ on $\Omega \times \Omega$. Then,

$$0 \leq K(x, x), \quad x \in \Omega, \tag{5.31}$$
$$|K(x, x)| \leq |K(x, x)|^{1/2} |K(x', x')|^{1/2}, \quad x, x' \in \Omega, \tag{5.32}$$

and for all orthonormal bases $\{e_m\}_{m\in\mathbb{N}}$ in $L^2(\Omega; d^n x)$,

$$\sum_{m\in\mathbb{N}} (e_m, K e_m)_{L^2(\Omega; d^n x)} = \int_\Omega d^n x\, K(x, x). \tag{5.33}$$

In particular, the finiteness of either side in (5.33) implies that of the other. Hence, $K \in \mathcal{B}_1\big(L^2(\Omega; d^n x)\big)$ if and only if $K(\,\cdot\,,\,\cdot\,) \in L^1(\Omega; d^n x)$, and, in this case,

$$\operatorname{tr}_{L^2(\Omega;d^n x)}(K) = \int_\Omega d^n x\, K(x,x) < \infty. \tag{5.34}$$

(For more details we refer, e.g., to [10, Proposition 5.6.9]. For more general measure spaces see, e.g., [54, pp. 65–66], [60, Sect. 3.11].)

Next, we introduce the family of self-adjoint projections $P_R$ in $L^2(\mathbb{R}_+; dx)$ via

$$(P_R f)(x) = \chi_{[0,R]}(x) f(x), \quad f \in L^2(\mathbb{R}_+; dx),\ R > 0, \tag{5.35}$$

with $\chi_{[0,R]}(\,\cdot\,)$ the characteristic function associated with the interval $[0, R]$, $R > 0$. ($P_R$ will play the role of $R_n$, $T_n$ in our application of Lemma 5.2 in the proof of Theorem 5.3 below.)

One then obtains the following results.

**Theorem 5.3** *Assume Hypothesis 5.1, $z, z_0 \in \rho(H_{+,\alpha})$, and $\alpha \in [0, \pi)$. Then,*

$$\big[(H_{+,\alpha} - zI_+)^{-1} - (H_{+,\alpha} - z_0 I_+)^{-1}\big] \in \mathcal{B}_1\big(L^2(\mathbb{R}_+; dx)\big), \tag{5.36}$$

*and*

$$\begin{aligned}
&\operatorname{tr}_{L^2(\mathbb{R}_+;dx)}\big((H_{+,\alpha} - zI_+)^{-1} - (H_{+,\alpha} - z_0 I_+)^{-1}\big)\\
&\quad = -\frac{d}{dz}\ln\big(\det{}_{2,L^2(\mathbb{R}_+;dx)}\big(I_+ - (z - z_0)(H_{+,\alpha} - z_0 I_+)^{-1}\big)\big)\\
&\quad = \frac{d}{dz}\ln\big(\sin(\alpha) f'_{+,1}(z,0,x_0) + \cos(\alpha) f_{+,1}(z,0,x_0)\big)\Big|_{z=z_0}\\
&\qquad - \frac{d}{dz}\ln\big(\sin(\alpha) f'_{+,1}(z,0,x_0) + \cos(\alpha) f_{+,1}(z,0,x_0)\big)\\
&\qquad + \frac{1}{2}\mathcal{I}(z, z_0, x_0),
\end{aligned} \tag{5.37}$$

*as well as,*

$$\begin{aligned}
&\det{}_{2,L^2(\mathbb{R}_+;dx)}\big(I_+ - (z - z_0)(H_{+,\alpha} - z_0 I_+)^{-1}\big)\\
&\quad = \left[\frac{\sin(\alpha) f'_{+,1}(z,0,x_0) + \cos(\alpha) f_{+,1}(z,0,x_0)}{\sin(\alpha) f'_{+,1}(z_0,0,x_0) + \cos(\alpha) f_{+,1}(z_0,0,x_0)}\right]\\
&\qquad \times \exp\left(-(z - z_0)\, \frac{\sin(\alpha) \dot{f}'_{+,1}(z_0,0,x_0) + \cos(\alpha) \dot{f}_{+,1}(z_0,0,x_0)}{\sin(\alpha) f'_{+,1}(z_0,0,x_0) + \cos(\alpha) f_{+,1}(z_0,0,x_0)}\right)\\
&\qquad \times \exp\left(-\frac{1}{2}\int_{z_0}^{z} d\zeta\, \mathcal{I}(\zeta, z_0, x_0)\right),
\end{aligned} \tag{5.38}$$

*where we abbreviated*

$$\mathcal{I}(z, z_0, x_0) = \int_{x_0}^{\infty} dx\{[q(x) - z]^{-1/2} - [q(x) - z_0]^{-1/2}\}. \tag{5.39}$$

*Proof* Since the resolvents of $H_{+,\alpha}$, $\alpha \in (0, \pi)$, and $H_{+,0}$ differ only by a rank-one operator, it suffices to choose $\alpha = 0$ when establishing (5.36).

We will first prove (5.36) for $z = \lambda < 0$, $z_0 = \lambda_0 < \lambda < 0$, and employ monotonicity of resolvents with respect to $\lambda < 0$ sufficiently negative, implying

$$0 \leq \big[(H_{+,0} - \lambda I_+)^{-1} - (H_{+,0} - \lambda_0 I_+)^{-1}\big], \quad \lambda_0 < \lambda < 0, \tag{5.40}$$

with $\lambda < 0$ sufficiently negative (the latter will be assumed for most of the remainder of this proof). Subsequently, we will apply (5.34) to $K$ given by the right-hand side of inequality (5.40).

Equations (5.23) and (5.24) yield for $\alpha = 0$,

$$\begin{aligned}\phi_0(\lambda, \cdot)\psi_{+,0}(\lambda, \cdot) &= f_{+,1}(\lambda, \cdot, x_0) f_{+,2}(\lambda, \cdot, x_0) \\ &\quad - f_{+,1}(\lambda, 0, x_0)^{-1} f_{+,2}(\lambda, 0, x_0) f_{+,1}(\lambda, \cdot, x_0)^2,\end{aligned} \tag{5.41}$$

and since by (5.21) for $j = 1$ integrability properties of (5.41) over $\mathbb{R}_+$ depend on those of $f_{+,1}(\lambda, \cdot, x_0) f_{+,2}(\lambda, \cdot, x_0)$, we now investigate the latter on $[x_0, \infty)$. Employing (5.21) once more then yields

$$\begin{aligned}0 &\leq [\phi_0(\lambda, x)\psi_{+,0}(\lambda, x) - \phi_0(\lambda_0, x)\psi_{+,0}(\lambda_0, x)] \\ &\underset{x\to\infty}{=} 2^{-1}\{[q(x) - \lambda]^{-1/2} - [q(x) - \lambda_0]^{-1/2}\}[1 + o(1)] \\ &\underset{x\to\infty}{=} 4^{-1}(\lambda - \lambda_0) q(x)^{-3/2}[1 + o(1)] \\ &\underset{x\to\infty}{=} 4^{-1}(\lambda - \lambda_0)\, C_0\, x^{-1-(3\varepsilon_0/2)}[1 + o(1)],\end{aligned} \tag{5.42}$$

according to (5.3), proving integrability of $[\phi_0(\lambda, \cdot)\psi_{+,0}(\lambda, \cdot) - \phi_0(\lambda_0, \cdot)\psi_{+,0}(\lambda_0, \cdot)]$ near $+\infty$. An application of (5.34) to $K$ given by the right-hand side of inequality (5.40) then yields (5.36) with $z = \lambda$, $z_0 = \lambda_0 < \lambda < 0$. Analytic continuation in $\lambda$ and subsequently in $\lambda_0$ proves (5.36).

By (2.7) with $p = 2$ this proves the first equality in (5.37).

To prove the second equality in (5.37), we now apply Lemma 5.2 in the trace class case $q = 1$ and combine it with (5.36) to arrive at

$$\begin{aligned}&\operatorname{tr}_{L^2(\mathbb{R}_+;dx)}\big((H_{+,\alpha} - \lambda I_+)^{-1} - (H_{+,\alpha} - \lambda_0 I_+)^{-1}\big) \\ &\quad = \lim_{R\to\infty} \operatorname{tr}_{L^2(\mathbb{R}_+;dx)}\big(P_R\big[(H_{+,\alpha} - \lambda I_+)^{-1} - (H_{+,\alpha} - \lambda_0 I_+)^{-1}\big]P_R\big) \\ &\quad = \lim_{R\to\infty} \int_0^R dx\, [\phi_\alpha(\lambda, x)\psi_{+,\alpha}(\lambda, x) - \phi_\alpha(\lambda_0, x)\psi_{+,\alpha}(\lambda_0, x)]\end{aligned}$$

$$
\begin{aligned}
&= \lim_{R\to\infty} \big[W\big(\phi_\alpha(\lambda_0,\,\cdot\,), \dot{\psi}_{+,\alpha}(\lambda_0,\,\cdot\,)\big)(R) - W\big(\phi_\alpha(\lambda,\,\cdot\,), \dot{\psi}_{+,\alpha}(\lambda,\,\cdot\,)\big)(R)\big] \\
&\quad + W\big(\phi_\alpha(\lambda,\,\cdot\,), \dot{\psi}_{+,\alpha}(\lambda,\,\cdot\,)\big)(0) - W\big(\phi_\alpha(\lambda_0,\,\cdot\,), \dot{\psi}_{+,\alpha}(\lambda_0,\,\cdot\,)\big)(0) \\
&= \lim_{R\to\infty} \big[W\big(\phi_\alpha(\lambda_0,\,\cdot\,), \dot{\psi}_{+,\alpha}(\lambda_0,\,\cdot\,)\big)(R) - W\big(\phi_\alpha(\lambda,\,\cdot\,), \dot{\psi}_{+,\alpha}(\lambda,\,\cdot\,)\big)(R)\big],
\end{aligned} \tag{5.43}
$$

since

$$
\begin{aligned}
W\big(\phi_\alpha(\lambda,\cdot), \dot{\psi}_{+,\alpha}(\lambda,\cdot)\big)(0) &= -\sin(\alpha)\dot{\psi}'_{+,\alpha}(\lambda,0) - \cos(\alpha)\dot{\psi}_{+,\alpha}(\lambda,0) \\
&= -\frac{d}{d\lambda}\big[\sin(\alpha)\psi'_{+,\alpha}(\lambda,0) + \cos(\alpha)\psi_{+,\alpha}(\lambda,0)\big] = 0.
\end{aligned} \tag{5.44}
$$

It remains to analyze the right-hand side of (5.43). To this end we note that

$$
\tau_+ \dot{f}_{+,1}(z,x,x_0) = z\dot{f}_{+,1}(z,x,x_0) + f_{+,1}(z,x,x_0), \tag{5.45}
$$

and hence

$$
\begin{aligned}
\dot{f}_{+,1}(z,x,x_0) &= c_1(z) f_{+,1}(z,x,x_0) + c_2(z) f_{+,2}(z,x,x_0) \\
&\quad + f_{+,1}(z,x,x_0) \int_0^x dx'\, f_{+,1}(z,x',x_0) f_{+,2}(z,x',x_0) \\
&\quad - f_{+,2}(z,x,x_0) \int_0^x dx'\, f_{+,1}(z,x',x_0)^2,
\end{aligned} \tag{5.46}
$$

$$
\begin{aligned}
\dot{f}'_{+,1}(z,x,x_0) &= c_1(z) f'_{+,1}(z,x,x_0) + c_2(z) f'_{+,2}(z,x,x_0) \\
&\quad + f'_{+,1}(z,x,x_0) \int_0^x dx'\, f_{+,1}(z,x',x_0) f_{+,2}(z,x',x_0) \\
&\quad - f'_{+,2}(z,x,x_0) \int_0^x dx'\, f_{+,1}(z,x',x_0)^2.
\end{aligned} \tag{5.47}
$$

Next, we claim that

$$
c_2(z) = \int_0^\infty dx'\, f_{+,1}(z,x',x_0)^2, \quad z \in \rho(H_{+,\alpha}), \tag{5.48}
$$

and hence (5.46), (5.47) simplify to

$$
\begin{aligned}
\dot{f}_{+,1}(z,x,x_0) &= c_1(z) f_{+,1}(z,x,x_0) \\
&\quad + f_{+,1}(z,x,x_0) \int_0^x dx'\, f_{+,1}(z,x',x_0) f_{+,2}(z,x',x_0) \\
&\quad + f_{+,2}(z,x,x_0) \int_x^\infty dx'\, f_{+,1}(z,x',x_0)^2,
\end{aligned} \tag{5.49}
$$

$$\begin{aligned}\dot{f}'_{+,1}(z,x,x_0) = c_1(z)f'_{+,1}(z,x,x_0)\\ &+ f'_{+,1}(z,x,x_0)\int_0^x dx'\, f_{+,1}(z,x',x_0)f_{+,2}(z,x',x_0)\\ &+ f'_{+,2}(z,x,x_0)\int_x^\infty dx'\, f_{+,1}(z,x',x_0)^2.\end{aligned} \tag{5.50}$$

To infer the necessity of (5.48), one can argue by contradiction as follows: If (5.48) does not hold, then integrating $\dot{f}_{+,1}(z,x)$ with respect to $z$ from $\lambda_0$ to $\lambda$ along the negative real axis on the left-hand side of (5.46) yields

$$\int_{\lambda_0}^{\lambda} dz\, \dot{f}_{+,1}(z,x,x_0) = f_{+,1}(\lambda,x,x_0) - f_{+,1}(\lambda_0,x,x_0) \underset{x\to\infty}{\longrightarrow} 0 \tag{5.51}$$

by the first asymptotic relation in (5.21). However, with (5.48) violated, integrating the right-hand side of (5.46) with respect to $z$ from $\lambda_0$ to $\lambda$ along the negative real axis now yields several contributions vanishing as $x \to \infty$ (again invoking (5.21)), but there will also be one integral of the type

$$\int_{\lambda_0}^{\lambda} dz\, f_{+,2}(z,x,x_0)A(z,x) \underset{x\to\infty}{\not\to} 0 \tag{5.52}$$

where $A(z,\,\cdot\,)$ is bounded with a finite nonzero limit, $\lim_{x\to\infty} A(z,x) = A(z,\infty) \neq 0$. Relation (5.52) contradicts (5.51), proving (5.48).

Investigating the asymptotics of the right-hand sides of (5.49), (5.50), invoking the leading asymptotic behavior (5.21), then shows that to obtain the leading asymptotic behavior of $\dot{f}_{+,1}(\lambda,x,x_0)$, $\dot{f}_{+,2}(\lambda,x,x_0)$ one can formally differentiate relations (5.21) with respect to $\lambda$ and hence obtains,

$$\begin{aligned}\dot{f}_{+,1}(\lambda,x,x_0) &\underset{x\to\infty}{=} 2^{-3/2}[q(x)-\lambda]^{-1/4}\int_{x_0}^x dx''\,[q(x'')-\lambda]^{-1/2}\\ &\quad\times\exp\bigg(-\int_{x_0}^x dx'[q(x')-\lambda]^{1/2}\bigg)[1+o(1)],\\ \dot{f}'_{+,1}(\lambda,x,x_0) &\underset{x\to\infty}{=} -2^{-3/2}[q(x)-\lambda]^{1/4}\int_{x_0}^x dx''\,[q(x'')-\lambda]^{-1/2}\\ &\quad\times\exp\bigg(-\int_{x_0}^x dx'[q(x')-\lambda]^{1/2}\bigg)[1+o(1)],\end{aligned} \tag{5.53}$$

for $\lambda < 0$ sufficiently negative according to our convention in this proof.

Next, one utilizes (5.23) and (5.24) and computes

$$
\begin{aligned}
& W\big(\phi_\alpha(\lambda,\,\cdot\,), \dot{\psi}_{+,\alpha}(\lambda,\,\cdot\,)\big)(R) \\
& \quad \underset{R\to\infty}{=} f_{+,2}(\lambda,R,x_0) \dot{f}'_{+,1}(\lambda,R,x_0) \\
& \qquad - f_{+,2}(\lambda,R,x_0) f'_{+,1}(\lambda,R,x_0) \frac{\sin(\alpha)\dot{f}'_{+,1}(\lambda,0,x_0)+\cos(\alpha)\dot{f}_{+,1}(\lambda,0,x_0)}{\sin(\alpha)f'_{+,1}(\lambda,0,x_0)+\cos(\alpha)f_{+,1}(\lambda,0,x_0)} \\
& \qquad - f'_{+,2}(\lambda,R,x_0)\dot{f}_{+,1}(\lambda,R,x_0) \\
& \qquad + f'_{+,2}(\lambda,R,x_0) f_{+,1}(\lambda,R,x_0) \frac{\sin(\alpha)\dot{f}'_{+,1}(\lambda,0,x_0)+\cos(\alpha)\dot{f}_{+,1}(\lambda,0,x_0)}{\sin(\alpha)f'_{+,1}(\lambda,0,x_0)+\cos(\alpha)f_{+,1}(\lambda,0,x_0)} \\
& \quad \underset{R\to\infty}{=} \dot{f}'_{+,1}(\lambda,R,x_0) f_{+,2}(\lambda,R,x_0) - \dot{f}_{+,1}(\lambda,R,x_0) f'_{+,2}(\lambda,R,x_0) \\
& \qquad + \frac{\sin(\alpha)\dot{f}'_{+,1}(\lambda,0,x_0)+\cos(\alpha)\dot{f}_{+,1}(\lambda,0,x_0)}{\sin(\alpha)f'_{+,1}(\lambda,0,x_0)+\cos(\alpha)f_{+,1}(\lambda,0,x_0)},
\end{aligned} \tag{5.54}
$$

again for $\lambda < 0$ sufficiently negative. Insertion of (5.21) and (5.53) into (5.54) finally implies

$$
\begin{aligned}
W\big(\phi_\alpha(\lambda,\,\cdot\,), \dot{\psi}_{+,\alpha}(\lambda,\,\cdot\,)\big)(R) & \underset{R\to\infty}{=} \frac{\sin(\alpha)\dot{f}'_{+,1}(\lambda,0,x_0)+\cos(\alpha)\dot{f}_{+,1}(\lambda,0,x_0)}{\sin(\alpha)f'_{+,1}(\lambda,0,x_0)+\cos(\alpha)f_{+,1}(\lambda,0,x_0)} \\
& \qquad - 2^{-1}\bigg(\int_{x_0}^{R} dx\,[q(x)-\lambda]^{-1/2}\bigg)[1+o(1)].
\end{aligned} \tag{5.55}
$$

Returning to (5.43) this yields

$$
\begin{aligned}
& \operatorname{tr}_{L^2(\mathbb{R}_+;dx)}\big((H_{+,\alpha}-\lambda I_+)^{-1} - (H_{+,\alpha}-\lambda_0 I_+)^{-1}\big) \\
& \quad = \lim_{R\to\infty} \big[W\big(\phi_\alpha(\lambda_0,\,\cdot\,),\dot{\psi}_{+,\alpha}(\lambda_0,\,\cdot\,)\big)(R) - W\big(\phi_\alpha(\lambda,\,\cdot\,),\dot{\psi}_{+,\alpha}(\lambda,\,\cdot\,)\big)(R)\big] \\
& \quad \underset{R\to\infty}{=} \frac{\sin(\alpha)\dot{f}'_{+,1}(\lambda_0,0,x_0)+\cos(\alpha)\dot{f}_{+,1}(\lambda_0,0,x_0)}{\sin(\alpha)f'_{+,1}(\lambda_0,0,x_0)+\cos(\alpha)f_{+,1}(\lambda_0,0,x_0)} \\
& \qquad - \frac{\sin(\alpha)\dot{f}'_{+,1}(\lambda,0,x_0)+\cos(\alpha)\dot{f}_{+,1}(\lambda,0,x_0)}{\sin(\alpha)f'_{+,1}(\lambda,0,x_0)+\cos(\alpha)f_{+,1}(\lambda,0,x_0)} \\
& \qquad + 2^{-1}\bigg(\int_{x_0}^{R} dx\,\big\{[q(x)-\lambda]^{-1/2} - [q(x)-\lambda_0]^{-1/2}\big\}\bigg)[1+o(1)] \\
& \quad = \frac{\sin(\alpha)\dot{f}'_{+,1}(\lambda_0,0,x_0)+\cos(\alpha)\dot{f}_{+,1}(\lambda_0,0,x_0)}{\sin(\alpha)f'_{+,1}(\lambda_0,0,x_0)+\cos(\alpha)f_{+,1}(\lambda_0,0,x_0)} \\
& \qquad - \frac{\sin(\alpha)\dot{f}'_{+,1}(\lambda,0,x_0)+\cos(\alpha)\dot{f}_{+,1}(\lambda,0,x_0)}{\sin(\alpha)f'_{+,1}(\lambda,0,x_0)+\cos(\alpha)f_{+,1}(\lambda,0,x_0)}
\end{aligned}
$$

$$+2^{-1}\bigg(\int_{x_0}^{\infty} dx\,\big\{[q(x)-\lambda]^{-1/2}-[q(x)-\lambda_0]^{-1/2}\big\}\bigg), \tag{5.56}$$

and hence (5.37) for $z=\lambda<0$, $z_0=\lambda_0<0$, both sufficiently negative. In this context one observes that for $x_0>0$ sufficiently large,

$$\begin{aligned} &2^{-1}\bigg(\int_{x_0}^{R} dx\,\big\{[q(x)-\lambda]^{-1/2}-[q(x)-\lambda_0]^{-1/2}\big\}\bigg) \\ &\underset{R\to\infty}{=} \frac{1}{4}(\lambda-\lambda_0)\bigg(\int_{x_0}^{R} dx\, q(x)^{-3/2}\bigg)[1+o(1)] \end{aligned} \tag{5.57}$$

with $q^{-3/2}\in L^1([x_0,\infty);dx)$ by Hypothesis (5.3).

Analytic continuation in $\lambda$ of both sides in (5.56) extends the latter to $z\in\rho(H_{+,\alpha})$. Similarly, analytic continuation in $\lambda_0$ of both sides in (5.56) extends the latter to $z_0\in\rho(H_{+,\alpha})$, completing the proof of (5.37).

Relation (5.38) then follows from integrating (5.37) with respect to the energy variable from $z_0$ to $z$. □

*Remark 5.4* Employing the resolvent equation,

$$\begin{aligned}(H_{+,0}-zI_+)^{-1}-(H_{+,0}-z_0I_+)^{-1}=(z-z_0)(H_{+,0}-zI_+)^{-1}(H_{+,0}-z_0I_+)^{-1},\\ z,z_0\in\rho(H_{+,0}),\end{aligned} \tag{5.58}$$

an alternative proof of relation (5.36) follows upon establishing

$$(H_{+,0}-zI_+)^{-1}\in\mathcal{B}_2\big(L^2(\mathbb{R}_+;dx)\big),\quad z\in\rho(H_{+,0}). \tag{5.59}$$

To prove (5.59) in turn it suffices to establish the Hilbert–Schmidt property for some $z=\lambda<0$ sufficiently negative, followed by analytic continuation with respect to $z\in\rho(H_{+,0})$. Given the Green's function of $H_{+,0}$ in (5.19), it thus suffices to prove that

$$\int_{\mathbb{R}_+}\int_{\mathbb{R}_+} dx\,dx'\,|\phi_0(\lambda,x)\,\psi_{+,0}(\lambda,x')|^2<\infty; \tag{5.60}$$

we omit further details at this point.

Next, we apply Theorem 5.3 to the following explicitly solvable example concerning the linear potential and denote by $Ai(\,\cdot\,)$, $Bi(\,\cdot\,)$ the Airy functions as discussed, for instance, in [1, Sect. 10.4].

*Example 5.5* Consider the special case $q(x)=x$, $x\in\mathbb{R}_+$, and $\alpha=0$. Then, for $x\in\mathbb{R}_+$, $z,z_0\in\rho(H_{+,0})$,

$$f_{+,1}(z,x,x_0) = (2\pi)^{1/2} e^{(2/3)(x_0-z)^{3/2}} Ai(x-z), \tag{5.61}$$

$$f_{+,2}(z,x,x_0) = (\pi/2)^{1/2} e^{-(2/3)(x_0-z)^{3/2}} Bi(x-z), \tag{5.62}$$

$$W(f_{+,1}(z,\cdot,x_0), f_{+,2}(z,\cdot,x_0)) = 1, \tag{5.63}$$

$$\phi_0(z,x) = \pi[Ai(-z)Bi(x-z) - Bi(-z)Ai(x-z)], \tag{5.64}$$

$$\psi_{+,0}(z,x) = Ai(x-z)/Ai(-z), \tag{5.65}$$

$$\begin{aligned} &W(\phi_0(z,\cdot), \dot{\psi}_{+,0}(z,\cdot))(x) \\ &\quad = \pi[Ai'(x-z)Bi'(x-z) - (x-z)Ai(x-z)Bi(x-z)] - [Ai'(-z)/Ai(-z)], \end{aligned} \tag{5.66}$$

$$\begin{aligned} \mathcal{I}(z,z_0,x_0) &= \int_{x_0}^{\infty} dx \big\{[x-z]^{-1/2} - [x-z_0]^{-1/2}\big\} \\ &= 2\big[(x_0-z_0)^{1/2} - (x_0-z)^{1/2}\big], \end{aligned} \tag{5.67}$$

$$\begin{aligned} &\operatorname{tr}_{L^2(\mathbb{R}_+;dx)}\big((H_{+,0}-zI_+)^{-1} - (H_{+,0}-z_0I_+)^{-1}\big) \\ &\quad = \psi'_{+,0}(z,0) - \psi'_{+,0}(z_0,0) = [Ai'(-z)/Ai(-z)] - [Ai'(-z_0)/Ai(-z_0)], \end{aligned} \tag{5.68}$$

$$\begin{aligned} &{\det}_{2,L^2(\mathbb{R}_+;dx)}\big(I_+ - (z-z_0)(H_{+,0}-z_0I_+)^{-1}\big) \\ &\quad = [Ai(-z)/Ai(-z_0)] \exp\big((z-z_0)[Ai'(-z_0)/Ai(-z_0)]\big). \end{aligned} \tag{5.69}$$

We note that (5.69) was recently considered in [48], but the exponential factor in (5.69) was missed in [48].

Finally, we generalize Theorem 5.3 to the following setting.

**Theorem 5.6** *Assume Hypothesis 5.1, $z \in \rho(H_{+,\alpha_2})$, $z_0 \in \rho(H_{+,\alpha_1})$, and $\alpha_1, \alpha_2 \in [0,\pi)$. Then,*

$$\big[(H_{+,\alpha_2}-zI_+)^{-1} - (H_{+,\alpha_1}-z_0I_+)^{-1}\big] \in \mathcal{B}_1\big(L^2(\mathbb{R}_+;dx)\big), \tag{5.70}$$

*and (cf. (5.39))*

$$\begin{aligned} &\operatorname{tr}_{L^2(\mathbb{R}_+;dx)}\big((H_{+,\alpha_2}-zI_+)^{-1} - (H_{+,\alpha_1}-z_0I_+)^{-1}\big) \\ &\quad = -\frac{d}{dz}\ln\left(\frac{\sin(\alpha_2)f'_{+,1}(z,0,x_0) + \cos(\alpha_2)f_{+,1}(z,0,x_0)}{\sin(\alpha_1)f'_{+,1}(z_0,0,x_0) + \cos(\alpha_1)f_{+,1}(z_0,0,x_0)}\right) \\ &\qquad + \frac{1}{2}\mathcal{I}(z,z_0,x_0). \end{aligned} \tag{5.71}$$

*Proof* Equation (5.70) is established exactly as in the proof of Theorem 5.3. Furthermore, as argued there one has

$$\begin{aligned} &\operatorname{tr}_{L^2(\mathbb{R}_+;dx)}\big((H_{+,\alpha_2}-\lambda I_+)^{-1} - (H_{+,\alpha_1}-\lambda_0 I_+)^{-1}\big) \\ &\quad = \lim_{R\to\infty}\big[W\big(\phi_{\alpha_1}(\lambda_0,\cdot), \dot{\psi}_{+,\alpha_1}(\lambda_0,\cdot)\big)(R) - W\big(\phi_{\alpha_2}(\lambda,\cdot), \dot{\psi}_{+,\alpha_2}(\lambda,\cdot)\big)(R)\big]. \end{aligned} \tag{5.72}$$

Using Eq. (5.55) then immediately implies (5.71). □

Setting $z = z_0$, we obtain in particular

$$\begin{aligned} &\operatorname{tr}_{L^2(\mathbb{R}_+;dx)}\big((H_{+,\alpha_2} - zI_+)^{-1} - (H_{+,\alpha_1} - zI_+)^{-1}\big) \\ &\quad = -\frac{d}{dz}\ln\left(\frac{\sin(\alpha_2)f'_{+,1}(z,0,x_0) + \cos(\alpha_2)f_{+,1}(z,0,x_0)}{\sin(\alpha_1)f'_{+,1}(z,0,x_0) + \cos(\alpha_1)f_{+,1}(z,0,x_0)}\right). \end{aligned} \tag{5.73}$$

*Remark 5.7* In order to prove Theorem 5.6, one could instead have proven the slightly simpler result (5.73) and then note that

$$\begin{aligned} &\operatorname{tr}_{L^2(\mathbb{R}_+;dx)}\big((H_{+,\alpha_2} - zI_+)^{-1} - (H_{+,\alpha_1} - z_0I_+)^{-1}\big) \\ &= \operatorname{tr}_{L^2(\mathbb{R}_+;dx)}\big((H_{+,\alpha_2} - zI_+)^{-1} - (H_{+,\alpha_1} - zI_+)^{-1}\big) \\ &\quad + \operatorname{tr}_{L^2(\mathbb{R}_+;dx)}\big((H_{+,\alpha_1} - zI_+)^{-1} - (H_{+,\alpha_1} - z_0I_+)^{-1}\big), \end{aligned} \tag{5.74}$$

which, using (5.73) together with Theorem 5.3 implies Theorem 5.6. ◇

**Acknowledgements** We are indebted to Christer Bennewitz for very helpful discussions.

## References


1. Abramowitz, M., Stegun, I.A.: Handbook of Mathematical Functions. Dover, New York (1972)
2. Akhiezer, N.I., Glazman, I.M.: Theory of Linear Operators in Hilbert Space, vol. II. Pitman, Boston (1981)
3. Atkinson, F.V., Mingarelli, A.B.: Asymptotics of the number of zeros and of the eigenvalues of general weighted Sturm-Liouville problems. J. reine angew. Math. **375/376**, 380–393 (1987)
4. Bennewitz, C.: Private communication
5. Burghelea, D., Friedlander, L., Kappeler, T.: On the determinant of elliptic boundary value problems on a line segment. Proc. Am. Math. Soc. **123**, 3027–3038 (1995)
6. Buslaev, V.S., Faddeev, L.D.: Formulas for traces for a singular Sturm-Liouville differential operator. Sov. Math. Dokl. **1**, 451–454 (1960)
7. Chadan, K., Sabatier, P.C.: Inverse Problems in Quantum Scattering Theory, 2nd edn. Springer, New York (1989)
8. Clark, S., Gesztesy, F., Mitrea, M.: Boundary data maps for Schrödinger operators on a compact interval. Math. Model. Nat. Phenom. **5**(4), 73–121 (2010)
9. Clark, S., Gesztesy, F., Nichols, R., Zinchenko, M.: Boundary data maps and Krein's resolvent formula for Sturm-Liouville operators on a finite interval. Oper. Matrices **8**, 1–71 (2014)
10. Davies, E.B.: Linear Operators and Their Spectra. Cambridge Studies in Advanced Mathematics, vol. 106. Cambridge University Press, Cambridge (2007)
11. Deift, P.A.: Applications of a commutation formula. Duke Math. J. **45**, 267–310 (1978)
12. Demirel, S., Usman, M.: Trace formulas for Schrödinger operators on the half-line. Bull. Math. Sci. **1**, 397–427 (2011)
13. Dreyfus, T., Dym, H.: Product formulas for the eigenvalues of a class of boundary value problems. Duke Math. J. **45**, 15–37 (1978)
14. Eastham, M.S.P.: The Asymptotic Solution of Linear Differential Systems. Application of the Levinson Theorem. Clarendon Press, Oxford (1989)

15. Falco, G.M., Fedorenko, A.A., Gruzberg, I.A.: On functional determinants of matrix differential operators with multiple zero modes. J. Phys. A **50**, 485201 (2017) 29 pp
16. Forman, R.: Functional determinants and geometry. Invent. Math. **88**, 447–493 (1987); Erratum **108**, 453–454 (1992)
17. Forman, R.: Determinants, finite-difference operators and boundary value problems. Commun. Math. Phys. **147**, 485–526 (1992)
18. Fucci, G., Graham, C., Kirsten, K.: Spectral functions for regular Sturm-Liouville problems. J. Math. Phys. **56**, 043503 (2015), 24 pp
19. Gesztesy, F., Kirsten, K.: Effective computation of traces, determinants, and $\zeta$-functions for Sturm–Liouville operators. J. Funct. Anal. **276**, 520–562 (2019)
20. Gesztesy, F., Kirsten, K.: On traces and modified Fredholm determinants for half-line Schrödinger operators with discrete spectra. Q. Appl. Math. (to appear), https://doi.org/10.1090/qam/1520. https://www.ams.org/journals/qam/0000-000-00/S0033-569X-2018-01520-3/
21. Gesztesy, F., Makarov, K.A.: (Modified) Fredholm determinants for operators with matrix-valued semi-separable integral kernels revisited. Integral Equ. Oper. Theory **47**, 457–497 (2003). (See also Erratum **48**, 425–426 (2004) and the corrected electronic only version in **48**, 561–602 (2004).)
22. Gesztesy, F., Nichols, R.: Weak convergence of spectral shift functions for one-dimensional Schrödinger operators. Math. Nachr. **285**, 1799–1838 (2012)
23. Gesztesy, F., Nichols, R.: An abstract approach to weak convergence of spectral shift functions and applications to multi-dimensional Schrödinger operators. J. Spectr. Theory **2**, 225–266 (2012)
24. Gesztesy, F., Simon, B.: Uniqueness theorems in inverse spectral theory for one-dimensional Schrödinger operators. Trans. Am. Math. Soc. **348**, 349–373 (1996)
25. Gesztesy, F., Weikard, R.: Floquet theory revisited. In: Knowles, I. (ed.) Differential Equations and Mathematical Physics, pp. 67–84. International Press, Boston (1995)
26. Gesztesy, F., Zinchenko, M.: Symmetrized perturbation determinants and applications to boundary data maps and Krein-type resolvent formulas. Proc. Lond. Math. Soc. **3**(104), 577–612 (2012)
27. Gesztesy, F., Latushkin, Y., Mitrea, M., Zinchenko, M.: Nonselfadjoint operators, infinite determinants, and some applications. Russ. J. Math. Phys. **12**, 443–471 (2005)
28. Gesztesy, F., Mitrea, M., Zinchenko, M.: Variations on a theme of Jost and Pais. J. Funct. Anal. **253**, 399–448 (2007)
29. Gohberg, I., Krein, M.G.: Introduction to the Theory of Linear Nonselfadjoint Operators. Translations of Mathematical Monographs, vol. 18. American Mathematical Society, Providence (1969)
30. Gohberg, I., Goldberg, S., Krupnik, N.: Traces and determinants of linear operators. Integral Equ. Oper. Theory **26**, 136–187 (1996)
31. Gohberg, I., Goldberg, S., Krupnik, N.: Hilbert-Carleman and regularized determinants for linear operators. Integral Equ. Oper. Theory **27**, 10–47 (1997)
32. Gohberg, I., Goldberg, S., Krupnik, N.: Traces and Determinants for Linear Operators, Operator Theory: Advances and Applications, vol. 116. Birkhäuser, Basel (2000)
33. Gohberg, I.C., Krein, M.G.: Theory and Applications of Volterra Operators in Hilbert Space. Translations of Mathematical Monographs, vol. 24. American Mathematical Society, Providence (1970)
34. Gradshteyn, I.S., Ryzhik, I.M.: Table of Integrals, Series, and Products, corrected and enlarged edition, prepared by A. Jeffrey. Academic, San Diego (1980)
35. Grümm, H.R.: Two theorems about $\mathcal{C}_p$. Rep. Math. Phys. **4**, 211–215 (1973)
36. Jost, R., Pais, A.: On the scattering of a particle by a static potential. Phys. Rev. **82**, 840–851 (1951)
37. Kato, T.: Wave operators and similarity for some non-selfadjoint operators. Math. Ann. **162**, 258–279 (1966)
38. Kirsten, K.: Spectral Functions in Mathematics and Physics. CRC Press, Boca Raton (2001)

39. Kirsten, K., McKane, A.J.: Functional determinants by contour integration methods. Ann. Phys. **308**, 502–527 (2003)
40. Kirsten, K., McKane, A.J.: Functional determinants for general Sturm-Liouville problems. J. Phys. A **37**, 4649–4670 (2004)
41. Kirsten, K., Loya, P., Park, J.: The very unusual properties of the resolvent, heat kernel, and zeta function for the operator $-d^2/dr^2 - 1/(4r^2)$. J. Math. Phys. **47**, 043506 (2006), 27 pp
42. Kuroda, S.T.: On a generalization of the Weinstein-Aronszajn formula and the infinite determinant. Sci. Pap. Coll. Gen. Educ. Univ. Tokyo **11**(1), 1–12 (1961)
43. Lesch, M.: Determinants of regular singular Sturm-Liouville operators. Math. Nachr. **194**, 139–170 (1998)
44. Lesch, M., Tolksdorf, J.: On the determinant of one-dimensional elliptic boundary value problems. Commun. Math. Phys. **193**, 643–660 (1998)
45. Lesch, M., Vertman, B.: Regular singular Sturm-Liouville operators and their zeta-determinants. J. Funct. Anal. **261**, 408–450 (2011)
46. Levit, S., Smilansky, U.: A theorem on infinite products of eigenvalues of Sturm-Liouville type operators. Proc. Am. Math. Soc. **65**, 299–302 (1977)
47. Levitan, B.M., Sargsjan, I.S.: Introduction to Spectral Theory, Translation of Mathematical Monographs, vol. 39. American Mathematical Society, Providence (1975)
48. Menon, G.: The Airy function is a Fredholm determinant. J. Dyn. Differ. Equ. **28**, 1031–1038 (2016)
49. Müller, W.: Relative zeta functions, relative determinants and scattering theory. Commun. Math. Phys. **192**, 309–347 (1998)
50. Muñoz-Castañeda, J.M., Kirsten, K., Bordag, M.: QFT over the finite line. Heat kernel coefficients, spectral zeta functions and selfadjoint extensions. Lett. Math. Phys. **105**, 523–549 (2015)
51. Naimark, M.A.: Linear Partial Differential Operators Part I. Elementary Theory of Linear Differential Operators. Translation by E.R. Dawson, English translation edited by W.N. Everitt, F. Ungar Publishing, New York (1967)
52. Newton, R.G.: Inverse scattering. I. One dimension. J. Math. Phys. **21**, 493–505 (1980)
53. Östensson, J., Yafaev, D.R.: A trace formula for differential operators of arbitrary order. In: Dym, H., Kaashoek, M.A., Lancaster, P., Langer, H., Lerer, L. (eds.) A Panorama of Modern Operator Theory and Related Topics. The Israel Gohberg Memorial Volume. Operator Theory: Advances and Applications, vol. 218, pp. 541–570. Birkhäuser, Springer, Boston, Berlin (2012)
54. Reed, M., Simon, B.: Methods of Modern Mathematical Physics. III: Scattering Theory. Academic, New York (1979)
55. Reed, M., Simon, B.: Methods of Modern Mathematical Physics. IV: Analysis of Operators. Academic, New York (1978)
56. Robert, D., Sordoni, V.: Generalized determinants for Sturm-Liouville problems on the real line. In: Greiner, P.C., Ivrii, V., Seco, L.A., Sulem, C. (eds.) Partial Differential Equations and Their Applications. CRM Proceedings and Lecture Notes, vol. 12, pp. 251–259. American Mathematical Society, Providence (1997)
57. Simon, B.: Quantum Mechanics for Hamiltonians Defined as Quadratic Forms. Princeton University Press, Princeton (1971)
58. Simon, B.: Notes on infinite determinants of Hilbert space operators. Adv. Math. **24**, 244–273 (1977)
59. Simon, B.: Trace Ideals and Their Applications. Mathematical Surveys and Monographs, vol. 120, 2nd edn. American Mathematical Society, Providence (2005)
60. Simon, B.: Operator Theory. A Comprehensive Course in Analysis, Part 4. American Mathematical Society, Providence (2015)
61. Spreafico, M.: Zeta function and regularized determinant on a disc and on a cone. J. Geom. Phys. **54**, 355–371 (2005)
62. Weidmann, J.: Verteilung der Eigenwerte für eine Klasse von Integraloperatoren in $L_2(a, b)$. J. reine angew. Math. **276**, 213–220 (1975)

63. Weidmann, J.: Lineare Operatoren in Hilberträumen. Teil I. Mathematische Leitfäden. Teubner, Stuttgart (2000)
64. Weidmann, J.: Lineare Operatoren in Hilberträumen. Teil II. Mathematische Leitfäden. Teubner, Stuttgart (2003)
65. Yafaev, D.R.: Mathematical Scattering Theory. General Theory. Translations of Mathematical Monographs, vol. 105. American Mathematical Society, Providence (1992)
66. Zettl, A.: Sturm-Liouville Theory. Mathematical Surveys and Monographs, vol. 121. American Mathematical Society, Providence (2005)

# On the Domain of a Magnetic Schrödinger Operator with Complex Electric Potential

**Bernard Helffer and Jean Nourrigat**

**Abstract** The aim of this paper is to review and compare the spectral properties of the Schrödinger operators $-\Delta + U$ $(U \geq 0)$ and $-\Delta + iV$ in $L^2(\mathbb{R}^d)$ for $C^\infty$ real potentials $U$ or $V$ with polynomial behavior. The case with magnetic field will be also considered. We present the existing criteria for essential self-adjointness, maximal accretivity, compactness of the resolvent, and maximal inequalities. Motivated by recent works with X. Pan, Y. Almog, and D. Grebenkov, we actually improve the known results in the case with purely imaginary potential.

## 1 Introduction

In this paper,[1], we will discuss the spectral theory for the Schrödinger operators $-\Delta + U$ $(U \geq 0)$ or $-\Delta + i\, V$ in $L^2(\mathbb{R}^d)$. More precisely, we present the state of the art (and sometime will go beyond) for the following properties:

- essential self-adjointness or maximal accretivity,
- compactness of the resolvent,
- maximal inequalities, i.e., the existence of $C > 0$ such that, $\forall u \in C_0^\infty(\mathbb{R}^d)$,

$$||u||^2_{H^2(\mathbb{R}^d)} + ||Uu||^2_{L^2(\mathbb{R}^d)} \leq C \left( ||(-\Delta + U)u||^2_{L^2(\mathbb{R}^d)} + ||u||^2_{L^2(\mathbb{R}^d)} \right), \quad (1)$$

or

[1]The results of this paper have been presented by the first author at the Kato centennial conference in Tokyo in September 2017.

B. Helffer (✉)
Laboratoire de Mathématiques Jean Leray, CNRS and Université de Nantes, 2 rue de la Houssinière, 44322 Nantes Cedex, France
e-mail: Bernard.Helffer@univ-nantes.fr

J. Nourrigat
LMR EA 4535 and FR CNRS 3399, Université de Reims Champagne-Ardenne, Moulin de la Housse, BP 1039 51687 REIMS Cedex 2, France
e-mail: jean-francois.nourrigat@univ-reims.fr

© Springer Nature Switzerland AG 2019

T. M. Rassias and V. A. Zagrebnov (eds.), *Analysis and Operator Theory*, Springer Optimization and Its Applications 146,
https://doi.org/10.1007/978-3-030-12661-2_8

$$||u||^2_{H^2} + ||Vu||^2 \leq C\left(||(-\Delta + i\,V)u||^2 + ||u||^2\right). \tag{2}$$

We will also discuss the case with magnetic field where the operator reads:

$$P_{\mathbf{A},W} = -\Delta_{\mathbf{A}} + W := \sum_{j=1}^{d}(D_{x_j} - A_j(x))^2 + W(x)\,,$$

where $D_{x_j} = -i\frac{\partial}{\partial x_j}$, $\mathbf{A} = (A_1, \dots, A_d)$ is a $C^\infty$ vector field on $\mathbb{R}^d$ and $W$ is a complex-valued $C^\infty$ potential, and the maximal regularity is expressed in terms of the magnetic Sobolev spaces:

$$\begin{aligned} &||(D-\mathbf{A})u||^2_{L^2(\mathbb{R}^d,\mathbb{C}^d)} + \textstyle\sum_{j,\ell}||(D_j - A_j)(D_\ell - A_\ell)u||^2_{L^2(\mathbb{R}^d)} + ||\,|W|\,u||^2_{L^2(\mathbb{R}^d)} \\ &\quad \leq C\left(||P_{\mathbf{A},W}u||^2_{L^2(\mathbb{R}^d)} + ||u||^2_{L^2(\mathbb{R}^d)}\right), \end{aligned} \tag{3}$$

The question of analyzing $-\Delta + i\,V$ or more generally $P_{\mathbf{A},iV} := -\Delta_{\mathbf{A}} + i\,V$ appears in many situations [1–3]. It seems therefore useful to present in a unified way, what is known on the subject in the self-adjoint case and try to go further in the accretive case, where much less is known. If we assume that the potential $V$ is $C^\infty$, we know that the operator is essentially self-adjoint starting from $C_0^\infty(\mathbb{R}^d)$ in the first situation and maximally accretive in the second case. Hence in the two cases, the closed operator in consideration is uniquely defined by its restriction to $C_0^\infty$.

At least for the self-adjoint case, the subject has a long story, in which T. Kato and his school plays an important role. We refer to [36] for a rather complete presentation with an exhaustive list of reference. One should also mention the work of Avron–Herbst–Simon [5] which popularizes the basic questions on the subject and in particular the magnetic bottles.

For the compactness of the resolvent, outside the easy case when $U \to +\infty$, the story starts around the 80s with the treatment of instructive examples (Simon [35], Robert [29]) and in the case with magnetic field [5] (the simplest example being for $d = 2$ and $U = 0$, when $B(x) \to +\infty$). In the polynomial case, many results are deduced as a byproduct of the analysis in Helffer–Nourrigat [15], at least in the case when $V$ is a sum of square of polynomials. Using Kohn's type inequality, Helffer and Morame (Mohamed) [14] obtain more general results which can be combined with the analysis of Iwatsuka [19]. Another family of results using the notion of capacity can be found in [21, 22] (see references therein).

T. Kato proves, for example, the inequality

$$||\Delta u||_{L^1} + ||Uu||_{L^1} \leq 3\,||(-\Delta + U)u||_{L^1}\,,\ \forall u \in C_0^\infty(\mathbb{R}^d)\,, \tag{4}$$

under the condition that $U \geq 0$ and $U \in L^1_{loc}$.

The generalization to the $L^p$ $(p > 1)$ is only possible under stronger conditions on $U$. We will mention some of these results but will focus on the $L^2$ estimates which are sometimes easier to obtain. In the case, when $U(x) = \sum_\ell U_\ell(x)^2$, the maximal $L^2$ estimate is obtained as a byproduct of the analysis of the hypoellipticity (see Hörmander [17], Rothschild–Stein [30] and the book Helffer–Nourrigat [16] (including polynomial magnetic potentials)). This was then generalized to the case when $V$ is a positive polynomial by J. Nourrigat in an unpublished paper [27] and used in the PhD of Guibourg [11, 12] defended in 1992, which considers the case when the electric potential $U \geq 0$ and the magnetic potential $\mathbf{A}$ are polynomials (one chapter treats a more general situation). In his thesis, Zhong [37] proves the same result by showing that $\nabla^2(-\Delta + U)^{-1}$ is a Calderon–Zygmund operator. Shen [31] generalizes the result to the case when $U$ is in the reverse Hölder class $RH_q$ $(q \geq \frac{d}{2})$, a class which contains the positive polynomials.

**Definition 1** A locally $L^q$ function $\omega$ and strongly positive almost everywhere belongs to $RH_q$ if there exists a constant $C > 0$ such that for any cube $Q$ in $\mathbb{R}^d$

$$\left(\frac{1}{|Q|}\int \omega^q \, dx\right)^{\frac{1}{q}} \leq C\left(\frac{1}{|Q|}\int \omega \, dx\right).$$

One should also mention the unpublished thesis of Mba-Yébé [23] defended in 1995. Together with the techniques developed by Guibourg, some of his techniques are useful for the improvements presented in the last section.

Z. Shen considers also the case with magnetic fields in 1996 [32]. Further progress is obtained in the thesis of Ben Ali [6], published in [4, 7, 8]. The methods applied by Shen and Auscher–Ben Ali include the Fefferman–Phong inequalities, the Calderón–Zygmund decompositions, and various techniques of interpolation. We will come back to one of these results in Sect. 2.3.

## 2 Kohn's Approach

This approach was mainly used for getting the compactness of the resolvent. Except in a few cases, these estimates do not lead to the maximal regularity but are sufficient for getting the compactness and we will see that surprisingly, they could also be a step for proving $L^2$-maximal estimates.

### *2.1 Self-adjoint Case*

Here we mainly refer to [14] (see also [15, 24]). We analyze the problem for the family of operators:

$$\mathscr{P}_{\mathbf{A},U} = \sum_{j=1}^{d}(D_{x_j} - A_j(x))^2 + \sum_{\ell=1}^{p} U_\ell(x)^2 . \tag{5}$$

Here, the magnetic potential $\mathbf{A}(x) = (A_1(x), A_2(x), \cdots, A_n(x))$ and the $U_j$ are assumed to be $C^\infty$. Under these conditions, the operator is essentially self-adjoint on $C_0^\infty(\mathbb{R}^d)$. We note also that it can be written in the form

$$\mathscr{P}_{\mathbf{A},U} = \sum_{j=1}^{d+p} X_j^2 = \sum_{j=1}^{d} X_j^2 + \sum_{\ell=1}^{p} Y_\ell^2 ,$$

with

$$X_j = (D_{x_j} - A_j(x)) , \; j = 1, \ldots, d , \; Y_\ell = U_\ell , \; \ell = 1, \ldots, p .$$

In particular, the magnetic field is recovered by observing that

$$B_{jk} = \frac{1}{i}[X_j, X_k] = \partial_j A_k - \partial_k A_j , \quad \text{for } j, k = 1, \ldots, d .$$

Of course, when $U \to +\infty$, it is well known that the operator has a compact resolvent. (see the argument below).

On the opposite, when $V = 0$ and $d = 2$ if $B(x) = B_{12} \geq 0$, one immediately deduces from the trivial inequality:

$$\int B(x)|u(x)|^2 dx \leq ||X_1 u||^2 + ||X_2 u||^2 = \langle \mathscr{P}_{\mathbf{A},U}\, u \mid u \rangle . \tag{6}$$

that $\lim_{|x| \to +\infty} B(x) = +\infty$ implies that the operator has a compact resolvent. A typical example is

$$A_1(x_1, x_2) = -x_2 x_1^2 , \; A_2(x_1, x_2) = +x_1 x_2^2 .$$

In order to treat more general situations, we introduce the quantities:

$$\check{m}_q(x) = \sum_{\ell} \sum_{|\alpha|=q} |\partial_x^\alpha U_\ell| + \sum_{j<k} \sum_{|\alpha|=q-1} |\partial_x^\alpha B_{jk}(x)| . \tag{7}$$

It is easy to reinterpret this quantity in terms of commutators of the $X_j$'s.

When $q = 0$, the convention is that

$$\check{m}_0(x) = \sum_{\ell} |U_\ell(x)| . \tag{8}$$

Let us also introduce

$$\check{m}^r(x) = 1 + \sum_{q=0}^{r} \check{m}_q(x) . \tag{9}$$

Then the criterion proven by Helffer–Mohamed in [14] reads:

**Theorem 1** *Let us assume that there exist r and C such that*

$$\check{m}_{r+1}(x) \leq C\, \check{m}^r(x)\,,\ \forall x \in \mathbb{R}^d\,, \tag{10}$$

*and*

$$\check{m}^r(x) \to +\infty\,,\ \text{as}\ |x| \to +\infty\,. \tag{11}$$

*Then, with* $U = \sum_\ell U_\ell^2$, $\mathscr{P}_{\mathbf{A},U}(h)$ *has a compact resolvent.*

*Remark 1* It is shown in [24], that one can get the same result as in Theorem 1 under the weaker assumption that

$$\check{m}_{r+1}(x) \leq C\, [\check{m}^r(x)]^{1+\delta}\,, \tag{12}$$

where $\delta = \frac{1}{2^{r+1}-3}$ $(r \geq 1)$. This result is optimal for $r = 1$. Iwatsuka [19] has indeed exhibited an example of a Schrödinger operator which has a non compact resolvent and such that $\sum_{j<k} |\nabla B_{jk}(x)|$ has the same order as $\sum_{j<k} |B_{jk}|^2$.

Other generalizations are given in [31] (Corollary 0.11) (see also references therein and [22] for a quite recent contribution including other references).

One can, for example, replace $\sum_j U_j^2$ by a more general $U$ and the conditions on the $m_j$'s can be reformulated in terms of the variation of $U$ and $B$ in suitable balls. In particular, Iwatsuka [19] showed that a necessary condition is

$$\int_{B(x,1)} \left( U(x) + \sum_{j<k} B_{jk}(x)^2 \right) dx \to +\infty \text{ as } |x| \to +\infty\,, \tag{13}$$

where $B(x, 1)$ is the ball of radius 1 centered at $x$.

## 2.2 The Accretive Case: Maximal Accretiveness

There is a general statement (see for example, [3, 15]) about the maximal accretiveness of $\mathscr{P}_{\mathbf{A},W} := -\Delta_{\mathbf{A}} + W$, when $U$ is non negative.

**Theorem 2** *Consider the magnetic Schrödinger operator* $\mathscr{P} := \mathscr{P}_{\mathbf{A},W}$ *defined on* $\mathbb{R}^d$ *with* $\mathbf{A} \in C^\infty(\mathbb{R}^d, \mathbb{R}^d)$ *and* $W = U + i\,V \in C^\infty(\mathbb{R}^d, \mathbb{C})$ *such that*

$$U(x) \geq 0\,. \tag{14}$$

*Then the operator* $\mathscr{P}$ *is maximally accretive. Moreover*

$$\mathscr{P}_{\mathbf{A},W} = (P_{\mathbf{A},\bar{W}})^*\,. \tag{15}$$

We extend Theorem 1 to the family of operators:

$$P_{\mathbf{A},W} = \sum_{j=1}^{d} (D_{x_j} - A_j(x))^2 + \sum_{\ell=1}^{p} U_\ell(x)^2 + i\,V(x)\,. \tag{16}$$

Here $U = \sum_{\ell=1}^{p} U_\ell(x)^2$ and $V$ is $C^\infty$. We note also that it has the form

$$P_{\mathbf{A},W} = \sum_{j=1}^{d+p} X_j^2 = \sum_{j=1}^{d} X_j^2 + \sum_{\ell=1}^{p} Y_\ell^2 + X_0\,,$$

with

$$X_j = (D_{x_j} - A_j(x))\,,\ j = 1, \dots, d\,,\ Y_\ell = U_\ell\,,\ \ell = 1, \dots, p\,,\ X_0 = i\,V\,.$$

We introduce the new quantity:

$$\check{m}_q(x) = \sum_{\ell} \sum_{|\alpha|=q} |\partial_x^\alpha U_\ell| + \sum_{j<k} \sum_{|\alpha|=q-1} |\partial_x^\alpha B_{jk}(x)| + \sum_{|\alpha|=q-1} |\partial_x^\alpha V|\,. \tag{17}$$

and keep for $\check{m}^r(x)$ the same definition as in previous subsection:

$$\check{m}^r(x) = 1 + \sum_{q=0}^{r} \check{m}_q(x)\,. \tag{18}$$

Then the criterion reads

**Theorem 3** *Let us assume that there exist $r$ and $C_0$ such that*

$$\check{m}_{r+1}(x) \le C_0\, \check{m}^r(x)\,,\ \forall x \in \mathbb{R}^d\,. \tag{19}$$

*In this case, we say that $(\mathbf{A}, W) \in \mathscr{T}(r, C_0)$.*

*Then there exist $\delta > 0$ and $C_1 := C_1(C_0)$ such that, $\forall u \in C_0^\infty(\mathbb{R}^d)$*

$$||(\check{m}^r(x))^\delta u||^2 \le C_1 \left(||P_{\mathbf{A},W} u||^2 + ||u||^2\right)\,. \tag{20}$$

*Remark 2* The proof, which was first given in a particular case in [3], will show that we can take $\delta = 2^{-r}$ which is in general not optimal. This $\delta$ can indeed be improved when $U = 0$ and $\mathbf{A} = 0$ (see [1]) but this improvement will not be used in this paper.

**Corollary 1** *Under the same assumptions, if*

$$\check{m}^r(x) \to +\infty\,,\ \text{as } |x| \to +\infty\,. \tag{21}$$

*Then $\mathscr{P}_{\mathbf{A},W}(h)$ has a compact resolvent.*

Before entering into the core of the proof, we observe that we can replace $\check{m}^r(x)$ by an equivalent $C^\infty$ function $\Psi(x)$ which has the property that there exist constants

$C_\alpha$ and $C > 0$ such that:

$$\begin{aligned} &\tfrac{1}{C}\Psi(x) \leq \check{m}^r(x) \leq C\Psi(x)\,, \\ &|D_x^\alpha \Psi(x)| \leq C_\alpha \Psi(x)\,. \end{aligned} \tag{22}$$

Indeed, it suffices to replace quantities like $\sum |u_k|$ by $(\sum |u_k|^2)^{1/2}$, in the definition (17) of $\check{m}_q$. The second condition is a consequence of (19).

In the same spirit as in Kohn's proof, let us introduce:

**Definition 2** For all $s > 0$, we denote by $M^s$ the space of $C^\infty$ real functions $T$ such that there exists $C_s$ such that:

$$||\Psi^{-1+s} T u||^2 \leq C_s \left(||P_{\mathbf{A},W} u||\, ||u|| + ||u||^2\right)\,,\ \forall u \in C_0^\infty(\mathbb{R}^d)\,. \tag{23}$$

We immediately verify that

$$U_\ell \in M^1\,. \tag{24}$$

We will then show the

**Lemma 1**

$$i\,[X_j, X_k] \in M^{\frac{1}{2}}\,,\ \forall j, k = 1, \ldots, d\,. \tag{25}$$

*and*

$$V \in M^{\frac{1}{2}}\,. \tag{26}$$

Another claim is contained in the

**Lemma 2** *If $T$ is in $M^s$ and $|\partial_x^\alpha T| \leq C_\alpha \Psi$ then $i[X_k, T] \in M^{\frac{s}{2}}$, when $|\alpha| = 1$ or $|\alpha| = 2$.*

Assuming these two lemmas, then it is clear that

$$\Psi(x) \in M^{2^{-r}}\,.$$

Lemma 2 and (24) lead to

$$\partial_x^\alpha U_\ell \in M^{2^{-|\alpha|}}\,,$$

and we deduce from Lemmas 1 and 2:

$$\partial_x^\alpha B_{jk} \in M^{2^{-(|\alpha|+1)}}\,.$$

The proof of Theorem 3 then becomes easy.

**Proof of Lemma** 1 We start from the identity (and observing that $X_j^* = X_j$):

$$\begin{aligned} ||\Psi^{-\frac{1}{2}}[X_j, X_k]u||^2 &= \langle (X_j X_k - X_k X_j)u \mid \Psi^{-1}[X_j, X_k]u\rangle \\ &= \langle X_k u \mid X_j \Psi^{-1}[X_j, X_k]u\rangle \end{aligned}$$

$$
\begin{aligned}
&-\langle X_j u \mid X_k \Psi^{-1}[X_j, X_k]u\rangle \\
&= \langle X_j u \mid \Psi^{-1}[X_k, X_j]X_k u\rangle \\
&\quad -\langle X_k u \mid \Psi^{-1}[X_k, X_j]X_k u\rangle \\
&\quad +\langle X_j u \mid [X_k, \Psi^{-1}[X_k, X_j]]u\rangle \\
&\quad -\langle X_k u \mid [X_j, \Psi^{-1}[X_k, X_j]]u\rangle\,.
\end{aligned}
$$

If we observe that $\Psi^{-1}[X_k, X_j]$ and $[X_k, \Psi^{-1}[X_k, X_j]]$ are bounded (look at the definition of $\Psi$), we obtain:

$$||\Psi^{-\frac{1}{2}}[X_j, X_k]u||^2 \leq C\left(||X_k u||^2 + ||X_j u||^2 + ||u||^2\right)\,.$$

We just observe that

$$\sum_j ||X_j u||^2 = \Re\langle P_{\mathbf{A},W} u \mid u\rangle\,. \tag{27}$$

This ends the proof of the first part of the lemma.

For the second part, we start from

$$\Im\left(\langle \Psi^{-1} P_{\mathbf{A},W} u\,,\, Qu\rangle\right) = ||\psi^{-\frac{1}{2}} Qu||^2 + \Im\left(\langle \Psi^{-1} \sum_j X_j^2 u\,,\, Vu\rangle\right)\,,$$

and observe that

$$\langle \Psi^{-1} X_j^2 u\,,\, Vu\rangle = \langle \Psi^{-1} V X_j u\; X_j u\rangle + \langle [X_j, \Psi^{-1} V] X_j u \mid u\rangle\,.$$

Hence

$$\Im\langle \Psi^{-1} X_j^2 u\,,\, Vu\rangle = \Im\langle [X_j, \Psi^{-1} V] X_j u \mid u\rangle\,.$$

Then using the property of $V$ and $\Psi$, we get the proof easily.

**Proof of Lemma** 2 Let $T \in M^s$. For each $k$, we can write:

$$
\begin{aligned}
||\Psi^{-1+\frac{s}{2}}[X_k, T]u||^2 &= \langle \Psi^{-1+s}(X_k T - T X_k)u \mid \Psi^{-1}[X_k, T]u\rangle \\
&= \langle \Psi^{-1+s} X_k T u \mid \Psi^{-1}[X_k, T]u\rangle \\
&\quad -\langle \Psi^{-1+s} T X_k u \mid \Psi^{-1}[X_k, T]u\rangle \\
&= \langle \Psi^{-1+s} T u \mid \Psi^{-1}[X_k, T]X_k u\rangle \\
&\quad -\langle X_k u \mid \Psi^{-1}[X_k, T]\Psi^{-1+s} T u\rangle \\
&\quad +\langle T u \mid [X_k, \Psi^{-2+s}[X_k, T]]u\rangle \\
&= \langle \Psi^{-1+s} T u \mid \Psi^{-1}[X_k, T]X_k u\rangle \\
&\quad -\langle X_k u \mid \Psi^{-1}[X_k, T]\Psi^{-1+s} T u\rangle \\
&\quad +\langle \Psi^{-1+s} T u \mid \Psi^{1-s}[X_k, \Psi^{-2+s}[X_k, T]]u\rangle\,.
\end{aligned}
$$

We now observe, according to the assumptions of the lemma and the properties of $\Psi$, that $\Psi^{1-s}[X_k, \Psi^{-2+s}[X_k, T]]$ and $\Psi^{-1}[X_k, T]$ are bounded. So finally we get:

$$||\Psi^{-\frac{1}{2}}[X_k, T]u||^2 \leq C\left(||\Psi^{-1+s}Tu||^2 + ||X_k u||^2 + ||u||^2\right).$$

This ends the proof of the lemma.

### 2.3 Nourrigat–Guibourg–Shen Results

The results where first obtained in the polynomial case ($V = 0$, $U \geq 0$), then for $\mathbf{A}$ and $U \geq 0$ satisfying a condition "à la Helffer-Mohamed" and then with condition of type Reverse-Hölder. We take a version presented in Shen [32] as a consequence of his Theorem 0.9 (see p. 820 lines –13 to –8) which in addition refers to [33]. We only write the $L^2$ criterion.

**Theorem 4** *For $d \geq 3$, let us assume that $\mathbf{A} \in C^{r+2}(\mathbb{R}^d)$, $U \in C^{r+2}(\mathbb{R}^d)$ and $U \geq 0$ and*

$$\sum_{|\beta|=r+1} |\partial_x^\beta B(x)| + \sum_{|\beta|=r+2} |\partial_x^\beta U(x)| \leq C\, \check{m}_{B,U}(x), \tag{28}$$

*where*

$$\check{m}_{B,U}(x) = \sum_{|\beta|\leq r} |\partial_x^\beta B(x)| + \sum_{|\beta|\leq r+1} |\partial_x^\beta U(x)| + 1. \tag{29}$$

*Then*

$$\sum_{1\leq j,\ell\leq d} ||(D_{x_j} - A_j(x))(D_{x_\ell} - A_\ell(x))u||_2 \leq C\left(||P_{\mathbf{A},U}u||_2 + ||u||_2\right), \ \forall u \in C_0^\infty(\mathbb{R}^d). \tag{30}$$

*Remark 3* As observed in [32], the statement appearing in the thesis of D. Guibourg (Chapter B) is a little weaker. When $d = 2$, according to [34], there is no particular difficulty. The assumption $(RH)_{n/2}$ appearing in [32] needs to be changed to $(RH)_p$ for some $p > 1$, and the size estimate for the fundamental solution should have a logarithm. Otherwise, all results should be true.

## 3 Nilpotent Approach

The basic idea is to start from the maximal hypoellipticity of the Hörmander operator $\mathscr{P} := \sum_j \check{X}_j^2 + \check{X}_0$ where the $\check{X}_j$'s are real vector fields satisfying the so called Hörmander condition. When considering the special case of a stratified group $\mathscr{G} = \mathscr{G}_1 \oplus \cdots \oplus \mathscr{G}_r$, with the $\check{X}_j$'s ($j = 1, \ldots, k$) being a basis of $\mathscr{G}_1$, $\check{X}_0 \in \mathscr{G}_2$ and the $\check{X}_j$ ($j = 0, \ldots, k$) generating the Lie algebra of rank $r$.

We get for any induced representation $\Pi$ of $\mathscr{G}$ in $\mathscr{H}_\Pi \sim L^2(\mathbb{R}^k)$ the maximal estimate

$$||\Pi(\check{X}_0)u||_{\mathscr{H}_\Pi} + \sum_{j,\ell=1,\dots,k} ||\Pi(\check{X}_j\check{X}_\ell)u||_{\mathscr{H}_\Pi} \leq C||\pi(\mathscr{P})u||_{\mathscr{H}_\Pi}, \ \forall u \in \mathscr{S}_\Pi. \tag{31}$$

We can then use a variant of Proposition 1.6.1 in [16], to get for any polynomial $V$ with $k$ variables the inequality

$$\sum_{j,\ell} ||\partial^2_{x_j x_\ell} u|| + ||V(x)u|| \leq C\,||(-\Delta + iV(x))u||, \ \forall u \in C_0^\infty(\mathbb{R}^k). \tag{32}$$

We can indeed find $\mathscr{G}$, a subalgebra $\mathscr{V}$ and $\ell \in \mathscr{G}^*$ such that $\Pi$ is unitary equivalent to the representation $\pi_{\ell,\mathscr{V}}$ with

$$\pi_{\ell,\mathscr{V}}(\check{X}_j) = \partial_{x_j} \text{ for } j = 1,\dots,k\,,\ \pi_{\ell,\mathscr{V}}(\check{X}_0) = i\,V(x)\,.$$

As observed a long time ago, if $U(x) = \sum_j U_j(x)^2$, we can use the maximal hypoellipticity of $\sum_j \check{X}_j^2$ and obtain

$$\sum_{j,\ell} ||\partial^2_{x_j x_\ell} u|| + ||U(x)u|| \leq C\,||(-\Delta + U(x))u||, \ \forall u \in C_0^\infty(\mathbb{R}^d). \tag{33}$$

It is then natural to ask if it is true for any polynomial $U \geq 0$ and more generally if we can relax the polynomial condition. Here, we refer to Guibourg or Nourrigat[2] [27, 28] for the first point (this is indeed true) and to Shen [31] for the second point.

There is indeed a more general class of reverse Hölder potentials:

**Definition 3** We say that a non negative $\omega$ belongs to the reverse Hölder class if there exists $C$ such that, $\forall x \in \mathbb{R}^d$, $\forall r > 0$,

$$\sup_{y \in B(x,r)} \omega(x) \leq \frac{C}{|B(x,r)|} \int_{B(x,r)} \omega(y)\,dy\,. \tag{34}$$

Note that a nonnegative polynomial satisfies this condition.

Then, other authors work on the subject with the aim of obtaining $L^p$ estimates [4, 6]. The case with magnetic fields is always considered (see [7, 8, 32, 33] and references therein).

---

[2]There are actually two different proofs proposed by J. Nourrigat a rather direct one and another based to the analysis of $\sum_j \check{X}_j^2 + i\check{X}_0$ the difficulty (but this was sometimes treated in [15]) that this operator is no more hypoelliptic.

# 4 Maximal Estimates for the Complex Schrödinger Operator with Magnetic Potentials (Non Necessarily Polynomial Case)

## 4.1 Main Statement

We consider as before

$$W = \sum_{\ell} U_\ell^2 + i\, V\,,$$

and the associated complex Schrödinger operator $P_{\mathbf{A},W}$.

**Theorem 5** *If* $(\mathbf{A}, W) \in \mathscr{T}(r, C_0)$*, there exists* $C > 0$ *such that, for all* $u \in C_0^\infty(\mathbb{R}^d)$*:*

$$\| |W| u\|^2 \le C \left(\| P_{\mathbf{A},W}\, u\|^2 + ||u||^2\right)\,. \tag{35}$$

## 4.2 Hörmander's Metrics and Partition of Unity

We introduce, for $t \in [0, 1]$ and $x \in \mathbb{R}^d$,

$$\Phi(x,t) = \sum_{\ell} \sum_{|\alpha|\le r} t^{|\alpha|+1}\, |\partial_x^\alpha U_\ell(x)| + \sum_{j<k} \sum_{|\alpha|\le r-1} t^{|\alpha|+2}\, |\partial_x^\alpha B_{jk}(x)| + \sum_{|\alpha|\le r-1} t^{|\alpha|+2} |\partial_x^\alpha V(x)|\,. \tag{36}$$

As in Mba-Yébé [23], we introduce a parameter $\mu \ge 1$ to be determined later and which is at the moment arbitrary and we define:

$$R(x,\mu) = \sup\{t \in [0,1], \quad \Phi(x,t) \le \mu\}\,.$$

**Proposition 1** *If* $(\mathbf{A}, W) \in \mathscr{T}(r, C_0)$ *(see condition (19), there exists* $C_2 > 1$ *such that, for all* $t \in (0, 1)$*, we have the implication:*

$$|y - x| \le t \Longrightarrow \Phi(y,t) \le C_2 \Phi(x,t) + C_2\, t^{r+1}\,.$$

*Proof* For all $x$ et $u$ in $\mathbb{R}^d$ such that $|u| = 1$, for all $t$ and $\theta$ such that $0 < \theta \le t \le 1$, let us introduce:

$$\Psi(x,u,\theta,t) = \Phi(x + \theta u, t)\,.$$

Using Taylor's formula with integral remainder, we can write, if $\theta \le t$,

$$\Psi(x,u,\theta,t) \le C\,\Phi(x,t) + C\mathbf{R}(t)\,,$$

with

$$\mathbf{R}(t) := t^{r+1} \sum_{\ell} \sum_{|\beta|=r+1} \int_0^{\theta} |\partial_x^{\beta} U_{\ell}(x + \sigma u)| \, d\sigma + \dots$$

$$\dots + C t^{r+1} \sum_{j<k} \sum_{|\beta|=r} \int_0^{\theta} |\partial_x^{\beta} B_{jk}(x + \sigma u)| \, d\sigma + C t^{r+1} \sum_{|\beta|=r} \int_0^{\theta} |\partial_x^{\beta} V(x + \sigma u)| d\sigma \, .$$

Using condition (19) and $\theta \leq t \leq 1$, there exists $C > 0$ such that

$$\mathbf{R}(t) \leq C \int_0^{\theta} \Psi(x, u, \sigma, t) \, d\sigma + C \, t^{r+1} \, .$$

We now apply Gronwall's Lemma and obtain the existence of $C_2 > 0$ such that, for $\theta \leq t \leq 1$,

$$\Psi(x, u, \theta, t) \leq C_2 \Phi(x, t) + C_2 t^{r+1} \, .$$

This achieves the proof of the proposition.

**Proposition 2** *Let $C_2 > 1$ the constant of Proposition 1. Then we have:*

$$|y - x| \leq \frac{R(x, \mu)}{2C_2} \Longrightarrow \frac{1}{2C_2} \leq \frac{R(y, \mu)}{R(x, \mu)} \leq 2C_2 \, .$$

*Proof* We apply Proposition 1 with $t_0 = R(x, \mu) \leq 1$.

If $|y - x| \leq R(x, \mu)$, we have

$$\Phi(y, t_0) \leq C_2(\Phi(x, t_0) + t_0^{r+1}) \leq 2C_2 \mu \, .$$

We have indeed $\Phi(x, t_0) \leq \mu$ and $t_0^{r+1} \leq 1 \leq \mu$. Consequently $t_1 = t_0/2C_2 = R(x, \mu)/(2C_2)$ satisfies $t_1 \leq 1$ and $\Phi(y, t_1) \leq \mu$. Therefore we get $t_1 \leq R(y, \mu)$, and consequently the first inequality above.

If now $|y - x| \leq R(x, \mu)/(2C_2)$, we deduce $|y - x| \leq R(y, \mu)$, and, permuting the roles of $x$ and $y$, we effectively get

$$R(y, \mu) \leq 2C_2 R(x, \mu) \, .$$

This achieves the proof of the proposition.

This proposition shows that the metric defined on $\mathbb{R}^d$ by $g_x(t) = |t|^2/R(x, \mu)^2$ ($x \in \mathbb{R}^d$, $t \in \mathbb{R}^d$), is slowly varying in the sense of Definition 18.4.1 in [18]. Moreover, the constant in the definition can be chosen independently of $\mu$. We deduce from Lemma 18.4.4 in [18] the following proposition.

**Proposition 3** *For any $\mu \geq 1$, there exist a sequence of real valued functions $(\varphi_m)$ in $C_0^{\infty}(\mathbb{R}^d)$, and a sequence $(x^m)$ in $\mathbb{R}^d$, such that:*

- $$\sum_m \varphi_m(x)^2 = 1\,,\ \forall x \in \mathbb{R}^d\,. \tag{37}$$

- $$\operatorname{supp} \varphi_m \subset B(x^m, R(x^m, \mu))\,. \tag{38}$$

- *For any multi-index $\alpha$, there exists $\hat{C}_\alpha > 0$, independent of $\mu$, such that*

$$\sum_m |\partial^\alpha \varphi_m(x)|^2 \le \frac{\hat{C}_\alpha}{R(x, \mu)^{2|\alpha|}}\,. \tag{39}$$

- *There exists $\hat{C} > 0$, independent of $\mu$, such that, for $k = 1, 2$, for any $u$ in $C_0^\infty(\mathbb{R}^d)$,*

$$\int_{\mathbb{R}^d} \frac{|u(x)|^2}{R(x, \mu)^{2k}} dx \le \hat{C} \sum_j \int_{\mathbb{R}^d} \frac{\varphi_m(x)^2 |u(x)|^2}{R(x^m, \mu)^{2k}} dx\,. \tag{40}$$

## 4.3 Proof of Theorem 5

Just observing that:

$$\Re < P_{U+iV} f, f >= ||(D - \mathbf{A}) f||^2 + \int U |f|^2 dx\,,$$

we obtain:

**Lemma 3** *For all $f \in C_0^\infty(\mathbb{R}^d)$, we have:*

$$\sum_j \|(D_{x_j} - A_j) f\|^2 + \sum_{\ell=1}^{p} \|U_\ell f\|^2 \le \|P_{\mathbf{A},W} f\|\ \|f\|\,. \tag{41}$$

**Proposition 4** *For any $\mu > 1$, let $(x^m)$ be the sequence of points in $\mathbb{R}^d$ constructed in Proposition 3 and let $(\mathbf{A}, W) \in \mathscr{T}(r, C_0)$. Then, there exist $\mu_0 > 1$ and $C_3$ (depending only on $r$ and $C_0$) such that, for any $\mu \ge \mu_0$, for any $m$ such that $R(x^m, \mu) \le 1/2$, and for any $f \in C_0^\infty(\mathbb{R}^d)$ supported in the ball $B_m = B(x^m, R(x^m, \mu))$,*

$$\frac{\mu^\delta}{R(x^m, \mu)^2} \|f\| + \frac{\mu^{\delta/2}}{R(x^m, \mu)} \|(D - \mathbf{A}) f\| \le C_3 \|P_{\mathbf{A},W} f\| \tag{42}$$

*where $\delta$ is the constant given by Theorem 3.*

*Proof* With $R_m := R(x^m, \mu)$, we introduce

$$V^{(m,loc)}(y) = R_m^2 V(x^m + R_m y), \qquad U_\ell^{(m,loc)}(y) = R_m U_\ell(x^m + R_m y),$$

$$A_k^{(m,loc)}(y) = R_m A_k(x^m + R_m y), \qquad B_{jk}^{(m,loc)}(y) = R_m^2 B_{jk}(x^m + R_m y).$$

If $R_m \leq 1$, one verifies that, for any $(\mathbf{A}, W) \in \mathscr{T}(r, C_0)$, the corresponding pair $(\mathbf{A}^{(m,loc)}, W^{(m,loc)})$ belongs to $\mathscr{T}(r, C_0)$.

If $R_m \leq 1/2$, we have $\Phi(x^m, R(x^m, \mu)) = \mu$. Applying Proposition 1 with $t = R_m = R(x^m, \mu) \leq 1$, we have, if $|y| \leq 1$,

$$\Phi(x^m, R_m) \leq C_2\, \Phi(x^m + R_m y, R_m) + C_2 \leq C_2\, \check{m}^{r,m,loc}(y),$$

where $\check{m}^{r,m,loc}(y)$ is the function associated, as in (17), with the localized operator $P_{\mathbf{A}^{m,loc}, W^{m,loc}}$ at the point $x^m$.

Consequently, we have $\mu \leq C_2 \check{m}^{r,m,loc}(y)$ for all $y \in \mathbb{R}^d$ such that $|y| \leq 1$.

By Theorem 3, for all $g \in C_0^\infty(\mathbb{R}^d)$ with support in $B(0,1)$, we have:

$$\mu^\delta (1/C_2)^\delta \|g\| \leq C_1(\|P_{\mathbf{A}^{m,loc}, W^{m,loc}} g\| + \|g\|)$$

where $C_1$ and $C_2$ depend only on $C_0$.

Then, one can find $\mu_0$ and $C_3$ with the same properties such that, for $\mu \geq \mu_0$,

$$\mu^\delta \|g\| \leq C_3 \|P_{\mathbf{A}^{m,loc}, W^{m,loc}} g\|. \tag{43}$$

If $f$ is supported in $B(x^m, R(x^m, \mu))$, we apply (43) to the function $g(y) = f(x^m + yR(x^m, \mu))$ and obtain for $\mu \geq \mu_0$

$$\frac{\mu^\delta}{R(x^m, \mu)^2} \|f\| \leq C_3\, \|P_{\mathbf{A}, W} f\|.$$

Inequality (41) leads to (42).

## 4.4 End of the Proof of Theorem 5

Let $u \in C_0^\infty(\mathbb{R}^d)$. For any $\mu \geq 1$, we apply (40) and get, distinguishing in the localization formula the $x^m$ such that $R(x^m, \mu) > \frac{1}{2}$ from the terms such that $R(x^m, \mu) \leq \frac{1}{2}$,

$$\int_{\mathbb{R}^d} \left[ \frac{|u(x)|^2}{R(x,\mu)^4} + \frac{|(D - \mathbf{A})u(x)|^2}{R(x,\mu)^2} \right] dx \leq C(\|u\|^2 + \|(D - \mathbf{A})u\|^2) + \mathbf{R},$$

$$\mathbf{R} := C \sum_{R(x^m,\mu) \leq 1/2} \frac{\|\varphi_m u\|^2}{R(x^m,\mu)^4} + \frac{\|(D - \mathbf{A})(\varphi_m u)\|^2}{R(x^m,\mu)^2}.$$

Here we have also used (39) for the control of commutators. If $\mu \geq \mu_0$ with $\mu_0$ large enough, and for any $m$ such that $R(x^m, \mu) \leq 1/2$, we apply Proposition 4 to the function $f = \varphi_m u$ and obtain:

$$\begin{aligned}\mathbf{R} &\leq C\mu^{-2\delta} \textstyle\sum_{R(x^m,\mu)\leq\frac{1}{2}} \|P_{\mathbf{A},W}(\varphi_m u)\|^2 \\ &\leq C\mu^{-2\delta}\|P_{\mathbf{A},W}u\|^2 + C\mu^{-2\delta}\textstyle\sum_m \left[\|\nabla\varphi_m\cdot(\nabla - i\mathbf{A})u\|^2 + \|u(\Delta\varphi_m)\|^2\right].\end{aligned}$$

From (39), we deduce:

$$\mathbf{R} \leq C\mu^{-2\delta}\|P_{\mathbf{A},W}u\|^2 + C\mu^{-2\delta}\int_{\mathbb{R}^d}\left[\frac{|u(x)|^2}{R(x,\mu)^4} + \frac{|(D-\mathbf{A})u(x)|^2}{R(x,\mu)^2}\right]dx\,.$$

There exists a possibly larger new $\mu_0$ such that, for $\mu \geq \mu_0$,

$$\int_{\mathbb{R}^d}\left[\frac{|u(x)|^2}{R(x,\mu)^4} + \frac{|(D-\mathbf{A})u(x)|^2}{R(x,\mu)^2}\right]dx \leq C\left(\|u\|^2 + \|(D-\mathbf{A})u\|^2\right) + C\,\mu^{-2\delta}\|P_{\mathbf{A},W}u\|^2\,.$$

Using again (41), we get:

$$\int_{\mathbb{R}^d}\left[\frac{|u(x)|^2}{R(x,\mu)^4} + \frac{|(D-\mathbf{A})u(x)|^2}{R(x,\mu)^2}\right]dx \leq C\,\|u\|^2 + C\,(1+\mu^{-2\delta})\|P_{\mathbf{A},W}u\|^2\,.$$

Theorem 5 follows since $\Phi(x, R(x,\mu)) \leq \mu$ and consequently

$$R(x,\mu)\sum_{\ell}|U_\ell(x)| + R(x,\mu)^2\sum_{j<k}|B_{jk}(x)| + R(x,\mu)^2|V(x)| \leq \mu\,.$$

*Remark 4* The proof of Theorem 5 does not give directly the maximal estimates as described in the introduction. But the control of the theorem reduces this question to the analysis for a self-adjoint operator which could be either the magnetic Laplacian ($W = 0$), or the magnetic Schrödinger operator $\mathscr{P}_{\mathbf{A},U}$ or the operator $\mathscr{P}_{\mathbf{A},U+\sqrt{1+V^2}}$. For these operators, one can, for example, use Shen's Theorem 4 (at least when $d \geq 3$, but see Remark 3 for $d = 2$). Note that to be in $\mathscr{T}(r, C_0)$ is usually not enough except if $\mathbf{A}$ is a polynomial. In this case, one can use the nilpotent approach for getting the maximal regularity of the magnetic Laplacian. Otherwise, one can, for example, prove the complete maximal estimates under the condition that ($\mathbf{A}$, $W = 0$) belongs to $\mathscr{T}(r', C_0')$.

**Acknowledgements** The authors would like to thank J. Camus for the transmission of [11] and Z. Shen for an enlightening clarification [34]. The first author would like to thank S. Nakamura and K. Yajima for their invitation to the Kato centennial conference and the RIMS Kyoto for its support.

## References


1. Almog, Y., Grebenkov, D., Helffer, B.: On a Schrödinger operator with a purely imaginary potential in the semiclassical limit. ArXiv 1703–07733, (2017)
2. Almog, Y., Helffer, B.: On the spectrum of non-selfadjoint Schrödinger operators with compact resolvent. Commun. PDE **40**(8), 1441–1466 (2015)
3. Almog, Y., Helffer, B., Pan, X.: Superconductivity near the normal state under the action of electric currents and induced magnetic fields in $\mathbb{R}^2$. Commun. Math. Phys. **300**(1), 147–184 (2010)
4. Auscher, P., Ben Ali, B.: Maximal inequalities and Riesz transform on $L^p$ space for Schrödinger operators with non negative potentials. Ann. Inst. Fourier **57**(6), 1975–2013 (2007)
5. Avron, Y., Herbst, I., Simon, B.: Schrödinger operators with magnetic fields I. General interactions. Duke Math. J. **45**(4), 847–883 (1978)
6. Ben Ali, B.: Inégalités maximales et estimations $L^p$ des transformées de Riesz des opérateurs de Schrödinger. Thèse de doctorat de l'université Paris-Sud (2008)
7. Ben Ali, B.: Maximal inequalities and Riesz transform estimates on $L^p$ spaces for magnetic Schrödinger operators I. J. Funct. Anal. **259**, 1631–1672 (2010)
8. Ben Ali, B.: Maximal inequalities and Riesz transform estimates on $L^p$ spaces for magnetic Schrödinger operators II. Math. Z. **274**, 85–116 (2013)
9. Folland, G.B.: On the Rothschild-Stein lifting theorem. Commun. PDE **212**, 165–191 (1977)
10. Fournais, S., Helffer, B.: Spectral Methods in Surface Superconductivity. Progress in Non-Linear PDE, vol. 77. Birkhäuser (2010)
11. Guibourg, D.: Inégalités maximales pour l'opérateur de Schrödinger, PhD Thesis, Université de Rennes 1 (1992)
12. Guibourg, D.: Inégalités maximales pour l'opérateur de Schrödinger. CRAS **316**, 249–252 (1993)
13. Helffer, B.: Spectral Theory and its Applications. Cambridge University Press, Cambridge (2013)
14. Helffer, B., Mohamed, A.: Sur le spectre essentiel des opérateurs de Schrödinger avec champ magnétique. Ann. Inst. Fourier **38**(2), 95–113 (1988)
15. Helffer, B., Nier, F.: Hypoelliptic Estimates and Spectral Theory for Fokker-Planck Operators and Witten Laplacians. Springer Lecture Note in Mathematics, vol. 1862 (2005)
16. Helffer, B., Nourrigat, J.: Hypoellipticité Maximale pour des Opérateurs Polynômes de Champs de Vecteurs. Progress in Mathematics. Birkhäuser, vol. 58 (1985)
17. Hörmander, L.: Hypoelliptic second order differential equations. Acta Math. **119**, 147–171 (1967)
18. Hörmander, L.: The Analysis of Linear Partial Differential Operators, vol. III. Springer, Berlin (1985)
19. Iwatsuka, A.: Magnetic Schrödinger operators with compact resolvent. J. Math. Kyoto Univ. **26**, 357–374 (1986)
20. Kohn, J.: Lectures on Degenerate Elliptic Problems. Pseudodifferential Operators with Applications, C.I.M.E., Bressanone, vol. 1977, pp. 89–151 (1978)
21. Kondratiev, V., Maz'ya, V., Shubin, M.: Gauge optimization and spectral properties of magnetic Schrödinger operators. Commun. Part. Differ. Equ. **34**(10–12), 1127–1146 (2009)
22. Kondratiev, V., Shubin, M.: Discreteness of spectrum for the magnetic Schrödinger operators. Commun. Part. Differ. Equ. **27**(3–4), 477–526 (2002)
23. Mba Yébé, J.T.: Réalisation et spectre d'opérateurs de Schrödinger et de Klein-Gordon avec des potentiels irréguliers. Thèse de Doctorat (Université de Reims) (1995)
24. Meftah, M.: Conditions suffisantes pour la compacité de la résolvante d'un opérateur de Schrödinger avec un champ magnétique. J. Math. Kyoto Univ. **31**(3), 875–880 (1991)
25. Mohamed, A., Nourrigat, J.: Encadrement du $N(\lambda)$ pour un opérateur de Schrödinger avec un champ magnétique et un potentiel électrique. Journal de Mathématiques Pures et Appliquées (9) 70, no 1 (1991), 87–99

26. Nourrigat, J.: Inégalités $L^2$ et représentations de groupes nilpotents. J. Funct. Anal. **74** (1987)
27. Nourrigat, J.: Une inégalité $L^2$. Unpublished manuscript (1990)
28. Nourrigat, J.: $L^2$ Inequalities and Representations of Nilpotent Groups. CIMPA School of Harmonic Analysis. Wuhan (China) (1991)
29. Robert, D.: Comportement asymptotique des valeurs propres d'opérateurs du type Schrödinger à potentiel dégénéré. J. Math. Pures Appl. (9) **61**(3), 275–300 (1982), (1983)
30. Rothschild, L.P., Stein, E.: Hypoelliptic operators and nilpotent groups. Acta Math. **137**, 248–315 (1977)
31. Shen, Z.: $L^p$ estimates for Schrödinger with certain potentials. Ann. Inst. Fourier **45**, 513–546 (1995)
32. Shen, Z.: Estimates in $L^p$ for magnetic Schrödinger operators. Indiana Univ. Math. J. **45**, 817–841 (1996)
33. Shen, Z.: Eigenvalue asymptotics and exponential decay of eigenfunctions for Schrödinger operators with magnetic fields. TAMS **348**, 4465–4488 (1996)
34. Shen, Z.: Personal Communication (2017)
35. Simon, B.: Some quantum operators with discrete spectrum but classically continuous spectrum. Ann. Phys. **146**, 209–220 (1983)
36. Simon, B.: Tosio Kato's Work on Non-relativistic Quantum Mechanics. ArXiv (2017)
37. Zhong, J.: The Sobolev estimates for some Schrödinger type operators. Math. Sci. Res. Hot-Line **3**(8), 1–48 (1999); (and Harmonic Analysis of some Schrödinger type Operators, PhD thesis, Princeton University, 1993)

# Matrix Limit Theorems of Kato Type Related to Positive Linear Maps and Operator Means

**Fumio Hiai**

**Abstract** We obtain limit theorems for $\Phi(A^p)^{1/p}$ and $(A^p\sigma B)^{1/p}$ as $p \to \infty$ for positive matrices $A, B$, where $\Phi$ is a positive linear map between matrix algebras (in particular, $\Phi(A) = KAK^*$) and $\sigma$ is an operator mean (in particular, the weighted geometric mean), which are considered as certain reciprocal Lie–Trotter formulas and also a generalization of Kato's limit to the supremum $A \vee B$ with respect to the spectral order.

**2010 Mathematics Subject Classification** Primary: 15A45 · 15A42 · 47A64

## 1 Introduction

For any matrices $X$ and $Y$, the well-known Lie–Trotter formula is the convergence

$$\lim_{n\to\infty} (e^{X/n}e^{Y/n})^n = e^{X+Y}.$$

The symmetric form with a continuous parameter is also well-known for positive semidefinite matrices $A, B \ge 0$ as

$$\lim_{p\searrow 0} (A^{p/2}B^pA^{p/2})^{1/p} = P_0 \exp(\log A \dot{+} \log B), \tag{1.1}$$

where $P_0$ is the orthogonal projection onto the intersection of the supports of $A, B$ and $\log A \dot{+} \log B$ is defined as $P_0(\log A)P_0 + P_0(\log B)P_0$. When $\sigma$ is an operator mean [13] corresponding to an operator monotone function $f$ on $(0, \infty)$ such that $\alpha := f'(1)$ is in $(0, 1)$, the operator mean version of the Lie–Trotter formula is also known to hold as

F. Hiai (✉)
Tohoku University (Emeritus), Hakusan 3-8-16-303, Abiko 270-1154, Japan
e-mail: hiai.fumio@gmail.com

© Springer Nature Switzerland AG 2019

T. M. Rassias and V. A. Zagrebnov (eds.), *Analysis and Operator Theory*,
Springer Optimization and Its Applications 146,
https://doi.org/10.1007/978-3-030-12661-2_9

$$\lim_{p\searrow 0}(A^p\sigma B^p)^{1/p} = P_0\exp((1-\alpha)\log A\dot{+}\alpha\log B) \tag{1.2}$$

for matrices $A, B \geq 0$. In particular, when $\sigma$ is the geometric mean $A\#B$ (introduced first in [15] and further discussed in [13]) corresponding to the operator monotone function $f(x) = x^{1/2}$, (1.2) yields

$$\lim_{p\searrow 0}(A^p\#B^p)^{2/p} = P_0\exp(\log A\dot{+}\log B), \tag{1.3}$$

which has the same right-hand side as (1.1). Due to the Araki–Lieb–Thirring inequality and the Ando–Hiai log-majorization [1, 4], it is worthwhile to note that $(A^{p/2}B^pA^{p/2})^{1/p}$ and $(A^p\#B^p)^{2/p}$ both tend to $P_0\exp(\log A\dot{+}\log B)$ as $p\searrow 0$, with the former decreasing and the latter increasing in the log-majorization order (see [1] for details on log-majorization for matrices).

In the previous paper [6], under the name "reciprocal Lie–Trotter formula", we considered the question complementary to (1.1) and (1.3), that is, about what happens to the limits of $(A^{p/2}B^pA^{p/2})^{1/p}$ and $(A^p\#B^p)^{2/p}$ as $p$ tends to $\infty$ instead of 0. If $A$ and $B$ are commuting, then $(A^{p/2}B^pA^{p/2})^{1/p} = (A^p\#B^p)^{2/p} = AB$, independently of $p > 0$. However, if $A$ and $B$ are not commuting, then the question is rather complicated. Indeed, although we can prove the existence of the limit $\lim_{p\to\infty}(A^{p/2}B^pA^{p/2})^{1/p}$, the description of the limit has a rather complicated combinatorial nature. Moreover, it is unknown so far whether the limit of $(A^p\#B^p)^{2/p}$ as $p\to\infty$ exists or not. In the present paper, we consider a similar (but seemingly a bit simpler) question about what happens to the limits of $(BA^pB)^{1/p}$ and $(A^p\#B)^{1/p}$ as $p$ tends to $\infty$, the case where $B$ is fixed without the $p$-power, in certain more general settings.

The rest of the paper is organized as follows. In Section 2, we first prove the existence of the limit of $(KA^pK^*)^{1/p}$ as $p\to\infty$ and give the description of the limit in terms of the diagonalization (eigenvalues and eigenvectors) data of $A$ and the images of the eigenvectors by $K$. We then extend the result to the limit of $\Phi(A^p)^{1/p}$ as $p\to\infty$ for a positive linear map $\Phi$ between matrix algebras. For instance, this limit is applied to the map $\Phi(A\oplus B) := (A+B)/2$ to reformulate Kato's limit theorem $((A^p+B^p)/2)^{1/p}\to A\vee B$ in [12]. Another application is given to find the limit formula as $\alpha\searrow 0$ of the sandwiched $\alpha$-Rényi divergence [14, 18], a new relative entropy relevant to quantum information theory. In Section 3, we discuss the limit behavior of $(A^p\sigma B)^{1/p}$ as $p\to\infty$ for operator means $\sigma$. To do this, we may assume without loss of generality that $B$ is an orthogonal projection $E$. Under a certain condition on $\sigma$, we prove that $(A^p\sigma E)^{1/p}$ is decreasing as $1\leq p\to\infty$, so that the limit as $p\to\infty$ exists. Furthermore, when $\sigma$ is the weighted geometric mean, we obtain an explicit description of the limit in terms of $E$ and the spectral projections of $A$.

It is worth noting that a limit formula in the same vein as those in [12] and this paper was formerly given in [2, 3] for the spectral shorting operation.

## 2 $\lim_{p\to\infty} \Phi(A^p)^{1/p}$ for Positive Linear Maps $\Phi$

For each $n \in \mathbb{N}$, we write $\mathbb{M}_n$ for the $n \times n$ complex matrix algebra and $\mathbb{M}_n^+$ for the set of positive semidefinite matrices in $\mathbb{M}_n$. When $A \in \mathbb{M}_n$ is positive definite, we write $A > 0$. We denote by Tr the usual trace functional on $\mathbb{M}_n$. For $A \in \mathbb{M}_n^+$, $\lambda_1(A) \geq \cdots \geq \lambda_n(A)$ are the eigenvalues of $A$ in decreasing order with multiplicities, and ran $A$ is the range of $A$.

Let $A \in \mathbb{M}_n^+$ be given, whose diagonalization is

$$A = V\mathrm{diag}(a_1, \ldots, a_n)V^* = \sum_{i=1}^{n} a_i |v_i\rangle\langle v_i| \tag{2.1}$$

with the eigenvalues $a_1 \geq \cdots \geq a_n$ and a unitary matrix $V = [v_1 \ \cdots \ v_n]$ so that $Av_i = a_i v_i$ for $1 \leq i \leq d$. Let $K \in \mathbb{M}_n$ and assume that $K \neq 0$ (our problem below is trivial when $K = 0$). Consider the sequence of vectors $Kv_1, \ldots, Kv_n$ in $\mathbb{C}^n$. Let $1 \leq l_1 < l_2 < \cdots < l_m$ be chosen so that if $l_{k-1} < i < l_k$ for $1 \leq k \leq m$, then $Kv_i$ is in $\mathrm{span}\{Kv_{l_1}, \ldots, Kv_{l_{k-1}}\}$ (this means, in particular, $Kv_i = 0$ if $i < l_1$). Then $\{Kv_{l_1}, \ldots, Kv_{l_m}\}$ is a linearly independent subset of $\{Kv_1, \ldots, Kv_n\}$, so we perform the Gram–Schmidt orthogonalization to obtain an orthonormal vectors $u_1, \ldots, u_m$ from $Kv_{l_1}, \ldots, Kv_{l_m}$. In particular, $u_1 = Kv_{l_1}/\|Kv_{l_1}\|$. The next theorem is our first limit theorem. This implicitly says that the right-hand side of (2.2) is independent of the expression of (2.1) (note that $v_i$'s are not unique for degenerate eigenvalues $a_i$).

**Theorem 2.1** *We have*

$$\lim_{p\to\infty} (KA^pK^*)^{1/p} = \sum_{k=1}^{m} a_{l_k} |u_k\rangle\langle u_k|, \tag{2.2}$$

*and in particular,*

$$\lim_{p\to\infty} \lambda_k((KA^pK^*)^{1/p}) = \begin{cases} a_{l_k}, & 1 \leq k \leq m, \\ 0, & m < k \leq n. \end{cases} \tag{2.3}$$

*Proof* Write $Z_p := (KA^pK^*)^{1/p}$ and $\lambda_i(p) := \lambda_i(Z_p)$ for $p > 0$ and $1 \leq i \leq n$. First we prove (2.3). Note that

$$\begin{aligned} Z_p^p = KA^pK^* &= KV\mathrm{diag}(a_1^p, \ldots, a_n^p)V^*K^* \\ &= \left[a_1^p Kv_1 \ a_2^p Kv_2 \ \cdots \ a_n^p Kv_n\right]\left[Kv_1 \ Kv_2 \ \cdots \ Kv_n\right]^*. \end{aligned} \tag{2.4}$$

Since

$$\begin{aligned}\lambda_1(p)^p \le \operatorname{Tr} Z_p^p &= \operatorname{Tr}\big[Kv_1 \ \cdots \ Kv_n\big]^*\big[a_1^p Kv_1 \ \cdots \ a_n^p Kv_n\big] \\ &= \sum_{i=1}^n a_i^p \langle Kv_i, Kv_i\rangle \le a_{l_1}^p \sum_{i=1}^n \|Kv_i\|^2,\end{aligned}$$

we have

$$\limsup_{p\to\infty} \lambda_1(p) \le a_{l_1}.$$

Moreover, since

$$n\lambda_1(p)^p \ge \operatorname{Tr} Z_p^p = \sum_{i=1}^n a_i^p \langle Kv_i, Kv_i\rangle \ge a_{l_1}^p \|Kv_{l_1}\|^2,$$

we have

$$\liminf_{p\to\infty} \lambda_1(p) \ge a_{l_1}.$$

Therefore, (2.3) holds for $k = 1$.

To prove (2.3) for $k > 1$, we consider the *antisymmetric tensor powers* $A^{\wedge k}$ and $K^{\wedge k}$ for each $k = 1, \dots, n$. Note that

$$Z_p^{\wedge k} = \big((K^{\wedge k})(A^{\wedge k})^p (K^{\wedge k})^*\big)^{1/p} \tag{2.5}$$

and

$$A^{\wedge k} = \sum_{1\le i_1<\cdots<i_k\le d} a_{i_1}\cdots a_{i_k} |v_{i_1}\wedge\cdots\wedge v_{i_k}\rangle\langle v_{i_1}\wedge\cdots\wedge v_{i_k}|.$$

The above case applied to $A^{\wedge k}$ and $K^{\wedge k}$ yields that

$$\begin{aligned}&\lim_{p\to\infty} \lambda_1(p)\lambda_2(p)\cdots\lambda_k(p) = \lim_{p\to\infty} \lambda_1\big(Z_p^{\wedge k}\big) \\ &\quad = \max\big\{a_{i_1}\cdots a_{i_k} : 1 \le i_1 < \cdots < i_k \le d,\ K^{\wedge k}(v_{i_1}\wedge\cdots\wedge v_{i_k}) \ne 0\big\}. \qquad (2.6)\end{aligned}$$

Since $K^{\wedge k}(v_{i_1}\wedge\cdots\wedge v_{i_k}) = Kv_{i_1}\wedge\cdots\wedge Kv_{i_k}$ is nonzero if and only if $\{Kv_{i_1}, \dots, Kv_{i_k}\}$ is linearly independent, it is easy to see that (2.6) is equal to $a_{l_1}\cdots a_{l_k}$ if $k \le m$ and equal to 0 if $k > m$. Therefore,

$$\lim_{p\to\infty} \lambda_1(p)\lambda_2(p)\cdots\lambda_k(p) = \begin{cases} a_{l_1}\cdots a_{l_k}, & 1 \le k \le m, \\ 0, & m < k \le n, \end{cases}$$

which implies (2.3).

Now, for $p > 0$ choose an orthonormal basis $\{u_1(p), \dots, u_n(p)\}$ of $\mathbb{C}^n$ for which $Z_p u_i(p) = \lambda_i(p)u_i(p)$ for $1 \le i \le n$. To prove (2.2), write $\tilde{a}_k := a_{l_k}$ for $1 \le k \le m$. If $\tilde{a}_1 = 0$ then it is obvious that $\lim_{p\to\infty} Z_p = 0$. So assume that $\tilde{a}_1 > 0$ and furthermore $\tilde{a}_1 > \tilde{a}_2$, i.e., $\lim_{p\to\infty} \lambda_1(p) > \lim_{p\to\infty} \lambda_2(p)$ at the moment. From (2.4) we have

$$
\begin{aligned}
Z_p^p &= \sum_{i=1}^{n} a_i^p |Kv_i\rangle\langle Kv_i| \\
&= \sum_{i=l_1}^{l_2-1} a_i^p |Kv_i\rangle\langle Kv_i| + \sum_{i=l_2}^{n} a_i^p |Kv_i\rangle\langle Kv_i| \\
&= \sum_{i=l_1}^{l_2-1} a_i^p \|Kv_i\|^2 |u_1\rangle\langle u_1| + \sum_{i=l_2}^{n} a_i^p |Kv_i\rangle\langle Kv_i|
\end{aligned}
$$

so that

$$
\left(\frac{Z_p}{\tilde{a}_1}\right)^p = \sum_{i=l_1}^{l_2-1} \left(\frac{a_i}{\tilde{a}_1}\right)^p \|Kv_i\|^2 |u_1\rangle\langle u_1| + \sum_{i=l_2}^{n} \left(\frac{a_i}{\tilde{a}_1}\right)^p |Kv_i\rangle\langle Kv_i| \longrightarrow \alpha |u_1\rangle\langle u_1|
$$

as $p \to \infty$ for some $\alpha > 0$, since $a_i/\tilde{a}_1 \le \tilde{a}_2/\tilde{a}_1 < 1$ for $i \ge l_2$. Hence, for any $p > 0$ sufficiently large, the largest eigenvalue of $(Z_p/\tilde{a}_1)^p$ is simple and the corresponding eigen projection converges to $|u_1\rangle\langle u_1|$ as $p \to \infty$. Since the eigen projection $E_1(p)$ of $Z_p$ corresponding to the largest eigenvalue $\lambda_1(p)$ (simple for any large $p > 0$) is the same as that of $(Z_p/\tilde{a}_1)^p$, we have

$$
E_1(p) = |u_1(p)\rangle\langle u_1(p)| \longrightarrow |u_1\rangle\langle u_1| \quad \text{as } p \to \infty.
$$

In the general situation, we assume that $\tilde{a}_1 \ge \cdots \ge \tilde{a}_k > \tilde{a}_{k+1}$ with $1 \le k \le m$, where $\tilde{a}_{k+1} = 0$ if $k = m$. From (2.3), note that

$$
\lim_{p\to\infty} \lambda_1(Z_p^{\wedge k}) = \tilde{a}_1 \cdots \tilde{a}_{k-1}\tilde{a}_k > \tilde{a}_1 \cdots \tilde{a}_{k-1}\tilde{a}_{k+1} = \lim_{p\to\infty} \lambda_2(Z_p^{\wedge k}).
$$

Hence, for any sufficiently large $p > 0$, the largest eigenvalue $\lambda_1(Z_p^{\wedge k}) = \lambda_1(p) \cdots \lambda_k(p)$ of $Z_p^{\wedge k}$ is simple, and from the above case applied to (2.5) it follows that

$$
|u_1(p) \wedge \cdots \wedge u_k(p)\rangle\langle u_1(p) \wedge \cdots \wedge u_k(p)| \longrightarrow |u_1 \wedge \cdots \wedge u_k\rangle\langle u_1 \wedge \cdots \wedge u_k| \quad \text{as } p \to \infty,
$$

since the vector in the present situation corresponding to $u_1 = Kv_{l_1}/\|Kv_{l_1}\|$ is

$$
\frac{K^{\wedge k}(v_{l_1} \wedge \cdots \wedge v_{l_k})}{\|K^{\wedge k}(v_{l_1} \wedge \cdots \wedge v_{l_k})\|} = \frac{Kv_{l_1} \wedge \cdots \wedge Kv_{l_k}}{\|Kv_{l_1} \wedge \cdots \wedge Kv_{l_k}\|} = u_1 \wedge \cdots \wedge u_k.
$$

(The last identity follows from the fact that, for linearly independent $w_1, \ldots, w_k$, $w_1 \wedge \cdots \wedge w_k/\|w_1 \wedge \cdots \wedge w_k\| = w_1' \wedge \cdots \wedge w_k'$ if $w_1', \ldots, w_k'$ are the Gram–Schmidt orthogonalization of $w_1, \ldots, w_k$.) By [6, Lemma 2.4], we see that the orthogonal projection $E_k(p)$ onto $\mathrm{span}\{u_1(p),\dots,u_k(p)\}$ converges to the orthogonal projection $E_k$ of $\mathrm{span}\{u_1,\dots,u_k\}$.

Finally, let $0=k_0<k_1<\cdots<k_{s-1}<k_s=m$ be such that

$$\tilde a_1=\cdots=\tilde a_{k_1}>\tilde a_{k_1+1}=\cdots=\tilde a_{k_2}>\cdots>\tilde a_{k_{s-1}+1}=\cdots=\tilde a_{k_s}.$$

The above argument says that, for every $r=1,\dots,s-1$, the orthogonal projection $E_{k_r}(p)$ onto $\mathrm{span}\{u_1(p),\dots,u_{k_r}(p)\}$ converges to the orthogonal projection $E_{k_r}$ onto $\mathrm{span}\{u_1,\dots,u_{k_r}\}$. When $\tilde a_{k_s}>0$, this holds for $r=s$ as well. Therefore, when $\tilde a_{k_s}>0$, we have

$$\begin{aligned}
Z_p&=\sum_{i=1}^n\lambda_i(p)|u_i(p)\rangle\langle u_i(p)|\\
&=\sum_{r=1}^s\sum_{i=k_{r-1}+1}^{k_r}\lambda_i(p)|u_i(p)\rangle\langle u_i(p)|+\sum_{i=k_s+1}^n\lambda_i(p)|u_i(p)\rangle\langle u_i(p)|\\
&=\sum_{r=1}^s\sum_{i=k_{r-1}+1}^{k_r}(\lambda_i(p)-\tilde a_i)|u_i(p)\rangle\langle u_i(p)|+\sum_{r=1}^s\tilde a_{k_r}(E_{k_r}(p)-E_{k_{r-1}}(p))\\
&\qquad+\sum_{i=k_s+1}^n\lambda_i(p)|u_i(p)\rangle\langle u_i(p)|\\
&\longrightarrow\sum_{r=1}^s\tilde a_{k_r}(E_{k_r}-E_{k_{r-1}})=\sum_{i=1}^m\tilde a_i|u_i\rangle\langle u_i|,\quad\text{where } E_0(p)=E_0=0.
\end{aligned}$$

When $\tilde a_{k_s}=0$, we may modify the above estimate as

$$\begin{aligned}
Z_p&=\sum_{r=1}^{s-1}\sum_{i=k_{r-1}+1}^{k_r}\lambda_i(p)|u_i(p)\rangle\langle u_i(p)|+\sum_{i=k_{s-1}+1}^n\lambda_i(p)|u_i(p)\rangle\langle u_i(p)|\\
&\longrightarrow\sum_{r=1}^{s-1}\tilde a_{k_r}(E_{k_r}-E_{k_{r-1}})=\sum_{i=1}^{k_{s-1}}\tilde a_i|u_i\rangle\langle u_i|=\sum_{i=1}^m\tilde a_i|u_i\rangle\langle u_i|.
\end{aligned}$$

□

The following corollary of Theorem 2.1 is an improvement of [7, Theorem 1.2].

**Corollary 2.2** *Let $A\in\mathbb{M}_n$ be positive definite. Then $\lim_{p\to\infty}\lambda_i((KA^pK^*)^{1/p})=a_i$ for all $i=1,\dots,n$ if and only if $\{Kv_1,\dots,Kv_n\}$ is linearly independent.*

**Remark 2.3** Note that Theorem 2.1 can easily extend to the case where $K$ is a rectangle $n'\times n$ matrix. In fact, when $n'<n$ we may apply Theorem 2.1 to $n\times n$ matrices $\begin{bmatrix}K\\O\end{bmatrix}$ and $A$, and when $n'>n$ we may apply to $n'\times n'$ matrices $\begin{bmatrix}K & O\end{bmatrix}$ and $A\oplus O_{n'-n}$.

A linear map $\Phi : \mathbb{M}_n \to \mathbb{M}_{n'}$ is said to be positive if $\Phi(A) \in \mathbb{M}_{n'}^+$ for all $A \in \mathbb{M}_n^+$, which is further said to be strictly positive if $\Phi(I_n) > 0$, that is, $\Phi(A) > 0$ for all $A \in \mathbb{M}_n$, $A > 0$. The following is an extended and refined version of Theorem 2.1.

**Theorem 2.4** *Let $\Phi : \mathbb{M}_n \to \mathbb{M}_{n'}$ be a positive linear map. Let $A \in \mathbb{M}_n^+$ be given as $A = \sum_{i=1}^n a_i |v_i\rangle\langle v_i|$ with $a_1 \ge \cdots \ge a_n$ and an orthonormal basis $\{v_1, \dots, v_n\}$ of $\mathbb{C}^n$. Then $\lim_{p\to\infty} \Phi(A^p)^{1/p}$ exists and*

$$\lim_{p\to\infty} \Phi(A^p)^{1/p} = \sum_{i=1}^n a_i P_{\mathcal{M}_i},$$

*where*

$$\mathcal{M}_1 := \operatorname{ran} \Phi(|v_1\rangle\langle v_1|),$$
$$\mathcal{M}_i := \bigvee_{j=1}^{i} \operatorname{ran} \Phi(|v_j\rangle\langle v_j|) \ominus \bigvee_{j=1}^{i-1} \operatorname{ran} \Phi(|v_j\rangle\langle v_j|), \qquad 2 \le i \le n,$$

*and $P_{\mathcal{M}_i}$ is the orthogonal projection onto $\mathcal{M}_i$ for $1 \le i \le n$.*

*Proof* Let $C^*(I, A)$ be the commutative $C^*$-subalgebra of $\mathbb{M}_n$ generated by $I, A$. We can consider the composition of the conditional expectation from $\mathbb{M}_n$ onto $C^*(I, A)$ with respect to Tr and $\Phi|_{C^*(I,A)} : C^*(I, A) \to \mathbb{M}_{n'}$ instead of $\Phi$, so we may assume that $\Phi$ is completely positive. By the Stinespring representation there are a $\nu \in \mathbb{N}$, a $*$-homomorphism $\pi : \mathbb{M}_n \to \mathbb{M}_{n\nu}$ and a linear map $K : \mathbb{C}^{n\nu} \to \mathbb{C}^{n'}$ such that $\Phi(X) = K\pi(X)K^*$ for all $X \in \mathbb{M}_n$. Moreover, since $\pi : \mathbb{M}_n \to \mathbb{M}_{n\nu}$ is represented, under a suitable change of an orthonormal basis of $\mathbb{C}^{n\nu}$, as $\pi(X) = I_\nu \otimes X$ for all $X \in \mathbb{M}_n$ under identification $\mathbb{M}_{n\nu} = \mathbb{M}_\nu \otimes \mathbb{M}_n$, we can assume that $\Phi$ is given (with a change of $K$) as

$$\Phi(X) = K(I_\nu \otimes X)K^*, \qquad X \in \mathbb{M}_n.$$

We then write

$$\begin{aligned} I_\nu \otimes A &= (I_\nu \otimes V)\operatorname{diag}\big(\underbrace{a_1, \dots, a_1}_{\nu}, \underbrace{a_2, \dots, a_2}_{\nu}, \dots, \underbrace{a_n, \dots, a_n}_{\nu}\big)(I_\nu \otimes V)^* \\ &= \sum_{i=1}^n a_i (|e_1 \otimes v_i\rangle\langle e_1 \otimes v_i| + \cdots + |e_\nu \otimes v_i\rangle\langle e_\nu \otimes v_i|). \end{aligned}$$

Now, we consider the following sequence of $n\nu$ vectors in $\mathbb{C}^{n'}$:

$$K(e_1 \otimes v_1), \dots, K(e_\nu \otimes v_1), K(e_1 \otimes v_2), \dots, K(e_\nu \otimes v_2), \dots, K(e_1 \otimes v_n), \dots, K(e_\nu \otimes v_n),$$

and if $K(e_j \otimes v_i)$ is a linear combination of the vectors in the sequence preceding it, then we remove it from the sequence. We write the resulting linearly independent subsequence as

$$K(e_j \otimes v_{l_1})\ (j \in J_1),\ \ K(e_j \otimes v_{l_2})\ (j \in J_2),\ \ \dots,\ \ K(e_j \otimes v_{l_m})\ (j \in J_m),$$

where $1 \le l_1 < l_2 < \cdots < l_m \le n$ and $J_1, \dots, J_m \subset \{1, \dots, \nu\}$. Furthermore, by performing the Gram–Schmidt orthogonalization to this subsequence, we end up making an orthonormal sequence of vectors in $\mathbb{C}^{n'}$ as follows:

$$u_j^{(l_1)}\ (j \in J_1),\ \ u_j^{(l_2)}\ (j \in J_2),\ \ \dots,\ \ u_j^{(l_m)}\ (j \in J_m).$$

Since

$$\Phi(A^p)^{1/p} = (K(I_\nu \otimes A)^p K^*)^{1/p},$$

Theorem 2.1 and Remark 2.3 imply that $\lim_{p\to\infty} \Phi(A^p)^{1/p}$ exists and

$$\lim_{p\to\infty} \Phi(A^p)^{1/p} = \sum_{k=1}^m a_{l_k} \Biggl(\sum_{j\in J_k} |u_j^{(l_k)}\rangle\langle u_j^{(l_k)}|\Biggr),$$

where $\sum_{j\in J_k} |u_j^{(l_k)}\rangle\langle u_j^{(l_k)}| = 0$ if $J_k = \emptyset$.

The next step of the proof is to find what is $\sum_{j\in J_k} |u_j^{(l_k)}\rangle\langle u_j^{(l_k)}|$ for $1 \le k \le m$. For this we first note that

$$\begin{aligned}\sum_{j=1}^{\nu} |K(e_j \otimes v_i)\rangle\langle K(e_j \otimes v_i)| &= K\Biggl(\sum_{j=1}^{\nu} |e_j \otimes v_i\rangle\langle e_j \otimes v_i|\Biggr)K^* \\ &= K(I_\nu \otimes |v_i\rangle\langle v_i|)K^* = \Phi(|v_i\rangle\langle v_i|).\end{aligned}$$

From Lemma 2.5 below this implies that

$$\mathcal{R}_i := \mathrm{ran}\,\Phi(|v_i\rangle\langle v_i|) = \mathrm{span}\{K(e_j \otimes v_i) : 1 \le j \le \nu\}.$$

Through the procedure of the Gram–Schmidt diagonalization we see that

$$\begin{aligned}
&\mathcal{R}_i = 0, \qquad 1 \le i < l_1, \\
&\mathcal{R}_{l_1} = \mathrm{span}\bigl\{u_j^{(l_1)} : j \in J_1\bigr\}, \\
&\mathcal{R}_i \subset \mathcal{R}_{l_1}, \qquad l_1 < i < l_2, \\
&(\mathcal{R}_{l_1} \vee \mathcal{R}_{l_2}) \ominus \mathcal{R}_{l_1} = \mathrm{span}\bigl\{u_j^{(l_2)} : j \in J_2\bigr\}, \\
&\mathcal{R}_i \subset \mathcal{R}_{l_1} \vee \mathcal{R}_{l_2}, \qquad l_2 < i < l_3, \\
&(\mathcal{R}_{l_1} \vee \mathcal{R}_{l_2} \vee \mathcal{R}_{l_3}) \ominus (\mathcal{R}_{l_1} \vee \mathcal{R}_{l_2}) = \mathrm{span}\bigl\{u_j^{(l_3)} : j \in J_3\bigr\}, \\
&\quad \vdots
\end{aligned}$$

$$\mathcal{R}_i \subset \mathcal{R}_{l_1} \vee \cdots \vee \mathcal{R}_{l_{m-1}}, \qquad l_{m-1} < i < l_m,$$
$$(\mathcal{R}_{l_1} \vee \cdots \vee \mathcal{R}_{l_m}) \ominus (\mathcal{R}_{l_1} \vee \cdots \vee \mathcal{R}_{l_{m-1}}) = \mathrm{span}\{u_j^{(l_m)} : j \in J_m\},$$
$$\mathcal{R}_i \subset \mathcal{R}_{l_1} \vee \cdots \vee \mathcal{R}_{l_m}, \qquad l_m < i \le n.$$

Now, let $P_{\mathcal{M}_i}$ be the orthogonal projections, respectively, onto the subspaces

$$\mathcal{M}_1 := \mathcal{R}_1, \quad \mathcal{M}_i := (\mathcal{R}_1 \vee \cdots \vee \mathcal{R}_i) \ominus (\mathcal{R}_1 \vee \cdots \vee \mathcal{R}_{i-1}), \quad 2 \le i \le n,$$

so that $P_{\mathcal{M}_i} = 0$ if $i \notin \{l_1, \ldots, l_m\}$ and $P_{\mathcal{M}_{l_k}}$ is the orthogonal projection onto $\mathrm{span}\{u_j^{(l_k)} : j \in J_k\}$ for $1 \le k \le m$. Therefore, we have

$$\lim_{p\to\infty} \Phi(A^p)^{1/p} = \sum_{k=1}^{m} a_{l_k} \Big( \sum_{j \in J_k} |u_j^{(l_k)}\rangle\langle u_j^{(l_k)}| \Big) = \sum_{k=1}^{m} a_{l_k} P_{\mathcal{M}_{l_k}} = \sum_{i=1}^{n} a_i P_{\mathcal{M}_i}.$$

□

**Lemma 2.5** *For any finite set* $\{w_1, \ldots, w_k\}$ *in* $\mathbb{C}^{n'}$, $\mathrm{span}\{w_1, \ldots, w_k\}$ *is equal to the range of* $|w_1\rangle\langle w_1| + \cdots + |w_k\rangle\langle w_k|$. *More generally, for every* $B_1, \ldots, B_k \in \mathbb{M}_{n'}^+$, $\bigvee_{j=1}^k \mathrm{ran}\, B_j$ *is equal to the range of* $B_1 + \cdots + B_k$.

*Proof* Let $Q := |w_1\rangle\langle w_1| + \cdots + |w_k\rangle\langle w_k|$. Since

$$Qx = \langle w_1, x\rangle w_1 + \cdots + \langle w_k, x\rangle w_k \in \mathrm{span}\{w_1, \ldots, w_k\}$$

for all $x \in \mathbb{C}^{n'}$, we have $\mathrm{ran}\, Q \subset \mathrm{span}\{w_1, \ldots, w_k\}$. Since $|w_i\rangle\langle w_i| \le Q$, we have

$$w_i \in \mathrm{ran}\, |w_i\rangle\langle w_i| \subset \mathrm{ran}\, Q, \quad 1 \le i \le k.$$

Hence, we have $\mathrm{span}\{w_1, \ldots, w_k\} \subset \mathrm{ran}\, Q$. The proof of the latter assertion is similar. □

Thanks to the lemma we can restate Theorem 2.4 as follows:

**Theorem 2.6** *Let* $\Phi : \mathbb{M}_n \to \mathbb{M}_{n'}$ *be a positive linear map. Let* $A \in \mathbb{M}_n^+$ *be given with the spectral decomposition* $A = \sum_{k=1}^m a_k P_k$, *where* $a_1 > a_2 > \cdots > a_m > 0$. *Define*

$$\mathcal{M}_1 := \mathrm{ran}\, \Phi(P_1),$$
$$\mathcal{M}_k := \mathrm{ran}\, \Phi(P_1 + \cdots + P_k) \ominus \mathrm{ran}\, \Phi(P_1 + \cdots + P_{k-1}), \quad 2 \le k \le m.$$

*Then*

$$\lim_{p\to\infty} \Phi(A^p)^{1/p} = \sum_{k=1}^{m} a_k P_{\mathcal{M}_k}.$$

**Example 2.7** Consider a linear map $\Phi : \mathbb{M}_{2n} \to \mathbb{M}_n$ given by

$$\Phi\left(\begin{bmatrix} X_{11} & X_{12} \\ X_{21} & X_{22} \end{bmatrix}\right) := \frac{X_{11} + X_{22}}{2}, \qquad X_{ij} \in \mathbb{M}_n.$$

Clearly, $\Phi$ is completely positive. For any $A, B \in \mathbb{M}_n^+$ and $p > 0$ we have

$$\Phi\left(\begin{bmatrix} A & 0 \\ 0 & B \end{bmatrix}^p\right)^{1/p} = \left(\frac{A^p + B^p}{2}\right)^{1/p}.$$

Thus, it is well-known [12] that

$$\lim_{p\to\infty} \Phi\left(\begin{bmatrix} A & 0 \\ 0 & B \end{bmatrix}^p\right)^{1/p} = \lim_{p\to\infty} \left(\frac{A^p + B^p}{2}\right)^{1/p} = A \vee B, \tag{2.7}$$

where $A \vee B$ is the supremum of $A, B$ in the spectral order. Here, let us show (2.7) from Theorem 2.6. The spectral decompositions of $A, B$ are given as

$$A = \sum_{i=1}^{m} a_i P_i, \qquad B = \sum_{j=1}^{m'} b_j Q_j,$$

where $a_1 > \cdots > a_m \ge 0$, $b_1 > \cdots > b_{m'} \ge 0$ and $\sum_{i=1}^m P_i = \sum_{j=1}^{m'} Q_j = I$. Then

$$A \oplus B = \sum_{k=1}^{l} c_k R_k,$$

where $\{c_k\}_{k=1}^l = \{a_i\}_{i=1}^m \cup \{b_j\}_{j=1}^{m'}$ with $c_1 > \cdots > c_l$ and

$$R_k = \begin{cases} P_i \oplus Q_j & \text{if } a_i = b_j = c_k, \\ P_i \oplus 0 & \text{if } a_i = c_k \text{ and } b_j \ne c_k \text{ for all } j, \\ 0 \oplus Q_j & \text{if } b_j = c_k \text{ and } a_i \ne c_k \text{ for all } i. \end{cases}$$

Note that

$$\Phi(R_1 + \cdots + R_k) = \frac{1}{2}\left(\sum_{i:a_i \ge c_k} P_i + \sum_{j:b_j \ge c_k} Q_j\right)$$

so that by Lemma 2.5 the support projection $F_k$ (i.e., the orthogonal projection onto the range) of $\Phi(R_1 + \cdots + R_k)$ is

$$F_k = \left(\sum_{i:a_i \ge c_k} P_i\right) \vee \left(\sum_{j:b_j \ge c_k} Q_j\right).$$

Theorem 2.6 implies that

$$\lim_{p\to\infty} \Phi\left(\begin{bmatrix} A & 0 \\ 0 & B \end{bmatrix}^p\right)^{1/p} = C := \sum_{k=1}^{l} c_k(F_k - F_{k-1}).$$

For every $x \in \mathbb{R}$, we denote by $E_{[x,\infty)}(A)$ the spectral projection of $A$ corresponding to the interval $[x, \infty)$, i.e.,

$$E_{[x,\infty)}(A) := \sum_{i:a_i \geq x} P_i,$$

and similarly for $E_{[x,\infty)}(B)$ and $E_{[x,\infty)}(C)$. If $c_k \geq x > c_{k+1}$ for some $1 \leq k < l$, then we have

$$E_{[x,\infty)}(C) = F_k = E_{[x,\infty)}(A) \vee E_{[x,\infty)}(B).$$

This holds also when $x > c_1$ and $x \leq c_l$. Indeed, when $x > c_1$, $E_{[x,\infty)}(C) = 0 = E_{[x,\infty)}(A) \vee E_{[x,\infty)}(B)$. When $x \leq c_l$, $E_{[x,\infty)}(C) = I = E_{[x,\infty)}(A) \vee E_{[x,\infty)}(B)$. This description of $C$ is the same as $A \vee B$ in [12], so we have $C = A \vee B$.

**Example 2.8** The example here is relevant to quantum information. For density matrices $\rho, \sigma \in \mathbb{M}_n$ (i.e., $\rho, \sigma \in \mathbb{M}_n^+$ with $\mathrm{Tr}\,\rho = \mathrm{Tr}\,\sigma = 1$) and for a parameter $\alpha \in (0, \infty) \setminus \{1\}$, the traditional *Rényi relative entropy* is

$$D_\alpha(\rho\|\sigma) := \begin{cases} \frac{1}{\alpha-1} \log\left[\mathrm{Tr}\,\rho^\alpha \sigma^{1-\alpha}\right] & \text{if } \rho^0 \leq \sigma^0 \text{ or } 0 < \alpha < 1, \\ +\infty & \text{otherwise,} \end{cases}$$

where $\rho^0$ denotes the support projection of $\rho$. On the other hand, the new concept was recently introduced and called the *sandwiched Rényi relative entropy* [14, 18] is

$$\widetilde{D}_\alpha(\rho\|\sigma) := \begin{cases} \frac{1}{\alpha-1} \log\left[\mathrm{Tr}\,\left(\sigma^{\frac{1-\alpha}{2\alpha}} \rho \sigma^{\frac{1-\alpha}{2\alpha}}\right)^\alpha\right] & \text{if } \rho^0 \leq \sigma^0 \text{ or } 0 < \alpha < 1, \\ +\infty & \text{otherwise.} \end{cases}$$

By taking the limit, we also consider

$$D_0(\rho\|\sigma) := \lim_{\alpha\searrow 0} D_\alpha(\rho\|\sigma) = -\log \mathrm{Tr}\,(\rho^0\sigma),$$

$$\widetilde{D}_0(\rho\|\sigma) := \lim_{\alpha\searrow 0} \widetilde{D}_\alpha(\rho\|\sigma) = -\log\left[\lim_{\alpha\searrow 0} \mathrm{Tr}\,\left(\sigma^{\frac{1-\alpha}{2\alpha}} \rho \sigma^{\frac{1-\alpha}{2\alpha}}\right)^\alpha\right].$$

(We remark that the notations $D_\alpha$ and $\widetilde{D}_\alpha$ are interchanged from those in [8].) Here, note that

$$\lim_{\alpha\searrow 0} \mathrm{Tr}\,\left(\sigma^{\frac{1-\alpha}{2\alpha}} \rho \sigma^{\frac{1-\alpha}{2\alpha}}\right)^\alpha = \lim_{p\to\infty} \mathrm{Tr}\,(\sigma^{p/2}\rho\sigma^{p/2})^{1/p} = \lim_{p\to\infty} \mathrm{Tr}\,(\rho^0\sigma^p\rho^0)^{1/p},$$

where the existence of $\lim_{p\to\infty} \mathrm{Tr}\,(\rho^0\sigma^p\rho^0)^{1/p}$ follows from the *Araki–Lieb–Thirring inequality* [4] (also [1]), and the latter equality above follows since $\lambda\rho^0 \le \rho \le \mu\rho^0$ for some $\lambda, \mu > 0$ and

$$\lambda^{1/p}\mathrm{Tr}\,(\sigma^{p/2}\rho^0\sigma^{p/2})^{1/p} \le \mathrm{Tr}\,(\sigma^{p/2}\rho\sigma^{p/2})^{1/p} \le \mu^{1/p}\mathrm{Tr}\,(\sigma^{p/2}\rho^0\sigma^{p/2})^{1/p}.$$

It was proved in [8] that

$$\widetilde{D}_0(\rho\|\sigma) \le D_0(\rho\|\sigma) \tag{2.8}$$

and equality holds in (2.8) if $\rho^0 = \sigma^0$. Let us here prove the following:

(1) $\widetilde{D}_0(\rho\|\sigma) = -\log \widetilde{Q}_0(\rho\|\sigma)$, where

$$\widetilde{Q}_0(\rho\|\sigma) := \max\bigl\{\mathrm{Tr}\,(P\sigma) : P \text{ an orthogonal projection}, \\ [P,\sigma]=0,\ (P\rho^0P)^0 = P\bigr\}.$$

(2) $\widetilde{D}_0(\rho\|\sigma) = D_0(\rho\|\sigma)$ holds if and only if $[\rho^0,\sigma]=0$. (Obviously, $[\rho^0,\sigma]=0$ if $\rho^0=\sigma^0$.)

Indeed, to prove (1), first note that $(P\rho^0P)^0 = P$ means that the dimension of $\mathrm{ran}\,\rho^0P$ is equal to that of $P$, that is, $\rho^0v_1,\dots,\rho^0v_d$ are linearly independent when $\{v_1,\dots,v_d\}$ is an orthonormal basis of $\mathrm{ran}\,P$. Choose $1 \le l_1 < l_2 < \cdots < l_m$ as in the first paragraph of this section (before Theorem 2.1) for $A=\sigma$ and $K=\rho^0$. Let $P_0$ be the orthogonal projection onto $\mathrm{span}\{v_{l_1},\dots,v_{l_m}\}$. Then $[P_0,\sigma]=0$, $(P_0\rho^0P_0)^0 = P_0$, and Theorem 2.1 gives

$$\lim_{p\to\infty} \mathrm{Tr}\,(\rho^0\sigma^p\rho^0)^{1/p} = \sum_{k=1}^m a_{l_k} = \mathrm{Tr}\,(P_0\sigma).$$

On the other hand, let $P$ be an orthogonal projection with $[P,\sigma]=0$ and $(P\rho^0P)^0 = P$. From $[P,\sigma]=0$, we may assume that $P=\sum_{k=1}^d |v_{i_k}\rangle\langle v_{i_k}|$ for some $1 \le i_1 < \cdots < i_d \le n$ (after, if necessary, changing $v_i$ for degenerate eigenvalues $a_i$). Since $(P\rho^0P)^0 = P$ implies that $\rho^0v_{i_1},\dots,\rho^0v_{i_d}$ are linearly independent, we have $d \le m$ and

$$\mathrm{Tr}\,(P\sigma) = \sum_{k=1}^d a_{i_k} \le \sum_{k=1}^m a_{l_k} = \mathrm{Tr}\,(P_0\sigma).$$

Next, to prove (2), note that $\mathrm{Tr}\,(\rho^0\sigma^p\rho^0)^{1/p}$ is increasing in $p>0$ by the Araki–Lieb–Thirring inequality mentioned above, which shows that

$$\mathrm{Tr}\,(\rho^0\sigma) \le \lim_{p\to\infty} \mathrm{Tr}\,(\rho^0\sigma^p\rho^0)^{1/p}.$$

This means inequality (2.8), and equality holds in (2.8) if and only if Tr $(\rho^0\sigma^p\rho^0)^{1/p}$ is constant for $p \geq 1$. By [10, Theorem 2.1] this is equivalent to the commutativity $\rho^0\sigma = \sigma\rho^0$.

Finally, we consider the complementary convergence of $\Phi(A^p)^{1/p}$ as $p \to -\infty$, or $\Phi(A^{-p})^{-1/p}$ as $p \to \infty$. Here, the expression $\Phi(A^{-p})^{-1/p}$ for $p > 0$ is defined in such a way that the $(-p)$-power of $A$ is restricted to the support of $A$, i.e., defined in the sense of the generalized inverse, and the $(-1/p)$-power of $\Phi(A^{-p})$ is also in this sense. The next theorem is the complementary counterpart of Theorem 2.6.

**Theorem 2.9** *Let $\Phi : \mathbb{M}_n \to \mathbb{M}_{n'}$ be a positive linear map. Let $A \in \mathbb{M}_n^+$ be given with the spectral decomposition $A = \sum_{k=1}^m a_k P_k$, where $a_1 > a_2 > \cdots > a_m > 0$. Define*

$$\begin{aligned}\widetilde{\mathcal{M}}_k &:= \operatorname{ran}\Phi(P_k + \cdots + P_m) \ominus \operatorname{ran}\Phi(P_{k+1} + \cdots + P_m), \quad 1 \leq i \leq m-1,\\ \widetilde{\mathcal{M}}_m &:= \operatorname{ran}\Phi(P_m).\end{aligned}$$

*Then*

$$\lim_{p\to\infty} \Phi(A^{-p})^{-1/p} = \sum_{k=1}^m a_k P_{\widetilde{\mathcal{M}}_k}.$$

*Proof* The proof is just a simple adaptation of Theorem 2.6. We can write for any $p > 0$

$$\Phi(A^{-p})^{-1/p} = \left\{\Phi((A^{-1})^p)^{1/p}\right\}^{-1},$$

where $A^{-1}$ and $\{\cdots\}^{-1}$ are defined in the sense of the generalized inverse so that

$$A^{-1} = \sum_{k=1}^m a_k^{-1} P_k = \sum_{k=1}^m a_{m+1-k}^{-1} P_{m+1-k}$$

with $a_m^{-1} > \cdots > a_1^{-1} > 0$. By Theorem 2.6 we have

$$\lim_{p\to\infty} \Phi((A^{-1})^p)^{1/p} = \sum_{k=1}^m a_{m+1-k}^{-1} P_{\widetilde{\mathcal{M}}_{m+1-k}} = \sum_{k=1}^m a_k^{-1} P_{\widetilde{\mathcal{M}}_k},$$

where

$$\begin{aligned}\widetilde{\mathcal{M}}_m &:= \operatorname{ran}\Phi(P_{m+1-1}) = \operatorname{ran}\Phi(P_m),\\ \widetilde{\mathcal{M}}_{m+1-k} &:= \operatorname{ran}\Phi(P_{m+1-1} + \cdots + P_{m+1-k}) \ominus \operatorname{ran}\Phi(P_{m+1-1} + \cdots + P_{m+1-(k-1)})\\ &= \operatorname{ran}\Phi(P_{m+1-k} + \cdots + P_m) \ominus \operatorname{ran}\Phi(P_{m+2-k} + \cdots + P_m), \quad 2 \leq k \leq m.\end{aligned}$$

According to the proofs of Theorems 2.1 and 2.4, we see that the $i$th eigenvalue $\lambda_i(p)$ of $\Phi((A^{-1})^p)^{1/p}$ converges to a positive real as $p \to \infty$, or otherwise, $\lambda_i(p) = 0$

for all $p > 0$. That is, $\lambda_i(p) \to 0$ as $p \to \infty$ occurs only when $\lambda_i(p) = 0$ for all $p > 0$. This implies that

$$\lim_{p\to\infty} \Phi(A^{-p})^{-1/p} = \left\{\lim_{p\to\infty} \Phi((A^{-1})^p)^{1/p}\right\}^{-1} = \sum_{k=1}^m a_k P_{\widetilde{\mathcal{M}}_k}.$$

□

**Remark 2.10** Assume that $\Phi : \mathbb{M}_n \to \mathbb{M}_{n'}$ is a *unital* positive linear map. Let $A \in \mathbb{M}_n$ be positive definite and $1 \le p < q$. Since $x^{p/q}$ and $x^{1/p}$ are operator monotone on $[0,\infty)$, we have $\Phi(A^q)^{p/q} \ge \Phi(A^p)$ and so $\Phi(A^q)^{1/q} \ge \Phi(A^p)^{1/p}$. Hence $\Phi(A^p)^{1/p}$ increases as $1 \le p \nearrow$. Similarly, $\Phi(A^{-q})^{p/q} \ge \Phi(A^{-p})$ and so $\Phi(A^{-q})^{-1/q} \le \Phi(A^{-p})^{-1/p}$ since $x^{-1/p}$ is operator monotone decreasing on $(0,\infty)$. Hence $\Phi(A^{-p})^{-1/p}$ decreases as $1 \le p \nearrow$. Moreover, since $x^{-1}$ is operator convex on $(0,\infty)$, we have $\Phi(A^{-1})^{-1} \le \Phi(A)$. (See [5, Theorem 2.1] for more details.) Combining altogether, when $A$ is positive definite, we have

$$\Phi(A^{-p})^{-1/p} \le \Phi(A^q)^{1/q}, \qquad p, q \ge 1, \tag{2.9}$$

and in particular,

$$\lim_{p\to\infty} \Phi(A^{-p})^{-1/p} \le \lim_{p\to\infty} \Phi(A^p)^{1/p}.$$

However, the latter inequality does not hold unless $\Phi$ is unital. For example, let

$$P_1 := \begin{bmatrix} 1 & 0 \\ 0 & 0 \end{bmatrix}, \quad P_2 := \begin{bmatrix} 0 & 0 \\ 0 & 1 \end{bmatrix}, \quad Q_1 := \begin{bmatrix} 1/2 & 1/2 \\ 1/2 & 1/2 \end{bmatrix}, \quad Q_2 := \begin{bmatrix} 1/2 & -1/2 \\ -1/2 & 1/2 \end{bmatrix},$$

and consider $\Phi : \mathbb{M}_2 \to \mathbb{M}_2$ given by

$$\Phi\left(\begin{bmatrix} a_{11} & a_{12} \\ a_{21} & a_{22} \end{bmatrix}\right) := a_{11}P_1 + a_{22}Q_1,$$

and $A := aP_1 + bP_2$ where $a > b > 0$. Since $P_{\mathrm{ran}\,\Phi(P_1+P_2)} = I$, $P_{\mathrm{ran}\,\Phi(P_1)} = P_1$ and $P_{\mathrm{ran}\,\Phi(P_2)} = Q_1$, Theorems 2.6 and 2.9 give

$$\lim_{p\to\infty} \Phi(A^p)^{1/p} = aP_1 + b(I - P_1) = aP_1 + bP_2,$$

$$\lim_{p\to\infty} \Phi(A^{-p})^{-1/p} = a(I - Q_1) + bQ_1 = aQ_2 + bQ_1.$$

We compute

$$(aP_1 + bP_2) - (aQ_2 + bQ_1) = \begin{bmatrix} \frac{a-b}{2} & \frac{a-b}{2} \\ \frac{a-b}{2} & \frac{b-a}{2} \end{bmatrix},$$

which is not positive semidefinite.

**Remark 2.11** We may always assume that $\Phi : \mathbb{M}_n \to \mathbb{M}_{n'}$ is strictly positive. Indeed, we may consider $\Phi$ as $\Phi : \mathbb{M}_n \to Q_0\mathbb{M}_{n'}Q_0 \cong \mathbb{M}_{n''}$, where $Q_0$ is the support projection of $\Phi(I_n)$. Under this convention, another reasonable definition of $\Phi(A^{-p})^{-1/p}$ for $p \ge 1$ is

$$\Phi(A^{-p})^{-1/p} := \lim_{\varepsilon\searrow 0} \Phi((A + \varepsilon I_n)^{-p})^{-1/p},$$

which is well defined since $\Phi((A + \varepsilon I)^{-p})$ is increasing so that $\Phi((A + \varepsilon I)^{-p})^{-1/p}$ is decreasing as $\varepsilon \searrow 0$. But this definition is different from the above definition of $\Phi(A^{-p})^{-1/p}$. For example, let $\Phi : \mathbb{M}_2 \to \mathbb{M}_2$ be given by $\Phi(A) := KAK^*$ with an invertible $K = \begin{bmatrix} a & b \\ c & d \end{bmatrix}$, and let $A = P = \begin{bmatrix} 1 & 0 \\ 0 & 0 \end{bmatrix}$. Then $A^{-p} = P$ (in the generalized inverse) so that

$$KA^{-p}K^* = \begin{bmatrix} |a|^2 & a\bar{c} \\ \bar{a}c & |c|^2 \end{bmatrix}$$

and so

$$(KA^{-p}K^*)^{-1/p} = \frac{1}{(|a|^2 + |c|^2)^{1-\frac{1}{p}}} \begin{bmatrix} |a|^2 & a\bar{c} \\ \bar{a}c & |c|^2 \end{bmatrix}. \tag{2.10}$$

On the other hand,

$$\begin{aligned} \lim_{\varepsilon\searrow 0}(K(A + \varepsilon I)^{-p}K^*)^{-1/p} &= \lim_{\varepsilon\searrow 0}(K^{*-1}(A + \varepsilon I)^{p}K^{-1})^{1/p} \\ &= (K^{*-1}A^pK^{-1})^{1/p} = (K^{*-1}PK^{-1})^{1/p} \end{aligned}$$

is equal to

$$\frac{1}{|ad - bc|^{2/p}(|b|^2 + |d|^2)^{1-\frac{1}{p}}} \begin{bmatrix} |d|^2 & -b\bar{d} \\ -\bar{b}d & |b|^2 \end{bmatrix}. \tag{2.11}$$

Hence we find that (2.10) and (2.11) are very different, even after taking the limits as $p \to \infty$.

Here is a simpler example. Let $\varphi : \mathbb{M}_2 \to \mathbb{C} = \mathbb{M}_1$ be a state (hence, unital) with density matrix $\begin{bmatrix} 1/2 & 1/2 \\ 1/2 & 1/2 \end{bmatrix}$, and let $A = \begin{bmatrix} 1 & 0 \\ 0 & 0 \end{bmatrix}$. For the first definition, we have

$$\lim_{p\to\infty} \varphi(A^{-p})^{-1/p} = \lim_{p\to\infty} 2^{1/p} = 1.$$

For the second definition,

$$\lim_{\varepsilon\searrow 0} \varphi((A + \varepsilon I)^{-p})^{-1/p} = \lim_{\varepsilon\searrow 0} \left\{ \frac{(1 + \varepsilon)^{-p} + \varepsilon^{-p}}{2} \right\}^{-1/p} = 0$$

for all $p > 0$. Moreover, since $\varphi(A^p)^{1/p} = 2^{-1/p}$ for $p > 0$, this example says also that (2.9) does not hold for general positive semidefinite $A$.

**Problem 2.12** It is also interesting to consider the limit of $(A^p B A^p)^{1/p}$ as $p \to \infty$ for $A, B \in \mathbb{M}_n^+$, a version different from the limit treated in Theorem 2.1. To consider $\lim_{p\to\infty}(A^p B A^p)^{1/p}$, we may assume without loss of generality that $B$ is an orthogonal projection $E$ (see the argument around (3.2) below). Since $(A^p E A^p)^{1/p} = (A^p E^{2p} A^p)^{1/p}$ converges as $p \to \infty$ by [6, Theorem 2.5], the existence of the limit $\lim_{p\to\infty}(A^p B A^p)^{1/p}$ follows. But it seems that the description of the limit is a combinatorial problem much more complicated than that in Theorem 2.1.

## 3 $\lim_{p\to\infty}(A^p \sigma B)^{1/p}$ for Operator Means $\sigma$

In theory of operator means due to Kubo and Ando [13], a main result says that each operator mean $\sigma$ is associated with a nonnegative operator monotone function $f$ on $[0, \infty)$ with $f(1) = 1$ in such a way that

$$A\sigma B = A^{1/2} f(A^{-1/2} B A^{-1/2}) A^{1/2}$$

for $A, B \in \mathbb{M}_n^+$ with $A > 0$, which is further extended to general $A, B \in \mathbb{M}_n^+$ as

$$A\sigma B = \lim_{\varepsilon\searrow 0}(A + \varepsilon I)\sigma(B + \varepsilon I).$$

We write $\sigma_f$ for the operator mean associated with $f$ as above. For $0 \le \alpha \le 1$, the operator mean corresponding to the function $x^\alpha$ $(x \ge 0)$ is the *weighted geometric mean* $\#_\alpha$, i.e.,

$$A\#_\alpha B = A^{1/2}(A^{-1/2} B A^{-1/2})^\alpha A^{1/2}$$

for $A, B \in \mathbb{M}_n^+$ with $A > 0$. In particular, $\# = \#_{1/2}$ is the so-called *geometric mean* first introduced by Pusz and Woronowicz [15].

The transpose of $f$ above is given by

$$\widetilde{f}(x) := x f(x^{-1}), \qquad x > 0,$$

which is again an operator monotone function on $[0, \infty)$ (after extending to $[0, \infty)$ by continuity) corresponding to the transposed operator mean of $\sigma_f$, i.e., $A\sigma_{\widetilde{f}} B = B\sigma_f A$. We also write

$$\widehat{f}(x) := \begin{cases} \widetilde{f}(x^{-1}) = f(x)/x & \text{if } x > 0, \\ 0 & \text{if } x = 0. \end{cases} \tag{3.1}$$

In the rest of the section, let $f$ be such an operator monotone function as above and $\sigma_f$ be the corresponding operator mean. We are concerned with the existence and the description of the limit $\lim_{p\to\infty}(A^p\sigma_f B)^{1/p}$, in particular, $\lim_{p\to\infty}(A^p\#_\alpha B)^{1/p}$ for $A, B \in \mathbb{M}_n^+$. For this, we may assume without loss of generality that $B$ is an orthogonal projection. Indeed, let $E$ be the support projection of $B$. Then we can choose $\lambda, \mu > 0$ with $\lambda < 1 < \mu$ such that $\lambda E \le B \le \mu E$. Thanks to monotonicity and positive homogeneity of $\sigma_f$, we have

$$\lambda(A^p\sigma_f E) = (\lambda A^p)\sigma_f(\lambda E) \le A^p\sigma_f B \le (\mu A^p)\sigma_f(\mu E) = \mu(A^p\,\sigma_f E).$$

Hence, for every $p \ge 1$, since $x^{1/p}$ $(x \ge 0)$ is operator monotone,

$$\lambda^{1/p}(A^p\sigma_f E)^{1/p} \le (A^p\sigma_f B)^{1/p} \le \mu^{1/p}(A^p\sigma_f E), \tag{3.2}$$

so that $\lim_{p\to\infty}(A^p\sigma_f B)^{1/p}$ exists if and only if $\lim_{p\to\infty}(A^p\sigma_f E)^{1/p}$ does, and in this case, both limits are equal. In particular, when $B > 0$, since $(A^p\sigma_f I)^{1/p} = \widetilde{f}(A^p)^{1/p}$, we note that

$$\lim_{p\to\infty}(A^p\sigma_f B)^{1/p} = \widetilde{f}^{(\infty)}(A)$$

whenever $\widetilde{f}^{(\infty)}(x) := \lim_{p\to\infty}\widetilde{f}(x^p)^{1/p}$ exists for all $x \ge 0$. For instance,

- if $f(x) = 1 - \alpha + \alpha x$ where $0 < \alpha < 1$, then $\sigma_f = \nabla_\alpha$, the $\alpha$-arithmetic mean $A\nabla_\alpha B := (1-\alpha)A + \alpha B$, and $\widetilde{f}^{(\infty)}(x) = \max\{x, 1\}$,
- if $f(x) = x^\alpha$ where $0 \le \alpha \le 1$, then $\sigma_f = \#_\alpha$ and $\widetilde{f}^{(\infty)}(x) = \widetilde{f}(x) = x^{1-\alpha}$,
- if $f(x) = x/((1-\alpha)x + \alpha)$ where $0 < \alpha < 1$, then $\sigma_f = !_\alpha$, the $\alpha$-harmonic mean $A\,!_\alpha B := (A^{-1}\nabla_\alpha B)^{-1}$, and $\widetilde{f}^{(\infty)}(x) = \min\{x, 1\}$.

But it is unknown to us that, for any operator monotone function $f$ on $[0, \infty)$, the limit $\lim_{p\to\infty} f(x^p)^{1/p}$ exists for all $x \ge 0$, while it seems so.

When $E$ is an orthogonal projection, the next proposition gives a nice expression for $A\sigma_f E$. This was shown in [11, Lemma 4.7], while the proof is given here for the convenience of the reader.

**Lemma 3.1** *Assume that $f(0) = 0$. If $A \in \mathbb{M}_n$ is positive definite and $E \in \mathbb{M}_n$ is an orthogonal projection, then*

$$A\sigma_f E = \widehat{f}(EA^{-1}E), \tag{3.3}$$

*where $\widehat{f}$ is given in* (3.1).

*Proof* For every $m = 1, 2, \ldots$ we have

$$A^{-1/2}(EA^{-1}E)^m A^{-1/2} = (A^{-1/2}EA^{-1/2})^{m+1}. \tag{3.4}$$

Note that the eigenvalues of $EA^{-1}E$ and those of $A^{-1/2}EA^{-1/2}$ are the same including multiplicities. Choose a $\delta > 0$ such that the positive eigenvalues of $EA^{-1}E$ and

$A^{-1/2}EA^{-1/2}$ are included in $[\delta, \delta^{-1}]$. Then, since $\widehat{f}(x)$ is continuous on $[\delta, \delta^{-1}]$, one can choose a sequence of polynomials $p_k(x)$ with $p_k(0) = 0$ such that $p_k(x) \to \widehat{f}(x)$ uniformly on $[\delta, \delta^{-1}]$ as $k \to \infty$. By (3.4) we have

$$A^{-1/2}p_k(EA^{-1}E)A^{-1/2} = A^{-1/2}EA^{-1/2}p_k(A^{-1/2}EA^{-1/2})$$

for every $k$. Since $\widehat{f}(0) = 0$ by definition, we have

$$p_k(EA^{-1}E) \longrightarrow \widehat{f}(EA^{-1}E)$$

and

$$A^{-1/2}EA^{-1/2}p_k(A^{-1/2}EA^{-1/2}) \longrightarrow A^{-1/2}EA^{-1/2}\widehat{f}(A^{-1/2}EA^{-1/2})$$

as $k \to \infty$. Since $f(0) = 0$ by assumption, we have $f(x) = x\widehat{f}(x)$ for all $x \in [0, \infty)$. This implies that

$$A^{-1/2}EA^{-1/2}\widehat{f}(A^{-1/2}EA^{-1/2}) = f(A^{-1/2}EA^{-1/2}).$$

Therefore,

$$A^{-1/2}\widehat{f}(EA^{-1}E)A^{-1/2} = f(A^{-1/2}EA^{-1/2})$$

so that we have $\widehat{f}(EA^{-1}E) = A^{1/2}f(A^{-1/2}EA^{-1/2})A^{1/2} = A\sigma_f E$, as asserted. □

Formula (3.3) can equivalently be written as

$$A\sigma_f E = \widetilde{f}((EA^{-1}E)^{-1}), \tag{3.5}$$

where $(EA^{-1}E)^{-1}$ is the inverse restricted to ran $E$ (in the sense of the generalized inverse) and $\widetilde{f}((EA^{-1}E)^{-1})$ is also restricted to ran $E$. In particular, if $f$ is symmetric (i.e., $f = \widetilde{f}$) with $f(0) = 0$, then

$$A\sigma_f E = f((EA^{-1}E)^{-1}).$$

**Example 3.2** Assume that $0 < \alpha \le 1$ and $A, E$ are as in Lemma 3.1.

(1) When $f(x) = x^\alpha$ and $\sigma_f = \#_\alpha$, $\widehat{f}(x) = x^{\alpha-1}$ for $x > 0$ so that

$$A\#_\alpha E = (EA^{-1}E)^{\alpha-1},$$

where the $(\alpha - 1)$-power in the right-hand side is defined with restriction to ran $E$.

(2) When $f(x) = x/((1-\alpha)x + \alpha)$ and $\sigma_f = !_\alpha$, $\widehat{f}(x) = (1 - \alpha + \alpha x)^{-1}$ for $x > 0$ so that

$$A\,!_\alpha E = \{(1-\alpha)E + \alpha EA^{-1}E\}^{-1} = \{E((1-\alpha)I + \alpha A^{-1})E\}^{-1},$$

where the inverse of $E((1-\alpha)I+\alpha A^{-1})E$ in the right-hand side is restricted to ran $E$.

(3) When $f(x)=(x-1)/\log x$ and so $\sigma_f$ is the logarithmic mean, $\widehat{f}(x)=(1-x^{-1})/\log x$ for $x>0$ so that

$$A\sigma_f E=(E-(EA^{-1}E)^{-1})(\log EA^{-1}E)^{-1},$$

where the right-hand side is defined with restriction to ran $E$.

**Theorem 3.3** *Assume that $f(0)=0$ and $f(x^r)\ge f(x)^r$ for all $x>0$ and all $r\in(0,1)$. Let $A\in\mathbb{M}_n^+$ and $E\in\mathbb{M}_n$ be an orthogonal projection. Then*

$$(A^p\,\sigma_f\,E)^{1/p}\ge(A^q\,\sigma_f\,E)^{1/q}\quad \text{if } 1\le p<q. \tag{3.6}$$

*Proof* First, note that $\widetilde{f}(x^r)=x^r f(x^{-r})\ge x^r f(x^{-1})^r=\widetilde{f}(x)^r$ for all $x>0$, $r\in(0,1)$. By replacing $A$ with $A+\varepsilon I$ and taking the limit as $\varepsilon\searrow 0$, we may assume that $A$ is positive definite. Let $1\le p<q$ and $r:=p/q\in(0,1)$. By (3.5) we have

$$(A^q\,\sigma_f\,E)^r=\widetilde{f}((EA^{-q}E)^{-1})^r\le\widetilde{f}((EA^{-q}E)^{-r}). \tag{3.7}$$

Since $x^r$ is operator monotone on $[0,\infty)$, we have by Hansen's inequality [9]

$$(EA^{-q}E)^r\ge EA^{-qr}E=EA^{-p}E$$

so that $(EA^{-q}E)^{-r}\le(EA^{-p}E)^{-1}$. Since $\widetilde{f}(x)$ is operator monotone on $[0,\infty)$, we have

$$\widetilde{f}((EA^{-q}E)^{-r})\le\widetilde{f}((EA^{-p}E)^{-1})=A^p\,\sigma_f\,E. \tag{3.8}$$

Combining (3.7) and (3.8) gives

$$(A^q\,\sigma_f\,E)^r\le A^p\,\sigma_f\,E.$$

Since $x^{1/p}$ is operator monotone on $[0,\infty)$, we finally have

$$(A^q\,\sigma_f\,E)^{1/q}\le(A^p\,\sigma_f\,E)^{1/p}.$$

□

**Corollary 3.4** *Assume that $f(0)=0$ and $f(x^r)\ge f(x)^r$ for all $x>0$, $r\in(0,1)$. Then for every $A,B\in\mathbb{M}_n^+$, the limit*

$$\lim_{p\to\infty}(A^p\,\sigma_f\,B)^{1/p}$$

*exists.*

*Proof* From the argument around (3.2) we may assume that $B$ is an orthogonal projection $E$. Then, Theorem 3.3 implies that $(A^p \sigma_f E)^{1/p}$ converges as $p \to \infty$. □

**Remark 3.5** Following [17], an operator monotone function $f$ on $[0, \infty)$ is said to be *power monotone increasing* (p.m.i. for short) if $f(x^r) \geq f(x)^r$ for all $x > 0$, $r > 1$ (equivalently, $f(x^r) \leq f(x^r)$ for all $x > 0, r \in (0, 1)$), and *power monotone decreasing* (p.m.d.) if $f(x^r) \leq f(x)^r$ for all $x > 0, r > 1$. These conditions play a role to characterize the operator means $\sigma_f$ satisfying Ando–Hiai's inequality [1], see [17, Lemmas 2.1, 2.2]. For instance, the p.m.d. condition is satisfied for $f$ in (1) and (2) of Example 3.2, while $f$ in Example 3.2 (3) does the p.m.i. condition. Hence, for any $\alpha \in [0, 1]$, $(A^p \#_\alpha E)^{1/p}$ and $(A^p\,!_\alpha E)^{1/p}$ converge decreasingly as $1 \leq p \nearrow \infty$. In fact, for the harmonic mean, we have the limit $A \wedge B := \lim_{p\to\infty}(A^p\,!\,B^p)^{1/p}$, the decreasing limit as $1 \leq p \nearrow \infty$ for any $A, B \geq 0$, which is the infimum counterpart of $A \vee B$ in [12] (see also Example 2.7). The reader might be wondering if the opposite inequality to (3.6) holds (i.e., $(A^p \sigma_f E)^{1/p}$ is increasing as $1 \leq p \nearrow \infty$) when $f$ satisfies the p.m.i. condition. Although this is the case when $\sigma = \nabla_\alpha$ the weighted arithmetic mean, it is not the case in general. In fact, if it were true, $(A^p \#_\alpha E)^{1/p}$ must be constant for $p \geq 1$ since $x^\alpha$ satisfies both p.m.i. and p.m.d. conditions, that is impossible.

Finally, for the weighted geometric mean $\#_\alpha$ we obtain the explicit description of $\lim_{p\to\infty}(A^p \#_\alpha E)^{1/p}$ for any $A \in \mathbb{M}_n^+$. For the trivial cases $\alpha = 0, 1$ note that $(A^p \#_0 E)^{1/p} = A$ and $(A^p \#_1 E)^{1/p} = E$ for all $p > 0$.

**Theorem 3.6** *Assume that $0 < \alpha < 1$. Let $A \in \mathbb{M}_n^+$ be given with the spectral decomposition $A = \sum_{k=1}^m a_k P_k$ where $a_1 > \cdots > a_m > 0$, and $E \in \mathbb{M}_n$ be an orthogonal projection. Then*

$$\lim_{p\to\infty} (A^p \#_\alpha E)^{1/p} = \sum_{k=1}^m a_k^{1-\alpha} Q_k, \tag{3.9}$$

*where*

$$\begin{aligned} Q_1 &:= P_1 \wedge E, \\ Q_k &:= (P_1 + \cdots + P_k) \wedge E - (P_1 + \cdots + P_{k-1}) \wedge E, \qquad 2 \leq k \leq m. \end{aligned}$$

*Proof* First, assume that $A$ is positive definite so that $P_1 + \cdots + P_m = I$. When $f(x) = x^\alpha$ with $0 < \alpha < 1$, formula (3.5) is given as

$$A \#_\alpha E = (EA^{-1}E)^{-(1-\alpha)}.$$

Since

$$\lim_{p\to\infty}(A^p\#_\alpha E)^{1/p}=\lim_{p\to\infty}(EA^{-p}E)^{-(1-\alpha)/p}=\lim_{p\to\infty}(E(A^{1-\alpha})^{-p}E)^{-1/p},$$

it follows from Theorem 2.9 that

$$\lim_{p\to\infty}(A^p\#_\alpha E)^{1/p}=\sum_{k=1}^{m}a_k^{1-\alpha}P_{\widetilde{\mathcal{M}}_k},$$

where

$$\widetilde{\mathcal{M}}_k:=\mathrm{ran}\,E(P_k+\cdots+P_m)E\ominus\mathrm{ran}\,E(P_{k+1}+\cdots+P_m)E,\quad 1\le k\le m-1,$$
$$\widetilde{\mathcal{M}}_m:=\mathrm{ran}\,EP_mE.$$

From Lemma 3.7 below, we have

$$\widetilde{\mathcal{M}}_1=\mathrm{ran}\,E\ominus\mathrm{ran}\,EP_1^\perp E=\mathrm{ran}\,P_1\wedge E,$$

and for $2\le k\le m$,

$$\begin{aligned}\widetilde{\mathcal{M}}_k&=\mathrm{ran}\,E(P_1+\cdots+P_{k-1})^\perp E\ominus\mathrm{ran}\,E(P_1+\cdots+P_k)^\perp E\\&=\big[\mathrm{ran}\,E\ominus\mathrm{ran}\,E(P_1+\ldots+P_k)^\perp E\big]\ominus\big[\mathrm{ran}\,E\ominus\mathrm{ran}\,E(P_1+\cdots+P_{k-1})^\perp E\big]\\&=\mathrm{ran}\,(P_1+\cdots+P_k)\wedge E\ominus\mathrm{ran}\,(P_1+\cdots+P_{k-1})\wedge E\\&=\mathrm{ran}\big[(P_1+\cdots+P_k)\wedge E-(P_1+\cdots+P_{k-1})\wedge E\big].\end{aligned}$$

Therefore, (3.9) is established when $A$ is positive definite.

Next, when $A$ is not positive definite, let $P_{m+1}:=(P_1+\cdots+P_m)^\perp$. For any $\varepsilon\in(0,a_m)$ define $A_\varepsilon:=A+\varepsilon P_{m+1}$. Then the above case implies that

$$\lim_{p\to\infty}(A_\varepsilon\#_\alpha E)^{1/p}=\sum_{k=1}^{m}a_k^{1-\alpha}Q_k+\varepsilon^{1-\alpha}Q_{m+1},$$

where

$$Q_{m+1}:=E-(P_1+\cdots+P_m)\wedge E.$$

Assume that $0<\varepsilon<\varepsilon'<a_m$. For every $p\ge 1$, since $A_\varepsilon^p\le A_{\varepsilon'}^p$, we have $A_\varepsilon^p\#_\alpha E\le A_{\varepsilon'}^p\#_\alpha E$ and hence $(A_\varepsilon^p\#_\alpha E)^{1/p}\le(A_{\varepsilon'}^p\#_\alpha E)^{1/p}$. Furthermore, since $A_\varepsilon^p\#_\alpha E\to A^p\#_\alpha E$ as $a_m>\varepsilon\searrow 0$, we have

$$(A^p\#_\alpha E)^{1/p}=\lim_{a_m>\varepsilon\searrow 0}(A_\varepsilon^p\#_\alpha E)^{1/p}\ \text{ decreasingly.}\tag{3.10}$$

Now, we can perform a calculation of limits as follows:

$$\begin{aligned}\lim_{1\le p\to\infty}(A^p\#_\alpha E)^{1/p} &= \lim_{1\le p\to\infty}\lim_{a_m>\varepsilon\searrow 0}(A_\varepsilon^p\#_\alpha E)^{1/p}\\ &= \lim_{a_m>\varepsilon\searrow 0}\lim_{1\le p\to\infty}(A_\varepsilon^p\#_\alpha E)^{1/p}\\ &= \lim_{a_m>\varepsilon\searrow 0}\left(\sum_{k=1}^m a_k^{1-\alpha}Q_k+\varepsilon^{1-\alpha}Q_{m+1}\right)\\ &= \sum_{k=1}^m a_k^{1-\alpha}Q_k.\end{aligned}$$

In the above, the second equality (the exchange of two limits) is confirmed as follows. Let $X_{p,\varepsilon} := (A_\varepsilon^p\#_\alpha E)^{1/p}$ for $p\ge 1$ and $0<\varepsilon<a_m$. Then $X_{p,\varepsilon}$ is decreasing as $1\le p\nearrow\infty$ by Theorem 3.3 and also decreasing as $a_m>\varepsilon\searrow 0$ as seen in (3.10). Let $X_p := \lim_\varepsilon X_{p,\varepsilon}$ $(= (A^p\#_\alpha E)^{1/p})$, $X_\varepsilon := \lim_p X_{p,\varepsilon}$, and $X := \lim_p X_p$. Since $X_{p,\varepsilon}\ge X_p$, we have $X_\varepsilon \ge X$ and hence $\lim_\varepsilon X_\varepsilon\ge X$. On the other hand, since $X_{p,\varepsilon}\ge X_\varepsilon$, we have $X_p\ge\lim_\varepsilon X_\varepsilon$ and hence $X\ge\lim_\varepsilon X_\varepsilon$. Therefore, $X=\lim_\varepsilon X_\varepsilon$, which gives the assertion. □

In particular, when $A=P$ is an orthogonal projection, we have $(A^p\#_\alpha E)^{1/p} = P\wedge E$ for all $p>0$ (see [13, Theorem 3.7]) so that both sides of (3.9) are certainly equal to $P\wedge E$.

**Lemma 3.7** *For every orthogonal projections $E$ and $P$,*

$$\mathrm{ran}\,EP^\perp E=\mathrm{ran}\,(E-P\wedge E),$$

*or equivalently,*

$$\mathrm{ran}\,P\wedge E=\mathrm{ran}\,E\ominus\mathrm{ran}\,EP^\perp E.$$

*Proof* According to the well-known representation of two projections (see [16, pp. 306–308]), we write

$$E = I\oplus I\oplus 0\oplus\begin{bmatrix} I & 0\\ 0 & 0\end{bmatrix}\oplus 0,$$

$$P = I\oplus 0\oplus I\oplus\begin{bmatrix} C^2 & CS\\ CS & S^2\end{bmatrix}\oplus 0,$$

where $0<C,S<I$ with $C^2+S^2=I$. We have

$$P\wedge E=I\oplus 0\oplus 0\oplus 0\oplus 0.$$

Since

$$P^\perp = 0\oplus I\oplus 0\oplus\begin{bmatrix} S^2 & -CS\\ -CS & C^2\end{bmatrix}\oplus I,$$

we also have

$$EP^{\perp}E = 0 \oplus I \oplus 0 \oplus \begin{bmatrix} S^2 & 0 \\ 0 & 0 \end{bmatrix} \oplus 0,$$

whose range is that of

$$0 \oplus I \oplus 0 \oplus \begin{bmatrix} I & 0 \\ 0 & 0 \end{bmatrix} \oplus 0 = E - P \wedge E,$$

which yields the conclusion. □

**Acknowledgements** This work was supported by JSPS KAKENHI Grant Number JP17K05266.

## References

1. Ando, T., Hiai, F.: Log majorization and complementary Golden-Thompson type inequalities. Linear Algebr. Appl. **197/198**, 113–131 (1994)
2. Antezana, J., Corach, G., Stojanoff, D.: Spectral shorted matrices. Linear Algebr. Appl. **381**, 197–217 (2004)
3. Antezana, J., Corach, G., Stojanoff, D.: Spectral shorted operators. Integr. Equ. Oper. Theory **55**, 169–188 (2006)
4. Araki, H.: On an inequality of Lieb and Thirring. Lett. Math. Phys. **19**, 167–170 (1990)
5. Audenaert, K.M.R., Hiai, F.: On matrix inequalities between the power means: counterexamples. Linear Algebr. Appl. **439**, 1590–1604 (2013)
6. Audenaert, K.M.R., Hiai, F.: Reciprocal Lie-Trotter formula. Linear Multilinear Algebr. **64**, 1220–1235 (2016)
7. Bourin, J.-C.: Convexity or concavity inequalities for Hermitian operators. Math. Inequalities Appl. **7**, 607–620 (2004)
8. Datta, N., Leditzky, F.: A limit of the quantum Rényi divergence. J. Phys. A: Math. Theor. **47**, 045304 (2014)
9. Hansen, F.: An operator inequality. Math. Ann. **246**, 249–250 (1980)
10. Hiai, F.: Equality cases in matrix norm inequalities of Golden-Thompson type. Linear Multilinear Algebr. **36**, 239–249 (1994)
11. Hiai, F.: A generalization of Araki's log-majorization. Linear Algebr. Appl. **501**, 1–16 (2016)
12. Kato, T.: Spectral order and a matrix limit theorem. Linear Multilinear Algebr. **8**, 15–19 (1979)
13. Kubo, F., Ando, T.: Means of positive linear operators. Math. Ann. **246**, 205–224 (1980)
14. Müller-Lennert, M., Dupuis, F., Szehr, O., Fehr, S., Tomamichel, M.: On quantum Rényi entropies: a new generalization and some properties. J. Math. Phys. **54**, 122203 (2013)
15. Pusz, W., Woronowicz, S.L.: Functional calculus for sesquilinear forms and the purification map. Rep. Math. Phys. **8**, 159–170 (1975)
16. Takesaki, M.: Theory of Operator Algebras I, Encyclopaedia of Mathematical Sciences, vol. 124. Springer, Berlin (2002)
17. Wada, S.: Some ways of constructing Furuta-type inequalities. Linear Algebr. Appl. **457**, 276–286 (2014)
18. Wilde, M.M., Winter, A., Yang, D.: Strong converse for the classical capacity of entanglement-breaking and Hadamard channels via a sandwiched Rényi relative entropy. Commun. Math. Phys. **331**, 593–622 (2014)

# The Howland–Kato Commutator Problem

**Ira Herbst and Thomas L. Kriete**

*Dedicated to the memory of Tosio Kato*

**Abstract** We investigate the following problem: For what bounded measurable real $f$ and $g$ is the commutator $i[f(P), g(Q)]$ positive? This problem originated in work of James Howland and was pursued by Tosio Kato who suggested what might be the answer. So far, there is no proof that Kato was correct but in this paper we discuss the problem and give some partial answers to the above question.

## 1 Introduction

In a paper on spectral theory [1], J. Howland used the positive commutator of two bounded functions of the Heisenberg operators $P$ and $Q$, $i[f(P), g(Q)]$, as a technical tool. Here, $P = -id/dx$ and $Q$ is multiplication by $x$ in $L^2(\mathbb{R})$. The functions $f$ and $g$ were specifically

$$f(t) = \tan^{-1}(t/2), g(t) = \tanh(t). \tag{1.1}$$

He sent his paper to T. Kato, his former thesis advisor, who got interested in the more general question: for what bounded real functions is the above commutator positive? Kato made much progress on this problem and in a beautiful paper, [2], he identified a very interesting class of pairs of such functions for which the commutator was

I. Herbst (✉) · T. L. Kriete
Department of Mathematics, University of Virginia, Charlottesville, VA 22904, USA
e-mail: iwh@virginia.edu; irah@virginia.edu

T. L. Kriete
e-mail: tlk8q@virginia.edu

© Springer Nature Switzerland AG 2019

T. M. Rassias and V. A. Zagrebnov (eds.), *Analysis and Operator Theory*, Springer Optimization and Its Applications 146,
https://doi.org/10.1007/978-3-030-12661-2_10

positive. In this paper, we will always assume that $f$ and $g$ are bounded measurable real functions. To state the main result of [2], for $a > 0$ define

$$K_a = \{f : \mathbb{R} \to \mathbb{R} |\ f \text{ is bounded and has an analytic continuation to the strip } |\mathrm{Im} z| < a \text{ with } \mathrm{Im} f(z)\mathrm{Im} z \geqslant 0\}.$$

Then

**Theorem 1.1** (Kato) *If $f \in K_a$, $g \in K_b$ with $ab \geqslant \pi/2$, then $i[f(P), g(Q)] \geqslant 0$.*

As we will see later, if $f$ and $g$ are two bounded real measurable functions for which $i[f(P), g(Q)] = C \geqslant 0$, then as noted by Kato, $C$ is trace class. It is thus natural to look at the case where $C$ is a nonzero, rank one operator. Kato does this in [2] and shows that in this case there exist $a$ and $b$ with $ab = \pi/2$ such that $\pm f \in K_a$ and $\pm g \in K_b$ (the signs are correlated). In fact $f(x) = c_1 + d_1 \tanh(\hat{a}(x - t_1))$ and $g(x) = c_2 + d_2 \tanh(\hat{b}(x - t_2))$ where $c_j$, $d_j$, and $t_j$ are real, $d_1 d_2 > 0$, and (following Kato) $\hat{a} = \pi/2a$ and $\hat{b} = \pi/2b$, so that $\hat{a} = b$ and $\hat{b} = a$. (Kato assumes that $f$ and $g$ are absolutely continuous with derivatives in $L^1(\mathbb{R})$ but this is not necessary as we will see later.) It is clear that from these functions, more pairs of functions with positive commutators can be constructed by convolution with a positive measure. In fact as Kato shows, the family of $f$ of the form

$$f(x) = \int_{\mathbb{R}} \tanh \hat{a}(x - t) d\nu(t) + c, \tag{1.2}$$

where $c$ is real and $\nu$ is a finite positive measure, exhausts all of $K_a$. These results led Kato to state (in [2]):

"In fact there is some reason to believe that these $[f \in K_a, g \in K_b$ with $ab \geqslant \pi/2]$ are the only solutions to $[i[f(P), g(Q)] \geqslant 0]$."

Note that since $K_a \supset K_c$ whenever $0 < a < c$, the meaning of Kato's statement just quoted, as well as the meaning of the statement of Theorem 1.1, is unchanged if the inequality $ab \geqslant \pi/2$ is replaced by the equality $ab = \pi/2$.

Kato also shows that if $i[f(P), g(Q)] = 0$ and both $f$ and $g$ are absolutely continuous with $L^1(\mathbb{R})$ derivatives then at least one of them is constant. In much of this paper, we will relax this assumption and only assume that $f$ and $g$ are bounded measurable real functions. This opens up another interesting possibility. We will show that for $f$ and $g$ bounded, real, and measurable,

**Theorem 1.2** *The commutator $[f(P), g(Q)] = 0$ if and only if either $f$ or $g$ is almost everywhere constant or both have periodic versions with periods $\tau_f$ and $\tau_g$ satisfying $\tau_f \tau_g = 2\pi$.*

Actually, the theorem is still true without the assumption that $f$ and $g$ are real.

There is a striking difference in the set of allowed $f$ and $g$ when $C = i[f(P), g(Q)] \geqslant 0$ and in addition we impose $C \neq 0$. We will show the following theorem.

**Theorem 1.3** *Suppose $C = i[f(P), g(Q)] \geqslant 0$ and in addition $C \neq 0$. Then there are versions of $f$ and $g$ which are both monotone (either both increasing or both decreasing).*

For convenience, let us formulate a conjecture which we shall call K (after Kato):

**Conjecture 1.4** *Suppose $C = i[f(P), g(Q)] \geqslant 0$ and in addition $C \neq 0$. Suppose both $f$ and $g$ are increasing. Then, there exist $a$ and $b$ with $ab = \pi/2$ such that $f \in K_a$ and $g \in K_b$.*

Unfortunately, we are far from proving K. But in the rest of this paper, we will give several results which illuminate the properties of the set of $f$ and $g$ for which the commutator is positive. In addition to the theorems of this Introduction and their proofs, see in particular the section on $2 \times 2$ positivity which gives inequalities which $f$ and $g$ must satisfy under a mild assumption.

**Theorem 1.5** *Suppose $C = i[f(P), g(Q)] \geqslant 0$ and in addition $C \neq 0$. Suppose both $f$ and $g$ are increasing. Then, both $f$ and $g$ have continuous versions which are strictly increasing. Taking $f$ and $g$ to be continuous, their inverse functions are absolutely continuous.*

A hint that K might be true is the following result:

**Theorem 1.6** *Suppose $g$ is nonconstant, lies in $K_b$, and has the integral representation*

$$g(x) = \int_{\mathbb{R}} \tanh \hat{b}(x - t) d\mu(t) + d \tag{1.3}$$

*where $d$ is real and $\mu$ is a finite positive measure such that*

$$\int |t| e^{2\hat{b}t} d\mu(t) < \infty$$

*or*

$$\int |t| e^{-2\hat{b}t} d\mu(t) < \infty.$$

*If $f \in L^\infty(\mathbb{R})$ and $i[f(P), g(Q)] \geqslant 0$ then $f \in K_{\hat{b}}$.*

There is a reason that a condition such as this assumption of exponential decay is required. Just because we have assumed $g \in K_b$ does not mean that $g$ is not in a smaller class $K_e$ with $e > b$. Suppose $c$ is the largest number $e$ such that $g \in K_e$. Then $\hat{c} = \pi/2c < \hat{b}$. We could not be able to prove that $f \in K_{\hat{b}}$ when only $f \in K_{\hat{c}}$ is required. To see that our exponential decay assumption eliminates this possibility, note that if $b$ and $c$ are as above, then besides (1.3) there is another representation of $g$ of the form

$$g(x) = \int_{\mathbb{R}} \tanh \hat{c}(x - t) d\mu_1(t) + d_1$$

where $\mu_1$ is a finite positive measure and $d_1$ is a real constant. We have an explicit formula for the imaginary part of $g$ on the line $z = x + ib$,

$$\mathrm{Im}g(x+ib) = \int_{\mathbb{R}} \sin(b\pi/c)(\cosh(2\hat{c}(x-t)) + \cos(b\pi/c))^{-1} d\mu_1(t) \geqslant k(\cosh 2\hat{c}x)^{-1}$$

for some positive $k$. If we note that (see [2]) $d\mu(t) = (2\pi)^{-1}\mathrm{Im}g(t+ib)dt$, it follows that $\int |t|e^{\pm 2\hat{b}t} d\mu(t) = \infty$.

The analyticity requirements of the conjecture K remind us of a theorem of Loewner [3]. To state this theorem, let $(a, b)$ be an open interval (finite or infinite). We say the continuous function $f : (a, b) \to \mathbb{R}$ is matrix monotone if for all $n$ and all self-adjoint $n \times n$ matrices $A$ and $B$ both with spectrum in $(a, b)$ satisfying $A \geqslant B$ we have $f(A) \geqslant f(B)$.

**Theorem 1.7** (Loewner) *$f$ is matrix monotone if and only if $f$ is the restriction to $(a, b)$ of a function analytic in $\{Imz \neq 0\} \cup (a, b)$ satisfying $Imf(z)Imz \geqslant 0$.*

It is not hard to see that if $f$ is matrix monotone on $(a, b)$ then for any two self-adjoint operators $A$ and $B$ on a Hilbert space with spectrum in $(a, b)$ and satisfying $A \geqslant B$ we have $f(A) \geqslant f(B)$.

We can formulate the positivity of the commutator $i[f(P), g(Q)]$ in a way that makes a connection with Loewner's theorem in the following way, at least formally (since we have not shown that $f$ is absolutely continuous, for example). Since

$$d/dt\left(e^{itf(P)}g(Q)e^{-itf(P)}\right) = e^{itf(P)}i[f(P), g(Q)]e^{-itf(P)},$$

if the commutator is positive and nonzero we have

$$g(e^{itf(P)}Qe^{-itf(P)}) = g(Q + tf'(P)) \geqslant g(Q)$$

for positive $t$. As we have seen we can assume that $f$ (and $g$) are increasing so that at least formally $Q + tf'(P) \geqslant Q$ for positive $t$. Thus with $A = Q + tf'(P)$ and $B = Q$ we have $A \geqslant B$ for $t$ positive while

$$g(A) \geqslant g(B) \text{ for all such } t$$

is the same as the positivity of the commutator.

In the following sections, we prove Theorems 1.2, 1.3, 1.5, and 1.6 and add some further information about this fascinating problem.

If $C$ is an operator and $C \geqslant 0$, we will say that $C$ is positive, although perhaps nonnegative would be more accurate. Similarly, we sometimes call a nondecreasing function an increasing function. For the inner product of two vectors $h$ and $k$ in Hilbert space, we write $(h, k)$, linear in $k$, and conjugate–linear in $h$. The Fourier transform of a function $h$ on $\mathbb{R}$ is denoted by $\hat{h}$ and includes the factor $(2\pi)^{-1/2}$, while $\check{h}$ denotes the inverse Fourier transform.

## 2 Finite Rank Commutators: $i[\tanh \alpha P, \tanh \beta Q]$

In this section, we consider the commutator $i[f(P), g(Q)]$ where $f$ and $g$ are the basic functions from which all functions in the Kato classes $K_a$ are constructed. We see that

$$i[\tanh \alpha P, \tanh \beta Q] = 4i[(1 + e^{2\alpha P})^{-1}, (1 + e^{2\beta Q})^{-1})]$$

so that the positivity of the commutator with tanh is just a statement about the positivity of the commutator of resolvents of the exponential function. For that reason it is interesting to note the easily verified fact that for $\alpha$ and $\beta$ real with $(2\alpha)(2\beta) = 2n\pi$ and $n \in \mathbb{Z}$

$$e^{2i\alpha P} e^{2i\beta Q} e^{-2i\alpha P} = e^{2i\beta(Q+2\alpha)} = e^{2i\beta Q}$$

and thus

$$[(\lambda + e^{2i\alpha P})^{-1}, (\lambda + e^{2i\beta Q})^{-1})] = 0,$$

for $|\lambda| \neq 1$. A formal calculation with unbounded operators would yield

$$e^{2\alpha P} e^{2\beta Q} e^{-2\alpha P} = e^{2\beta(Q-2i\alpha)} = e^{2\beta Q}$$

thus leading to

$$[(1 + e^{2\alpha P})^{-1}, (1 + e^{2\beta Q})^{-1})] = 0$$

if $(2\alpha)(2\beta) = 2n\pi$. But this is incorrect as we see from the next proposition.

**Proposition 2.1** *The integral kernel of the commutator* $i[\tanh \alpha P, \tanh \beta Q] = 4i[(1 + e^{2\alpha P})^{-1}, (1 + e^{2\beta Q})^{-1}]$, *for* $\alpha$ *and* $\beta$ *real and* $(2\alpha)(2\beta) = 2\pi n > 0$, *is given by*

$$(\beta/n\pi) \sum_{k=0}^{n-1} \psi_k(x)\phi_k(y) \tag{2.1}$$

*where*

$$\psi_k(x) = (\cosh \beta x)^{-1} e^{(n-1-2k)\beta x/n}$$

*and*

$$\phi_k(x) = (\cosh \beta x)^{-1} e^{-(n-1-2k)\beta x/n}$$

Thus $i[\tanh \alpha P, \tanh \beta Q]$ is a rank $n$ operator if $(2\alpha)(2\beta) = 2\pi n$. Note that for the rank one case (the one considered by Kato) where $(2\alpha)(2\beta) = 2\pi$, the commutator is positive. We have assumed $n > 0$. Reversing the sign of $n$ just amounts to a sign change in the commutator. If $n > 1$ this operator is not positive. This can be seen by looking at the matrix $K_{ij} := K(x_i, x_j)$ in the $2 \times 2$ case. We have

$$K_{ij} = (\beta/n\pi)(\cosh\beta x_i \cosh\beta x_j)^{-1} \sum_{k=0}^{n-1} f_k(x_i - x_j) \tag{2.2}$$

where $f_k(x) = e^{(n-(2k+1))\beta x/n}$ and thus in the $2 \times 2$ case

$$\det(K_{ij}) = K_{11}K_{22} - K_{12}^2 = \tag{2.3}$$

$$(\beta/n\pi)^2(\cosh\beta x_1 \cosh\beta x_2)^{-2}(n^2 - |\sum_{k=0}^{n-1} f_k(x_1 - x_2)|^2) \tag{2.4}$$

Using the fact that for a positive number $a \neq 1$ we have $a + a^{-1} > 2$, we see that if $x_1 - x_2 \neq 0$ and $n > 1$, $\sum_{k=0}^{n-1} f_k(x_1 - x_2) > n$ and thus the determinant is negative.

*Remark 2.2* The noncommutativity of $e^{\alpha P}$ and $e^{\beta Q}$ for $\alpha\beta = \pm 2\pi$ has been a source of counterexamples for the uniqueness of the representation of the canonical commutation relations [4] and the hypotheses under which the so-called virial theorem is true [5]. In the latter reference, one can also see in what sense these operators commute.

Before proving Proposition 2.1 we need the following lemma.

**Lemma 2.3** *If $f$ and $g$ are bounded, $f$ is monotone increasing, and $g$ is smooth with bounded derivatives then for $\psi \in L^2(\mathbb{R})$*

$$(i[f(P), g(Q)]\psi)(x) = \int K(x, y)\psi(y)dy \tag{2.5}$$

*where*

$$K(x, y) = \frac{1}{\sqrt{2\pi}} \frac{g(x) - g(y)}{x - y} \widehat{df}(y - x). \tag{2.6}$$

*If $g$ is smooth with bounded derivatives, and $f$ is bounded then if $\psi \in \mathcal{S}(\mathbb{R})$ (the Schwartz space) we have*

$$(i[f(P), g(Q)]\psi)(x) = \int \frac{1}{\sqrt{2\pi}} \frac{g(x) - g(y)}{x - y} \widehat{f'}(y - x)\psi(y)dy$$

*where $f$ is considered a tempered distribution.*

*Remarks 2.4* If $f$ is monotone increasing, we take the version of $f$ which is right continuous in the definition of $df$. See Lemma 5.1 for a more complete result.

*Proof* We first assume $\psi \in \mathcal{S}(\mathbb{R})$.

$$(i[f(P), g(Q)]\psi)(x) = i \int \left( \int f(\xi)(g(y) - g(x))\psi(y)e^{-i\xi y}dy \right) e^{i\xi x} d\xi/2\pi.$$

Let

$$h_x(y) = \frac{g(y+x) - g(x)}{y} \psi(y+x)/\sqrt{2\pi}.$$

Then

$$(i[f(P), g(Q)]\psi)(x) = i \int f(\xi) i(\widehat{h_x})'(\xi) d\xi = \int \widehat{h_x}(\xi) df(\xi) = \int h_x(y) \widehat{df}(y) dy$$

where in the second equality we have used the integration by parts formula [6] in the case where we assume $f$ is monotone. After a change of variable, this gives the first result (2.5) for $\psi \in \mathcal{S}(\mathbb{R})$. A limiting argument gives (2.5) for $\psi \in L^2(\mathbb{R})$. If we only assume $f$ is bounded, the first equality in the last equation gives our result when we note that $h_x \in \mathcal{S}(\mathbb{R})$ and remember the definition of the derivative and the Fourier transform of a tempered distribution. □

*Proof of Proposition* 2.1 The kernel $K(x, y)$ is given by

$$(2\pi)^{-1/2} \frac{\tanh \beta x - \tanh \beta y}{x - y} \hat{f}'(y - x)$$

with $f(\xi) = \tanh \alpha\xi$. Using the known Fourier transform of $(\cosh x)^{-2}$ (see (6.11)), we obtain

$$K(x, y) = (\beta/n\pi) \frac{\tanh \beta x - \tanh \beta y}{\sinh(\beta(x - y)/n)}.$$

We calculate

$$\tanh \beta x - \tanh \beta y = \sinh \beta(x - y)(\cosh \beta x \cosh \beta y)^{-1}.$$

Using

$$a^n - b^n = (a - b)(a^{n-1} + a^{n-2}b + \cdots + b^{n-1})$$

with $a = e^{\beta x/n}$, $b = e^{-\beta x/n}$, we obtain

$$\frac{\sinh \beta x}{\sinh \beta x/n} = \sum_{k=0}^{n-1} e^{(n-(2k+1))\beta x/n}$$

which gives the result. □

## 3 Theorem 1.2—The Case $[f(P), g(Q)] = 0$

In this section, we consider the case where for real bounded measurable functions $f$ and $g$, we have $C = i[f(P), g(Q)] = 0$.

*Proof of Theorem* 1.2 If $f$ is periodic with period $a = \tau_f$ then $e^{iaQ} f(P) e^{-iaQ} = f(P-a) = f(P)$, so $e^{iaQ}$ commutes with $f(P)$. If $g$ is periodic with period $\tau_g = 2\pi/\tau_f$, then $g(Q)$ is a function of $e^{iaQ}$ and thus commutes with $f(P)$.

For the converse first assume $f$ and $g$ are $C^\infty$ with bounded derivatives and $f(P)$ and $g(Q)$ commute. Suppose $f$ is not constant and suppose $k_1$ and $k_2$ are two points so that $f(k_1) \neq f(k_2)$. Choose open neighborhoods $N_j$ of $k_j$ with $\overline{N_j}$ compact such that $f(\overline{N_1}) \cap f(\overline{N_2}) = \emptyset$. Choose a bounded continuous function $F$ with $F = 0$ on $f(\overline{N_1})$ and $F = 1$ on $f(\overline{N_2})$. Note that $(F \circ f)(P)$ commutes with $g(Q)$ so that if $\widehat{\phi_j} \in C_0^\infty(N_j)$

$$0 = (\phi_1, [F \circ f(P), g(Q)]\phi_2) = -(\phi_1, g(Q)\phi_2).$$

Considering $g$ as a tempered distribution and denoting $\psi_1 = \overline{\hat{\phi}_1} \in C_0^\infty(N_1)$, $\psi_2 = \hat{\phi}_2 \in C_0^\infty(N_2)$, we have

$$0 = g(\overline{\phi_1}\phi_2) = \hat{g}_r(\widehat{\overline{\phi_1}\phi_2}) = \int \hat{g}_r(\psi_2(\xi + \cdot)\psi_1(\xi) d\xi/\sqrt{2\pi}$$

where $h_r(x) = h(-x)$. Since this is true for all $\psi_1 \in C_0^\infty(N_1)$, $\hat{g}_r(\psi_2(\cdot + \xi)) = 0$ for all $\xi \in N_1$ and thus $\hat{g}_r(\psi) = 0$ for all $\psi \in C_0^\infty(N_2 - \xi)$ for all $\xi \in N_1$ or more concisely $\hat{g}_r(\psi) = 0$ for all $\psi \in C_0^\infty(N_2 - N_1)$. Thus $(\operatorname{supp} \hat{g}) \cap (N_1 - N_2) = \emptyset$. We have thus proved that

$$f(k_1) \neq f(k_2) \text{ implies } k_1 - k_2 \notin \operatorname{supp} \hat{g}$$

or

$$k_1 - k_2 \in \operatorname{supp} \hat{g} \text{ implies } f(k_1) = f(k_2)$$

or

$$\operatorname{supp} \hat{g} \subset P_f := \text{ the set of periods of } f \tag{3.1}$$

It follows that either $\operatorname{supp} \hat{g} = \{0\}$ in which case $g$ is constant, or $f$ is periodic. We have already assumed that $f$ is not constant; now assume $g$ is not constant. Then $f$ has a smallest positive period, $\tau_f$, and

$$\operatorname{supp} \hat{g} \subset \tau_f \mathbb{Z}.$$

If we choose $\phi \in C_0^\infty((-\tau_f/2, \tau_f/2))$ with $\phi = 1$ in a neighborhood of the origin, then there is a sequence of integers $\{n_j\}$ such that $\hat{g} = \sum_j \phi_j \hat{g}$ where $\phi_j(\xi) = \phi(\xi - n_j \tau_f)$. Thus $\hat{g} = \sum_j c_j \delta_{n_j \tau_f}$ with $\delta_k$ the Dirac delta at $k$. It follows that $g = \sum_j c_j e^{i n_j \tau_f x}/\sqrt{2\pi}$ which implies $g$ has period $\tau_g = 2\pi/\tau_f$. This proves the result when $f$ and $g$ are smooth with bounded derivatives. In the general case convolve $f$ and $g$ with the Gaussian $\delta_0^\sigma(x) = \frac{1}{\sqrt{2\pi}\sigma} e^{-|x|^2/2\sigma^2}$. The resulting functions of $P$ and $Q$ still commute. Since

$$\operatorname{supp}(\widehat{\delta_0^\sigma * f}) = \operatorname{supp} \hat{f}$$
$$\operatorname{supp}(\widehat{\delta_0^\sigma * g}) = \operatorname{supp} \hat{g},$$

it follows that $\delta_0^\sigma * f$ and $f$ have the same periods as do $\delta_0^\sigma * g$ and $g$. This completes the proof. □

*Remark 3.1* Actually, the result is true without the requirement that the functions $f$ and $g$ be real. The complex case follows from the result just proved and a result of Fuglede [7] (see also [8] for a very simple proof) which states that if a normal operator commutes with another operator then so does its adjoint. Thus the real and imaginary parts of $f(P)$ and $g(Q)$ commute with one another. With this information the proof is straightforward.

## 4 Theorem 1.3—Monotonicity

If we assume that $f$ and $g$ are absolutely continuous with derivatives in $L^1(\mathbb{R})$, the proof of monotonicity is given by Kato in [2] and is straightforward. We give a sketch of the proof: The integral kernel of the positive operator $i[f(P), g(Q)]$ is

$$K(x, y) = (\sqrt{2\pi})^{-1} \frac{g(x) - g(y)}{x - y} \widehat{f'}(y - x).$$

With $[f] = f(\infty) - f(-\infty)$ the condition of positivity implies $K(x, x) = g'(x)$ $[f]/2\pi \geqslant 0$ and $K(x, x)K(y, y) \geqslant |K(x, y)|^2$ or $g'(x)g'(y)[f]^2 \geqslant (2\pi)^2|K(x, y)|^2$. Thus, unless $K$ is identically zero, $[f] \neq 0$ which implies $g$ is monotone. Using complex conjugation, $C_0$, and the Fourier transform, $\mathcal{F}$, we note that

$$\mathcal{F}^{-1} C_0 i[f(P), g(Q)] C_0^{-1} \mathcal{F} = i[g(P), f(Q)] \tag{4.1}$$

which is therefore positive. Thus, the same argument gives that $f$ is also monotone and it follows they are both either increasing or both decreasing.

*Proof of Theorem* 1.3 We first assume that $f$ and $g$ are infinitely differentiable with bounded derivatives. In the proof, we use the distribution kernel of the operator $C = i[f(P), g(Q)]$ given by

$$K(x, y) = (\sqrt{2\pi})^{-1} \frac{g(x) - g(y)}{x - y} \widehat{f'}(y - x)$$

We do not know that $f'$ is integrable which accounts for the distribution nature of the kernel.

Since the continuity of the kernel is not yet known, we use a smeared out version of the inequality $|K(x_1, x_2)|^2 \leqslant K(x_1, x_1)K(x_2, x_2)$. Put

$$\delta_a^\sigma(x) = \frac{1}{\sqrt{2\pi}\sigma} e^{-|x-a|^2/2\sigma^2}$$

and note that

$$\delta_a^\sigma(x)\delta_a^\sigma(y) = \delta_0^{\sqrt{2}\sigma}(x-y)\delta_a^{\sigma/\sqrt{2}}((x+y)/2).$$

For $\phi \in C_0^\infty(\mathbb{R})$, let

$$I_\sigma = (\delta_0^\sigma * \phi, C\delta_0^\sigma * \phi) = \int (\delta_{x_1}^\sigma, C\delta_{x_2}^\sigma)\overline{\phi(x_1)}\phi(x_2)dx_1dx_2 \tag{4.2}$$

and note

$$|(\delta_{x_1}^\sigma, C\delta_{x_2}^\sigma)| \leqslant (\delta_{x_1}^\sigma, C\delta_{x_1}^\sigma)^{1/2}(\delta_{x_2}^\sigma, C\delta_{x_2}^\sigma)^{1/2}$$

so that

$$\begin{aligned} I_\sigma &\leqslant \Big(\int (\delta_x^\sigma, C\delta_x^\sigma)^{1/2}|\phi(x)|dx\Big)^2 \leqslant \int_{\operatorname{supp}\phi} 1dx \int (\delta_x^\sigma, C\delta_x^\sigma)|\phi(x)|^2dx \\ &= c\int \delta_0^{\sqrt{2}\sigma}(x-y)K(x,y)\delta_0^{\sigma/\sqrt{2}}((x+y)/2 - x_1)|\phi(x_1)|^2dx_1dxdy \\ &= c\int \delta_0^{\sqrt{2}\sigma}(x-y)K(x,y)\delta_0^{\sigma/\sqrt{2}} * |\phi|^2((x+y)/2)dxdy \\ &=: cJ_\sigma/\sqrt{2\pi} \end{aligned} \tag{4.3}$$

where $c = \int_{\operatorname{supp}\phi} 1dx$. We abbreviate $\psi_\sigma = \delta_0^{\sigma/\sqrt{2}} * |\phi|^2$ and use Taylor's theorem to write

$$\frac{g(y+t) - g(y)}{t} = g'(y) + tG_1(y,t) \text{ where} \tag{4.4}$$

$$G_1(y,t) = \int_0^1 g^{(2)}(y+\theta t)(1-\theta)d\theta. \tag{4.5}$$

Thus, we obtain

$$\begin{aligned} J_\sigma = &\int \delta_0^{\sqrt{2}\sigma}(t)g'(y)\widehat{f}'(-t)\psi_\sigma(y+t/2)dtdy \\ &+ \int t\delta_0^{\sqrt{2}\sigma}(t)G_1(y,t)\widehat{f}'(-t)\psi_\sigma(y+t/2)dtdy \end{aligned} \tag{4.6}$$

and replacing $\psi_\sigma(y+t/2)$ by $\psi_\sigma(y)$ we pick up another error term so that

$$J_\sigma = \int \delta_0^{\sqrt{2}\sigma}(t) g'(y) \widehat{f}'(-t)\psi_\sigma(y)dtdy + \int t\delta_0^{\sqrt{2}\sigma}(t) G_2^\sigma(y,t)\widehat{f}'(-t)dtdy \text{ where} \tag{4.7}$$

$$G_2^\sigma(y,t) = g'(y)(\psi_\sigma(y+t/2) - \psi_\sigma(y))/t + G_1(y,t)\psi_\sigma(y+t/2). \tag{4.8}$$

We now have

$$\begin{aligned}
J_\sigma &= J_\sigma^1 + J_\sigma^2 \\
J_\sigma^1 &= \left(\int \delta_0^{\sqrt{2}\sigma}(t)\widehat{f}'(-t)dt\right)\left(\int g'(y)\psi_\sigma(y)dy\right) \\
J_\sigma^2 &= \int I_2^\sigma(y)dy \\
I_2^\sigma(y) &= \int G_2^\sigma(y,t)t\delta_0^{\sqrt{2}\sigma}(t)\widehat{f}'(-t)dt = \int F_y^\sigma(\xi)f(\xi)d\xi \\
F_y^\sigma(\xi) &= (-i/\sqrt{2\pi})\int G_2^\sigma(y,t)t^2\delta_0^{\sqrt{2}\sigma}(t)e^{i\xi t}dt.
\end{aligned} \tag{4.9}$$

We estimate $D^m\psi_\sigma = \delta_0^{\sigma/\sqrt{2}} * D^m|\phi|^2$ to find for all $m$ and $n$

$$|D^m\psi_\sigma(x)| \leqslant c_{n,m}(1+|x|)^{-n}$$

uniformly for $\sigma \in (0,1)$. We use $G_2^\sigma(y,t) = \int_0^{1/2} \psi_\sigma'(y+\theta t)d\theta g'(y) + G_1^\sigma(y,t)\psi_\sigma(y+t/2)$ and obtain

$$|G_2^\sigma(y,t)| \leqslant c_n((1+|y+t/2|)^{-n} + (1+|y|)^{-n}).$$

Thus for $\sigma \in (0,1)$

$$\begin{aligned}
|F_y^\sigma(\xi)| &\leqslant c_n' \int ((1+|y+t/2|)^{-n} + (1+|y|)^{-n})t^2\delta_0^{\sqrt{2}\sigma}(t)dt \\
&\leqslant d_n\sigma^2(1+|y|)^{-n}.
\end{aligned} \tag{4.10}$$

We also need some decay of $F_y^\sigma(\xi)$ in $\xi$. Thus

$$|(1+\xi^2)F_y^\sigma(\xi)| \leqslant c\int |(1-d^2/dt^2)G_2^\sigma(y,t)t^2\delta_0^{\sqrt{2}\sigma}(t)|dt. \tag{4.11}$$

We easily see that $|d^m/dt^m G_2^\sigma(y,t)| \leqslant c_m'$ uniformly for $\sigma \in (0,1)$. Thus differentiating $t^2\delta_0^{\sqrt{2}\sigma}(t)$ and integrating in (4.11), we find

$$(1+\xi^2)|F_y^\sigma(\xi)| \leqslant c'. \tag{4.12}$$

Interpolating between (4.10) and (4.12), we obtain for $\theta \in (0, 1)$

$$|F_y^\sigma(\xi)| \leqslant c_n''(1+\xi^2)^{-(1-\theta)}\sigma^{2\theta}(1+|y|)^{-n\theta}.$$

It thus follows that for $\theta \in (0, 1/2)$ and any $n$

$$|I_2^\sigma| \leqslant c_{n,\theta}\sigma^{2\theta}(1+|y|)^{-n\theta}$$

which gives

$$|J_2^\sigma| \leqslant c_\theta \sigma^{2\theta}.$$

We have thus shown

$$J_\sigma/\sqrt{2\pi} = \int (\delta_x^\sigma, C\delta_x^\sigma)|\phi(x)|^2 dx = \left(\int \delta_0^{\sqrt{2}\sigma}(t)\widehat{f}'(-t)dt\right)\left(\int g'(y)\psi_\sigma(y)dy\right) + O(\sigma^{2\theta}) \tag{4.13}$$

for $\theta \in (0, 1/2)$. Notice that

$$\beta(\sigma) := \sqrt{2\pi}\int \delta_0^{\sqrt{2}\sigma}(t)\widehat{f}'(-t)dt = \int e^{-(\sigma\xi)^2} f'(\xi)d\xi \tag{4.14}$$

is independent of $\phi$. Let us assume that $g'$ is not of constant sign. Choose $\phi_1 \in C_0^\infty(\mathbb{R})$ so that $\int g'(y)|\phi_1(y)|^2 dy > 0$. Then with $\psi_\sigma^1 = \delta_0^{\sigma/\sqrt{2}} * |\phi_1|^2$, we have $\lim_{\sigma\to 0}\int g'(y)\psi_\sigma^1(y)dy = \int g'(y)|\phi_1(y)|^2 dy > 0$ and thus

$$\beta(\sigma) = J_\sigma\Big(\int g'(y)\psi_\sigma^1(y)dy\Big)^{-1} + O(\sigma^{2\theta}) \tag{4.15}$$

which implies that for small $\sigma$, $\beta(\sigma) \geqslant -c_1\sigma^{2\theta}$ for some $c_1 > 0$ since $J_\sigma \geqslant 0$. Similarly choosing $\phi_2 \in C_0^\infty(\mathbb{R})$ so that $\int g'(y)|\phi_2(y)|^2 dy < 0$, we find for small $\sigma$ $\beta(\sigma) \leqslant c_2\sigma^{2\theta}$ for some $c_2 > 0$. It follows that for small $\sigma$, for any $\phi \in C_0^\infty(\mathbb{R})$, $J_\sigma = O(\sigma^{2\theta})$. But since $I_\sigma = (\delta_0^\sigma * \phi, C\delta_0^\sigma * \phi) \leqslant cJ_\sigma/\sqrt{2\pi}$, we can take the limit $\sigma \to 0$ and conclude $(\phi, C\phi) = 0$ for all $\phi \in C_0^\infty(\mathbb{R})$ and thus $C = 0$. Hence, if $C \geqslant 0$ and $C \neq 0$, $g$ is a monotone function. By (4.1) the same holds for $f$. Combining (4.13) and (4.14), we see that

$$\begin{aligned} I_\sigma &= (\delta_0^\sigma * \phi, C\delta_0^\sigma * \phi) \leqslant cJ_\sigma/\sqrt{2\pi} \\ &\leqslant c(1/\sqrt{2\pi})\left(\int e^{-(\sigma\xi)^2} f'(\xi)d\xi\right)\left(\int g'(y)\psi_\sigma(y)dy\right) + O(\sigma^{2\theta}). \end{aligned}$$

Taking $\sigma \to 0$, we obtain

$$(\phi, C\phi) \leqslant c(1/\sqrt{2\pi}) \int f'(\xi)d\xi \int g'(y)|\phi(y)|^2 dy$$

(where again $c = \int_{\text{supp}\,\phi} 1dx$) so that $g'(x)f'(\xi) \geqslant 0$.

To deal with the general case where neither $f$ nor $g$ is known to be smooth note that if $i[f(P), g(Q)] \geqslant 0$ and nonzero, the same is true if we replace $g$ with $g * \rho$ where $\rho$ is a smooth nonnegative approximation to the identity. Similarly for $f$ replaced with $f * \zeta$. It then follows that both $g * \rho$ and $f * \zeta$ are monotone. By taking $\rho_n$ a suitable sequence, $g * \rho_n \to g$ on a set $E$ of full Lebesgue measure and thus $g$ is monotone on $E$. If we take $G(x) = \lim_{u \in E, u \uparrow x} g(u)$, then $G$ is monotone and equals $g$ a.e. Clearly the same idea works for $f$. It is easy to see that $G$ and $F$, the corresponding version of $f$, are either both increasing or both decreasing. □

## 5 Theorem 1.5—Continuity of g, Absolute Continuity of the Inverse of g

In this section, we assume that the commutator $i[f(P), g(Q)]$ is nonzero and positive.

**Lemma 5.1** *Suppose $i[f(P), g(Q)] = C$ where $C$ is positive and $f$ and $g$ are monotone nondecreasing and bounded. Then for all $\psi \in L^2(\mathbb{R})$, $(i[f(P), g(Q)]\psi)(x) = \int K(x, y)\psi(y)dy$ where*

$$K(x, y) = \frac{1}{\sqrt{2\pi}} \frac{g(x) - g(y)}{x - y} \widehat{df}(y - x). \tag{5.1}$$

*The operator $C$ is trace class with* $\operatorname{tr} C = [f][g]/2\pi$, *and the kernel $K$ is square integrable.*

*Proof* Let $g_t = \phi_t * g$ where $\phi$ is a nonnegative smooth function of compact support whose integral is 1 and $\phi_t(x) = t^{-1}\phi(t^{-1}x)$. Then

$$i[f(P), g_t(Q)] = \int_{\mathbb{R}} C(a)\phi_t(a)da$$

where $C(a) = e^{-iPa}Ce^{iPa}$. The commutator $i[f(P), g_t(Q)]$ is a positive operator with continuous kernel so we can calculate its trace which is easily seen to be $[f][g]/2\pi$. We have

$$\operatorname{tr} \int_{\mathbb{R}} C(a)\phi_t(a)da = \int (\operatorname{tr} e^{-iPa}Ce^{iPa})\phi_t(a)da = \int \operatorname{tr}(C)\phi_t(a)da = \operatorname{tr} C$$

which shows $\operatorname{tr} C = [f][g]/2\pi$. Since $i[f(P), g_t(Q)]$ has constant trace its Hilbert–Schmidt norm is uniformly bounded. Thus if we abbreviate $K_t(x, y) = \int K(x-a, y-a)\phi_t(a)da$, we have

$$\infty > \liminf_{t\downarrow 0} \int |K_t(x, y)|^2 dxdy \geqslant \int |K(x, y)|^2 dxdy$$

by Fatou's lemma. Here, we have used that $K_t \to K$ a.e. It follows easily that $K_t \to K$ in $L^2(\mathbb{R}^2)$. For an $L^2(\mathbb{R})$ function $\psi$, let $F_t = \int K_t(x, y)\psi(y)dy$ and $F = \int K(x, y)\psi(y)dy$. We have $||F_t - F|| \leqslant ||K_t - K||_{L^2(\mathbb{R}^2)}||\psi||$. But $F_t \to C\psi$ strongly which gives (5.1). This finishes the proof. □

**Lemma 5.2** *Suppose $i[f(P), g(Q)] = C$ where $C$ is nonzero and positive, while $f$ and $g$ are bounded and monotone increasing. Then $g$ is continuous.*

*Proof* We use the square integrability of the kernel $K$. Suppose $g$ has a jump of $k'$ at 0. Then for all $\epsilon > 0$

$$\int_{x<-\epsilon, y>\epsilon} |(g(y) - g(x))/(y - x)|^2 |\widehat{df}(y - x)|^2 dxdy < k$$

for some $k$. Now $|\widehat{df}(y - x)| \geqslant \delta > 0$ for $y - x$ small, while $g(y) - g(x) \geqslant k' > 0$ in the range of integration. This contradicts the above inequality. □

**Lemma 5.3** *Suppose $i[f(P), g(Q)] = C$ where $C$ is positive and $f$ and $g$ are monotone nondecreasing and bounded. Then $g$ is strictly increasing.*

*Proof* Suppose $g$ is constant on $I = (a, b)$. Then by the explicit form of the kernel $K$ of the commutator $C = i[f(P), g(Q)]$, we see that for $\psi \in C_0^\infty(I)$, $(\psi, C\psi) = 0$. Since $C \geqslant 0$, this means $C\psi = 0$ which in turn implies $(\frac{g(x)-g(y)}{x-y})\widehat{df}(y - x) = 0$ for $y \in I$ and $x \in \mathbb{R}$. But $\widehat{df}$ is continuous and nonzero at 0, so nonzero in an interval $J = (-c, c)$, $c > 0$. Thus $g$ is constant on the interval $I + J$. The result follows by induction. □

We give a quick proof of Putnam's theorem [9, 10] on positive commutators:

**Theorem 5.4** (Putnam) *Suppose $A$ and $H$ are self-adjoint operators with $A$ bounded. Suppose $i[H, A] = C$ with $C$ self-adjoint and positive. We assume this equality is true in the sense that for each $\psi \in D(H)$*

$$i(H\psi, A\psi) - i(A\psi, H\psi) = ||C^{1/2}\psi||^2.$$

*Then, the absolutely continuous subspace of $H$, $\mathcal{H}_{ac}(H) \supset \overline{\operatorname{Ran} C}$.*

*Remark 5.5* It will be clear in the proof that $A$ need not be bounded but it rather suffices that $D(A)$ contain all vectors in the range of $E_H(J)$ for all finite intervals $J$. (Here $E_H(\cdot)$ is the projection-valued spectral measure associated with the operator $H$.) This generalization may have limited applicability so we do not include a proof (although it should be obvious).

*Proof* For a finite interval $I = (a, b)$, let $\lambda = (a + b)/2$, and $E = E_H(I)$. We use the notation $|J|$ for the Lebesgue measure of the Borel set $J$. Then note

$$\begin{aligned} |(E\psi, i[H, A]E\psi)| &= |(E\psi, i[H - \lambda I, A]E\psi)| \\ &= |((H - \lambda)E\psi, AE\psi) - (AE\psi, (H - \lambda)E\psi)| \\ &\leqslant 2(|I|/2)||A||||\psi||^2. \end{aligned}$$

Thus

$$||ECE|| \leqslant |I|||A||.$$

We also have

$$||ECE|| = ||(EC^{1/2})(EC^{1/2})^*|| = ||(EC^{1/2})^*(EC^{1/2})|| = ||C^{1/2}EC^{1/2}||$$

so that if $||\psi|| = 1$,

$$(C^{1/2}\psi, E_H(I)C^{1/2}\psi) \leqslant |I|||A||.$$

This inequality extends from finite open intervals, $I$, to arbitrary Borel sets and shows that $\mathcal{H}_{ac}(H) \supset \text{Ran}\, C^{1/2}$. But it is easily seen that $\overline{\text{Ran}\, C^{1/2}} = \overline{\text{Ran}\, C}$ which proves the theorem. □

We use Putnam's theorem to show absolute continuity of the inverse of the function $g$.

**Proposition 5.6** *Suppose $i[f(P), g(Q)] = C$, where $f$ and $g$ are real increasing bounded functions and $C$ is positive and nonzero. Then the inverse function $g^{-1}$ is absolutely continuous.*

*Proof* Let $M_g$ be multiplication by $g$, that is, $M_g = g(Q)$. From Putnam's theorem, we have that $\mathcal{H}_{ac}(M_g) \supset \text{Ran}\, C$ and since for $a \in \mathbb{R}$

$$i[f(P - aI), g(Q)] = e^{iaQ}Ce^{-iaQ}$$

it follows that $\mathcal{H}_{ac}(M_g) \supset e^{iaQ}\text{Ran}\, C$. Suppose $(\phi, e^{iaQ}C\psi) = 0$ for all $a \in \mathbb{R}$ and all $\psi \in L^2(\mathbb{R})$. It follows that $\phi(x)(C\psi)(x) = 0$ a.e. Choose $a > 0$ so that $\text{Re}\, \widehat{df}(x) > 0$ for $x \in I = (-a, a)$. Choose a nonnegative $\psi \in C_0^\infty(\frac{1}{2}I)$ with $\psi(0) > 0$. Then for any $x_0 \in \mathbb{R}$, $\text{Re}(C\psi(x)) = (2\pi)^{-1/2} \int \frac{g(x)-g(y)}{x-y} \text{Re}(\widehat{df}(y - x))\psi(y - x_0)dy > 0$ for $x \in \frac{1}{2}I + x_0$. Thus for all $x_0$, $\phi(x) = 0$ for a.e. $x \in \frac{1}{2}I + x_0$, or in other words $\phi$ is the zero vector in $L^2(\mathbb{R})$. It follows that $\mathcal{H}_{ac}(M_g) = L^2(\mathbb{R})$.

With $H = M_g$ we have that $E_H(J)$ is multiplication by $1_{\{x:g(x)\in J\}} = 1_{g^{-1}(J)}$. So $M_g$ being absolutely continuous as an operator means that if the Borel set $J$ has Lebesgue measure 0 then $E_{M_g}(J) = 0$ and thus $g^{-1}(J)$ has Lebesgue measure 0. Thus the strictly increasing function $g^{-1}$ is absolutely continuous. (Note that even with the fact that $g$ is strictly increasing, the absolute continuity of $M_g$ as an operator alone does not imply that $g$ is continuous.) □

We remark that because of (4.1) all the results of this section apply to $f$ as well as $g$.

## 6 Theorem 1.6—$(K_b, L^\infty) \overset{?}{\to} (K_b, K_{\hat{b}})$

*Proof of Theorem* 1.6 We assume $g$ has the integral representation (1.3) where $\mu$ is a nonzero finite positive measure with $\int |t|e^{2\hat{b}t}d\mu(t) < \infty$. The proof where $\int |t|e^{-2\hat{b}t}d\mu(t) < \infty$ is similar. Assuming the commutator $i[f(P), g(Q)]$ is non-negative, we are aiming to prove that $f$ has the integral representation

$$f(x) = \int \tanh b(x-t)d\nu(t) + d$$

for some finite positive measure $\nu$ and with $b\hat{b} = \pi/2$. We first assume that $f$ is as above but with $d\nu(t)$ replaced with $w(t)dt$ where $w \in L^1(\mathbb{R}) \cap L^\infty(\mathbb{R})$ and where $w$ is real and continuous but not necessarily nonnegative. This will be justified at the end of the proof. Our first task is to prove that $w \geqslant 0$. For the purpose of obtaining a contradiction, suppose that is not the case. Then the sets $E_1 = \{s : w(s) > 0\}$ and $E_2 = \{s : w(s) < 0\}$ are both non-empty. For $j = 1, 2$, put $w_j = 1_{E_j} \cdot |w|$, where $1_F$ denotes the characteristic function of the set $F$. Clearly, $w = w_1 - w_2$. A computation based on (2.5) shows that if $\psi$ is in $L^2(\mathbb{R})$, then

$$(\psi, i[f(P), g(Q)]\psi) = \frac{\hat{b}}{\pi}\int\int\left|\int \frac{e^{-isx}}{\cosh \hat{b}(x-t)}\psi(x)dx\right|^2 d\mu(t)\, w(s)ds \geqslant 0, \tag{6.1}$$

or equivalently

$$\int\int\left|\int \frac{e^{isx}}{\cosh \hat{b}(x-t)}\psi(x)dx\right|^2 w_2(s)ds\, d\mu(t) \le \int\int\left|\int \frac{e^{isx}}{\cosh \hat{b}(x-t)}\psi(x)dx\right|^2 w_1(s)ds\, d\mu(t),$$

for all $\psi \in L^2(\mathbb{R})$ (we find it convenient to change signs in the complex exponential in (6.1), and so have replaced $\psi$ by $\bar{\psi}$). The last inequality can be written as

$$\|A_2\psi\| \le \|A_1\psi\|, \tag{6.2}$$

with $A_j : L^2(\mathbb{R}) \to L^2(\nu_j \times \mu)$ defined by

$$(A_j\psi)(x,t) = \int_{-\infty}^{\infty} \frac{e^{isx}}{\cosh \hat{b}(x-t)}\psi(x)dx$$

for $j = 1$ or 2, where $d\nu_j(s) = w_j(s)ds$. These operators are in fact bounded, as one can see by applying (6.1) with $\nu = \nu_1$ or $\nu = \nu_2$ and with $f$ correspondingly modified. We see (writing $A$ for $A_1$ or $A_2$ as the case may be)

$$\begin{aligned}\frac{\hat{b}}{\pi}\|A\psi\|^2 &= (\bar{\psi}, i[f(P), g(Q)]\bar{\psi}) \\ &\leq 2\|f\|_\infty\|g\|_\infty\|\psi\|^2.\end{aligned}$$

Let us turn to the inequality (6.2), which can be restated as $A_2^*A_2 \leq A_1^*A_1$. According to a lemma of Douglas [11], $A_2 = CA_1$ for some contraction operator $C : L^2(\nu_1 \times \mu) \to L^2(\nu_2 \times \mu)$, so $A_2^* = A_1^*C^*$ and Ran $A_2^* \subset$ Ran $A_1^*$. We will show that this range inclusion cannot occur, and to this end look closely at the operators $A_j^*$. Note that if $h \in L^2(\nu_j \times \mu)$,

$$(A_j^*h)(x) = \int\int \frac{e^{-isx}}{\cosh \hat{b}(x-t)} h(s,t)w_j(s)ds\,d\mu(t).$$

It follows that

$$2^{-1}e^{\hat{b}x}(A_j^*h)(x) = \int\int e^{-isx}(1 - e^{2\hat{b}(t-x)}(1+e^{2\hat{b}(t-x)})^{-1})h(s,t)w_j(s)dse^{\hat{b}t}d\mu(t).$$

Further

$$e^{2\hat{b}(t-x)}(1+e^{2\hat{b}(t-x)})^{-1} = 1_{(-\infty,t)}(x) - U(x-t), \text{ with } |U(x)| \leqslant e^{-2\hat{b}|x|}.$$

We see therefore that

$$2^{-1}e^{\hat{b}x}(A_j^*h)(x) = I_j(x) + II_j(x) + III_j(x)$$

where

$$I_j(x) = \int\int e^{-isx}h(s,t)w_j(s)dse^{\hat{b}t}d\mu(t) \tag{6.3}$$

$$II_j(x) = -\int\int e^{-isx}1_{(-\infty,t)}(x)h(s,t)w_j(s)dse^{\hat{b}t}d\mu(t) \tag{6.4}$$

$$III_j(x) = \int\int e^{-isx}U(x-t)h(s,t)w_j(s)dse^{\hat{b}t}d\mu(t). \tag{6.5}$$

First consider $I_j$. We define

$$q_j(s) = \Big(\int h(s,t)e^{\hat{b}t}d\mu(t)\Big)w_j(s).$$

Then $q_j \in L^2(\mathbb{R}) \cap L^1(\mathbb{R})$. Indeed

$$\begin{aligned}\int |q_j(s)|^2 ds &= \int \Big|\int h(s,t)e^{\hat{b}t}d\mu(t)\Big|^2 w_j(s)^2 ds \\ &\leqslant \|w\|_\infty \int e^{2\hat{b}t}d\mu(t) \int\int |h(s,t)|^2 w_j(s)dsd\mu(t) < \infty\end{aligned}$$

and

$$\int |q_j(s)|ds \leqslant$$

$$\int\int |h(s,t)|w_j(s)^{1/2}e^{\hat{b}t}w_j(s)^{1/2}d\mu(t)ds$$

$$\leqslant \left(\int\int |h(s,t)|^2 w_j(s)d\mu(t)ds\right)^{1/2}\left(\int\int e^{2\hat{b}t}w_j(s)d\mu(t)ds\right)^{1/2}$$

$$= ||w_j||_1^{1/2}\left(\int e^{2\hat{b}t}d\mu(t)\right)^{1/2}\left(\int\int |h(s,t)|^2 w_j(s)dsd\mu(t)\right)^{1/2} < \infty.$$

Here, we have used the Schwarz inequality and the facts that $w_j \in L^1(\mathbb{R}) \cap L^\infty(\mathbb{R})$, $\int e^{2\hat{b}t}d\mu(t) < \infty$, and $h \in L^2(\nu_j \times \mu)$. We see that $I_j = \sqrt{2\pi}\hat{q}_j$ which implies $I_j \in L^2(\mathbb{R})$ as well as in $C_0(\mathbb{R})$, the space of continuous functions on $\mathbb{R}$ vanishing at infinity.

We turn to $II_j$. We first write $g(t,x) = \int e^{-isx}h(s,t)w_j(s)ds$ and note that by the Plancherel theorem

$$\int |g(t,x)|^2 dxd\mu(t) \leqslant 2\pi||w_j||_\infty ||h||^2_{L^2(\nu_j\times\mu)} \tag{6.6}$$

We have

$$\begin{aligned} II_j(x) = &-\int_{(-\infty,0)} 1_{(-\infty,t)}(x)g(t,x)e^{\hat{b}t}d\mu(t) \\ &-\int_{[0,\infty)} 1_{(-\infty,0)}(x)g(t,x)e^{\hat{b}t}d\mu(t) \\ &-\int_{[0,\infty)} 1_{[0,t)}(x)g(t,x)e^{\hat{b}t}d\mu(t) \end{aligned}$$

Clearly from (6.6) and the Schwarz inequality, each of these three terms is in $L^2(\mathbb{R})$. Evidently, the first two terms vanish for $x > 0$ so that their sum can be written as a Fourier transform, $\hat{v}_j$, with $v_j \in H^2_-$, the Hardy space of the lower half plane [12]. (Here, we follow the convention of identifying elements of $H^2_-$, which are analytic in the open lower half plane, with their boundary value functions which lie in $L^2(\mathbb{R})$.) Using the condition $\int_{[0,\infty)} te^{2\hat{b}t}d\mu(t) < \infty$ and the Schwarz inequality, we see that the third piece of $II_j$ is also in $L^1(\mathbb{R})$, thus of the form $\hat{p}_j$ with $p_j \in C_0(\mathbb{R}) \cap L^2(\mathbb{R})$. In summary, $II_j = \hat{v}_j + \hat{p}_j$ with $v_j \in H^2_-$ and $p_j \in C_0(\mathbb{R}) \cap L^2(\mathbb{R})$. Finally, $III_j \in L^2(\mathbb{R})$ by essentially the same argument that shows each term of $II_j \in L^2(\mathbb{R})$. Moreover, $III_j \in L^1(\mathbb{R})$ which can be seen as follows. On rewriting $III_j(x)$ in terms of $g(t,x)$ defined above, we have

$$|III_j(x)|^2 \leqslant \int |U(x-t)|^2 e^{2\hat{b}t} d\mu(t) \int |g(t,x)|^2 d\mu(t)$$

and thus

$$\int |III_j(x)|dx \leqslant ||U||_2 \left(2\pi ||w_j||_\infty ||h||^2_{L^2(\nu_j\times\mu)} \int e^{2\hat{b}t} d\mu(t)\right)^{1/2}$$

where we have used (6.6). Thus $III_j = \hat{u}_j$ where $u_j \in C_0(\mathbb{R}) \cap L^2(\mathbb{R})$.

On putting the above together, we have

$$2^{-1} e^{\hat{b}x}(A_j^* h)(x) = (\hat{q}_j + \hat{v}_j + \hat{p}_j + \hat{u}_j)(x). \tag{6.7}$$

To obtain a contradiction we now use the fact that

$$\text{Ran} A_2^* \subset \text{Ran} A_1^*. \tag{6.8}$$

Choose $h_2 \in L^2(\nu_2 \times \mu)$ as $h_2(s,t) = 1_F(s)$, where $F$ is a nondegenerate compact interval contained in the (assumed) non-empty open subset $E_2$. By (6.8), there exists $h_1 \in L^2(\nu_1 \times \mu)$ with

$$A_2^* h_2 = A_1^* h_1. \tag{6.9}$$

For $j = 1$ or 2, associate $q_j$, $v_j$, $p_j$, and $u_j$ to $h_j$ as they are associated to $h$ in (6.7). By (6.9) we have

$$\hat{q}_2 + \hat{v}_2 + \hat{p}_2 + \hat{u}_2 = \hat{q}_1 + \hat{v}_1 + \hat{p}_1 + \hat{u}_1,$$

or applying the inverse Fourier transform and rearranging

$$q_2 - q_1 + p_2 - p_1 + u_2 - u_1 = v_1 - v_2. \tag{6.10}$$

Clearly,

$$q_2(s) = 1_F(s) w_2(s) \int e^{\hat{b}t} d\mu(t)$$

has nontrivial jump discontinuities at the end points of $F$ since $w_2$ is positive and continuous on $E_2$. Let $J \subset E_2$ be a compact interval containing one of the endpoints of $F$, $x_0$, in its interior. Since $q_1 \equiv 0$ on $E_2$ and $p_1, p_2, u_1, u_2$ are continuous, $v_1 - v_2$ has a jump discontinuity at $x_0$. However, this will contradict a theorem of Lindelöf, since $v_1 - v_2 \in H^2_-$. Indeed, the Poisson integral provides the analytic extension of $v_1 - v_2$ into the lower half plane. The left side of (6.10), and thus $v_1 - v_2$ are bounded in a neighborhood of $J$ in $\mathbb{R}$, so this analytic extension is bounded in the open half disk $B$ in the lower half plane with the interior of $J$ as its diameter. It follows that if $\phi$ is a conformal map of the open unit disk $\mathbb{D}$ onto $B$ (such a $\phi$ must extend to a homeomorphism of $\bar{\mathbb{D}}$ onto $\bar{B}$) then $(v_1 - v_2) \circ \phi$ is a bounded analytic function in $\mathbb{D}$

whose boundary value function has a jump discontinuity at $\phi^{-1}(x_0)$. This contradicts Lindelöf's theorem [13] (see also [14]) and shows that indeed $w \geqslant 0$.

Finally, we consider general $f$ in $L^\infty(\mathbb{R})$. Suppose our hypothesis $i[f(P), g(Q)] \geq 0$ is in effect. Let

$$u_r(x) = \frac{1}{r\pi}\left(\frac{\sin rx}{x}\right)^2,$$

an approximate identity on $\mathbb{R}$ as $r \to \infty$. Since $u_r \geqslant 0$ it follows that $i[f * u_r(P), g(Q)] \geq 0$. Let $f_r = f * u_r$. Since $f_r$ is smooth, it also follows from Theorem 1.3 that $(f_r)' \geq 0$. Since $f_r$ is bounded, we then see $(f_r)'$ is in $L^1(\mathbb{R})$. By direct computation $\widehat{u_r}$ has compact support, and so $\widehat{f_r}$ (taken in the sense of tempered distributions) does as well. The same must be true for the Fourier transform of $(f_r)'$ which is a continuous function.

Now we let $\phi(x) = \tanh bx$ and note

$$\widehat{\phi'}(y) = \hat{b}\sqrt{\frac{2}{\pi}}\frac{y}{\sinh(\hat{b}y)}. \tag{6.11}$$

Let $\sigma$ be a real even function in the Schwartz space agreeing with $1/\widehat{\phi'}$ on the support of $\widehat{f_r'}$. Then $\widehat{f_r'} = \widehat{\phi'} \cdot \widehat{f_r'} \cdot \sigma$, whence

$$f_r' = \frac{1}{2\pi}\phi' * f_r' * \check{\sigma}. \tag{6.12}$$

We put $w_r = \frac{1}{2\pi} f_r' * \check{\sigma}$, a convolution of real $L^1(\mathbb{R})$-functions, and so in $L^1(\mathbb{R})$ and real valued. In particular,

$$\begin{aligned}\|w_r\|_1 &\leq \frac{1}{2\pi}\left\{\int_{-\infty}^{\infty}(f_r)'(x)dx\right\}\|\check{\sigma}\|_1\\ &= \frac{1}{2\pi}\{(f * u_r)(\infty) - (f * u_r)(-\infty)\}\|\check{\sigma}\|_1\\ &\leq \frac{1}{\pi}\|f\|_\infty\|\check{\sigma}\|_1.\end{aligned}$$

Further, $w_r$ is continuous and tends to zero at $\pm\infty$, hence $w_r \in L^\infty(\mathbb{R})$. On integrating equation (6.12) (that is $f_r' = \phi' * w_r$), we have

$$f_r(x) = \int_{-\infty}^{\infty}\tanh b(x - s)w_r(s)ds + d_r, \tag{6.13}$$

where $d_r$ is a real constant. Since $i[f * u_r(P), g(Q)] \geq 0$, our previous results show that $w_r \geq 0$ a.e. On letting $x \to \pm\infty$ in (6.13), we find by the dominated convergence theorem that

$$f_r(\infty) + f_r(-\infty) = 2d_r,$$

and

$$f_r(\infty) - f_r(-\infty) = 2||w_r||_1$$

so that $|d_r| \le \|f\|_\infty$ for $r > 0$. Since $f_r(\pm\infty) = f(\pm\infty)$ we have $\int w_r(s)ds = [f]/2$. Now consider $w_r(s)ds$ as a measure on the two-point compactification $[-\infty, \infty]$ of $\mathbb{R}$. Then there exists a finite positive measure $\nu$ on $\mathbb{R}$ and $\epsilon_1, \epsilon_2 \ge 0$, a real number $d$ and a sequence $r_n \to \infty$ such that

$$w_{r_n}(s)ds \to \nu + \epsilon_1\delta_{-\infty} + \epsilon_2\delta_{+\infty},$$

in the weak-$*$ topology and $d_{r_n} \to d$ as $n \to \infty$; here, $\delta_{\pm\infty}$ denotes the unit point mass at $\pm\infty$. Then taking a limit along $r_n$ in (6.13) yields

$$\begin{aligned} f(x) &= \lim_{n\to\infty} f_{r_n}(x) \\ &= \int_{-\infty}^{\infty} \tanh b(x-s)d\nu(s) + \epsilon_1 - \epsilon_2 + d \end{aligned}$$

for all $x \in \mathbb{R}$. It follows that $f \in K_{\hat{b}}$, and the proof is complete. □

## 7 Finite Rank Positive Commutators

Considering that the positive commutator is trace class, it is interesting to consider the case where the positive commutator has finite rank. The next lemma shows that, in particular, in considering the rank one case, Kato's assumption that $f$ and $g$ are absolutely continuous with $L^1(\mathbb{R})$ derivatives can be proved.

**Lemma 7.1** *Suppose the commutator $C = i[f(P), g(Q)]$ is positive, nonzero, and has finite rank. Then $f$ and $g$ can be taken differentiable with continuous derivatives. These derivatives are everywhere nonzero.*

*Proof* We can assume that $f$ and $g$ are increasing. The kernel of the commutator $C = i[f(P), g(Q)]$, given by

$$K(x, y) = \frac{1}{\sqrt{2\pi}} \frac{g(x) - g(y)}{x - y} \widehat{df}(y - x),$$

has the form

$$K(x, y) = \sum_{j=1}^{N} \phi_j(x)\overline{\phi_j}(y) \tag{7.1}$$

where $\{\phi_1, \phi_2, \ldots, \phi_N\}$ is a linearly independent set of $L^2(\mathbb{R})$ functions. Let $x_0 \in \mathbb{R}$ and $I$ a small open interval centered at $x_0$. Denote the complement of $I$ as $J$ and the restriction of $\phi_i$ to $J$ as $\phi_{i,J}$. The $N \times N$ matrix with elements $(\phi_{i,J}, \phi_{k,J})$ is continuous in $|I|$ so we can choose a small such interval $I$ such that the determinant of this matrix is nonzero, and thus the vectors $\phi_{1,J}, \phi_{2,J}, \ldots, \phi_{N,J}$ are linearly independent. We can thus find $N$ functions $\psi_i \in C_0^\infty(J^o)$ so that the matrix with matrix elements $(\phi_{i,J}, \psi_k) = (\phi_i, \psi_k)$ has nonzero determinant. If we define $\gamma_j(x)$ for all $x \in \mathbb{R}$ by

$$\gamma_j(x) = \sum_i (\phi_i, \psi_j)\phi_i(x),$$

the $\phi_i$ are thus linear combinations of the $\gamma_j$ and at least for $x$ outside the support of the $\psi_j$ we have

$$\gamma_j(x) = \int K(x, y)\psi_j(y)dy.$$

It follows from the explicit form of $K$ that the $\phi_i$ can be taken continuous in a neighborhood of $\overline{I}$. Since we can cover $\mathbb{R}$ with a countable number of such intervals, we can assume all the $\phi_i$ are continuous functions. In addition, since $\widehat{df}(u)$ is continuous and $\widehat{df}(0) = [f]/\sqrt{2\pi}$ is nonzero, we can see from the explicit form of $K$ that $g$ is $C^1$ with

$$g'(x) = 2\pi \sum_j |\phi_j(x)|^2/[f].$$

If $\phi_j(x) = 0$ for all $j$, this contradicts the fact that $g$ is strictly increasing and that $\widehat{df}(y - x)$ is nonzero for $|y - x|$ small. Thus $g'(x)$ is positive for all $x$.

Since $\mathcal{F} f(P)\mathcal{F}^{-1} = f(Q)$ and $\mathcal{F} g(Q)\mathcal{F}^{-1} = g(-P)$, $i[-g(-P), f(Q)]$ has the integral kernel

$$\tilde{K}(\xi, \eta) = \sum_j \widehat{\phi}_j(\xi)\overline{\widehat{\phi}_j(\eta)}.$$

It follows that what we have proved for $g$ is also true for $f$. In particular, $f \in C^1(\mathbb{R})$ with

$$f'(\xi) = 2\pi \sum_j |\widehat{\phi}_j(\xi)|^2/[g]. \tag{7.2}$$

□

We can actually prove much more:

**Theorem 7.2** *Suppose the commutator $C = i[f(P), g(Q)]$ is positive, nonzero, and finite rank. Then the functions $f'$ and $g'$ are in $\mathcal{S}(\mathbb{R})$.*

*Proof* We write

$$K(x, y) = \sum_{j=1}^{N} \phi_j(x)\overline{\phi_j}(y)$$

where the functions $\phi_j$ are in $L^2(\mathbb{R})$ and linearly independent. We have shown that these functions can be assumed continuous. Let $v(y) =< \phi_1(y), \cdots, \phi_N(y) > \in \mathbb{C}^N$. The span of $\{v(y) : y \in \mathbb{R}\}$ is $\mathbb{C}^N$ for otherwise there is a nonzero vector $c \in \mathbb{C}^N$ perpendicular to this span. This implies $\sum \overline{c}_j\phi_j(y) = 0$ contradicting the fact that $C$ has rank $N$. Choose $N$ points $y_j$ so that $v(y_1), \cdots, v(y_N)$ are linearly independent. Let $\zeta \in C_0^\infty(-1, 1)$ be a real nonnegative function with $\int \zeta(y)dy = 1$. Let $\zeta_t(y) = t^{-1}\zeta(y/t)$ and $\psi_j(y) = \zeta_t(y_j - y)$. Then $v_j := \zeta_t * v(y_j) = \int \psi_j(y)v(y)dy =< \int \psi_j(y)\phi_1(y)dy, \cdots, \int \psi_j(y)\phi_N(y)dy >$. We choose $t > 0$ small enough so that the $v_j$ span $\mathbb{C}^N$. Let

$$\gamma_j(x) = \sum_k \phi_k(x)(\phi_k, \psi_j) = (v_j, v(x)).$$

The $\phi_j(x)$ are linear combinations of the $\gamma_j(x)$. In fact, a standard linear algebra calculation gives

$$v(x) = \sum_{j,l}(A^{-1})_{jl}\gamma_l(x)v_j,$$

where $A_{ij} = (v_i, v_j)$.

We will show that $||D^m\phi_j||_2 + ||D^m\widehat{\phi_j}||_2 < \infty$ for all $m$. We proceed to show $D^m\phi_j \in L^2(\mathbb{R}) \cap L^\infty(\mathbb{R})$ by induction. For $x \notin I := \cup_j(y_j - t, y_j + t)$ we have

$$\gamma_j(x) = \int_{\mathbb{R}} \frac{g(y) - g(x)}{y - x}\widehat{f}'(y - x)\psi_j(y)dy$$

or changing variables

$$\gamma_j(x) = \int_{\mathbb{R}} \frac{g(x + u) - g(x)}{u}\widehat{f}'(u)\psi_j(x + u)du.$$

We note that for all $i$, $\psi_i = 0$ in a neighborhood of $I^c$ so that $|\gamma_j(x)| \leqslant c \int |\psi_j(y)|/|x - y|dy$ and thus $\phi_j(x)$ and thus $g'(x)$ are bounded in a neighborhood of $I^c$. Then

$$\gamma_j'(x) = \int_{\mathbb{R}} [\frac{g'(x + u) - g'(x)}{u}\psi_j(x + u) + \frac{g(x + u) - g(x)}{u}\psi_j'(x + u)]\widehat{f}'(u)du$$

or

$$|\gamma_j'(x)| \leqslant c \int \frac{|g'(y) - g'(x)|}{|y - x|}|\psi_j(y)|dy + c \int \frac{|g(y) - g(x)|}{|y - x|}|\psi_j'(y)|dy$$

$$\leqslant C(|x|+1)^{-1}$$

for $x$ in a neighborhood of $I^c$. It follows that $\phi_j' \in L^2(I^c) \cap L^\infty(I^c)$.

We now choose a different set of $y_j$, call them $\tilde{y}_j$, and replace $I$ with $J = \cup_j(\tilde{y}_j - t, \tilde{y}_j + t)$. If, for example, $0 < |\tilde{y}_j - y_j|$ is small enough, the $v(\tilde{y}_j)$ will be linearly independent and if $t$ is small enough the $v_j$ will also be linearly independent and $\overline{I} \cap \overline{J} = \emptyset$.

Using the same technique, we find $\phi_j' \in L^2(J^c) \cap L^\infty(J^c)$. Thus, we finally conclude the first step: The $\phi_j$ are differentiable and $\phi_j, \phi_j' \in L^2(\mathbb{R}) \cap L^\infty(\mathbb{R})$. We now suppose $D^n\phi_j \in L^2(\mathbb{R}) \cap L^\infty(\mathbb{R})$ for $1 \leqslant n \leqslant m$ and $1 \leqslant j \leqslant N$. Using $g'(x) = 2\pi(\sum_j |\phi_j(x)|^2)/[f]$, $c = 2\pi/[f]$, we calculate

$$D^{n+1}g(x) = c\sum_{j,k}\binom{n}{k}[D^k\phi_j(x)D^{n-k}\overline{\phi}_j(x)].$$

This gives $D^{n+1}g \in L^1(\mathbb{R}) \cap L^\infty(\mathbb{R})$ for $1 \leqslant n \leqslant m$. Again using

$$\gamma_j(x) = \int_{\mathbb{R}} \frac{g(x+u)-g(x)}{u}\widehat{f}'(u)\psi_j(x+u)du,$$

we see that $\gamma_j$ is $m+1$ times continuously differentiable in a neighborhood of $I^c$ with

$$D^{m+1}\gamma_j(x) = \sum_k \binom{m+1}{k}\int \frac{D^kg(x+u)-D^kg(x)}{u}\widehat{f}'(u)D^{m+1-k}\psi_j(x+u)du$$

which gives

$$|D^{m+1}\gamma_j(x)| \leqslant C\sum_k \int \frac{|D^kg(x+u)-D^kg(x)|}{|u|}|D^{m+1-k}\psi_j(x+u)|du$$

$$= C\sum_k \int \frac{|D^kg(y)-D^kg(x)|}{|y-x|}|D^{m+1-k}\psi_j(y)|dy \leqslant C'(|x|+1)^{-1}$$

in a neighborhood of $I^c$. So $D^{m+1}\gamma_j$, and hence the $D^{m+1}\phi_k$ are in $L^2(I^c) \cap L^\infty(I^c)$. Using $J$ instead of $I$, we complete the induction step to learn $D^{m+1}\phi_j \in L^2(\mathbb{R}) \cap L^\infty(\mathbb{R})$.

As before $\mathcal{F}f(P)\mathcal{F}^{-1} = f(Q)$ and $\mathcal{F}g(Q)\mathcal{F}^{-1} = g(-P)$, so $i[-g(-P), f(Q)]$ has the integral kernel

$$\tilde{K}(\xi,\eta) = \sum_j \widehat{\phi}_j(\xi)\overline{\widehat{\phi}_j}(\eta).$$

It follows that what we have proved for $\phi_j$ is also true for $\widehat{\phi}_j$. This completes the proof. □

In case the commutator $C$ has finite rank with kernel as in (7.1), an important part of the kernel can be expressed entirely in terms of the $\phi_j$'s. In particular, we have

$$\widehat{f}'(y-x) = (2\pi/[g]) \sum_j \int \phi_j(x+u)\overline{\phi_j}(y+u)du.$$

There is another way of seeing this but let us calculate directly. Using the formula (7.2) for $f'$, we should calculate

$$\begin{aligned}\int e^{-i(y-x)\xi}|\widehat{\phi}(\xi)|^2 d\xi/\sqrt{2\pi} &= \int\int\int e^{i(x-y)\xi}\phi(x_1)e^{-i\xi x_1}e^{i\xi x_2}\overline{\phi}(x_2)dx_1dx_2d\xi(2\pi)^{-3/2} \\ &= \int \phi(x-y+x_2)\overline{\phi}(x_2)/\sqrt{2\pi} \\ &= \int \phi(x+u)\overline{\phi}(y+u)du/\sqrt{2\pi}.\end{aligned}$$

## 8 Operator Monotone Functions

The following proposition gives another connection between operator monotone functions and positive commutators.

**Proposition 8.1** *Suppose $f$ and $g$ are bounded measurable functions with $i[f(P), g(Q)] \geqslant 0$. Suppose $F$ is an operator monotone function on an open interval containing the range of $f$. Then $i[F(f(P)), g(Q)] \geqslant 0$.*

*Proof* Looking at the derivative, we see that if $t > 0$, $e^{-itg(Q)}f(P)e^{itg(Q)} \geqslant f(P)$. It follows that $e^{-itg(Q)}F(f(P))e^{itg(Q)} = F(e^{-itg(Q)}f(P)e^{itg(Q)}) \geqslant F(f(P))$. The result now follows from the fact that the derivative of the left side of the inequality at $t = 0$ is thus positive. □

## 9 Two Additional Representations of the Commutator

The integral kernel of the commutator $i[f(P), g(Q)]$ given in (2.6) is clearly not very symmetric in the functions $f$ and $g$ although there is much symmetry in the operator (see (4.1) for example). In this section, we remedy this by giving two symmetrical representations.

The integral kernel of our first symmetrical representation of $i[f(P), g(Q)]$ which is on $L^2(\mathbb{R}^2)$ appears in the next lemma.

**Lemma 9.1** *Suppose $i[f(P), g(Q)] = C$, where $f$ and $g$ are real measurable bounded functions and $C$ is positive and nonzero. Then a representation of the commutator on $L^2(\mathbb{R}^2)$ has integral kernel*

$$K(x_1, x_2; y_1, y_2) = \widehat{df}(\sqrt{2}(y_1 - x_1))\widehat{dg}(\sqrt{2}(y_2 - x_2))e^{i(x_1 y_2 - x_2 y_1)} \frac{\sin(y_1 - x_1)(y_2 - x_2)}{(y_1 - x_1)(y_2 - x_2)} / \pi$$

*Proof* We define a different representation of the canonical commutation relations. We set $p_j = -i\partial_{x_j}$ and

$$\begin{aligned} p &= (p_1 - x_2)/\sqrt{2}, x = (x_1 + p_2)/\sqrt{2} \\ a &= (x + ip)/\sqrt{2} \\ p' &= (p_2 - x_1)/\sqrt{2}, x' = (p_1 + x_2)/\sqrt{2} \\ a' &= (x' + ip')/\sqrt{2} \end{aligned}$$

We have

$$\begin{aligned} [a, a^*] &= [a', a'^*] = 1 \\ [a, a'] &= [a, a'^*] = 0 \end{aligned}$$

For $f, g \in L^1(\mathbb{R})$ we have

$$\begin{aligned} f(p)\psi(x_1, x_2) &= \pi^{-1/2} \int \hat{f}(\sqrt{2}(y_1 - x_1))e^{ix_2(x_1 - y_1)}\psi(y_1, x_2)dy_1 \\ g(x)\psi(x_1, x_2) &= \pi^{-1/2} \int \hat{g}(\sqrt{2}(y_2 - x_2))e^{-ix_1(x_2 - y_2)}\psi(x_1, y_2)dy_2. \end{aligned}$$

Thus

$$f(p)g(x)\psi(x_1, x_2) =$$

$$\begin{aligned} &\pi^{-1} \int \hat{f}(\sqrt{2}(y_1 - x_1))\hat{g}(\sqrt{2}(y_2 - x_2))e^{ix_2(x_1 - y_1) - iy_1(x_2 - y_2)}\psi(y_1, y_2)dy_1 dy_2 \\ &= \pi^{-1} \int \hat{f}(\sqrt{2}(y_1 - x_1))\hat{g}(\sqrt{2}(y_2 - x_2))e^{i(x_2 - y_2)(x_1 - y_1) + i(x_1 y_2 - x_2 y_1)}\psi(y_1, y_2)dy_1 dy_2 \end{aligned}$$

and

$$g(x)f(p)\psi(x_1, x_2) =$$

$$\begin{aligned} &\pi^{-1} \int \hat{f}(\sqrt{2}(y_1 - x_1))\hat{g}(\sqrt{2}(y_2 - x_2))e^{-ix_1(x_2 - y_2) + iy_2(x_1 - y_1)}\psi(y_1, y_2)dy_1 dy_2 \\ &= \pi^{-1} \int \hat{f}(\sqrt{2}(y_1 - x_1))\hat{g}(\sqrt{2}(y_2 - x_2))e^{-i(x_2 - y_2)(x_1 - y_1) + i(x_1 y_2 - x_2 y_1)}\psi(y_1, y_2)dy_1 dy_2. \end{aligned}$$

It follows that

$$i[f(p), g(x)] = \int K(x_1, x_2; y_1, y_2)\psi(y_1, y_2)dy_1dy_2$$

where

$$\begin{aligned} &K(x_1, x_2; y_1, y_2) \\ &= -2\pi^{-1}\hat{f}(\sqrt{2}(y_1 - x_1))\hat{g}(\sqrt{2}(y_2 - x_2)) \sin(x_1 - y_1)(x_2 - y_2)e^{i(x_1y_2 - x_2y_1)} \\ &= \pi^{-1}\widehat{df}(\sqrt{2}(y_1 - x_1))\widehat{dg}(\sqrt{2}(y_2 - x_2))\frac{\sin(x_1 - y_1)(x_2 - y_2)}{(x_1 - y_1)(x_2 - y_2)}e^{i(x_1y_2 - x_2y_1)} \end{aligned}$$

If $f$, $g$ are increasing and bounded, it is not hard to make sense of these manipulations using distributions. This proves the result. □

*Remark 9.2* Note that if $\Omega$ is the vacuum vector ($a\Omega = a'\Omega = 0$), then $L^2(\mathbb{R}^2) \simeq \mathcal{F}_1 \otimes \mathcal{F}_2$, where $\mathcal{F}_j$ are the Fock spaces which are the spans of $\{(a^*)^n\Omega : n = 0, 1, 2, ...\}$ and $\{(a'^*)^n\Omega : n = 0, 1, 2, ...\}$ and $i[f(P), g(Q)] = k \otimes I$ for some operator $k$ acting in $\mathcal{F}_1$.

Another representation on $L^2(\mathbb{R})$:

**Lemma 9.3** $i[f(P), g(Q)] = (2\pi)^{-1} \int e^{i(\xi Q + uP)}\widehat{df}(u)\widehat{dg}(\xi)\frac{\sin(u\xi/2)}{u\xi/2}dud\xi.$

*Proof* We calculate

$$\begin{aligned} g(\infty) - g(x) &= \int_0^\infty dg(t + x) = \lim_{\epsilon\downarrow 0} \int_0^\infty e^{-\epsilon t}dg(t + x) \\ &= \lim_{\epsilon\downarrow 0}(2\pi)^{-1/2} \int_0^\infty \int e^{-\epsilon t}e^{i(t+x)\xi}\widehat{dg}(\xi)d\xi dt \\ &= \lim_{\epsilon\downarrow 0}(2\pi)^{-1/2} \int e^{ix\xi}\frac{\widehat{dg}(\xi)}{\epsilon - i\xi}d\xi \end{aligned}$$

and

$$\begin{aligned} [f(P), e^{iQ\xi}] &= e^{iQ\xi}(f(P + \xi) - f(P)) \\ &= e^{ix\xi}\int_0^1 f'(P + s\xi)\xi ds = (2\pi)^{-1/2}\xi \int_0^1 \int e^{iQ\xi}e^{i(P+s\xi)u}\widehat{df}(u)duds \end{aligned}$$

and thus

$$\begin{aligned} i[f(P), g(Q)] &= \lim_{\epsilon\downarrow 0} -i(2\pi)^{-1} \int \int_0^1 \int \frac{\xi}{\epsilon - i\xi}e^{iQ\xi}e^{i(P+s\xi)u}\widehat{df}(u)\widehat{dg}(\xi)dudsd\xi \\ &= (2\pi)^{-1} \int \int_0^1 \int e^{iQ\xi}e^{i(P+s\xi)u}\widehat{df}(u)\widehat{dg}(\xi)dudsd\xi \end{aligned}$$

$$= (2\pi)^{-1} \int\int_0^1\int e^{i(Q\xi+Pu)} e^{i(s-1/2)\xi u} \widehat{df}(u)\widehat{dg}(\xi) du ds d\xi$$
$$= \pi^{-1} \int\int e^{i(Q\xi+Pu)} \frac{\sin(\xi u/2)}{\xi u} \widehat{df}(u)\widehat{dg}(\xi) du d\xi.$$

Since we do not know a priori that $f$ is differentiable, we should first mollify our bounded $f$ and then take a limit at the end of the calculation. □

As an application of this representation, we calculate expectations of the commutator in coherent states: With $a = (Q + iP)/\sqrt{2}$ and $\Omega = \pi^{-1/4} e^{-x^2/2}$ we have $a\Omega = 0$. Let $\psi(z) = e^{za^*}\Omega$. Then $a\psi(z) = z\psi(z)$. Let $\zeta = (\xi + iu)/\sqrt{2}$. We calculate

$$e^{i(\xi Q+uP)} = e^{i(\zeta a^* + \bar{\zeta} a)} = e^{|\zeta|^2/2} e^{i\bar{\zeta} a} e^{i\zeta a^*}$$

$$(\psi(w), \psi(z)) = e^{\bar{w}z}.$$

Thus

$$(\psi(w), i[f(P), g(Q)]\psi(z)) = \pi^{-1} \int e^{|\zeta|^2/2} e^{(\overline{w - i\zeta})(z+i\zeta)} \frac{\sin(\xi u/2)}{\xi u} \widehat{df}(u)\widehat{dg}(\xi) du d\xi$$
$$= \pi^{-1} e^{\bar{w}z} \int e^{-|\zeta|^2/2 + i(\zeta\bar{w} + \bar{\zeta}z)} \frac{\sin(\xi u/2)}{\xi u} \widehat{df}(u)\widehat{dg}(\xi) du d\xi \tag{9.1}$$

Putting $z = w$ we obtain

$$\int e^{-(\xi^2+u^2)/4} e^{i(\xi x+uy)} \frac{\sin(\xi u/2)}{\xi u} \widehat{df}(u)\widehat{dg}(\xi) du d\xi \geqslant 0 \tag{9.2}$$

for all $z = (x + iy)/\sqrt{2}$.

## 10 Some Results that Follow from 2 × 2 Positivity

We assume that $i[F(P), G(Q)] = C \geqslant 0$, $C \neq 0$. It then follows that $F$ and $G$ can be taken strictly increasing, continuous, with inverses which are absolutely continuous. We now look more carefully at the condition that the kernel

$$H_{xy} = \frac{G(x) - G(y)}{x - y} \widehat{dF}(y - x)$$

gives a positive semidefinite 2 by 2 matrix for any pair $(x, y)$. In this section only, we drop the factor of $\sqrt{2\pi}$ and use $\widehat{dF}(x) = \int e^{-ix\xi} dF(\xi)$. Let us restrict to $F$ with $\int dF = 1$. An important parameter will make its appearance, namely,

$$\sigma^2 = \inf\{\int \xi^2 df(\xi) - (\int \xi df(\xi))^2 : i[f(P), G(Q)] = C \geqslant 0, C \neq 0, \int df = 1\}$$

which we believe must be related to the Kato class of $G$ (see some discussion below). We assume $\sigma^2 < \infty$. Let $\hat{a} = \sqrt{3}\sigma$.

**Theorem 10.1** *Suppose $i[F(P), G(Q)] = C \geqslant 0, C \neq 0$ where $F$ and $G$ are real bounded measurable functions. Then, both are monotone increasing or monotone decreasing and continuous. Suppose there exists an increasing $f$ with $i[f(P), G(Q)] \geqslant 0$ and with $\int \xi^2 df(\xi) < \infty$, $\int df = 1$. Then in fact $G$ is $C^1$ with a Lipschitz derivative and thus $G'$ is absolutely continuous. $G$ satisfies the following estimates:*

$$G'(x_0)e^{-2\hat{a}|(x-x_0)|} \leqslant G'(x) \leqslant G'(x_0)e^{2\hat{a}|(x-x_0)|} \tag{10.1}$$

$$|G''(x)| \leqslant 2\hat{a}G'(x), \ \text{a.e. and both} \tag{10.2}$$

$$G'(x) \leqslant 2\hat{a}|G(\pm\infty) - G(x)|. \tag{10.3}$$

*If $G$ is odd then*

$$G'(x) \geqslant \frac{G'(0)}{(\cosh(\hat{a}x))^2}. \tag{10.4}$$

This inequality is an equality in the rank one case [2] and if K is true the rank, one case is the case of minimal variance (see Eq. 10.12).

The proof will require some preliminary work. Let $g_t = \phi_t * G$ where $\phi$ is a nonnegative smooth function of compact support whose integral is 1 and $\phi_t(x) = t^{-1}\phi(t^{-1}x)$. We take $f$ to satisfy $i[f(P), G(Q)] = C \geqslant 0$, $C \neq 0$ and $\int df = 1$ and additionally $\int \xi^2 df(\xi) < \infty$. The idea is to get uniform bounds on the derivatives of $g_t$ and then take limits to learn about $G$. In the following, we drop the subscript $t$.

**Lemma 10.2** *The function $g'$ is positive. The function $\psi - (g')^{-1/2}$ satisfies*

$$-\psi'' + \hat{a}^2\psi \geqslant 0 \tag{10.5}$$

*Proof* Let $y = x + h$ with $h$ small and nonzero. Then

$$\frac{g(x) - g(y)}{x - y} = (g(x+h) - g(x))/h = g'(x) + hg''(x)/2 + h^2 g'''(x)/6 + o(h^2) \tag{10.6}$$

$$\widehat{df}(y - x) = \int e^{-ih\xi} df(\xi) = e^{-ih\langle\xi\rangle} \int e^{-ih(\xi - \langle\xi\rangle)} df(\xi)$$
$$= e^{-ih\langle\xi\rangle}(1 - \frac{h^2}{2}(\langle\xi^2\rangle - \langle\xi\rangle^2) + o(h^2)) \tag{10.7}$$

where we have written $\int \xi^n df(\xi) = \langle \xi^n \rangle$. We have

$$H_{xx}H_{yy} - |H_{xy}|^2 =$$
$$g'(x)g'(y) - (\frac{g(x)-g(y)}{x-y})^2 |\widehat{df}(y-x)|^2$$
$$= g'(x)(g'(x) + hg''(x) + h^2 g'''(x)/2) - (g'(x) + hg''(x)/2 + h^2 g'''(x)/6)^2 +$$
$$(g'(x) + hg''(x)/2 + h^2 g'''(x)/6)^2 (1 - |\widehat{df}(y-x)|^2) + o(h^2)$$
$$= h^2 g' g'''/6 - h^2 (g'')^2/4 + (g')^2 (1 - [1 - h^2(\langle \xi^2 \rangle - \langle \xi \rangle^2)]) + o(h^2)$$
$$= \frac{h^2}{6}[g'g''' - 3(g'')^2/2 + 6(g')^2(\langle \xi^2 \rangle - \langle \xi \rangle^2)] + o(h^2) \tag{10.8}$$

We thus find

$$g'(x)g'''(x) - 3(g''(x))^2/2 + 6g'(x)^2(\langle \xi^2 \rangle - \langle \xi \rangle^2) \geqslant 0 \tag{10.9}$$

Since this is true for all monotone $f$ with $\int df = 1, \int \xi^2 df < \infty$ and $i[f(P), G(Q)]$ positive semidefinite, we take the infimum of the variance of $df$ over all such $f$ to find

$$g'(x)g'''(x) - 3(g''(x))^2/2 + 6\sigma^2 g'(x)^2 \geqslant 0 \tag{10.10}$$

Since $g'(x) > 0$ (see the formula for the determinant above and use the fact that $g$ is strictly increasing) we can introduce $\psi = (g')^{-1/2}$. Then recalling $\hat{a}^2 = 3\sigma^2$, (10.10) becomes

$$-\psi'' + \hat{a}^2 \psi \geqslant 0 \tag{10.11}$$

□

To see why we have used the abbreviation $\hat{a}^2 = 3\sigma^2$, suppose we assume K and take $g$ in its smallest Kato class, $K_a$. Let us find $\sigma^2$. We must have $f \in K_{\hat{a}}$, viz., $f(\xi) = \tanh a\xi * \mu/2 + c$ for some positive measure $\mu$ and where $\hat{a} = \pi/2a$. Then if $\mu$ is a probability measure, $\int f'(\xi)d\xi = 1$. We compute

$$\langle \xi^2 \rangle - \langle \xi \rangle^2 = a^{-2}[\frac{\int \xi^2 \cosh^{-2}(\xi)d\xi}{\int \cosh^{-2}(\xi)d\xi}] + \int t^2 d\mu - (\int t d\mu)^2 \tag{10.12}$$

This is minimized by taking $\mu$ a point measure. We compute with the help of Gradshteyn and Ryzhik 3527.3, $\int_{-\infty}^{\infty} x^2 \cosh^{-2}(x)dx = \pi^2/6$ and thus

$$\sigma^2 = \pi^2/12a^2 = \hat{a}^2/3 \tag{10.13}$$

Continuing with the proof of Theorem 10.1 we set $u = \psi'/\psi$ and obtain the Riccati-type inequality

$$u' \leqslant \hat{a}^2 - u^2.$$

**Lemma 10.3** $|\psi'(x)/\psi(x)| < \hat{a}$.

*Proof* If $u(x_0) > \hat{a}$ then as long as $u(x) \geqslant \hat{a}$, $\hat{a}^2 - u(x)^2 \leqslant -(u(x) - \hat{a})^2$ so that if $w = u - \hat{a}$, $w' \leqslant -w^2$. Thus $w$ tends to $+\infty$ for some $x < x_0$, a contradiction. If $u(x_0) < -\hat{a}$ then as long as $u(x) \leqslant -\hat{a}$, $u' \leqslant -(u+b)^2$. This leads to $u(x)$ tending to $+\infty$ for some $x > x_0$, another contradiction. Suppose $u(x_0) = \hat{a}$. We know that $|u(x)| \leqslant \hat{a}$ for all $x$. If $\epsilon > 0$ let $\hat{a}' = \hat{a} + \epsilon$. We know that $-\psi'' + \hat{a}'^2\psi \geqslant 0$. A computation shows that $\frac{\hat{a}'+u}{\hat{a}'-u}e^{-2\hat{a}'x}$ is monotone decreasing. Thus if $x < x_0$, $\frac{\hat{a}'+u(x)}{\hat{a}'-u(x)}e^{-2\hat{a}'x} \geqslant \frac{\hat{a}'+\hat{a}}{\epsilon}e^{-2\hat{a}x_0}$. If we take $\epsilon \to 0$ we find $u(x) = \hat{a}$. Integrating we find $\psi(x) = ce^{\hat{a}x}$ for $x < x_0$ or $g'(x) = e^{-2\hat{a}x}/c^2$. This contradicts the boundedness of $g$. Similarly if $u(x_0) = -\hat{a}$, $u(x) = -\hat{a}$ for all $x > x_0$ and this also contradicts the boundedness of $g$. So we have shown $-\hat{a} < \psi'/\psi < \hat{a}$ for all $x$. □

Let $m(t) = \frac{\hat{a}+u}{\hat{a}-u}$, $p(t) = \frac{1}{2}\log m(t)$. Then for any $x_0 \in \mathbb{R}$,

**Lemma 10.4**

$$g'(x) = g'(x_0)e^{-2\hat{a}\int_{x_0}^{x}\tanh(p(t))dt}; \quad p'(t) \leqslant \hat{a}. \tag{10.14}$$

$$\begin{aligned} &g'(x_0)e^{-2\hat{a}|(x-x_0)|} \leqslant g'(x) \leqslant g'(x_0)e^{2\hat{a}|(x-x_0)|} \\ &|g''(x)| \leqslant 2\hat{a}g'(x), \ \textit{and both} \\ &g'(x) \leqslant 2\hat{a}|g(\pm\infty) - g(x)|. \end{aligned} \tag{10.15}$$

*Proof* The fact that $p'(t) \leqslant \hat{a}$ is a computation which uses $u' \leqslant \hat{a}^2 - u^2$. We compute $u = \frac{m-1}{m+1}\hat{a} = -\frac{1}{2}(\log g')'$. Integration gives (10.14). The first two lines of (10.15) follow directly from (10.14). To prove the last inequalities first note that $\lim_{x\to\infty} g'(x) = 0$. To see this suppose the contrary. Then there is a sequence $x_n$ with $x_{n+1} > x_n + 1$ so that $g'(x_n) \geqslant \delta > 0$. But then from the first inequality of the lemma $g'(x) \geqslant \delta e^{-2\hat{a}}$ for $x \in [x_n, x_n + 1]$. This contradicts the integrability of $g'$. Similarly $\lim_{x\to-\infty} g'(x) = 0$. From the second line of (10.15),

$$\begin{aligned} &-g''(x) \leqslant 2\hat{a}g'(x) \text{ or} \\ &(g'(x) + 2\hat{a}g(x))' \geqslant 0. \end{aligned}$$

Integrating from $x$ to infinity gives one of the last inequalities. The other follows in a similar way. □

To make a connection with the conjecture K, consider the case where $g'$ is even.

**Corollary 10.5**

$$g'(x) \geqslant \frac{g'(0)}{(\cosh \hat{a}x)^2} \tag{10.16}$$

*Proof* This follows directly from $p'(t) \leqslant \hat{a}$ and the formula in (10.14). This inequality is an equality in the rank one case [2] and if K is true the rank one case is the case of minimal variance (see Eq. 10.12). □

*Proof of Theorem* 10.1 We have $g = g_t = \phi_t * G$ satisfying $0 < g_t'(x) \leqslant 2\hat{a}[G]$, $|g_t''(x)| \leqslant (2\hat{a})^2[G]^2$. By the Arzela–Ascoli theorem there is a sequence $g_n = g_{t_n}, t_n \downarrow 0$, such that $g_n \to G_1$ and $g_n' \to G_2$ uniformly on compact subsets of $\mathbb{R}$. Clearly, $G_1 = G$ and $g_n(x) - g_n(y) = \int_x^y g_n'(t)dt$ so that $G(x) - G(y) = \int_x^y G_2(t)dt$. Since $G_2$ is continuous it follows that $G$ is $C^1$ and $g_n' \to G'$ uniformly on compacts. We also have $|g_n'(x) - g_n'(y)| = |\int_x^y g_n''(t)dt| \leqslant (2\hat{a})^2[G]^2|x-y|$ giving $|G'(x) - G'(y)| \leqslant (2\hat{a})^2[G]^2|x-y|$. Thus $G'$ is absolutely continuous. We have $g_n'(x+h) - g_n'(x) = \int_x^{x+h} g_n''(u)du \leqslant \int_x^{x+h} 2\hat{a}g_n'(u)du$. Thus $G'(x+h) - G'(x) \leqslant \int_x^{x+h} 2\hat{a}G'(u)du$. It follows that $G''(x) \leqslant 2\hat{a}G'(x)$. Similarly, $-G''(x) \leqslant 2\hat{a}G'(x)$. We have used that $g_n(\pm\infty) = G(\pm\infty)$. We take $\phi$ to be even and then $g = g_t$ is odd. Thus, the last inequality follows simply from Corollary 10.5. □

If we now saturate (10.5) by making the inequality an equality, we obtain $g'^{-1/2} = \psi = be^{jx} + ce^{-jx}$ where $j = \sqrt{3}\sigma$. Taking $b$ and $c$ positive to insure that $g'$ does not blow up at $\pm\infty$, we can write $\psi = \hat{a}\cosh(j(x-t))$ for some $\hat{a} > 0$ and $t \in \mathbb{R}$. This gives $g(x) = \hat{a}^{-2}\tanh(j(x-t)) + c$ for some constant $c$. We can check that indeed $j = \hat{a}$ so that $g \in K_a$. This is to be expected from Theorem 1.6 which in particular states that if $i[f(P), g(Q)] \geqslant 0$ for all $f \in K_{\hat{a}}$, then $g \in K_a$. But it does show that the inequality (10.5) is in some sense sharp, so that cases of equality occur if the Kato conjecture is correct.

## 11 An Interesting Formula

Here is an interesting formula: Suppose $g$ is a bounded real function with a bounded analytic continuation to a strip of width $> 2\lambda$. Consider

$$\begin{aligned}
&i[\tanh\lambda P, g(Q)] = -2i[(1+e^{2\lambda P})^{-1}, g(Q)]\\
&= 2i(1+e^{2\lambda P})^{-1}[e^{2\lambda P}, g(Q)](1+e^{2\lambda P})^{-1}\\
&= 2i(1+e^{2\lambda P})^{-1}\left(e^{\lambda P}(e^{\lambda P}g(Q)e^{-\lambda P})e^{\lambda P} - e^{\lambda P}(e^{-\lambda P}g(Q)e^{\lambda P})e^{\lambda P}\right)(1+e^{2\lambda P})^{-1}\\
&= (i/2)(\cosh\lambda P)^{-1}\left(e^{\lambda P}g(Q)e^{-\lambda P} - e^{-\lambda P}g(Q)e^{\lambda P}\right)(\cosh\lambda P)^{-1}\\
&= (i/2)(\cosh\lambda P)^{-1}(g(Q-i\lambda) - g(Q+i\lambda))(\cosh\lambda P)^{-1}\\
&= (\cosh\lambda P)^{-1}\mathrm{Im}g(Q+i\lambda)(\cosh\lambda P)^{-1} \qquad (11.1)
\end{aligned}$$

Given the problems which occurred with manipulating unbounded functions of $P$ in Section 2, we assure the reader that the computations above can be made rigorous.

Perhaps (11.1) makes Theorem 1.6 intuitive.

**Acknowledgements** We are grateful to Brian Hall for many useful conversations about this problem.

## References


1. Howland, J.: Perturbation theory of dense point spectrum. J. Funct. Anal. **94**, 52–80 (1987)
2. Kato, T.: Positive commutators $i[f(P), g(Q)]$. J. Funct. Anal. **96**, 117–129 (1991)
3. Löwner, K.: Über monotone Matrixfunktionen. Math. Z. **38**, 177–216 (1934)
4. Fuglede, B.: On the relation $PQ - QP = -iI$. Math. Scand. **20**, 79–88 (1967)
5. Georgescu, V., Gerard, C.: On the virial theorem in quantum mechanics. Commun. Math. Phys. **208**, 275–281 (1999)
6. Folland, G.: Real Analysis, p. 107. Wiley, New York (1999)
7. Fuglede, B.: A commutativity theorem for normal operators. Proc. Natl. Acad. Sci. **36**, 35–40 (1950)
8. Rosenblum, M.: On a theorem of Fuglede and Putnam. J. Lond. Math. Soc. **33**, 376–377 (1958)
9. Cycon, H., Froese, R., Kirsch, W., Simon, B.: Schrödinger Operators with Applications to Quantum Mechanics and Global Geometry. Texts and Monographs in Physics, p. 61. Springer, Berlin (1987)
10. Putnam, C.: Commutation Properties of Hilbert Space Operators and Related Topics, p. 20. Springer, Berlin (1967)
11. Douglas, R.: On majorization, factorization and range inclusion of operators in Hilbert space. Proc. Am. Math. Soc. **17**, 413–415 (1966)
12. Duren, P.: Theory of $H^p$ Spaces, pp. 196–197. Academic, New York (1970)
13. Lindelöf, E.: Sur un principe gènèral de l'Analyse. Acta Societatis Scientiarum Finnicae **66**, 1–35 (1915)
14. Garnett, J.: Bounded Analytic Functions, p. 92. Academic, New York (1981). (Problem 7)

# Pointwise Exponential Decay of Bound States of the Nelson Model With Kato-Class Potentials

**Fumio Hiroshima**

**Abstract** Pointwise exponential decay of bound states of the so-called Nelson model with Kato-class potentials is shown by constructing a martingale derived from the semigroup generated by the Nelson Hamiltonian.

## 1 Introduction

To show the existence of the ground state of a model in quantum field theory has been a crucial issue. In particular, the so-called infrared regular condition is a critical condition for a scalar model to have the ground state. In this article, we are concerned with the so-called Nelson model [16, 17] describing an interaction between nonrelativistic nucleons and spinless scalar mesons. The time evolution of the nonrelativistic matters studied in this article is given by a Schrödinger operator. Then the model can be regarded as Schrödinger operator coupled to a quantum field.

The existence of the ground state of this kind of model has been shown under some general conditions so far. Next interesting issue concerning the ground state is to make properties of the ground state clear, which includes to estimate the number of bosons in the ground state and the decay properties on both field variable $\phi$ and matter variable $x$. In this article, we treat Kato-class potentials $V$, which were introduced and studied by Aizenman and Simon [1] and the definition of Kato-class potentials is based on a condition considered by Kato in [11]. The Nelson Hamiltonian $H$ with Kato-class potential is defined via functional integrations. The main purpose of this article is to show *pointwise* exponential decay of bound states $\Phi$ of the Nelson Hamiltonian:

$$\|\Phi(x)\|_{\mathscr{F}} \leq Ce^{-c|x|}, \quad a.e. x \in \mathbb{R}^d. \tag{1}$$

The strategy is an extension of Carmona, Masters and Simon [4]. Let $H\Phi = E\Phi$. Then $\Phi = e^{+tE}e^{-tH}\Phi$. By the functional integration of $e^{-tH}$, we can express $\Phi$ as

F. Hiroshima (✉)
Faculty of Mathematics, Kyushu University, Fukuoka, Japan
e-mail: hiroshima@math.kyushu-u.c.jp

© Springer Nature Switzerland AG 2019

T. M. Rassias and V. A. Zagrebnov (eds.), *Analysis and Operator Theory*, Springer Optimization and Its Applications 146,
https://doi.org/10.1007/978-3-030-12661-2_11

$$\Phi(x) = e^{tE}\mathbb{E}^x[\xi\Phi(B_t)]$$

for each $x \in \mathbb{R}^d$, where

$$\xi = e^{-\int_0^t V(B_s)ds} \mathrm{I}_0^* e^{-\phi(K)} \mathrm{I}_t \tag{2}$$

is the integral kernel and $\mathbb{E}$ denotes the expectation with respect to Brownian motion $(B_t)_{t\geq 0}$. See (18) below. We estimate $\mathbb{E}^x[\xi\Phi(B_t)]$ to get the bound (1).

Statement (1) is stronger rather than localization: $\int_{\mathbb{R}^d} dx \|e^{C|x|}\Phi(x)\|^2_{\mathscr{F}} < \infty$, which can be shown by IMS localization [5, Theorem 3.2] in, e.g., [2, 6, 8]. The pointwise exponential decay is also shown for the so-called semi-relativistic Pauli–Fierz model in quantum electrodynamics by the author oneself [9]. Here, the integral kernel of the semigroup generated by the semi-relativistic Pauli–Fierz Hamiltonian is of the form

$$\xi_{\mathrm{PF}} = e^{-\int_0^t V(X_s)ds} \mathrm{J}_0^* e^{-iA(K_{PF})} \mathrm{J}_t,$$

where $(X_t)_{t\geq 0}$ is a Lévy process and $\mathrm{J}_t$ the family of isometries and $e^{-iA(K_{PF})}$ a unitary operator. Hence, $\mathrm{J}_0^* e^{-iA(K_{PF})}\mathrm{J}_t$ is contractive, i.e., $\|\mathrm{J}_0^* e^{-iA(K_{PF})}\mathrm{J}_t\| \leq 1$. However, the integral kernel of the semigroup generated by the Nelson Hamiltonian is of the form (2), where $\mathrm{I}_t$ is also a family of isometries but $e^{-\phi(K)}$ is unbounded. To see the bound $\|\mathrm{I}_0^* e^{-\phi(K)}\mathrm{I}_t\|$, one can apply the hypercontractivity [19] of $\mathrm{I}_0^*\mathrm{I}_t$ for massive cases, but it is nontrivial to have a bound for massless cases. In this article, we derive the operator bound

$$\|\mathrm{I}_0^* e^{-\phi(K)}\mathrm{I}_t\| \leq e^{tE(\hat{\varphi})}$$

for massless cases in Corollary 4.6 but under infrared-regular condition, and consequently show pointwise exponential decay (1).

## 2 Basic Facts on Fock Space

### 2.1 Boson Fock Space

Let $\mathscr{H}$ be a separable Hilbert space over $\mathbb{C}$. Define $\mathscr{F}^{(n)} = \otimes_s^n \mathscr{H}$, where $\otimes_s^n \mathscr{H}$ denotes the $n$-fold symmetric tensor product with $\otimes_s^0 \mathscr{H} = \mathbb{C}$. The space

$$\mathscr{F} = \oplus_{n=0}^\infty \mathscr{F}^{(n)}(\mathscr{H})$$

is called boson Fock space over $\mathscr{H}$. The Fock space $\mathscr{F}$ can be identified with the space of $\ell_2$-sequences $(\Psi^{(n)})_{n\in\mathbb{N}}$ such that $\Psi^{(n)} \in \mathscr{F}^{(n)}$ and

$$\|\Psi\|^2_{\mathscr{F}} = \sum_{n=0}^\infty \|\Psi^{(n)}\|^2_{\mathscr{F}^{(n)}} < \infty. \tag{3}$$

Set $(\Psi, \Phi)_{\mathscr{F}} = \sum_{n=0}^{\infty}(\Psi^{(n)}, \Phi^{(n)})_{\mathscr{F}^{(n)}}$. The vector $\Omega = (1, 0, 0, \dots)$ is called the Fock vacuum. The creation operator denoted by $a^*(f)$ and the annihilation operator by $a(f)$ are defined by

$$(a^*(f)\Psi)^{(n)} = \sqrt{n} S_n(f \otimes \Psi^{(n-1)}), \quad n \geq 1,$$
$$(a^*(f)\Psi)^{(0)} = 0$$

with domain $D(a^*(f)) = \left\{ (\Psi^{(n)})_{n\geq 0} \in \mathscr{F} \,\middle|\, \sum_{n=1}^{\infty} n \| S_n(f \otimes \Psi^{(n-1)}) \|^2_{\mathscr{F}^{(n)}} < \infty \right\}$ and $a(f) = (a^*(\bar{f}))^*$. Furthermore, since both operators are closable and we denote their closed extensions by the same symbols. The space

$$\mathscr{F}_{\text{fin}} = \{(\Psi^{(n)})_{n\geq 0} \in \mathscr{F} \mid \Psi^{(m)} = 0 \text{ for all } m \geq M \text{ with some } M\}$$

is called finite particle subspace. Operators $a, a^*$ leave $\mathscr{F}_{\text{fin}}$ invariant and satisfy the canonical commutation relations on $\mathscr{F}_{\text{fin}}$:

$$[a(f), a^*(g)] = (\bar{f}, g), \quad [a(f), a(g)] = 0, \quad [a^*(f), a^*(g)] = 0.$$

Given a contraction operator $T$ on $\mathscr{H}$, the second quantization of $T$ is defined by

$$\Gamma(T) = \oplus_{n=0}^{\infty}(\otimes^n T).$$

Here $\otimes^0 T = 1$. For a self-adjoint operator $h$ on $\mathscr{H}$, $\{\Gamma(e^{ith}) : t \in \mathbb{R}\}$ is a strongly continuous one-parameter unitary group on $\mathscr{F}$. Then by Stone's theorem, there exists a unique self-adjoint operator $\mathrm{d}\Gamma(h)$ on $\mathscr{F}$ such that $\Gamma(e^{ith}) = e^{it\mathrm{d}\Gamma(h)}$. Let $N = \mathrm{d}\Gamma(1)$. To obtain the commutation relations between $a^\sharp(f)$ and $\mathrm{d}\Gamma(h)$, suppose that $f \in D(h)$. Then

$$[\mathrm{d}\Gamma(h), a^*(f)] = a^*(hf), \quad [\mathrm{d}\Gamma(h), a(f)] = -a(\overline{h\bar{f}}), \tag{4}$$

for $\Psi \in D(\mathrm{d}\Gamma(h)^{3/2}) \cap \mathscr{F}_{\text{fin}}$. The Segal field $\Phi(f)$ on the boson Fock space $\mathscr{F}(\mathscr{H})$ is defined by

$$\Phi(f) = \frac{1}{\sqrt{2}}(a^*(f) + a(\bar{f})), \quad f \in \mathscr{H}.$$

Here $\bar{f}$ denotes the complex conjugate of $f$. Field operator $\Phi(f)$, $f \in \mathscr{H}$, is a self-adjoint operator, but $a^*(f)$ and $a(f)$ are not. Nevertheless, we can define $e^{a^*(f)}$ and $e^{a(f)}$ by a geometric series. Let $f \in \mathscr{H}$ and we define the exponential of creation operators $F_f$ by

$$F_f = \sum_{n=0}^{\infty} \frac{1}{n!} a^*(f)^n$$

and $D(F_f) = \left\{ \Phi \in \cap_{n=1}^\infty D(a^*(f)^n) \middle| \sum_{n=0}^\infty \frac{1}{n!}\|a^*(f)^n\Phi\| < \infty \right\}$. Let $\Phi \in \mathscr{F}^{(m)}$. Thus we have

$$\|F_f\Phi\| \leq \|\Phi\| + \sum_{n=1}^\infty \frac{\sqrt{m+n-1}\cdots\sqrt{m}}{n!}\|f\|^n\|\Phi\| < \infty.$$

Then $\mathscr{F}_{\rm fin} \subset D(F_f)$ follows. We also define the exponential of annihilation operators by

$$G_f = \sum_{n=0}^\infty \frac{1}{n!}a(f)^n$$

with $D(G_f) = \left\{ \Phi \in \cap_{n=1}^\infty D(a(f)^n) \middle| \sum_{n=0}^\infty \frac{1}{n!}\|a(f)^n\Phi\| < \infty \right\}$. We simply write $F_f = e^{a^*(f)}$ and $G_f = e^{a(f)}$. Then we can see that $(e^{a^*(f)})^* \supset e^{a(\bar{f})}$ and this implies that $e^{a^*(f)}$ is closable. The closure of $e^{a^*(f)}$ is denoted by the same symbol. Similarly, the closure of $e^{a(f)}$ is denoted by the same symbol. We can represent $e^{\Phi(f)}$ in terms of both $e^{a^*(f)}$ and $e^{a(f)}$. Let $\mathscr{D}_{\rm b} = {\rm L.H.}\{C(g), \Phi \,|\, g \in \mathscr{H}, \Phi \in \mathscr{F}_{\rm fin}\}$.

**Proposition 2.1** (Baker–Campbell–Hausdorff formula) *Let* $f \in \mathscr{H}$ *and* $\alpha \in \mathbb{C}$. *Then it holds on* $\mathscr{D}_{\rm b}$ *that*

$$e^{\alpha\Phi(f)} = e^{\alpha a^*(f)/\sqrt{2}}e^{\alpha a(\bar{f})/\sqrt{2}}e^{\frac{1}{2}\alpha^2\|f\|^2}. \tag{5}$$

*Proof* We shall show (5) on $C(g)$. The proof of (5) on $\mathscr{F}_{\rm fin}$ is similar. We have

$$e^{\alpha a^*(f)}e^{\alpha a(\bar{f})}C(g) = e^{\alpha(f,g)}C(\alpha f + g). \tag{6}$$

Let $\psi(f) = a^*(f) + a(\bar{f})$. Then $\psi(f)$ is self-adjoint and it holds that

$$e^{\alpha\psi(f)} = \sum_{n=0}^\infty \frac{\alpha^n\psi(f)^n}{n!} \tag{7}$$

on $\mathscr{F}_{\rm fin}$. Let $C_m(g) = \sum_{n=0}^m \frac{a^*(g)^n}{n!}\Omega$. By using the expansion (7), we can compute as

$$e^{\alpha\psi(f)}C_m(g) = \sum_{n=0}^m \frac{(a^*(g) + \alpha(f,g))^n}{n!}e^{\alpha\psi(f)}\Omega.$$

Together with $e^{\alpha\psi(f)}\Omega = e^{\frac{1}{2}\alpha^2\|f\|^2}e^{\alpha a^*(f)}\Omega$, we see that

$$e^{\alpha\psi(f)}C_m(g) = \sum_{n=0}^m \frac{(a^*(g) + \alpha(f,g))^n}{n!}e^{\frac{1}{2}\alpha^2\|f\|^2}e^{\alpha a^*(f)}\Omega.$$

Hence we have

$$e^{\psi(f)}C(g) = e^{\alpha(f,g)}e^{\frac{1}{2}\alpha^2\|f\|^2}C(f+g). \tag{8}$$

By (6) and (8) the proposition follows. □

## 2.2 Bounds

In this section, we show several bounds concerning the exponential of annihilation operators and the creation operators. We learned all these bounds from the papers by Guneysu, Matte and Møller [7] and Matte and Møller [15]. We consider the case where $\mathscr{H} = L^2(\mathbb{R}^d)$. In this case, for $n \in \mathbb{N}$, the space $\mathscr{F}^{(n)}$ can be identified with the set of symmetric functions on $L^2(\mathbb{R}^{dn})$. The creation and annihilation operators act as

$$(a(f)\Psi)^{(n)}(k_1, \ldots, k_n) = \sqrt{n+1}\int_{\mathbb{R}^d} f(k)\Psi^{(n+1)}(k, k_1, \ldots, k_n)dk, \quad n \geq 0,$$

$$(a^*(f)\Psi)^{(n)}(k_1, \ldots, k_n) = \frac{1}{\sqrt{n}}\sum_{j=1}^{n} f(k_j)\Psi^{(n-1)}(k_1, \ldots, \hat{k}_j, \ldots, k_n),, \quad n \geq 1,$$

with $(a^*(f)\Psi)^{(0)} = 0$. Let $H_\mathrm{f} = d\Gamma(\omega)$ with $\omega = \omega(k) = \sqrt{|k|^2 + \nu^2}$. We have

$$(H_\mathrm{f}\Psi)^{(n)}(k_1, \ldots, k_n) = \left(\sum_{j=1}^{n} \omega(k_j)\right)\Psi^{(n)}(k_1, \ldots, k_n).$$

We can see the lemma below:

**Lemma 2.2** *Let $h : \mathbb{R}^d \to \mathbb{C}$ be measurable, and $g_j \in D(h)$ for $j = 1, ...., m$. Then for every $\Psi \in D(\mathrm{d}\Gamma(|h|^2)^{m/2})$, we have $\Psi \in D(\prod_{j=1}^m a(hg_j))$, and it follows that*

$$\left\|\prod_{j=1}^{m} a(hg_j)\Psi\right\| \leq \left(\prod_{j=1}^{m}\|g_j\|\right)\|\mathrm{d}\Gamma(|h|^2)^{m/2}\Psi\|. \tag{9}$$

*In particular*

$$\left\|\prod_{j=1}^{n} a(g_j)\Phi\right\| \leq \left(\prod_{j=1}^{n}\|g_j/\sqrt{\omega}\|\right)\|H_\mathrm{f}^{n/2}\Phi\|, \quad \Phi \in D(H_\mathrm{f}^{n/2}).$$

*Proof* Let $\Psi \in D(\mathrm{d}\Gamma(|h|^2)^m)$. First note that

$$(\Psi^{(n)}, (\mathrm{d}\Gamma(|h|^2)^m \Psi)^{(n)}) = \int_{\mathbb{R}^{nd}} |\Psi^{(n)}(k_1, \cdots, k_n)|^2 \left(\sum_{i=1}^n |h(k_i)|^2\right)^m dk_1 \cdots dk_n.$$

By the symmetry, we can replace $\left(\sum_{i=1}^n |h(k_i)|^2\right)^m$ with $C(n,m)\prod_{j=1}^m |h(k_j)|^2$. Here $C(n,m) = n(n-1)\cdots(n-m+1)$. Then we have

$$(\Psi^{(n)}, (\mathrm{d}\Gamma(|h|^2)^m \Psi)^{(n)}) = C(n,m)\int_{\mathbb{R}^{nd}} |\Psi^{(n)}(k_1, \cdots, k_n)|^2 \prod_{j=1}^m |h(k_j)|^2 dk_1 \cdots dk_n.$$

On the other hand by the definition of annihilation operators, we have

$$\begin{aligned} &\left\| (\prod_{j=1}^m a(hg_j)\Psi)^{(n-m)} \right\|^2 \\ &\le C(n,m)\left(\prod_{j=1}^m \|g_j\|^2\right) \int_{\mathbb{R}^{nd}} \left(\prod_{j=1}^m |h(k_j)|^2\right) |\Psi^{(n)}(k_1, \cdots, k_n)|^2 dk_1 \cdots dk_n \\ &= \left(\prod_{j=1}^m \|g_j\|^2\right) (\Psi^{(n)}, (\mathrm{d}\Gamma(|h|^2)^m \Psi)^{(n)}) \end{aligned}$$

and summation over $n$ gives (9). By the closedness of both operators $\mathrm{d}\Gamma(|h|^2)^{m/2}$ and $\prod_{j=1}^m a(hg_j)$, we can extend to $\Psi \in D(\mathrm{d}\Gamma(|h|^2)^{m/2})$. □

Next we estimate $\left\|\prod_{j=1}^n a^*(f_j)\Phi\right\|$.

**Lemma 2.3** *Let $f_i, g_j \in D(1/\sqrt{\omega})$ for $i, j = 1, \ldots, n$ and $\Phi \in D(H_{\mathrm{f}}^{n/2})$. Then*

$$\left|\left(\prod_{j=1}^n a^*(g_j)\Phi, \prod_{j=1}^n a^*(f_j)\Phi\right)\right| \le n!2^n \left(\prod_{l=1}^n \|f_l\|_\omega \|g_l\|_\omega\right) \sum_{m=0}^n \frac{1}{m!}\|H_{\mathrm{f}}^{m/2}\Phi\|^2,$$

*where $\|f\|_\omega = \|f\| + \|f/\sqrt{\omega}\|$.*

*Proof* Let $\Phi \in \mathscr{F}_{\mathrm{fin}}$ and $f_i, g_j \in \mathscr{H}$ for $i, j = 1, \ldots, m$. Then

$$\prod_{j=1}^n a(\bar{g}_j) \prod_{j=1}^n a^*(f_j)\Phi$$

$$= \sum_{m=0}^{n} \sum_{C_m \ni A} \sum_{C_{n-m} \ni B} \sum_{\substack{\sigma : A^c \to B \\ \text{bijection}}} \left( \prod_{l \in A^c} (g_l, f_{\sigma(l)}) \right) \left( \prod_{p \in B^c} a^*(f_p) \right) \left( \prod_{q \in A} a(\bar{g}_q) \right) \Phi. \quad (10)$$

Here $C_k = \{A \subset \{1, ..., n\} | \#A = k\}$, $C_0 = \emptyset$, and $\sum_{\substack{\sigma : A^c \to B \\ \text{bijection}}}$ is understood to take summation over all bijections from $A^c$ to $B$. In particular

$$\left[ \prod_{j=1}^{n} a(\bar{g}_j), \prod_{j=1}^{n} a^*(f_j) \right]$$

$$= \sum_{m=0}^{n-1} \sum_{C_m \ni A} \sum_{C_{n-m} \ni B} \sum_{\substack{\sigma : A^c \to B \\ \text{bijection}}} \left( \prod_{l \in A^c} (g_l, f_{\sigma(l)}) \right) \left( \prod_{p \in B^c} a^*(f_p) \right) \prod_{q \in A} a(\bar{g}_q).$$

By this formula we have

$$\left( \prod_{j=1}^{n} a^*(g_j) \Phi, \prod_{j=1}^{n} a^*(f_j) \Phi \right)$$

$$= \sum_{m=0}^{n} \sum_{C_m \ni A} \sum_{C_{n-m} \ni B} \sum_{\substack{\sigma : A^c \to B \\ \text{bijection}}} \left( \Phi, \prod_{l \in A^c} (g_l, f_{\sigma(l)}) \prod_{p \in B^c} a^*(f_p) \prod_{q \in A} a(\bar{g}_q) \Phi \right). \quad (11)$$

By $\| \prod_{j=1}^{n} a(h_j) \Phi \| \le \prod_{j=1}^{n} \|h_j\| \|H_{\mathrm{f}}^{n/2} \Phi\|$ and $\#B^c = m = \#A$, the right-hand side of (11) can be estimated as

$$\left| \left( \Phi, \prod_{p \in B^c} a^*(f_p) \prod_{q \in A} a(\bar{g}_q) \Phi \right) \right| < \left( \prod_{p \in B^c} \|f_p / \sqrt{\omega}\| \right) \left( \prod_{q \in A} \|g_q / \sqrt{\omega}\| \right) \|H_{\mathrm{f}}^{m/2} \Phi\|^2.$$

Since $\|f\| \le \|f\|_\omega$ and $\|f/\sqrt{\omega}\| \le \|f\|_\omega$, we have

$$\left| \left( \Phi, \prod_{l \in A^c} (g_l, f_{\sigma(l)}) \prod_{p \in B^c} a^*(f_p) \prod_{q \in A} a(\bar{g}_q) \Phi \right) \right| \le \left( \prod_{l=1}^{n} \|g_l\|_\omega \|f_l\|_\omega \right) \|H_{\mathrm{f}}^{m/2} \Phi\|^2. \quad (12)$$

Hence by (11) and (12)

$$\left| \left( \Phi, \prod_{j=1}^{n} a(\bar{g}_j) \prod_{j=1}^{n} a^*(f_j) \Phi \right) \right| \le n! 2^n \sum_{m=0}^{n} \frac{1}{m!} \left( \prod_{l=1}^{n} \|g_l\|_\omega \|f_l\|_\omega \right) \|H_{\mathrm{f}}^{m/2} \Phi\|^2.$$

Then the lemma follows. □

By Lemma 2.3, we have bounds for products of annihilation operators and creation operators. We summarize them as follows. Suppose that $f_j \in D(1/\sqrt{\omega})$ for $j = 1, \dots, n$. By introducing a scaling parameter $0 < s < 1$, we also have

$$\left\| \prod_{j=1}^{n} a(f_j)\Phi \right\| \le s^{-n/2} \left( \prod_{j=1}^{n} \|f_j/\sqrt{\omega}\| \right) \|(sH_{\mathrm{f}})^{n/2}\Phi\|, \tag{13}$$

$$\left\| \prod_{j=1}^{n} a^*(f_j)\Phi \right\| \le \sqrt{n!}2^{n/2}s^{-n/2} \left( \prod_{l=1}^{n} \|f_l\|_\omega \right) \left( \sum_{m=0}^{n} \frac{1}{m!} \|(sH_{\mathrm{f}})^{m/2}\Phi\|^2 \right)^{1/2}. \tag{14}$$

Although exponential operator $e^{a^*(f)}$ is unbounded, it can be seen in the proposition below that $e^{a^*(f)}e^{-\frac{t}{2}H_{\mathrm{f}}}$ is bounded for any $t > 0$.

**Proposition 2.4** *Let $t > 0$ and $f \in D(1/\sqrt{\omega})$. Then both $e^{a^*(f)}e^{-\frac{t}{2}H_{\mathrm{f}}}$ and $\overline{e^{-\frac{t}{2}H_{\mathrm{f}}}e^{a(f)}}$ are bounded.*

*Proof* Let $\Psi \in \cap_{n=1}^{\infty} D(H_{\mathrm{f}}^n)$. Suppose that $t < 1$. By (14) for any $s < t$ we have

$$\left\| \sum_{n=0}^{m} \frac{1}{n!} a^*(f)^n e^{-\frac{t}{2}H_{\mathrm{f}}}\Phi \right\| \le \sum_{n=0}^{m} \frac{1}{\sqrt{n!}} 2^{n/2}s^{-n/2}\|f\|_\omega^n \left( \sum_{k=0}^{n} \frac{1}{k!} \|(sH_{\mathrm{f}})^{k/2}e^{-\frac{t}{2}H_{\mathrm{f}}}\Phi\|^2 \right)^{1/2}.$$

We can see that sequence $\{\sum_{n=0}^{m} \frac{1}{n!}a^*(f)^n e^{-\frac{t}{2}H_{\mathrm{f}}}\Phi\}_{m=0}^{\infty}$ is a Cauchy sequence in $\mathscr{F}$. Hence $e^{-\frac{t}{2}H_{\mathrm{f}}}\Phi \in D(e^{a^*(f)})$ and as $m \to \infty$ on both sides above, we have

$$\|e^{a^*(f)}e^{-\frac{t}{2}H_{\mathrm{f}}}\Phi\| \le A(f,s)\|e^{-\frac{1}{2}(t-s)H_{\mathrm{f}}}\Phi\|,$$

where $A(f,s) = \sum_{n=0}^{\infty} \frac{1}{\sqrt{n!}}2^{n/2}s^{-n/2}\|f\|_\omega^n$. Choosing $s$ such that $s < t$, we can see that $\|e^{-\frac{1}{2}(t-s)H_{\mathrm{f}}}\Phi\| \le \|\Phi\|$ and $e^{a^*(f)}e^{-\frac{t}{2}H_{\mathrm{f}}}$ for $t < 1$ is bounded. Suppose $1 \le t$. Choosing $s = 1$ in the above discussion, we have

$$\|e^{a^*(f)}e^{-\frac{t}{2}H_{\mathrm{f}}}\Phi\| \le A(f,1)\|e^{-\frac{1}{2}(t-1)H_{\mathrm{f}}}\Phi\| \le A(f,1)\|\Phi\|.$$

Thus $e^{a^*(f)}e^{-\frac{t}{2}H_{\mathrm{f}}}$ for $t \ge 1$ is bounded. Finally since $\left(e^{-\frac{t}{2}H_{\mathrm{f}}}e^{a(f)}\right)^* \supset e^{a^*(\bar{f})}e^{-\frac{t}{2}H_{\mathrm{f}}}$, the second statement follows. Then the lemma follows. □

**Corollary 2.5** *Let $f \in D(1/\sqrt{\omega})$. Then*

$$\|e^{a^*(f)}e^{-\frac{t}{2}H_{\mathrm{f}}}\| \le \sqrt{2}e^{(2/s)\|f\|_\omega^2}\|e^{-\frac{1}{2}(t-s)H_{\mathrm{f}}}\|, \quad 0 < s < t < 1,$$
$$\|e^{a^*(f)}e^{-\frac{t}{2}H_{\mathrm{f}}}\| \le \sqrt{2}e^{2\|f\|_\omega^2}\|e^{-\frac{1}{2}(t-1)H_{\mathrm{f}}}\|, \quad 1 \le t.$$

*In particular, we have*

$$\|e^{a^*(f)}e^{-tH_f}e^{a(\bar{f})}\| \leq 2e^{(4/s)\|f\|_\omega^2},\ 0 < s < t < 1,$$
$$\|e^{a^*(f)}e^{-tH_f}e^{a(\bar{f})}\| \leq 2e^{4\|f\|_\omega^2},\quad 1 \leq t.$$

*Proof* We can estimate $A(f, s)$ as

$$A(f,s) \leq \left(\sum_{n=0}^{\infty} \frac{1}{n!} 8^n \|f\|_\omega^{2n} s^{-n} 2^{-n}\right)^{1/2} \left(\sum_{n=0}^{\infty} 1^2 \cdot 2^{-n}\right)^{1/2} \leq \sqrt{2} e^{(2/s)\|f\|_\omega^2}.$$

Then (1) follows from Proposition 2.4. □

## 3 Definition of the Nelson model

Let $\mathscr{S}'(\mathbb{R}^d)$ be the tempered distribution on $\mathbb{R}^d$. Let $s \in \mathbb{R}$ and $H^s(\mathbb{R}^d)$ be the inhomogeneous Sobolev space, i.e.,

$$H^s(\mathbb{R}^d) = \{u \in \mathscr{S}'(\mathbb{R}^d) | \hat{u} \in L^1_{\text{loc}}(\mathbb{R}^d), (1 + |k|^2)^{s/2} \hat{u} \in L^2(\mathbb{R}^d)\}.$$

Here $\hat{u}$ describes the Fourier transform on $\mathscr{S}'(\mathbb{R}^d)$. Homogeneous Sobolev space $\dot{H}^s(\mathbb{R}^d)$ is defined by

$$\dot{H}^s(\mathbb{R}^d) = \{u \in \mathscr{S}'(\mathbb{R}^d) | \hat{u} \in L^1_{\text{loc}}(\mathbb{R}^d), |k|^s \hat{u} \in L^2(\mathbb{R}^d)\}.$$

The scalar product on $\dot{H}^s(\mathbb{R}^d)$ is defined by $(f, g)_{\dot{H}^s(\mathbb{R}^d)} = (|k|^s \hat{f}, |k|^s \hat{g})_{L^2(\mathbb{R}^d)}$. It is known that $\dot{H}^s(\mathbb{R}^d)$ is a Hilbert space if and only if $s < d/2$. Now we modify $H^s(\mathbb{R}^d)$ to apply quantum field theory. We define $H^s_\nu(\mathbb{R}^d)$ by $H^s(\mathbb{R}^d)$ with $(1 + |k|^2)^{1/2}$ replaced by $\omega$. Hence $H^s(\mathbb{R}^d)$ and $H^s_\nu(\mathbb{R}^d)$ are equivalent for $\nu > 0$, and $\dot{H}^s(\mathbb{R}^d) = H^s_\nu(\mathbb{R}^d)$ for $\nu = 0$. We set

$$\mathscr{H}_s(\mathbb{R}^d) = \begin{cases} H^s_\nu(\mathbb{R}^d), & \nu > 0, \\ \dot{H}^s(\mathbb{R}^d), & \nu = 0. \end{cases}$$

We set $\mathscr{H}_M = \mathscr{H}_{-1/2}(\mathbb{R}^d)$ and $\mathscr{H}_E = \mathscr{H}_{-1}(\mathbb{R}^{d+1})$. We define the Fourier transform (in the sense of tempered distribution) of $\mathscr{H}_M$ and $\mathscr{H}_E$ by $\hat{\mathscr{H}}_M$ and $\hat{\mathscr{H}}_E$, respectively. Although $\mathscr{H}_M$, $\hat{\mathscr{H}}_M$, $\mathscr{H}_E$ and $\hat{\mathscr{H}}_E$ depend on the space dimension and $\nu \geq 0$, we do not write the dependence explicitly. We also define real Hilbert spaces below:

(1) $\mathscr{M} = \{f \in \mathscr{H}_M | f \text{ is real valued}\}$,
(2) $\mathscr{E} = \{f \in \mathscr{H}_E | f \text{ is real valued}\}$.

Both $\mathscr{M}$ and $\mathscr{E}$ are Hilbert spaces over $\mathbb{R}$, and note that $\mathscr{M}_\mathbb{C} = \mathscr{H}_M$ and $\mathscr{E}_\mathbb{C} = \mathscr{H}_E$.

Hilbert space $L^2(\mathbb{R}^d)$ describes the state space of the nonrelativistic matter, and $\mathscr{F}_\text{N} = \mathscr{F}(\hat{\mathscr{H}}_M)$ that of the scalar bose field. The joint state space is described by the tensor product

$$\mathscr{H} = L^2(\mathbb{R}^d) \otimes \mathscr{F}_{\mathrm{N}}.$$

The free particle Hamiltonian is described by the Schrödinger operator

$$H_{\mathrm{p}} = -\frac{1}{2}\Delta + V$$

acting in $L^2(\mathbb{R}^d)$. We introduce a class $\mathscr{R}$ of potentials. Let $V$ be relatively bounded with respect to $-(1/2)\Delta$ with a relative bound strictly smaller than one, i.e., $D(V) \subset D(-(1/2)\Delta)$ and

$$\|Vf\| \le a\| - (1/2)\Delta f\| + b\|f\|$$

for $f \in D(V)$ with some $a < 1$ and $b \ge 0$. Then we say $V \in \mathscr{R}$. We introduce Assumption 3.1.

**Assumption 3.1** The following conditions hold:

(1) $\varphi \in \mathscr{S}'(\mathbb{R}^d)$, $\overline{\hat{\varphi}(k)} = \hat{\varphi}(-k)$ and $\hat{\varphi}/\omega, \hat{\varphi}/\sqrt{\omega} \in L^2(\mathbb{R}^d)$.
(2) $V \in \mathscr{R}$.

The free field Hamiltonian $H_{\mathrm{f}} = \mathrm{d}\Gamma(\omega)$ on $\mathscr{F}_{\mathrm{N}}$ accounts for the energy carried by the field configuration. The matter -field interaction Hamiltonian $H_{\mathrm{I}}$ acting on the Hilbert space $\mathscr{H}$ describes then the interaction energy between the bose field and the matter . To give a definition of this operator we identify $\mathscr{H}$ as the space of $\mathscr{F}_{\mathrm{N}}$-valued $L^2$-functions on $\mathbb{R}^d$:

$$\mathscr{H} \cong \int_{\mathbb{R}^d}^{\oplus} \mathscr{F}_{\mathrm{N}} dx = \left\{ F : \mathbb{R}^d \to \mathscr{F}_{\mathrm{N}} \,\middle|\, \int_{\mathbb{R}^d} \|F(x)\|^2_{\mathscr{F}_{\mathrm{N}}} dx < \infty \right\}.$$

For each $x \in \mathbb{R}^d$, $H_{\mathrm{I}}(x)$ is defined by

$$H_{\mathrm{I}}(x) = \frac{1}{\sqrt{2}} \left( a_M^*(\hat{\varphi} e^{-ikx}) + a_M(\tilde{\hat{\varphi}} e^{ikx}) \right).$$

Here $a_M$ and $a_M^*$ denote the creation operator and the annihilation operator in the boson Fock space $\mathscr{F}(\hat{\mathscr{H}}_M)$, respectively. Since $\overline{\hat{\varphi}(k)} = \hat{\varphi}(-k)$, $H_{\mathrm{I}}(x)$ is symmetric, and it can be shown by using Nelson's analytic vector theorem that $H_{\mathrm{I}}(x)$ is essentially self-adjoint on $\mathscr{F}_{\mathrm{fin}}(\hat{\mathscr{H}}_M)$ of $\mathscr{F}_{\mathrm{N}}$. We denote the self-adjoint extension of $H_{\mathrm{I}}(x)$ by $\overline{H_{\mathrm{I}}(x)}$. The interaction $H_{\mathrm{I}}$ is then defined by the self-adjoint operator

$$H_{\mathrm{I}} = \int_{\mathbb{R}^d}^{\oplus} \overline{H_{\mathrm{I}}(x)} dx.$$

Under the conditions of Assumption 3.1 the operator

$$H = H_{\mathrm{p}} \otimes \mathbb{1} + \mathbb{1} \otimes H_{\mathrm{f}} + H_{\mathrm{I}} \tag{15}$$

acting in $\mathscr{H}$ is called the Nelson Hamiltonian. Let

$$H_0 = H_p \otimes \mathbb{1} + \mathbb{1} \otimes H_f, \quad D(H_0) = D(H_p \otimes \mathbb{1}) \cap D(\mathbb{1} \otimes H_f).$$

Then $H_0$ is self-adjoint on $D(H_0)$ and bounded below. Suppose Assumption 3.1. Then $H$ is also self-adjoint on $D(H_0)$ and bounded below, furthermore, it is essentially self-adjoint on any core of $H_0$. This follows from Kato–Rellich theorem [10].

# 4 Pointwise Exponential Decay

## 4.1 *Integral Kernels*

We review a family of Gaussian random variables indexed by a real vector space $\mathscr{M}$. We say that $(\phi(f), f \in \mathscr{M})$ is a family of Gaussian random variables on a probability space $(\mathscr{Q}, \Sigma, \mu)$ indexed by a real inner product space $\mathscr{M}$ whenever

(1) $\phi : \mathscr{M} \ni f \mapsto \phi(f)$ is a map from $\mathscr{M}$ to a Gaussian random variable on $(\mathscr{Q}, \Sigma, \mu)$ with $\mathbb{E}_\mu[\phi(f)] = 0$ and covariance $\mathbb{E}_\mu[\phi(f)\phi(g)] = \frac{1}{2}(f, g)_{\mathscr{M}}$,
(2) $\phi(\alpha f + \beta g) = \alpha\phi(f) + \beta\phi(g)$, $\alpha, \beta \in \mathbb{R}$,
(3) $\Sigma$ is the completion of the minimal $\sigma$-field generated by $\{\phi(f) | f \in \mathscr{M}\}$.

Also let $(\phi_E(f), f \in \mathscr{E})$ be a family of Gaussian random variables on a probability space $(\mathscr{Q}_E, \Sigma_E, \mu_E)$ indexed by a real inner product space $\mathscr{E}$. Let $\mathscr{O} \subset \mathbb{R}$ and put

$$\mathscr{E}(\mathscr{O}) = \{f \in \mathscr{E} | \mathrm{supp} f \subset \mathscr{O} \times \mathbb{R}^d\}$$

and the projection $\mathscr{E} \to \mathscr{E}(\mathscr{O})$ is denoted by $e_{\mathscr{O}}$. Let

$$\Sigma_{\mathscr{O}} = \sigma\left(\{\phi_E(f) | f \in \mathscr{E}(\mathscr{O})\}\right).$$

Define

$$\mathscr{E}_{\mathscr{O}} = \{\Phi \in L^2(\mathscr{Q}_E) | \Phi \text{ is } \Sigma_{\mathscr{O}}\text{-measurable}\}.$$

Let $e_t = \tau_t\tau_t^*$, $t \in \mathbb{R}$. Then $\{e_t\}_{t\in\mathbb{R}}$ is a family of projections from $\mathscr{E}$ to $\mathrm{Ran}(\tau_t)$. Let $\Sigma_t$, $t \in \mathbb{R}$, be the minimal $\sigma$-field generated by $\{\phi_E(f) | f \in \mathrm{Ran}(e_t)\}$. Define

$$\mathscr{E}_t = \{\Phi \in L^2(\mathscr{Q}_E) | \Phi \text{ is } \Sigma_t\text{-measurable}\}, \quad t \in \mathbb{R}.$$

We will see below that $F \in \mathscr{E}_{[a,b]}$ can be characterized by $\mathrm{supp} F \subset [a, b] \times \mathbb{R}^d$.

**Lemma 4.1** *(1)* $\mathscr{E}(\{t\}) = \mathrm{Ran}(e_t)$ *and any* $f \in \mathrm{Ran}(e_t)$ *can be expressed as* $f = \delta_t \otimes g$ *for some* $g \in \mathscr{M}$. *In particular,* $e_{\{t\}} = e_t$.
*(2)* $\mathscr{E}([a, b]) = \overline{\mathrm{L.H.}\{f \in \mathscr{E} | f \in \mathrm{Ran}(e_t), a \le t \le b\}}^{\|\cdot\|_{-1}}$ *holds.*

*Proof* Refer to see [18]. □

We will define a family of transformations $\mathrm{I}_t$ from $L^2(\mathscr{Q})$ to $L^2(\mathscr{Q}_\mathrm{E})$ through the second quantization of a specific transformation $\tau_t$ from $\mathscr{M}$ to $\mathscr{E}$. Define $\tau_t : \mathscr{M} \to \mathscr{E}$ by $\tau_t : f \mapsto \delta_t \otimes f$. Here $\delta_t(x) = \delta(x-t)$ is the delta function with mass at $t$. Note that $\overline{\delta_t \otimes f} = \delta_t \otimes f$, which implies that $\tau_t$ preserves realness. It follows that

$$\tau_s^* \tau_t = e^{-|s-t|\hat{\omega}}, \quad s, t \in \mathbb{R}.$$

In particular, $\tau_t$ is isometry between $\mathscr{M}$ and $\mathscr{E}$ for each $t \in \mathbb{R}$. Let $\mathrm{I}_t = \Gamma(\tau_t) : L^2(\mathscr{Q}) \to L^2(\mathscr{Q}_\mathrm{E})$, $t \in \mathbb{R}$, be the family of isometries:

$$\mathrm{I}_t \mathbb{1}_\mathrm{M} = \mathbb{1}_\mathrm{E}, \quad \mathrm{I}_t{:}\phi(f_1)\cdots\phi(f_n){:} = {:}\phi_\mathrm{E}(\delta_t \otimes f_1)\cdots\phi_\mathrm{E}(\delta_t \otimes f_n){:}.$$

Let $\hat{\omega} = \omega(-i\nabla) = \sqrt{-\Delta + \nu^2}$. The self-adjoint operator

$$\hat{H}_\mathrm{f} = \mathrm{d}\Gamma(\hat{\omega})$$

is called the free field Hamiltonian in $L^2(\mathscr{Q})$. It follows that $\hat{H}_\mathrm{f} = \theta_\mathrm{W} H_\mathrm{f} \theta_\mathrm{W}^{-1}$. From the identity $\tau_s^* \tau_t = e^{-|s-t|\hat{\omega}}$, it follows that

$$\mathrm{I}_t^* \mathrm{I}_s = e^{-|s-t|\hat{H}_\mathrm{f}}, \quad s, t \in \mathbb{R}. \tag{16}$$

On $L^2(\mathbb{R}^d) \otimes L^2(\mathscr{Q})$, we define the Nelson Hamiltonian by

$$\hat{H} = H_\mathrm{p} \otimes \mathbb{1} + \mathbb{1} \otimes \hat{H}_\mathrm{f} + g\hat{H}_\mathrm{I},$$

where $\hat{H}_\mathrm{I} = \int_{\mathbb{R}^d}^{\oplus} \hat{H}_\mathrm{I}(x)dx$ with $\hat{H}_\mathrm{I}(x) = \phi(\tilde{\varphi}(\cdot - x))$. Since $H$ and $\hat{H}$ are unitarily equivalent, we denote $H$ for $\hat{H}$ for simplicity in what follows. Let $(B_t)_{t\ge 0}$ be the brownian motion on the probability space $(\mathscr{X}, B(\mathscr{X}), \mathscr{W})$.

**Proposition 4.2** *Suppose Assumption 3.1. Then for $t \ge 0$ and $F, G \in \mathscr{H}$,*

$$(F, e^{-tH}G)_\mathscr{H} = \int_{\mathbb{R}^d} dx \mathbb{E}^x\left[e^{-\int_0^t V(B_s)ds}(F(B_0), \mathrm{I}_0^* e^{-\phi_\mathrm{E}(\int_0^t \delta_s \otimes \varphi(\cdot - B_s)ds)} \mathrm{I}_t G(B_t))_{L^2(\mathscr{Q})}\right]. \tag{17}$$

*Here $F, G \in \mathscr{H}$ are regarded as $L^2(\mathscr{Q})$-valued $L^2$-functions on $\mathbb{R}^d$.*

*Proof* See Appendix and see [14]. □

We call

$$\mathrm{I}_{[0,t]} = \mathrm{I}_0^* e^{-\phi_\mathrm{E}(\int_0^t \delta_s \otimes \varphi(\cdot - B_s)ds)} \mathrm{I}_t$$

the integral kernel of the semigroup generated by $H$. Thus

$$(F, e^{-tH}G)_\mathscr{H} = \int_{\mathbb{R}^d} dx \mathbb{E}^x\left[e^{-\int_0^t V(B_s)ds}(F(B_0), \mathrm{I}_{[0,t]} G(B_t))_{L^2(\mathscr{Q})}\right]$$

and we have

$$e^{-tH}G(x) = \mathbb{E}^x\left[e^{-\int_0^t V(B_s)ds}\mathrm{I}_{[0,t]}G(B_t)\right].$$

Moreover if $HG = EG$ we have

$$G(x) = e^{tE}\mathbb{E}^x\left[e^{-\int_0^t V(B_s)ds}\mathrm{I}_{[0,t]}G(B_t)\right]. \tag{18}$$

We discuss the boundedness of the norm of $\mathrm{I}_a^* e^{\phi_{\mathrm{E}}(f)}\mathrm{I}_b$ for not only massive case but also massless case. We see some intertwining properties of $\mathrm{I}_t$ and $\tau_t$. We can identify $\phi_{\mathrm{E}}(f)$ (resp.$\phi(f)$) with $\frac{1}{\sqrt{2}}(a_E^*(\hat{f}) + a_E(\tilde{\hat{f}}))$ (rep. $\frac{1}{\sqrt{2}}(a_M^*(\hat{f}) + a_M(\tilde{\hat{f}}))$). Under this identification $\mathrm{I}_t$ can be recognized as a map from $\mathscr{F}(\hat{\mathscr{H}}_M)$ to $\mathscr{F}(\hat{\mathscr{H}}_E)$, i.e.,

$$\mathrm{I}_t\prod_{j=1}^n a_M^*(\hat{f}_j)\Omega = \prod_{j=1}^n a_E^*(\hat{\tau}_t f_j)\Omega$$

and $\mathrm{I}_t^*$ as that from $\mathscr{F}(\hat{\mathscr{H}}_E)$ to $\mathscr{F}(\hat{\mathscr{H}}_M)$. It follows that on the finite particle subspace,

$$\mathrm{I}_t a_M^*(\hat{f}) = a_E^*(\widehat{\tau_t f})\mathrm{I}_t, \quad \mathrm{I}_t a_M(\hat{f}) = a_E(\widehat{\tau_t f})\mathrm{I}_t, \tag{19}$$

$$\mathrm{I}_t^* a_E^*(\hat{f}) = a_M^*(\widehat{\tau_t^* f})\mathrm{I}_t^*, \quad \mathrm{I}_t^* a_E(\widehat{e_t f}) = a_M(\widehat{\tau_t^* f})\mathrm{I}_t^*, \tag{20}$$

where $e_t = \tau_t\tau_t^*$. In particular

$$\mathrm{I}_t a_M^*(\widehat{\tau_t^* f}) = a_E^*(\widehat{e_t f})\mathrm{I}_t, \quad \mathrm{I}_t a_M(\widehat{\tau_t^* f}) = a_E(\widehat{e_t f})\mathrm{I}_t = a_E(\hat{f})\mathrm{I}_t, \tag{21}$$

$$\mathrm{I}_t^* a_E^*(\widehat{\tau_t f}) = a_M^*(\hat{f})\mathrm{I}_t^*, \quad \mathrm{I}_t^* a_E(\widehat{\tau_t f}) = a_M(\hat{f})\mathrm{I}_t^*. \tag{22}$$

**Theorem 4.3** *Suppose that $\hat{f} \in \hat{\mathscr{H}}_E$ and $\widehat{\tau_t^* f}/\sqrt{\omega} \subset \hat{\mathscr{H}}_M$ for $t = a, b$ with $a \neq b$. Then $\mathrm{I}_a^* e^{\phi_{\mathrm{E}}(f)}\mathrm{I}_b$ is bounded and*

$$\|\mathrm{I}_a^* e^{\phi_{\mathrm{E}}(f)}\mathrm{I}_b\| \leq 2\exp\left(\frac{1}{4}\|\hat{f}\|_{\hat{\mathscr{H}}_E} + \left(1 \vee \frac{1}{|a-b|}\right)(\|\widehat{\tau_a^* f}\|_\omega^2 + \|\widehat{\tau_b^* f}\|_\omega^2)\right).$$

*Here $x \vee y = \max\{x, y\}$ and*

$$\|\widehat{\tau_a^* f}\|_\omega = \|\widehat{\tau_a^* f}/\sqrt{\omega}\|_{L^2(\mathbb{R}^d)} + \|\widehat{\tau_a^* f}/\omega\|_{L^2(\mathbb{R}^d)}.$$

*Proof* By Baker–Campbell–Hausdorff formula, we have

$$e^{\phi_{\mathrm{E}}(f)} = e^{\frac{1}{\sqrt{2}}a_E^*(\hat{f})}e^{\frac{1}{\sqrt{2}}a_E(\tilde{\hat{f}})}e^{\frac{1}{4}\|\hat{f}\|^2_{\hat{\mathscr{H}}_E}}.$$

The intertwining property yields that

$$\mathrm{I}_a^* e^{\phi_{\mathrm{E}}(f)} \mathrm{I}_b = e^{\frac{1}{\sqrt{2}} a_M^*(\widehat{\tau_a^* f})} e^{-|a-b| H_{\mathrm{f}}} e^{\frac{1}{\sqrt{2}} a_M(\widetilde{\widehat{\tau_b^* f}})} e^{\frac{1}{4}\|\hat{f}\|^2_{\hat{\mathscr{H}}_E}}. \tag{23}$$

Since $e^{\frac{1}{\sqrt{2}} a_M^*(\widehat{\tau_a^* f})} e^{-\frac{|a-b|}{2} H_{\mathrm{f}}}$ and $e^{-\frac{|a-b|}{2} H_{\mathrm{f}}} e^{\frac{1}{\sqrt{2}} a_M(\widetilde{\widehat{\tau_b^* f}})}$ are bounded operators which operator bounds are given by

$$\|e^{\frac{1}{\sqrt{2}} a_M^*(\widehat{\tau_a^* f})} e^{-\frac{|a-b|}{2} H_{\mathrm{f}}}\| \le \sqrt{2} \exp\left\{\left(1 \vee \frac{1}{|a-b|}\right)\left(\|\widehat{\tau_a^* f}\|^2_{\hat{\mathscr{H}}_M} + \|\widehat{\tau_a^* f}/\sqrt{\omega}\|^2_{\hat{\mathscr{H}}_M}\right)\right\},$$

$$\|e^{-\frac{|a-b|}{2} H_{\mathrm{f}}} e^{\frac{1}{\sqrt{2}} a_M(\widetilde{\widehat{\tau_b^* f}})}\| \le \sqrt{2} \exp\left\{\left(1 \vee \frac{1}{|a-b|}\right)\left(\|\widehat{\tau_b^* f}\|^2_{\hat{\mathscr{H}}_M} + \|\widehat{\tau_b^* f}/\sqrt{\omega}\|^2_{\hat{\mathscr{H}}_M}\right)\right\}.$$

Hence together with them, we have

$$\|\mathrm{I}_a^* e^{\phi_{\mathrm{E}}(f)} \mathrm{I}_b\| \le 2 \exp\left(\frac{1}{4}\|\hat{f}\|^2_{\hat{\mathscr{H}}_{\mathrm{E}}} + \left(1 \vee \frac{1}{|a-b|}\right)\left(\|\widehat{\tau_a^* f}\|^2_\omega + \|\widehat{\tau_b^* f}\|^2_\omega\right)\right).$$

Then the proof is complete. □

**Corollary 4.4** (Integral kernel) *The integral kernel is given by*

$$\mathrm{I}_{[0,t]} = e^{\frac{1}{\sqrt{2}} a_M^*(U_t)} e^{-t H_{\mathrm{f}}} e^{\frac{1}{\sqrt{2}} a_M(\tilde{U}_t)} e^{\frac{1}{4} W},$$

*where* $U_t = \int_0^t ds e^{-s\omega(k)} e^{-ikB_s} \hat{\varphi}(k)/\sqrt{\omega(k)}$, $U_t = \int_0^t ds e^{-s\omega(k)} e^{ikB_s} \hat{\varphi}(k)/\sqrt{\omega(k)}$ *and*

$$W = \int_0^t dr \int_0^t ds \int_{\mathbb{R}^d} dk \frac{|\hat{\varphi}(k)|^2}{\omega(k)} e^{-|s-r|\omega(k)} e^{-ik(B_s - B_r)}.$$

*Proof* This follows from (23) and the definition of $\mathrm{I}_{[0,t]}$. □

We consider special cases of Theorem 4.3.

**Corollary 4.5** *Let $T \ge 0$. Let $f \in \mathscr{H}_M$, i.e., $\hat{f}/\sqrt{\omega} \in L^2(\mathbb{R}^d)$. We set*

$$\Phi_X = \phi_{\mathrm{E}}\left(\int_0^T (\tau_s f)(\cdot - B_s) ds\right).$$

*Then (1) and (2) follow.*
(1) *Suppose that $\hat{f}/\omega, \hat{f}/\sqrt{\omega^3} \in L^2(\mathbb{R}^d)$. Then*

$$\|\mathrm{I}_0^* e^{\Phi_X} \mathrm{I}_T\| \le 2 \exp\left(\frac{T}{2}\|\hat{f}/\omega\|^2 + 2(T \vee 1)(\|\hat{f}/\omega\|^2 + \|\hat{f}/\sqrt{\omega^3}\|^2)\right). \tag{24}$$

(2) *Suppose that $\hat{f}/\omega \in L^2(\mathbb{R}^d)$. Then*

$$\|\mathrm{I}_0^* e^{\Phi_X} \mathrm{I}_T\| \le 2 \exp\left(\frac{T}{2}\|\hat{f}/\omega\|^2 + 2T(T \vee 1)(\|\hat{f}/\sqrt{\omega}\|^2 + \|\hat{f}/\omega\|^2)\right). \tag{25}$$

*Proof* We see that

$$\left\| \int_0^T \tau_s f ds \right\|^2_{\hat{\mathscr{H}}_E} = \int_0^T ds \int_0^T dt (e^{-|s-t|\omega} \hat{f}, \hat{f})_{\hat{\mathscr{H}}_M} \leq 2T \|\hat{f}/\omega\|^2.$$

We can also see that

$$\left\| \tau_T^* \int_0^T \tau_s f ds \right\|^2_{\mathscr{H}_M} \leq T \|\hat{f}/\omega\|^2, \quad \left\| \tau_T^* \int_0^T \tau_s f ds/\sqrt{\omega} \right\|^2_{\mathscr{H}_M} \leq T \|\hat{f}/\sqrt{\omega^3}\|^2,$$
$$\left\| \tau_0^* \int_0^T \tau_s f ds \right\|^2_{\mathscr{H}_M} \leq T \|\hat{f}/\omega\|^2, \quad \left\| \tau_0^* \int_0^T \tau_s f ds/\sqrt{\omega} \right\|^2_{\mathscr{H}_M} \leq T \|\hat{f}/\sqrt{\omega^3}\|^2.$$

Then (1) follows from Theorem 4.3. For (2) we can estimate as

$$\left\| \tau_T^* \int_0^T \tau_s f ds \right\|^2_{\mathscr{H}_M} \leq T^2 \|\hat{f}/\sqrt{\omega}\|^2, \quad \left\| \tau_T^* \int_0^T \tau_s f ds/\sqrt{\omega} \right\|^2_{\mathscr{H}_M} \leq T^2 \|\hat{f}/\omega\|^2,$$
$$\left\| \tau_0^* \int_0^T \tau_s f ds \right\|^2_{\mathscr{H}_M} \leq T^2 \|\hat{f}/\sqrt{\omega}\|^2, \quad \left\| \tau_0^* \int_0^T \tau_s f ds/\sqrt{\omega} \right\|^2_{\mathscr{H}_M} \leq T^2 \|\hat{f}/\omega\|^2.$$

Then (2) follows. □

We can plug (24) and (25). Let

$$E(\hat{f}) = \max \left\{ \begin{array}{l} \frac{1}{2} \|\hat{f}/\omega\|^2 + 2(\|\hat{f}/\omega\|^2 + \|\hat{f}/\sqrt{\omega^3}\|^2) \\ \frac{1}{2} \|\hat{f}/\omega\|^2 + 2(\|\hat{f}/\sqrt{\omega}\|^2 + \|\hat{f}/\omega\|^2) \end{array} \right\}. \tag{26}$$

**Definition 4.1** (*Infrared regular condition*) The condition

$$\int_{\mathbb{R}^d} \frac{|\hat{\varphi}(k)|^2}{\omega(k)^3} dk < \infty$$

is called the *infrared regular condition*.

**Corollary 4.6** (Integral kernel under infrared regular condition) *Let $\Phi_X$ be in Corollary 4.5. Suppose that $\hat{f}/\sqrt{\omega}$, $\hat{f}/\omega \in L^2(\mathbb{R}^d)$ and $\int_{\mathbb{R}^d} |\hat{\varphi}(k)|^2/\omega(k)^3 dk < \infty$. Then it follows that*

$$\|\mathrm{I}_0^* e^{\Phi_X} \mathrm{I}_T\| \leq 2e^{TE(\hat{\varphi})}, \quad T \geq 0. \tag{27}$$

*Proof* We have

$$\|\mathrm{I}_0^* e^{\Phi_X} \mathrm{I}_T\| \leq 2 \exp\left( \frac{T}{2} \|\hat{f}/\omega\|^2 + 2T(\|\hat{f}/\omega\|^2 + \|\hat{f}/\sqrt{\omega^3}\|^2) \right)$$

for $T \geq 1$, and

$$\|\mathrm{I}_0^* e^{\Phi_x} \mathrm{I}_T\| \leq 2\exp\left(\frac{T}{2}\|\hat{f}/\omega\|^2 + 2T(\|\hat{f}/\sqrt{\omega}\|^2 + \|\hat{f}/\omega\|^2)\right)$$

for $T \leq 1$. Then (27) is shown. □

## 4.2 Kato-class Potentials

**Definition 4.2** (*Kato-class potentials* [1, 11])

(1) $V : \mathbb{R}^d \to \mathbb{R}$ is called a Kato-class potential whenever

$$\lim_{r\to 0} \sup_{x\in\mathbb{R}^d} \int_{B_r(x)} |g(x-y)V(y)|\, dy = 0$$

holds, where $B_r(x)$ is the closed ball of radius $r$ centered at $x$, and

$$g(x) = \begin{cases} |x|, & d = 1, \\ -\log|x|, & d = 2, \\ |x|^{2-d}, & d \geq 3. \end{cases}$$

We denote this linear space by $\mathscr{K}(\mathbb{R}^d)$.

(2) $V$ is Kato-decomposable whenever $V = V_+ - V_-$ with $V_+ \in L^1_{\mathrm{loc}}(\mathbb{R}^d)$ and $V_- \in \mathscr{K}(\mathbb{R}^d)$.

**Lemma 4.7** ([3]) *Let* $0 \leq V \in \mathscr{K}(\mathbb{R}^d)$. *Then there exist* $\beta, \gamma > 0$ *such that*

$$\sup_{x\in\mathbb{R}^d} \mathbb{E}^x[e^{\int_0^t V(B_s)ds}] < \gamma e^{\beta t}. \tag{28}$$

*Furthermore, if* $V \in L^p(\mathbb{R}^d)$ *with* $p > d/2$ *and* $1 \leq p < \infty$, *then*

$$\beta \leq c(p)^{1/\varepsilon}\Gamma(\varepsilon)^{1/\varepsilon}\|V\|_p^{1/\varepsilon}, \tag{29}$$

*where* $\varepsilon = 1 - \frac{d}{2p}$ *and*

$$c(p) = \begin{cases} (2\pi)^{-d/2} & p = 1, \\ (2\pi)^{-d/2p} q^{d/(2q)} & p > 1 \end{cases}$$

*with* $\frac{1}{p} + \frac{1}{q} = 1$. *In particular* $L^p(\mathbb{R}^d) \subset \mathscr{K}(\mathbb{R}^d)$ *for* $p > d/2$ *and* $1 \leq p < \infty$.

*Proof* There exists $t^* > 0$ such that $\alpha_t = \sup_{x\in\mathbb{R}^d} \mathbb{E}^x[\int_0^t V(B_s)ds] < 1$, for all $t \le t^*$, and $\alpha_t \to 0$ as $t \to 0$. By Khasminskii's lemma we have

$$\sup_{x\in\mathbb{R}^d} \mathbb{E}^x[e^{\int_0^t V(B_s)ds}] < \frac{1}{1-\alpha_t} \tag{30}$$

for all $t \le t^*$. We obtain

$$\mathbb{E}^x[e^{\int_0^{2t^*} V(B_s)ds}] = \mathbb{E}^x[e^{\int_0^{t^*} V(B_s)ds}\mathbb{E}^{B_{t^*}}[e^{\int_0^{t^*} V(B_s)ds}]] \le \left(\frac{1}{1-\alpha_{t^*}}\right)^2.$$

Repeating this procedure we see that

$$\sup_{x\in\mathbb{R}^d} \mathbb{E}^x[e^{\int_0^t V(B_s)ds}] \le \left(\frac{1}{1-\alpha_{t^*}}\right)^{[t/t^*]+1} \tag{31}$$

for all $t > 0$, where $[z] = \max\{w \in \mathbb{Z} | w \le z\}$. Setting $\gamma = (\frac{1}{1-\alpha_{t^*}})$ and $\beta = \log\left\{(\frac{1}{1-\alpha_{t^*}})^{1/t^*}\right\}$. This proves (28). Next we prove (29). Suppose $V \in L^p(\mathbb{R}^d)$ with $p > d/2$ and $1 \le p < \infty$. We let $p > 1$. By Schwarz inequality, we have

$$\mathbb{E}^x[V(B_t)] \le (2\pi t)^{-d/2}\left(\int_{\mathbb{R}^d} e^{-|x-y|^2 q/(2t)}\right)^{1/q} \|V\|_p = (2\pi t)^{-d/(2p)} q^{d/(2q)}\|V\|_p.$$

In particular, we have

$$\|\mathbb{E}^x[V(B_t)]\|_\infty \le c(p) t^{-d/(2p)}\|V\|_p, \quad p > 1.$$

We introduce a Mittag-Leffler function which is defined by $m_b(x) = \sum_{k=0}^\infty \frac{1}{\Gamma(1+kb)} x^k$, where $\Gamma$ denotes the Gamma function, $x \in \mathbb{R}$ and $b > 0$. It is known that Mittag-Leffler function $m_b(x)$ satisfies that $\lim_{x\to\infty}(m_b(x) - \frac{1}{b}e^{x^{1/b}}) = 0$ and there exists $k_b > 0$ such that $m_b(x) \le k_b e^{x^{1/b}}$ for all $x > 0$. Let $0 \le s_1 \le s_2 \le \cdots \le s_k$. By the Markov property of Brownian motion, we have

$$\mathbb{E}^x[V(B_{s_1})\cdots V(B_{s_k})] \le c(p)^k \|V\|_p^k s_1^{-d/(2p)}(s_2-s_1)^{-d/(2p)}\cdots(s_k-s_{k-1})^{-d/(2p)}.$$

Then

$$\begin{aligned}
&\mathbb{E}^x\left[\frac{1}{k!}\left[\int_0^t V(B_s)ds\right]^k\right] \\
&\le c(p)^k\|V\|_p^k \int_0^t ds_1 \cdots \int_{s_{k-1}}^t ds_k s_1^{-d/(2p)}(s_2-s_1)^{-d/(2p)}\cdots(s_k-s_{k-1})^{-d/(2p)}
\end{aligned}$$

$$= \frac{(c(p)\|V\|_p t^\varepsilon \Gamma(\varepsilon))^k}{\Gamma(1+k\varepsilon)},$$

where $\varepsilon = 1 - \frac{d}{2p} > 0$. Then it can be derived that

$$\sup_{x\in\mathbb{R}^d} \mathbb{E}^x[e^{\int_0^t V(B_s)ds}] \le m_\varepsilon\left(c(p)\|V\|_p \Gamma(\varepsilon)t^\varepsilon\right) \le k_\varepsilon e^{c(p)^{1/\varepsilon}\|V\|_p^{1/\varepsilon}\Gamma(\varepsilon)^{1/\varepsilon}t}.$$

Then (29) is proven. Let $p = 1$. Then $d = 1$ and it follows directly that

$$\mathbb{E}^x\left[V(B_t)\right]ds \le (2\pi t)^{-1/2}\|V\|_1 = c(1)t^{-1/2}\|V\|_1.$$

Hence (29) is proven in a similar way to the case of $p > 1$. □

We introduce Assumption 4.1.

**Assumption 4.1** The following conditions hold:

(1) $\varphi \in \mathscr{S}'(\mathbb{R}^d)$, $\overline{\hat{\varphi}(k)} = \hat{\varphi}(-k)$ and $\hat{\varphi}/\omega, \hat{\varphi}/\sqrt{\omega} \in L^2(\mathbb{R}^d)$.
(2) $V$ is Kato-decomposable.

Suppose Assumption 4.1 and define the family of operators

$$(T_t F)(x) = \mathbb{E}^x[e^{-\int_0^t V(B_r)dr} \mathrm{I}_{[0,t]} F(B_t)].$$

**Lemma 4.8** *$T_t$ is bounded.*

*Proof* Let $F \in \mathscr{H}_\mathrm{N}$. Since $\|\mathrm{I}_{[0,t]}F(B_t)\| \le 2e^{E(t)}\|F(B_t)\|$, where $E(t)$ is given by

$$E(t) = \frac{t}{2}\|\hat{\varphi}/\sqrt{\omega}\|^2 + 2t(1\vee t)(\|\hat{\varphi}/\sqrt{\omega}\|^2 + \|\hat{\varphi}/\omega\|^2),$$

from

$$\|T_t F\|^2_{\mathscr{H}_\mathrm{N}} = \int_{\mathbb{R}^d} dx \|\mathbb{E}^x[e^{-\int_0^t V(B_r)dr} \mathrm{I}_{[0,t]} F(B_t)]\|^2_{L^2(\mathscr{Q})}$$

it follows that

$$\|T_t F\|^2_{\mathscr{H}_\mathrm{N}} \le 2\int_{\mathbb{R}^d} dx \mathbb{E}[e^{-2\int_0^t V(B_r+x)dr}]\mathbb{E}[\|F(B_t+x)\|^2]e^{2E(t)}.$$

Since $V$ is Kato-decomposable, we have $\sup_{x\in\mathbb{R}^d}\mathbb{E}[e^{-2\int_0^t V(B_r+x)dr}] = C < \infty$, and thus $\|T_t F\|^2_{\mathscr{H}_\mathrm{N}} \le 4Ce^{2E(t)}\|F\|^2_{\mathscr{H}_\mathrm{N}}$ follows. □

We note that if infrared regular condition $\int_{\mathbb{R}^d} |\hat{\varphi}(k)|^2/\omega(k)^3 dk < \infty$ holds, then $\|\mathrm{I}_{[0,t]}\| \le 2e^{E(t)}$ can be replaced with

$$\|\mathrm{I}_{[0,t]}\| \le 2e^{tE(\hat{\varphi})}. \tag{32}$$

In what follows, we show that $\{T_t : t \geq 0\}$ is a symmetric $C_0$-semigroup. To do that, we introduce the time-shift operator $u_t$ on $L^2(\mathbb{R}^d)$ by

$$u_t f(x) = f(x_0 - t, \mathbf{x}), \quad x = (x_0, \mathbf{x}) \in \mathbb{R} \times \mathbb{R}^d.$$

It is straightforward that $u_t^* = u_{-t}$ and $u_t^* u_t = 1$. We denote the second quantization of $u_t$ by $U_t = \Gamma_{\mathrm{E}}(u_t)$ which acts on $L^2(\mathscr{Q}_{\mathrm{E}})$ and is a unitary map. We can see that $U_t \mathrm{I}_s = \mathrm{I}_{s+t}$. We set

$$K_t = \int_0^t \delta_s \otimes \varphi(\cdot - B_s) ds.$$

**Lemma 4.9** *$T_s T_t = T_{s+t}$ holds for $s, t \geq 0$.*

*Proof* By the definition of $T_t$ we have

$$T_s T_t F = \mathbb{E}^x[e^{-\int_0^s V(B_r)dr} \mathrm{I}_{[0,s]} \mathbb{E}^{B_s}[e^{-\int_0^t V(B_r)dr} \mathrm{I}_{[0,t]} F(B_t)]]. \tag{33}$$

By the formulae $\mathrm{I}_s \mathrm{I}_0^* = \mathrm{I}_s \mathrm{I}_s^* U_{-s}^* = \mathrm{E}_s U_{-s}^*$ and $\mathrm{I}_t = U_{-s} \mathrm{I}_{t+s}$, Eq. (33) is equal to

$$\mathbb{E}^x[e^{-\int_0^s V(B_r)dr} \mathrm{I}_0^* e^{-\phi_{\mathrm{E}}(K_s)} \mathrm{E}_s \mathbb{E}^{B_s}[e^{-\int_0^t V(B_r)dr} U_{-s}^* e^{-\phi_{\mathrm{E}}(K_t)} U_{-s} \mathrm{I}_{t+s} F(B_t)]]. \tag{34}$$

Since $U_s$ is unitary, we have $U_{-s}^* e^{-\phi_{\mathrm{E}}(K_t)} U_{-s} = e^{-\phi_{\mathrm{E}}(u_{-s}^* K_t)}$ as an operator. The test function of the exponent $u_{-s}^* K_t$ is given by

$$u_{-s}^* K_t = \int_0^t \delta_{r+s} \otimes \tilde{\varphi}(\cdot - B_r) dr.$$

Moreover by the Markov property of $\mathrm{E}_t$, $t \in \mathbb{R}$, we may neglect $\mathrm{E}_s$ in (34), and by the Markov property of $(B_t)_{t \geq 0}$ we have

$$\begin{aligned} T_s T_t F &= \mathbb{E}^x[e^{-\int_0^s V(B_r)dr} \mathrm{I}_0^* e^{-\phi_{\mathrm{E}}(K_s)} \mathbb{E}^x[e^{-\int_s^{s+t} V(B_r)dr} e^{-\phi_{\mathrm{E}}(K_s^{s+t})} \mathrm{I}_{s+t} F(B_{s+t}) | \mathscr{F}_s]] \\ &= \mathbb{E}^x[e^{-\int_0^{s+t} V(B_r)dr} \mathrm{I}_0^* e^{-\phi_{\mathrm{E}}(K_{s+t})} \mathrm{I}_{s+t} F(B_{s+t})] = T_{s+t} F, \end{aligned}$$

where $K_s^{s+t} = \int_s^{s+t} \delta_r \otimes \tilde{\varphi}(\cdot - B_r) dr$ and $(\mathscr{F}_t)_{t \geq 0}$ denotes the natural filtration of $(B_t)_{t \geq 0}$. □

The strong continuity of the map $t \mapsto T_t$ on $\mathscr{H}_{\mathrm{N}}$ can be checked, while $T_0 = 1$ is trivial.

**Theorem 4.10** *Semigroup $\{T_t : t \geq 0\}$ is a symmetric $C_0$-semigroup.*

*Proof* Since it was shown that $\{T_t : t \geq 0\}$ is a $C_0$-semigroup, it is enough to show that $T_t$ is symmetric, i.e., $(F, T_t G) = (T_t F, G)$. Let $R = \Gamma_{\mathrm{E}}(r)$ be the second quantization of the reflection $r$, and $U_t = \Gamma_{\mathrm{E}}(u_t)$. Then we have

$$(F, T_t G) = \int_{\mathbb{R}^d} dx \mathbb{E}^x \left[ e^{-\int_0^t V(B_s)ds} (\mathrm{I}_t F(B_0), e^{-\phi_{\mathrm{E}}(u_t r K_t)} \mathrm{I}_0 G(B_t)) \right],$$

where $u_t r K_t = \int_0^t \delta_{t-s} \otimes \tilde{\varphi}(\cdot - B_s) ds$. Noticing that $\dot{B}_s = B_{t-s} - B_t \stackrel{\mathrm{d}}{=} B_s$ for $0 \leq s \leq t$. Thus we can replace $B_s$ with $\dot{B}_s$. Thus

$$(F, T_t G) = \int_{\mathbb{R}^d} dx \mathbb{E}[e^{-\int_0^t V(\dot{B}_s + x)ds} (\mathrm{I}_t F(\dot{B}_0 + x), e^{-\phi_{\mathrm{E}}(u_t r \dot{K}_t)} \mathrm{I}_0 G(\dot{B}_t + x))].$$

Here $\dot{K}_t = \int_0^t \delta_s \otimes \tilde{\varphi}(\cdot - \dot{B}_s) ds$. Exchanging $\int_{\mathbb{R}^d} dx$ and $\int_{\mathscr{X}} d\mathscr{W}^0$, and changing variable $-B_t + x$ to $y$, we have

$$(F, T_t G) = \int_{\mathbb{R}^d} dy \mathbb{E}[e^{-\int_0^t V(B_{t-s} + y)ds} (\mathrm{I}_t F(B_t + y), e^{-\phi_{\mathrm{E}}(\tilde{K}_t)} \mathrm{I}_0 G(B_0 + y))].$$

Here $\tilde{K}_t = \int_0^t \delta_{t-s} \otimes \tilde{\varphi}(\cdot - B_{t-s} - y) ds$. Thus we conclude that $(F, T_t G) = (T_t F, G)$ and the theorem follows. □

By Theorem 4.10, there exists a self-adjoint operator $H_{\mathrm{Kato}}$ such that $T_t = e^{-t H_{\mathrm{Kato}}}$ for $t \geq 0$.

**Definition 4.3** (*Nelson Hamiltonian with Kato-class potential*) Let $V$ be a Kato - decomposable potential. Then we call the self-adjoint operator $H_{\mathrm{Kato}}$ Nelson Hamiltonian with Kato-class potential $V$.

In what follows notational simplicity, we also write $H$ for $H_{\mathrm{Kato}}$. That is, the Nelson Hamiltonian is written by $H$ but with Kato-decomposable potential is understood as $H_{\mathrm{Kato}}$.

## 4.3 Martingales

Let us consider the Schrödinger operator $H_{\mathrm{p}} = -\frac{1}{2}\Delta + V$. Let $f$ be an eigenvector of $H_{\mathrm{p}}$; $H_{\mathrm{p}} f = E_{\mathrm{p}} f$. Then we define the random process

$$h_t(x) = e^{t E_{\mathrm{p}}} e^{-\int_0^t V(B_r + x)dr} f(B_t + x), \quad t \geq 0.$$

Note that $\mathbb{E}[h_t(x)] = f(x)$ for all $t \geq 0$. We see that the random process $(h_t(x))_{t \geq 0}$ is martingale with respect to $(\mathscr{F}_t)_{t \geq 0}$. We extend this to the Nelson Hamiltonian. Suppose that $E$ is an eigenvalue associated with a bound state $\Phi$;

$$H\Phi = E\Phi.$$

Define

$$H_t(x) = e^{tE} e^{-\int_0^t V(B_r+x)dr} e^{-\phi_{\mathrm{E}}(\int_0^t \delta_r \otimes \varphi(\cdot - x - B_r)dr)} \mathrm{I}_t \Phi(B_t + x). \tag{35}$$

Then $(H_t(x))_{t\geq 0}$ is a random process on $(\mathscr{X} \times \mathscr{Q}_{\mathrm{E}}, \mathscr{B}(\mathscr{X}) \times \Sigma_{\mathrm{E}}, \mathscr{W} \times \mu_{\mathrm{E}})$ for each $x \in \mathbb{R}^d$. By the functional integral representation we have

$$(\Psi, \Phi) = (\Psi, e^{-t(H-E)}\Phi) = \int_{\mathbb{R}^d} dx \mathbb{E}_{\mu_{\mathrm{E}}}[\bar{\Psi}(x)\mathbb{E}[\mathrm{I}_0^* H_t(x)]].$$

Hence it follows that $\Phi(x) = (e^{-t(H-E)}\Phi)(x) = \mathbb{E}[\mathrm{I}_0^* H_t(x)]$. Define the filtration $(\mathscr{M}_t)_{t\geq 0}$ by

$$\mathscr{M}_t = \mathscr{F}_t \times \Sigma_{(-\infty,t]}, \quad t \geq 0.$$

**Theorem 4.11** *Suppose Assumption 4.1. Then $(H_t(x))_{t\geq 0}$ is martingale with respect to $(\mathscr{M}_t)_{t\geq 0}$.*

*Proof* Let us set

$$K(0, s) = e^{-\int_0^s V(B_r+x)dr} e^{-\phi_{\mathrm{E}}(\int_0^s \delta_r \otimes \varphi(\cdot - x - B_r)dr)}.$$

We have

$$\mathbb{E}_{\mu_{\mathrm{E}}}\mathbb{E}[H_t(x)|\mathscr{M}_s] = e^{tE} K(0, s)\mathbb{E}_{\mu_{\mathrm{E}}}\mathbb{E}[K(s, t)\Phi(B_t + x)|\mathscr{M}_s]$$

We compute the conditional expectation of the right-hand side above. From Markov property of $(B_t)_{t\geq 0}$, it follows that

$$\begin{aligned}
&\mathbb{E}_{\mu_{\mathrm{E}}}\mathbb{E}[K(s, t)\mathrm{I}_t\Phi(B_t + x)|\mathscr{M}_s] \\
&= \mathbb{E}_{\mu_{\mathrm{E}}}\left[\mathbb{E}^{B_s}\left[e^{-\int_0^{t-s} V(B_r+x)dr} e^{-\phi_{\mathrm{E}}(\int_s^t \delta_r \otimes \varphi(\cdot - x - B_{r-s})dr)} \mathrm{I}_t \Phi(B_{t-s} + x)|\Sigma_{(-\infty,s]}\right]\right].
\end{aligned}$$

From the Markov property of projection $\mathrm{E}_s = \mathrm{I}_s\mathrm{I}_s^*$ it furthermore follows that

$$\begin{aligned}
&= \mathbb{E}_{\mu_{\mathrm{E}}}\left[\mathbb{E}^{B_s}\left[e^{-\int_0^{t-s} V(B_r+x)dr} e^{-\phi_{\mathrm{E}}(\int_s^t \delta_r \otimes \varphi(\cdot - x - B_{r-s})dr)} \mathrm{I}_t \Phi(B_{t-s} + x)|\Sigma_s\right]\right] \\
&= \mathrm{E}_s\mathbb{E}^{B_s}\left[e^{-\int_0^{t-s} V(B_r+x)dr} e^{-\phi_{\mathrm{E}}(\int_s^t \delta_r \otimes \varphi(\cdot - x - B_{r-s})dr)} \mathrm{I}_t \Phi(B_{t-s} + x)\right].
\end{aligned}$$

Let $U_s = \Gamma(u_s)$, where $u_s : \mathscr{E} \to \mathscr{E}$ denotes the time-shift operator defined by $u_s f(t, x) = f(t + s, x)$. Since the shift operator and the projection are related to $E_s = \mathrm{I}_s\mathrm{I}_0^* U_{-s}$, we have

$$\begin{aligned}
&= \mathrm{I}_s\mathrm{I}_0^* U_{-s}\mathbb{E}^{B_s}\left[e^{-\int_0^{t-s} V(B_r+x)dr} e^{-\phi_{\mathrm{E}}(\int_s^t \delta_r \otimes \varphi(\cdot - x - B_{r-s})dr)} \mathrm{I}_t \Phi(B_{t-s} + x)\right] \\
&= \mathrm{I}_s\mathrm{I}_0^*\mathbb{E}^{B_s}\left[e^{-\int_0^{t-s} V(B_r+x)dr} e^{-\phi_{\mathrm{E}}(\int_s^t \delta_{r-s} \otimes \varphi(\cdot - x - B_{r-s})dr)} \mathrm{I}_{t-s} \Phi(B_{t-s} + x)\right] \\
&= \mathrm{I}_s\mathrm{I}_0^*\mathbb{E}^{B_s}\left[K(0, t-s)\mathrm{I}_{t-s}\Phi(B_{t-s} + x)\right].
\end{aligned}$$

Hence we conclude that

$$\mathbb{E}_{\mu_E}\mathbb{E}[H_t(x)|\mathcal{M}_s] = e^{sE}K(0,s)e^{-(t-s)(H-E)}\Phi(B_s+x) = H_s(x).$$

Then $(H_t(x))_{t\geq 0}$ is martingale. □

Let $\tau$ be a stopping time with respect to $\mathcal{M}_t$. Then $H_{t\wedge\tau}(x)$ is also martingale. In particular $\mathbb{E}_{\mu_E}\mathbb{E}[H_t(x)] = \mathbb{E}_{\mu_E}\mathbb{E}[H_{t\wedge\tau}(x)]$.

## 4.4 Main Theorem

In this section we assume that $\Phi$ is a bound state of $H$: $H\Phi = E\Phi$.

**Lemma 4.12** *Suppose Assumption 4.1. Then* $\|\Phi(\cdot)\|_{L^2(\mathscr{Q})} \in L^\infty(\mathbb{R}^d)$.

*Proof* It follows that $\Phi(x) = \mathbb{E}_{\mu_E}\mathbb{E}[\mathrm{I}_0^* H_t(x)]$ for an arbitrary $t > 0$. Since we can see that $\mathrm{I}_{[0,t]}$ is bounded and $\|\mathrm{I}_{[0,t]}\| \leq 2e^{E(t)}$, we obtain

$$\|\Phi(x)\| \leq 2e^{tE}e^{E(t)}\left(\mathbb{E}^x[e^{-2\int_0^t V(B_r)dr}]\right)^{1/2}\left(\mathbb{E}^x[\|\Phi(B_t)\|^2]\right)^{1/2}.$$

Since $\sup_{x\in\mathbb{R}^d}\mathbb{E}^x[e^{-2\int_0^t V(B_r)dr}] < \infty$ and $\mathbb{E}^x[\|\Phi(B_t)\|^2] \leq C\|\Phi\|$, the lemma follows. □

Now we state the main theorem in this article.

**Theorem 4.13** (Pointwise exponential decay) *Suppose Assumption 4.1 and infrared regular condition* $\int_{\mathbb{R}^d}|\hat{\varphi}(k)|^2/\omega(k)^3dk < \infty$. *Assume either (1) or (2):*

*(1)* $\lim_{|x|\to\infty} V(x) = \infty$,
*(2)* $\lim_{|x|\to\infty} V_-(x) + E + E(\hat{\varphi}) < 0$.

*Then there exist constants c and C such that*

$$\|\Phi(x)\|_{L^2(\mathscr{Q})} \leq Ce^{-c|x|}.$$

*Proof* Suppose (1). Let $\tau_R = \inf\{t||B_t| > R\}$. Then $\tau_R$ is a stopping time with respect to the natural filtration of the Brownian motion $(B_t)_{t\geq 0}$. Let

$$W_R(x) = \inf\{V(y)||x-y| < R\}.$$

Note that $W_R(x) \leq V(x+y)$ for $|y| < R$, and we see that

$$X_R(x) = W_R(x) - E - E(\hat{\varphi}) \to \infty \quad (|x| \to \infty).$$

Let $\Psi \in L^2(Q)$. Then $\mathrm{I}_0\Psi \cdot H_t(x)$ is also martingale. Actually we can see that

$$\mathbb{E}_{\mu_{\mathrm{E}}}\mathbb{E}[\mathrm{I}_0\Psi \cdot H_t(x)|\mathscr{M}_s] = \mathrm{I}_0\Psi \cdot \mathbb{E}_{\mu_{\mathrm{E}}}\mathbb{E}[H_t(x)|\mathscr{M}_s] = \mathrm{I}_0\Psi \cdot H_s(x).$$

Hence, it follows that

$$(\mathrm{I}_0\Psi, \Phi(x))_{L^2(Q)} = \mathbb{E}_{\mu_{\mathrm{E}}}\mathbb{E}[\mathrm{I}_0\Psi \cdot H_0(x)] = \mathbb{E}_{\mu_{\mathrm{E}}}\mathbb{E}[\mathrm{I}_0\Psi \cdot H_{t\wedge\tau_R}(x)].$$

Boud $\|\mathrm{I}_{[0,t]}\| \le 2e^{tE(\hat{\varphi})}$ is derived from the infrared regular condition and the exponent is linear in $t$. Then

$$\begin{aligned}\|\Phi(x)\| &= \sup_{\Psi\in L^2(\mathscr{Q}),\|\Psi\|=1} |(\mathrm{I}_0\Psi, \Phi(x))| = |\mathbb{E}_{\mu_{\mathrm{E}}}\mathbb{E}[\mathrm{I}_0\Psi \cdot H_{t\wedge\tau_R}(x)]| \\ &\le \|\mathbb{E}[\mathrm{I}_0^* H_{t\wedge\tau_R}(x)]\| \le 2\mathbb{E}[e^{-\int_0^{t\wedge\tau_R}(V(B_r+x)-E-E(\hat{\varphi}))dr}] \sup_{x\in\mathbb{R}^d} \|\Psi_{\mathrm{g}}(x)\|.\end{aligned}$$

Then it is enough to estimate $\mathbb{E}[e^{-\int_0^{t\wedge\tau_R}(V(B_r+x)-E-E(\hat{\varphi}))dr}]$. Let

$$\mathbb{E}[e^{-\int_0^{t\wedge\tau_R}(V(B_r+x)-E-E(\hat{\varphi}))dr}] \le \mathbb{E}[\mathbb{1}_{\{\tau_R<t\}}e^{-(t\wedge\tau_R)X_R(x)}] + \mathbb{E}[\mathbb{1}_{\{\tau_R\ge t\}}e^{-(t\wedge\tau_R)X_R(x)}].$$

It is trivial to see that $\mathbb{E}[\mathbb{1}_{\{\tau_R\ge t\}}e^{-(t\wedge\tau_R)X_R(x)}] \le e^{-tX_R(x)}$. We also see that

$$\mathbb{E}[\mathbb{1}_{\{\tau_R<t\}}e^{-(t\wedge\tau_R)X_R(x)}] \le \mathbb{E}[\mathbb{1}_{\{|B_t|\ge R\}}] = \frac{S_{d-1}}{(2\pi)^{d/2}}\int_{R/\sqrt{t}}^\infty e^{-r^2/2}r^{d-1}dr \le c_1e^{-c_2R^2/t}.$$

Let $R = p|x|$ $(0 < p < 1)$ and $t = \delta|x|$, where we assume that $\delta$ is a positive constant. Since $X_{p|x|}(x) \to \infty$ as $|x| \to \infty$, we have

$$\mathbb{E}[e^{-(t\wedge\tau_R)X_R(x)}] \le e^{-\delta|x|c} + c_1e^{-c_2(p^2/\delta)|x|}.$$

Then the theorem follows. Next suppose (2). Let $\tau_R(x) = \inf\{t \ge 0||B_t + x| < R\}$. Similarly to (1) it is enough to estimate $\mathbb{E}[e^{\int_0^{t\wedge\tau_R(x)}(V_-(B_r+x)+E+E(\hat{\varphi}))dr}]$. By the assumption on $V_-$, there exist $\varepsilon > 0$ and $R > 0$ such that for all $|x| > R$ it holds that $V_-(x) + E + E(\hat{\varphi}) < -\varepsilon < 0$. By the definition of stopping time $|B_r + x| \ge R$ for $r \le t \wedge \tau_R(x)$. Then the integrand in $\int_0^{t\wedge\tau_R}(V_-(B_r + x) + E + E(\hat{\varphi}))dr$ is less than $-\varepsilon$. Hence

$$\begin{aligned}&\mathbb{E}[e^{\int_0^{t\wedge\tau_R(x)}(V_-(B_r+x)+E+E(\hat{\varphi}))dr}] \le \mathbb{E}[e^{-(t\wedge\tau_R(x))\varepsilon}] \\ &= \mathbb{E}[\mathbb{1}_{\{t<\tau_R(x)\}}e^{-t\varepsilon}] + \mathbb{E}[\mathbb{1}_{\{t\ge\tau_R(x)\}}e^{-\tau_R(x)\varepsilon}] \le e^{-t\varepsilon} + \mathbb{E}^x[\mathbb{1}_{\{t\ge\tau_R(0)\}}].\end{aligned}$$

We have $\mathbb{E}^x[\mathbb{1}_{\{t\ge\tau_R(0)\}}] = (2\pi)^{-d/2}\int_{|u|\le R/t} e^{-\frac{1}{2}(|u|^2-2\frac{u\cdot x}{\sqrt{|x|}}+\frac{|x|^2}{t})}du$. Set $t = R = |x|$. Then

$$\mathbb{E}^x[\mathbb{1}_{\{t\ge\tau_R(0)\}}] \le (2\pi)^{-d/2}e^{\sqrt{|x|}-\frac{1}{2}|x|}\int_{|u|\le 1} e^{-\frac{1}{2}|u|^2}du \le c_1e^{-c_2|x|}$$

and

$$\mathbb{E}[e^{\int_0^{t\wedge\tau_R(x)}(V_-(B_r+x)+E+E(\hat{\varphi}))dr}] \leq e^{-\varepsilon|x|} + c_1 e^{-c_2|x|}.$$

Thus the proof is complete. □

## 5 Proof of Proposition 4.2

*Proof* For $\varepsilon \geq 0$ let $\hat{H}_{\rm I}^{\varepsilon} = \rho_\varepsilon(\hat{H}_{\rm I})$, where $\rho_\varepsilon(X) = X + \varepsilon X^2$. Then $\hat{H}_{\rm I}^{\varepsilon}$ is bounded below for $\varepsilon > 0$. We can also see that $e^{-tH^\varepsilon} \to e^{-tH}$ strongly as $\varepsilon \downarrow 0$. We shall construct a functional integral representation of $(F, e^{-tH^\varepsilon}G)$ for $\varepsilon > 0$ and by a limiting argument we construct that of $(F, e^{-tH}G)$ for $\varepsilon = 0$. Assume that $V \in C_0^\infty(\mathbb{R}^d)$. Let $h = (-1/2)\Delta$. By the Trotter–Kato product formula [12, 13] and $e^{-|s-t|\hat{H}_{\rm f}} = {\rm I}_s^*{\rm I}_t$, we have

$$(F, e^{-tH^\varepsilon}G) = \lim_{n\to\infty}\left({\rm I}_0 F, \left(\prod_{j=0}^{n-1} {\rm I}_{\frac{jt}{n}} e^{-\frac{t}{n}\hat{H}_{\rm I}^{\varepsilon}} e^{-\frac{t}{n}h} e^{-\frac{t}{n}V} {\rm I}^*_{\frac{jt}{n}}\right){\rm I}_t G\right).$$

Here $\prod_{j=1}^n t_j = t_1 \cdots t_n$. Using the identity ${\rm I}_s e^{-\hat{H}_{\rm I}^{\varepsilon}}{\rm I}_s = E_s e^{-\hat{H}_{\rm I}^{\varepsilon}(s)}E_s$ for $s \in \mathbb{R}$, where

$$\hat{H}_{\rm I}^{\varepsilon}(s) = \int_{\mathbb{R}^d}^{\otimes} \rho_\varepsilon\left(\phi_{\rm E}(\delta_s \otimes \varphi(\cdot - x))\right) dx$$

and $E_s = {\rm I}_s{\rm I}_s^*$ is a projection, we can see that

$$(F, e^{-tH^\varepsilon}G) = \lim_{n\to\infty}\left({\rm I}_0 F, \left(\prod_{j=0}^{n-1} E_{\frac{jt}{n}} e^{-\frac{t}{n}\hat{H}_{\rm I}^{\varepsilon}(\frac{jt}{n})} e^{-\frac{t}{n}h} e^{-\frac{t}{n}V} E_{\frac{jt}{n}}\right){\rm I}_t G\right).$$

By the Markov property of $E_s$'s, we can neglect $E_s$'s. Then

$$(F, e^{-tH^\varepsilon}G) = \lim_{n\to\infty}\left({\rm I}_0 F, \left(\prod_{j=0}^{n-1} e^{-\frac{t}{n}\hat{H}_{\rm I}^{\varepsilon}(\frac{jt}{n})} e^{-\frac{t}{n}h} e^{-\frac{t}{n}V}\right){\rm I}_t G\right).$$

The right-hand side above can be represented in terms of the Wiener measure as

$$(F, e^{-tH^\varepsilon}G) = \lim_{n\to\infty}\int_{\mathbb{R}^d} dx \mathbb{E}^x\left[e^{-\sum_{j=0}^{n-1} V(B_{\frac{jt}{n}})}\left({\rm I}_0 F(B_0) e^{-\sum_{j=0}^{n-1}\frac{t}{n}\hat{H}_{\rm I}^{\varepsilon}(\frac{jt}{n})}{\rm I}_t G(B_t)\right)\right].$$

Note that $s \mapsto \delta_s \otimes \varphi(\cdot - B_s)$ is strongly continuous as a map $\mathbb{R} \to \mathscr{E}$, almost surely. Hence, $s \mapsto \phi_{\mathrm{E}}(\delta_s \otimes \varphi(\cdot - B_s))$ is also strongly continuous as a map $\mathbb{R} \to L^2(\mathscr{Q}_E)$. Then we can compute the limit and the result is

$$(F, e^{-tH^\varepsilon} G) = \int_{\mathbb{R}^d} dx \mathbb{E}^x \left[ e^{-\int_0^t V(B_s)ds} \left( \mathrm{I}_0 F(B_0), e^{-\varepsilon Q_t - \phi_{\mathrm{E}}(\int_0^t \delta_s \otimes \varphi(\cdot - B_s)ds)} \mathrm{I}_t G(B_t) \right) \right].$$

Here $Q_t = \int_0^t \phi_{\mathrm{E}}(\delta_s \otimes \varphi(\cdot - B_s))^2 ds$. Take $\varepsilon \downarrow 0$ on both sides above we have

$$(F(B_0), e^{-tH} G) = \int_{\mathbb{R}^d} dx \mathbb{E}^x \left[ e^{-\int_0^t V(B_s)ds} \left( \mathrm{I}_0 F(B_0), e^{-\phi_{\mathrm{E}}(\int_0^t \delta_s \otimes \varphi(\cdot - B_s)ds)} 8 \mathrm{I}_t G(B_t) \right) \right].$$

Then the theorem follows for $V \in C_0^\infty(\mathbb{R}^d)$. By a simple limiting argument, we can prove (17) for $V \in \mathscr{R}$. □

**Acknowledgements** The author thanks Oliver Matte for useful discussions and comments for bounds in Section 2.2, and also thanks organizers: Kenji Yajima, Arne Jensen, Hisashi Okamoto, Yoshio Tsutsumi, Shu Nakamura, Keiichi Kato, Norikazu Saito, and Fumihiko Nakano for inviting him to "Tosio Kato Centennial Conference" held in the University of Tokyo at September 4–8, 2017.
Finally, this work is financially supported by JSPS KAKENHI 16H03942, CREST JPMJCR14D6 and JSPS open partnership joint research with Denmark 1007867.

## References


1. Aizenman, M., Simon, B.: Brownian motion and Harnack's inequality for Schrödinger operators. Commun. Pure Appl. Math. **35**, 209–271 (1982)
2. Bach, V., Fröhlich, J., Sigal, I.M.: Renormalization group analysis of spectral problems in quantum field theory. Adv. Math. **137**, 205–298 (1998)
3. Carmona, R.: Pointwise bounds for Schrödinger eigenstates. Commun. Math. Phys. **62**, 97–106 (1978)
4. Carmona, R., Masters, W.C., Simon, B.: Relativistic Schrödinger operators: Asymptotic behavior of the eigenfunctions. J. Funct. Anal. **91**, 117–142 (1990)
5. Cycon, H.L., Froese, R.G., Kirsch, W., Simon, B.: Schrödinger Operators With Application to Quantum Mechanics and Global Geometry. Springer, Berlin (1987)
6. Griesemer, M., Lieb, E., Loss, M.: Ground states in non-relativistic quantum electrodynamics. Invent. Math. **145**, 557–595 (2001)
7. Guneysu, B., Matte, O., Møller, J.: Stochastic differential equations for models of non-relativistic matter interacting with quantized radiation fields. Probab. Theory Relat. Fields **167**, 817–915 (2017)
8. Hirokawa, M., Hiroshima, F., Spohn, H.: Ground state for point particles interacting through a massless scalar bose field. Adv. Math. **191**, 339–392 (2005)
9. Hiroshima, F.: Functional integral approach to semi-relativistic Pauli-Fierz models. Adv. Math. **259**, 784–840 (2014)
10. Kato, T.: Fundamental properties of Hamiltonian operators of Schrödinger type. Trans. Am. Math. Soc. **70**, 195–211 (1951)
11. Kato, T.: Schrödinger operators with singular potentials. Isr. J. Math. **13**, 135–148 (1972)
12. Kato, T.: On the Trotter-Lie product formula. Proc. Jpn. Acad. **50**, 694–698 (1974)
13. Kato, T., Masuda, K.: Trotter's product formula for nonlinear semigroups generated by the subdifferentiables of convex functionals. J. Math. Soc. Jpn. **30**, 169–178 (1978)

14. Lőrinczi, J., Hiroshima, F., Betz, V.: Feynman-Kac Type Theorems and Its Applications. De Gruyter (2011)
15. Matte, O., Møller, J.: Feynman-Kac Formulas for the Ultra-violet Renormalized Nelson Model (2017). arXiv:1701.02600
16. Nelson, E.: Interaction of nonrelativistic particles with a quantized scalar field. J. Math. Phys. **5**, 1990–1997 (1964)
17. Nelson, E.: Schrödinger particles interacting with a quantized scalar field. In: Martin, W.T., Segal, I. (eds.) Proceedings of a conference on analysis in function space, p. 87. MIT Press (1964)
18. Simon, B.: The $P(\phi)_2$ Euclidean (Quantum) Field Theory. Princeton University Press (1974)
19. Simon, B., Høegh-Krohn, R.: Hypercontractive semigroup and two dimensional self-coupled bose fields. J. Funct. Anal. **9**, 121–180 (1972)

# Regular KMS States of Weakly Coupled Anharmonic Crystals and the Resolvent CCR Algebra

**Tomohiro Kanda and Taku Matsui**

**Abstract** We consider equilibrium states of weakly coupled anharmonic quantum oscillators(= anharmonic crystal) on an integer lattice $\mathbf{Z}$. We employed standard functional analytic methods for Schrödinger operators and we show existence of the infinite volume limit of equilibrium states, and uniqueness of the regular KMS (Kubo–Martin–Schwinger) states in the frame of Resolvent CCR Algebra introduced by D. Buchholz and H. Grundling.

## 1 Introduction

In this article, we consider equilibrium states of weakly coupled anharmonic quantum oscillators on $\mathbf{Z}$. We first consider finite quantum systems of particles located in sites $\Lambda_L := \{-L+1, -L+2, \cdots L-1, L\} \subset \mathbf{Z}$. Let $H_L$ be the Schrödinger operator (defined the domain of the quantum harmonic oscillator) on $L_2(\mathbf{R}^{2L})$ defined by

$$H_L = \sum_{k=-L+1}^{L} \left\{p_k^2 + \omega^2 x_k^2 + V(x_k)\right\} + \sum_{k=-L+1}^{L-1} \varphi(x_k - x_{k+1}) \tag{1}$$

where the potential $V(x)$ and $\varphi(x)$ are rapidly decreasing smooth functions defined on $\mathbf{R}$ and $p_k$ is the quantum mechanical momentum operator $p_k = -i\frac{\partial}{\partial x_k}$ as usual.

Bose particles are supposed to be fixed at sites in $\mathbf{Z}$, and we consider the nearest-neighbor interaction. Due to the presence of the term $V(x)$ in (1), the case of one-site double-well potentials is within reach of our study while $\varphi(x)$ gives rise to the (weak)

T. Kanda · T. Matsui (✉)
Faculty of Mathematics, Kyushu University, Fukuoka, Japan
e-mail: matsui@math.kyushu-u.ac.jp

T. Kanda
e-mail: t-kanda@math.kyushu-u.ac.jp

© Springer Nature Switzerland AG 2019

T. M. Rassias and V. A. Zagrebnov (eds.), *Analysis and Operator Theory*, Springer Optimization and Its Applications 146,
https://doi.org/10.1007/978-3-030-12661-2_12

interaction between Bose particles. By equilibrium states for finite quantum systems, we mean the state $\psi_{L,\beta}$ is defined by

$$\psi_{L,\beta}(Q) = \frac{tr(e^{-\beta H_L} Q)}{tr(e^{-\beta H_L})}$$

where $Q$ is any element in an algebra $\mathscr{R}_L$ of bounded operators we introduce below. Our interest is to examine the infinite volume limit $\lim_{L\to\infty} \psi_{L,\beta}$ and uniqueness of KMS states. We will not discuss the decay of correlation nor analyticity of correlation functions as these topics were already investigated before. To this end, we proceed with two steps.
(a-1) The first step is to find the physically reasonable choice of an algebra $\mathscr{R}_L$ of operators. More precisely, we require that the multiplication operators $f(x_k)$ of the variable $x_k$ and the function $g(p_k)$ of momentum operators $p_k$ $(-L+1 \le k \le L)$ are contained in $\mathscr{R}_L$ where $f(x)$ and $g(x)$ are rapidly decreasing smooth functions.
(a-2) We require that

$$\alpha_t^L(Q) = e^{itH_L} Q e^{-itH_L}, \quad Q \in \mathfrak{R}_L \tag{2}$$

gives rise to a one-parameter group of automorphisms on $\mathfrak{R}_L$, hence $Q \in \mathfrak{R}_L$ should imply $\alpha_t^L(Q) \in \mathfrak{R}_L$.
(b) $\mathfrak{R}_L$ can be embedded into $\mathfrak{R}_{L+1}$ so that the infinite volume limit $\lim_{L\to\infty}$ is well defined and we can consider the uniqueness of the infinite volume states (the states obtained by the infinite volume limit) satisfying the KMS conditions.

At a first look, these questions are mathematical problems of operator algebras, however, real mathematical contents are functional analysis related to Schrödinger operators as we see below. The conditions (a-1) and (a-2) are satisfied if we employ the resolvent CCR algebra introduced by D. Buchholz and H. Grundling in their investigation of supersymmetric quantum field theories. (c.f. [7–9] and D. Buchholz considered KMS states of lattice models recently in [10]). To show the uniqueness of the infinite volume states, we first show the existence of a state satisfying regularity and KMS conditions using the Trotter–Kato formula for the heat kernel associated with $H_L$ and next, we apply known results on relative entropy due to H. Araki, F. Hiai and A. Uhlmann in [26].

To state our results more precisely, we define the resolvent CCR algebra now.

Let $\mathfrak{R}_L$ be the unital C*-algebra on $L_2(\mathbf{R}^{2L})$ generated by the unit and operators

$$g(s_{-L+1}x_{-L+1} + \cdots + s_L x_L + t_{-L+1}p_{-L+1} + \cdots + t_L p_L)$$

where $g(x)$ is a continuous function (with a single variable) vanishing at infinity, $s_k, t_l (k, l \in \Lambda_L)$ are real constants. The sub-algebra generated by

$$g(s_{-L+2}x_{-L+2} + \cdots + s_{L-1}x_{L-1} + t_{-L+2}p_{-L+2} + \cdots + t_{L-1}p_{L-1})$$

can be identified with $\mathfrak{R}_{L-1}$, in fact, the tensor product structure of

$$L_2(\mathbf{R}^{2M}) = L_2(\mathbf{R}^{2(M-L)}) \otimes L_2(\mathbf{R}^{2L})$$

induces the natural inclusion $\mathfrak{R}_L \subset \mathfrak{R}_M$ if $L < M$. The inductive limit C*-algebra of $\cup_L \mathfrak{R}_L$ is denoted by $\mathfrak{R}$. Both $\mathfrak{R}_L$ and $\mathfrak{R}$ are referred to as the resolvent CCR algebras. D. Buchholz and H. Grundling named $\mathfrak{R}$ the resolvent algebra because they defined $\mathfrak{R}$ as the universal C*-algebra generated by resolvents

$$\frac{1}{(s_{-L+1}x_{-L+1} + \cdots + s_L x_L + t_{-L+1}p_{-L+1} + \cdots + t_L p_L) + i\lambda}$$

where $\lambda$ is an arbitrary real number. The definition of the resolvent algebra due to D. Buchholz and H. Grundling coincides with the one we introduced here.

The quantum mechanical Hamiltonian of the harmonic oscillators is denoted by $H_L^{\text{har}}$

$$H_L^{\text{har}} = \sum_{k=-L+1}^{L} \left\{ p_k^2 + \omega^2 x_k^2 \right\}, \qquad \alpha_t^{\text{har},L}(Q) = e^{itH_L^{\text{har}}} Q e^{-itH_L^{\text{har}}}, \quad Q \in \mathfrak{R}_L.$$

By definition, $\alpha_t^{\text{har},L}(Q) \in \mathfrak{R}_L$ if $Q \in \mathfrak{R}_L$.

**Theorem 1** *Let $H_L$ be the Schrödinger operator defined in (1) where the potential $V$ and $\varphi$ are rapidly decreasing smooth functions. Then,*

$$\alpha_t^L(Q) = e^{itH_L} Q e^{-itH_L}, \quad Q \in \mathfrak{R}_L$$

*gives rise to a one-parameter group of automorphisms on $\mathfrak{R}_L$.*

The above result was shown by D. Buchholz and H. Grundling in [8] for the case of $\omega = 0$. The key point of their argument is based on the fact that $e^{ith}e^{-ith_0} - 1$ is a compact operator for Schrödinger operators $h$ and $h_0$ defined by $h = p^2 + V(x)$, $h_0 = p^2$. We can show Theorem 1 in the same manner for the case of $\omega \neq 0$.

Combined with Lieb-Robinson bound techniques, (c.f. [13, 15, 17–19, 23]) existence of the limit $\lim_{L\to\infty} \alpha_t^L(Q)$ can be proved. Application of the techniques to anharmonic crystals was investigated by B. Nachtergaele and R. Sims in [21, 23], by B. Nachtergaele, B. Schlein, H. Raz, R. Sims, S. Starr in [20] and by B. Nachtergaele, B. Schlein, R. Sims, S. Starr and V. Zagrebnov in [22]. Even though they did not consider the case of resolvent CCR algebras, there is no essential difference between our case and that of papers cited above, hence we omit the detailed proof of the existence of the infinite volume limit of the Heisenberg time evolution.

**Theorem 2** *The infinite volume limit*

$$\alpha_t(Q) = \lim_{L\to\infty} \alpha_t^L(Q) \tag{3}$$

*exists in the norm topology of* $\mathfrak{R}$. *Let* $H^{\text{free}}$ *be the Hamiltoninan of decoupled oscillators*

$$H^{\text{free}} = \sum_{k=-\infty}^{\infty} \{p_k^2 + \omega^2 x_k^2 + V(x_k)\ \} \tag{4}$$

*and set* $\alpha_t^{\text{free}}(Q) = e^{itH^{\text{free}}} Q e^{-itH^{\text{free}}}$. $\alpha_t \circ \alpha_{-t}^{\text{free}}(Q)$ *and* $\alpha_{-t}^{\text{free}} \circ \alpha_t(Q)$ *are continuous in* $t$ *for the norm topology of* $\mathfrak{R}$.

Thus, we obtain the Heisenberg time evolution for weakly coupled anharmonic oscillators on the infinite lattice **Z**.

Now we state our main result. Uniqueness of the KMS states for one-dimensional quantum spin systems is a well-known fact, however, in our case, the time evolution $\alpha_t$ is not norm continuous in $t$. $\alpha_t$ is continuous in weak topology on the GNS representations which is locally quasi-equivalent to the standard Fock representation. We will see that there exists an infinite volume limit of $\psi_{L,\beta}$ that the restriction $\psi$ to each $\mathfrak{R}_L$ is obtained by a density matrix on $L_2(\mathbf{R}^{2L})$. In another word, the restriction $\psi$ to $\mathfrak{R}_L$ is normal to the Fock representation and $\psi(Q\alpha_t(R))$ is continuous in $t$. By regularity of states, we mean states locally normal to the standard Fock state.

**Definition 1** Suppose that $\beta > 0$. Let $\alpha_t$, $t \in \mathbf{R}$, be the one-parameter group of $*$-automorphism above on $\mathfrak{R}$. The state $\psi$ is an $(\alpha_t, \beta)$-KMS state, if $\psi$ is $\alpha_t$ invariant state, $\psi(\alpha_t(Q)) = \psi(Q)$, $Q \in \mathfrak{R}$, and $\psi(Q\alpha_t(R))$ is a continuous function in $t \in \mathbf{R}$ for any $Q, R \in \mathfrak{R}$ satisfying the KMS boundary condition, namely there exists a holomorphic function $F_{Q,R}(t)$ in the strip $I_\beta = \{z \in \mathbf{C}\ , |\ 0 < Imz < \beta\}$ of the complex plain, which is bounded continuous in the closure of $I_\beta$ such that

$$F_{Q,R}(t) = \psi(Q\alpha_t(R)), \quad F_{Q,R}(t + i\beta) = \psi(\alpha_t(R)Q) \tag{5}$$

for any $Q, R \in \mathfrak{R}$.

By a direct computation, we can verify that $\psi_{L,\beta}$ satisfies the KMS condition for $\alpha_t^L$ and for $Q, R \in \mathfrak{R}_L$.

**Theorem 3** *The unique regular KMS state,* $\psi_\beta$ *associated with the Hamiltonian (1) exists, and* $\psi_\beta$ *is obtained by the infinite volume limit* $\lim_{L\to\infty} \psi_{L,\beta} = \psi_\beta$ *of* $\psi_{L,\beta}$

Statistical mechanics of anharmonic crystals with the Hamiltonian (6) defined below has been extensively studied by several people. (c.f. [1, 16] and the references therein)

$$H = \sum_{j\in\mathbf{Z}^d} \{p_j^2 + V(x_j)\} + \sum_{j,i\in\mathbf{Z}^d, ||i-j||=1} |x_i - x_j|^2 \tag{6}$$

where $V$ is a polynomial giving rise to a double well potential.

Results obtained so far are based mainly on perturbation theories. If $\beta$ is small, it is possible to show exponential decay and and analyticity of correlation functions by using cluster expansion due to R. Minlos, A. Verbeure and V. Zagrebnov in

[16], however, at the low-temperature regime, it seems that different techniques are required. In [16], the confining potential has the form $ax^4 + bx^2$ and decay property of eigenfunctions of the anharmonic oscillators of [16] is different from our perturbed harmonic potential. In fact, R. Minlos, A. Verbeure and V. Zagrebnov used the fact of (each decoupled single) anharmonic oscillator that the ratio $\psi_n(x)/\psi_0(x)$ of the eigenfunction $\psi_n(x)$ for the $n$th eigenvalue and the ground state eigenfunction $\psi_0(x)$ is bounded as a function of $x$. Obviously, this property does not hold for our case even though our weakly coupled model looks mathematically simpler to handle. With suitable modification, the methods of [16] may be applied at low temperatures regime for our model but in this article, we concentrate on (non-perturbative) methods of operator theories.

We obtained proof of existence of the Heisenberg time evolution for the same Hamiltonian (6), with the potential $V(x) = \omega^2 x^2 + v(x)$ and $v(x)$ is smooth, rapidly decreasing, and we consider the application elsewhere.

## 2 Weyl and Resolvent CCR Algebras

For our proof of results, we use the Weyl CCR algebra generated by unitaries

$$e^{i(s_{-L+1}x_{-L+1}+\cdots+s_L x_L + t_{-L+1}p_{-L+1}+\cdots+t_L p_L)}$$

as well. Let $f = (s_k, t_m)$ be a pair of real valued functions $s_k, t_m$ with finite supports defined on $\mathbf{Z}$. As both $s_k$ and $t_m$ have a finite support, the following sum is finite.

$$\Psi(f) = \sum_{k=-\infty}^{\infty} s_k x_k + \sum_{l=-\infty}^{\infty} t_l p_l$$

We regard the domain of $\Psi(f)$ is the set of rapidly decreasing smooth functions and by abuse of notation, we denote its selfadjoint extension by the same symbol $\Psi(f)$. The Weyl CCR algebra is the C*-algebra generated by unitaries $W(f)$ where $W(f)$ is written as

$$W(f) = e^{i\Psi(f)}$$

(For a precise definition of the Weyl CCR algebra, see e.g., [6, Theorem 5.2.8.].) The Weyl CCR algebra is denoted by $\mathscr{W}$ and $\mathscr{W}_L$ is defined as the unital sub-algebra of $\mathscr{W}$ generated by $W(f)$ where $f$ has its support in $\Lambda_L$.

We also set

$$R(\lambda, f) = \frac{1}{i\lambda + \Psi(f)}$$

For any positive integer $L$, we set $\Lambda_L = \{j \in \mathbf{Z} \mid -L < j \leq L\}$. For simplicity, we set $\mathfrak{R}_L$ as the unital sub-algebra of $\mathfrak{R}$ generated by $R(\lambda, f)$ where $f$ has its support in $\Lambda_L$ and $\lambda$ is any real number.

Next, we consider regular states of Weyl and Resolvent CCR algebras.

Let $\psi$ be a state of a C*-algebra $\mathfrak{A}$ and by $\{\pi_\psi(\cdot), \Omega_\psi, \mathfrak{H}_\psi\}$, we denote the GNS representation associated with $\psi$ where $\pi_\psi(\cdot)$ is the representation on the GNS Hilbert space $\mathfrak{H}_\psi$ and $\Omega_\psi$ is the GNS cyclic vector in $\mathfrak{H}_\psi$ such that $\psi(Q) = \left(\Omega_\psi, \pi_\psi(Q)\Omega_\psi\right), \ Q \in \mathfrak{A}$.

Any state $\psi$ of $\mathfrak{R}_L$ is regular if the GNS representation $\{\pi_\psi(\cdot), \Omega_\psi, \mathfrak{H}_\psi\}$ is quasi-equivalent to the Fock representation. A state $\psi$ of $\mathfrak{R}$ is referred to as a regular state if the restriction of $\psi$ to $\mathfrak{R}_L$ is regular for any $L$.

In the same manner, a state $\psi$ of $\mathscr{W}_L$ regular if the GNS representation $\{\pi_\psi(\cdot), \Omega_\psi, \mathfrak{H}_\psi\}$ is quasi-equivalent to the Fock representation. A state $\psi$ of $\mathscr{R}$ is referred to as a regular state if the restriction of $\psi$ to $\mathscr{W}_L$ is regular for any $L$. D. Buchholz and H. Grundling proved that the regularity of a state $\psi$ of $\mathfrak{R}_L$ is equivalent to the condition if and only if $\ker(\pi_\psi(R(\lambda, f))) = \{0\}$ for any $f$ and $\lambda \in \mathbf{R}\backslash\{0\}$. In another word, $\pi_\psi(R(\lambda, f))$ is the resolvent of a closed operator if $\psi$ is regular. Note that, there is a one-to-one correspondence between a regular state of $\mathfrak{R}$ and that of $\mathscr{W}$. And by abuse of notations, we employ the same notation, $\psi$ or $\varphi$, etc., for the regular states of $\mathfrak{R}$ and $\mathscr{W}$. (See Definition 4.3 and Corollary 4.4 of [8].)

**Lemma 1** *We identify $\mathfrak{R}_L$ with operators on $L^2(\mathbf{R}^{2L})$ and let $\mathscr{K}_L$ be the algebra of compact operators on $L^2(\mathbf{R}^{2L})$ which we regard as a sub-algebra of $\mathfrak{R}_L$.*

*If $\psi$ is a regular state on $\mathfrak{R}_L$ ( $L \in \mathbf{N}$), $\pi_\psi(\mathscr{K}_L)$ is weekly dense in $\pi_\psi(\mathfrak{R}_L)''$.*

**Proposition 1** *Let $\psi$ be a regular state on $\mathfrak{R}_L$, $L \in \mathbf{N}$. Then, $\psi(\alpha_t^L(Q)R)$ is continuous on $t \in \mathbf{R}$ for any $Q, R \in \mathfrak{R}_L$ where $\alpha_t^L$ is defined in* (2).

For our resolvent CCR algebra, there exists the trivial (= of no interest) state $\psi_{trivial}$ defined by

$$\psi_{trivial}(R(\lambda_1, f_1)R(\lambda_2, f_2)\cdots R(\lambda_k, f_k)) = 0 \tag{7}$$

for any $\lambda_i \in \mathbf{C}$ and $f_j$.

For any finite system, we have a canonical decomposition of the KMS state into the regular part and the singular part.

**Lemma 2** *As before, we identify $\mathfrak{R}_L$ with operators in Fock representation on $L^2(\mathbf{R}^{2L})$ and let $\mathscr{K}_L$ be the algebra of compact operators on $L^2(\mathbf{R}^{2L})$. Let $H$ be a positive selfadjoint operator on $L^2(\mathbf{R}^{2L})$ satisfying the following conditions:*

*(a) $e^{itH}Qe^{-itH}$ ($Q \in \mathfrak{R}_L$) gives rise to a one-parameter group of automorphisms of $\mathfrak{R}_L$ denoted by $\gamma_t(Q)$. $\gamma_t(Q) = e^{itH}Qe^{-itH}$.*

*(b) $e^{-\beta H_{L,\psi}}$ is a trace class operator on $L^2(\mathbf{R}^{2L})$.*

*Suppose a $\beta$-KMS state $\psi_\beta$ for $\gamma_t$ is given. There exist $\beta$-KMS states $\psi_s$ and $\psi_r$ satisfying the following properties.*

*(i) The kernel of the GNS representation for $\psi_s$ contains the compact operator algebra $\mathscr{K}_L$ on $L^2(\mathbf{R}^{2L})$.*

*(ii)* $\psi_r$ *is the regular KMS state defined by*

$$\psi_r(Q) = \frac{tr_{L^2(\mathbf{R}^{2L})}(e^{-\beta H}Q)}{tr_{L^2(\mathbf{R}^{2L})}(e^{-\beta H})}, \quad Q \in \mathfrak{R}_L. \tag{8}$$

*(iii)* $\psi_\beta$ *is a convex combination of* $\psi_s$ *and* $\psi_r$,

$$\psi_\beta = \lambda\psi_r + (1-\lambda)\psi_s$$

*for some a positive real number* $\lambda$ $0 \leq \lambda \leq 1$.

*Proof* Let $p_j$ $(j = 0, 1, 2, \cdots)$ be the mutually orthogonal rank one projections in $\mathfrak{R}_L$ associated with an eigenvector for an eigenvalue $\varepsilon_j$ of $H$, hence the spectral decomposition of $H$ is written as follows:

$$H = \sum_j \varepsilon_j \pi_{\Lambda_L}(p_j).$$

Set $P_m = \sum_{j=1}^m p_j$ and $P = w - lim_{m\to\infty}\pi_{\psi_\beta}(P_m)$ on the GNS representation associated with $\psi_\beta$. We claim that the projection $P$ is in the center of the von Neumann algebra $\pi_{\psi_\beta}(\mathfrak{R}_L)''$. In fact, by definition $P$ commutes with any elements in $\pi_{\psi_\beta}(\mathcal{K}_L)$ and for any $Q \in \mathfrak{R}_L$ $QP_m$ is of finite rank in the Schrödinger representation, $\pi_{\psi_\beta}(QP_m)$ and its weak limit commutes with $\pi_{\psi_\beta}(\mathfrak{R}_L)$.

Set $\lambda = \lim_{m\to\infty} \psi(P_m)$ and

$$\psi_r(Q) = \lim_{m\to\infty} \psi_\beta(QP_m), \quad \psi_s(Q) = \lim_{m\to\infty} \psi_\beta(Q(1-P_m))$$

for any $Q \in \mathfrak{R}_L$. As $P$ is in the center of $\pi_{\psi_\beta}(\mathfrak{R}_L)''$, $\psi_r$ and $\psi_s$ are $\beta-$KMS states.

For any $Q \in \mathcal{K}_L$, $\psi_s(Q^*Q) = \lim_{m\to\infty} \psi_\beta(Q^*Q(1-P_m)) = 0$ as $\{P_m\}$ is an approximate unit for $Q \in \mathcal{K}_L$. As the GNS vector of the KMS state $\psi_s$ is separating for $\pi_{\psi_s}(\mathfrak{R}_L)$, the kernel of $\pi_{\psi_s}$ contains $\mathcal{K}_L$, $\pi_{\psi_s}(\mathcal{K}_L) = 0$.

Now suppose that $\lambda = \lim_{m\to\infty} \psi(P_m) \neq 0$. Then, $\psi(p_j) = \psi_r(p_j) \neq 0$ for any $j$. As $\lambda$ does not vanish, there is at least $j$ satisfying $\psi_r(p_j) \neq 0$. On the other hand, for a matrix unit system $p_{ij}$ of $\mathcal{K}_L$ satisfying $p_{ij}p_{ji} = p_i$, the KMS condition implies

$$\psi_r(p_i) = \psi_r(p_{ij}p_{ji}) = e^{\varepsilon_j - \varepsilon_i}\psi_r(p_j), \quad \psi_r(p_{kl}) = 0(k \neq l)$$

which shows that $\psi_r(p_j) \neq 0$ for any $j$.

These equations tell us that (8) for $Q = AP_m$ $(A \in \mathfrak{R}_L)$. By taking the limit $m \to \infty$, we obtain (8) holds for any $Q \in \mathfrak{R}_L$. □

In the case of a single harmonic oscillator, the singular part of any KMS state is $\psi_{trivial}$. We denote the time evolution of a single harmonic oscillator by $\alpha_t^{har}$ and the resolvent CCR algebra for one degree of freedom by $\mathfrak{R}$.

**Lemma 3** *For any $\beta > 0$, if $\psi$ is a $(\alpha_t^{har}, \beta)$-KMS state of a single harmonic oscillator, $\psi_s = \psi_{trivial}$ where $\psi_{trivial}$ is defined in (7).*

*Proof* Let $\varphi$ be a KMS state for the single harmonic oscillator $\alpha_t^{har}$ such that the kernel of the GNS representation for $\varphi$ contains the compact operator algebra. Let $\{\pi_\varphi(\cdot), \Omega, \mathfrak{H}\}$ be the GNS triple associated with $\varphi$.

Note that the quotient $\tilde{\mathfrak{R}} = \mathfrak{R}/\mathscr{K}_1$ is an abelian algebra.

Assuming $\pi_\varphi(Q) = 0$, $Q \in \mathscr{K}_L$, we show $\pi_\varphi(R(\lambda, f)) = 0$. If $\pi_\varphi(Q) = 0$, $Q \in \mathscr{K}_L$, $\varphi$ gives rise to the KMS state $\tilde{\varphi}$ of the quotient algebra $\tilde{\mathfrak{R}} = \mathfrak{R}/\mathscr{K}_1$ for the time evolution $\tilde{\alpha}_t^{har}$ induced by $\alpha_t^{har}$.

Let $Q$ and $R$ be entire analytic elements in $\tilde{R}$. Due to the KMS boundary condition and commutativity of $\tilde{R}$, $\tilde{\varphi}(Q\sigma_t(R))$ is bounded on the whole complex plane and is entire, so $\tilde{\varphi}(Q\sigma_t(R))$ is a constant,

$$\tilde{\varphi}(Q\sigma_t(R)) = \tilde{\varphi}(QR).$$

We set $Q = R = \pi_\varphi(f(x))$ where $f$ is a real continuous function with one variable vanishing at infinity. ($x$ is the position operator.) As $\alpha_t^{har}(f(x)) = f(\cos\omega t \cdot x + \sin\omega t \cdot p)$, for $t = \pi/(2\omega)$ $f(x)\alpha_t^{har}(f(x))$ is a compact operator. Thus,

$$\varphi(f(x)^2) = 0.$$

As $\Omega$ is separating for $\pi_\varphi(\tilde{R})'$, $\pi_\varphi(f(x)) = 0$. It turns out

$$\pi_\varphi(\alpha_t^{har,L}(f(x))) = \pi_\varphi(f(\cos\omega t \cdot x + \sin\omega t \cdot p))$$

which shows that $\varphi = \psi_{trivial}$. □

We are not certain that $\psi_s = \psi_{trivial}$ holds for more general finite quantum systems, though, the physical meaning of singular KMS states is not clear and we shall consider only regular KMS states.

## 3 Existence of Regular KMS States

In this section, we establish the existence of regular KMS states for infinite systems.

**Lemma 4** *For any positive integers $L < L'$ and any positive function $F \in \otimes_{k\in\Lambda_L} L^\infty(\mathbf{R})$, the following estimates are valid:*

$$e^{-2\beta\|\varphi\|_\infty}\mathrm{Tr}_{L'\setminus L}(e^{-\beta H_{L'\setminus L}})\mathrm{Tr}_L(e^{-\beta H_L}M_F) \le \mathrm{Tr}_{L'}(e^{-\beta H_{L'}}M_F) \tag{9}$$

$$\mathrm{Tr}_{L'}(e^{-\beta H_{L'}}M_F) \le e^{2\beta\|\varphi\|_\infty}\mathrm{Tr}_{L'\setminus L}(e^{-\beta H_{L'\setminus L}})\mathrm{Tr}_L(e^{-\beta H_L}M_F) \tag{10}$$

*where we set $\Lambda_L = \{\ j \in \mathbf{Z}\ \mid -L < j \le L\ \} \subset \mathbf{Z}$ and*

$$H_{L'\setminus L} = \sum_{k\in\Lambda_{L'}\setminus\Lambda_L} (p_k^2 + \omega^2 x_k^2 + V(x_k)) + \sum_{k,k+1\in\Lambda_{L'}\setminus\Lambda_L} \varphi(x_k - x_{k+1})$$

*and* $M_F$ *is the multiplication operator of* $F$ *on* $L^2(\mathbf{R}^{2L})$ *and* $\|\varphi\|_\infty$ *is the supremum norm of* $\varphi$.

*Proof* Note that for $\beta > 0$, $e^{-\beta H_L}$ is a positive trace class operator. $e^{-\beta H_L}$ is an integral operator with its kernel $e^{-\beta H_L}(x, y)$ satisfying $\int_{\mathbf{R}^{|\Lambda_L|}} \int_{\mathbf{R}^{|\Lambda_L|}} e^{-\beta H_L}(x, y)^2 dxdy < \infty$. For $L < L'$, we have

$$\begin{aligned}
\Upsilon_{L'} &= \sum_{\Lambda\subset\Lambda_L} \Phi(\Lambda) = \sum_{k\in\Lambda_{L'}} V(x_k) + \sum_{k,k+1\in\Lambda_{L'}} \varphi(x_k - x_{k+1}) \\
&\geq \sum_{k\in\Lambda_{L'}} V(x_k) + \sum_{k,k+1\in\Lambda_L} \varphi(x_k - x_{k+1}) + \sum_{k,k+1\in\Lambda_{L'}\setminus\Lambda_L} \varphi(x_k - x_{k+1}) - 2\|\varphi\|_\infty \\
&= \sum_{\Lambda\subset\Lambda_{L'}} \Phi(\Lambda) + \sum_{\Lambda\subset\Lambda_{L'}\setminus\Lambda_L} \Phi(\Lambda) - 2\|\varphi\|_\infty \\
&= \Upsilon_L + \Upsilon_{L'\setminus L} - 2\|\varphi\|_\infty
\end{aligned}$$

and

$$\Upsilon_{L'} \leq \Upsilon_L + \Upsilon_{L'\setminus L} + 2\|\varphi\|_\infty.$$

$e^{-\beta H_{L'}^h}$ is the Mehler kernel $k_\beta^h(x, y) \in \mathscr{S}(\mathbf{R}^{2L'})$,

$$k_\beta^h(x, y) = \left(\frac{\omega}{2\pi \sinh(2\omega\beta)}\right)^{\frac{n}{2}} \prod_{k\in\Lambda_{L'}} \exp\left(-\frac{\omega(x_k^2 + y_k^2)\coth(2\omega\beta) - 2\mathrm{cosech}(2\omega\beta)x_k y_k}{2}\right)$$

for $x = (x_{-L'+1}, \cdots, x_{-1}, x_0, x_1, \cdots, x_{L'})$, $y = (y_{-L'+1}, \cdots, y_{L'}) \in \mathbf{R}^{2L'}$.

For any positive functions $f, g \in \mathscr{S}(\mathbf{R}^{2L'})$, we have

$$\begin{aligned}
&\left\langle f, (e^{-\frac{\beta H_{L'}^{\mathrm{har}}}{n}} e^{-\frac{\beta\Upsilon_{L'}}{n}})^n g\right\rangle_{L^2} \\
&= \int_{\mathbf{R}^{2L'}} f(w) \int_{\mathbf{R}^{2L'}} k_{\beta/n}^h(w, z_1) e^{-\frac{\beta\Upsilon_{L'}}{n}}(z_1) \int_{\mathbf{R}^{2L'}} k_{\beta/n}^h(z_1, z_2) e^{-\frac{\beta\Upsilon_{L'}}{n}}(z_2) \\
&\quad \times\cdots\times \int_{\mathbf{R}^{2L'}} k_{\beta/n}^h(z_{n-1}, z_n) e^{-\frac{\beta\Upsilon_{L'}}{n}}(z_n) g(z_n) dz_n \cdots dz_2 dz_1 dw \\
&\geq e^{-2\beta\|\varphi\|_\infty} \int_{\mathbf{R}^{2L'}} f(w) \int_{\mathbf{R}^{2L'}} k_{\beta/n}^h(w, z_1) e^{-\beta\frac{\Upsilon_L + \Upsilon_{L'\setminus L}}{n}}(z_1) \\
&\quad \times\cdots\times \int_{\mathbf{R}^{2L'}} k_{\beta/n}^h(z_{n-1}, z_n) e^{-\beta\frac{\Upsilon_L + \Upsilon_{L'\setminus L}}{n}}(z_n) g(z_n) dz_n \cdots dz_2 dz_1 dw \\
&= e^{-2\beta\|\varphi\|_\infty} \left\langle f, (e^{-\frac{\beta H_{L'}^{\mathrm{har}}}{n}} e^{-\beta\frac{\Upsilon_L + \Upsilon_{L'\setminus L}}{n}})^n g\right\rangle_{L^2}.
\end{aligned}$$

As a consequence,

$$\begin{aligned}\left\langle f, e^{-\beta H_{L'}} g\right\rangle_{L^2} &= \lim_{n\to\infty}\left\langle f, (e^{-\frac{\beta H^{\mathrm{har}}_{L'}}{n}} e^{-\frac{\beta \Upsilon_{L'}}{n}})^n g\right\rangle_{L^2} \\ &\geq \lim_{n\to\infty} e^{-2\beta\|\varphi\|_\infty}\left\langle f, (e^{-\frac{\beta H^{\mathrm{har}}_{L'}}{n}} e^{-\frac{\beta\Upsilon_L+\beta\Upsilon_{L'\setminus L}}{n}})^n g\right\rangle_{L^2} \\ &= e^{-2\beta\|\varphi\|_\infty}\left\langle f, e^{-\beta H_L-\beta H_{L'\setminus L}} g\right\rangle_{L^2}.\end{aligned}$$

Then,

$$e^{-\beta H_{L'}}(x, y) \geq e^{-2\beta\|\varphi\|_\infty} e^{-\beta H_L-\beta H_{L'\setminus L}}(x, y), \quad x, y \in \mathbf{R}^{2L'}.$$

Since $e^{-\beta H_{L'}}$ is a trace class operator, $\int_{R^{2L'}} e^{-\beta H_{L'}}(x, x)dx < \infty$. Thus, we obtain the following estimates for any positive function $F \in \bigotimes_{k\in\Lambda_L} L^\infty(\mathbf{R})$:

$$\begin{aligned}\mathrm{Tr}_{L'}(e^{-\beta H_L} M_F) &= \int_{\mathbf{R}^{2L}} e^{-\beta H_L}(x, x)F(x)dx \\ &\geq e^{-2\beta\|\varphi\|_\infty}\int_{\mathbf{R}^{2L'}} e^{-\beta(H_{L'\setminus L}+H_L)}(x, x)F(x)dx \\ &= e^{-2\beta\|\varphi\|_\infty}\mathrm{Tr}_{L'}(e^{-\beta H_{L'\setminus L}} e^{-\beta H_L} M_F) \\ &= e^{-2\beta\|\varphi\|_\infty}\mathrm{Tr}_{L'\setminus L}(e^{-\beta H_{L'\setminus L}})\mathrm{Tr}_L(e^{-\beta H_L} M_F)\end{aligned}$$

and

$$\begin{aligned}\mathrm{Tr}_{L'}(e^{-\beta H_{L'}} M_F) &\leq e^{2\beta\|\varphi\|_\infty}\mathrm{Tr}_{L'}(e^{-\beta H_{L'\setminus L}} e^{-\beta H_L} M_F) \\ &= e^{2\beta\|\varphi\|_\infty}\mathrm{Tr}_{L'\setminus L}(e^{-\beta H_{L'\setminus L}})\mathrm{Tr}_L(e^{-\beta H_L} M_F).\end{aligned}$$

We obtain (9) and (10). □

**Proposition 2** *For any positive integers $L \leq L'$ and any $F \in \mathfrak{R}_L$ such that $\pi_0(F)$ is a positive multiplication operator on $\mathfrak{H}_{\Lambda_L}$, the following estimate hold:*

$$e^{-4\beta\|\varphi\|_\infty}\psi_L(F) \leq \psi_{L'}(F) \leq e^{4\beta\|\varphi\|_\infty}\psi_L(F), \tag{11}$$

*where $\psi_L$ and $\psi_{L'}$ are KMS states of finite systems $\mathfrak{R}_L$ and $\mathfrak{R}_{L'}$.*

*Proof* By (9) and (10), we obtain the following inequalities:

$$\begin{aligned}&e^{-2\beta\|\varphi\|_\infty}\mathrm{Tr}_{L\setminus L'}(e^{-\beta H_{L\setminus L'}})\mathrm{Tr}_{L'}(e^{-\beta H_{L'}} M_F) \leq \mathrm{Tr}_L(e^{-\beta H_L} M_F), \\ &\mathrm{Tr}_L(e^{-\beta H_L} M_F) \leq e^{2\beta\|\varphi\|_\infty}\mathrm{Tr}_{L\setminus L'}(e^{-\beta H_{L\setminus L'}})\mathrm{Tr}_{L'}(e^{-\beta H_{L'}} M_F), \\ &e^{-2\beta\|\varphi\|_\infty}\mathrm{Tr}_{L\setminus L'}(e^{-\beta H_{L\setminus L'}})\mathrm{Tr}_{L'}(e^{-\beta H_{L'}}) \leq \mathrm{Tr}_L(e^{-\beta H_L}), \\ &\mathrm{Tr}_L(e^{-\beta H_L}) \leq e^{2\beta\|\varphi\|_\infty}\mathrm{Tr}_{L\setminus L'}(e^{-\beta H_{L\setminus L'}})\mathrm{Tr}_{L'}(e^{-\beta H_{L'}}).\end{aligned}$$

Thus, we obtain (11). □

Note that $\psi_L$ can be regarded as a state on $\mathscr{W}_L$, i.e.,

$$\psi_L(W) = \frac{\mathrm{Tr}_L(e^{-\beta H_L}\pi_0(W))}{\mathrm{Tr}_L(e^{-\beta H_L})}, \quad W \in \mathscr{W}_L. \tag{12}$$

Also, for the Weyl CCR algebra, the following statement follows.

**Proposition 3** *For any positive integers $L \leq L'$ and any $F \in \mathscr{W}_L$ such that $\pi_0(F)$ is a positive multiplication operator on $\mathfrak{H}_{\Lambda_L}$, the following estimate hold:*

$$e^{-4\beta\|\varphi\|_\infty}\psi_L(F) \leq \psi_{L'}(F) \leq e^{4\beta\|\varphi\|_\infty}\psi_L(F) \tag{13}$$

*where $\psi_L$ and $\psi_{L'}$ are states on $\mathscr{W}_L$ and $\mathscr{W}_{L'}$ defined in (12), respectively.*

Since $e^{-\beta(p^2+\omega^2x^2)}$ is a trace class operator on $L^2(\mathbf{R}, dx)$ and by [6, Proposition 5.2.27], $e^{-\beta d\Gamma(p^2+\omega^2x^2)}$ is a trace class on $\mathscr{F}_+(L^2(\mathbf{R}, dx))$, where $d\Gamma(p^2+\omega^2x^2)$ is the second quantization of $p^2+\omega^2x^2$ and $\mathscr{F}_+(L^2(\mathbf{R}, dx))$ is the Boson Fock space of $L^2(\mathbf{R}, dx)$ (see also [6, Section 5.2.1]). Set

$$\widetilde{\psi}_L := \psi_L \otimes \frac{\mathrm{Tr}_{\mathscr{F}_+(L^2(\mathbf{R},dx))}(e^{-\beta H^{\mathrm{har}}(\mathbf{Z}\setminus\Lambda_L)}\pi_0(\cdot))}{\mathrm{Tr}_{\mathscr{F}_+(L^2(\mathbf{R},dx))}(e^{-\beta H^{\mathrm{har}}(\mathbf{Z}\setminus\Lambda_L)})},$$

then $\widetilde{\psi}_L$ is a regular state on $\mathscr{W}$ and $\widetilde{\psi}_L|_{\mathscr{W}_L} = \psi_L$. Thus, the regular state $\psi_L$ on $\mathscr{W}_L$ can extend to a regular state on $\mathscr{W}$. Since $\mathscr{W}$ is a unital C*-algebra, the family of states $\{\widetilde{\psi}_L\}_{L\in\mathbf{N}}$ has at least one cluster point $\psi$. Next, we show that $\psi$ is a regular state.

**Theorem 4** *The state $\psi$ defined in the above is a regular state on $\mathscr{W}$.*

*Proof* To show regularity of $\psi$, we show that for $t$ in $|t| \leq \delta$

$$\widetilde{\psi}_L(QW_0(t))$$

is equicontinuous with respect to $L$ where $Q = Q(x)$ is an arbitrary essentially bounded function on $\mathbf{R}^{2L}$ and $(W_0(t)) = e^{itp_0}$. (We can show the continuity of $\lim_L \widetilde{\psi}_L(QW_0(t))$ for the general $W = e^{i\sum_{k=-L+1}^{L} t_k p_k}$ in the same way.)

For simplicity of presentation, we consider the case $\omega = 1$ here. The case $\omega \neq 1$ can be shown analogously.

Note that

$$\widetilde{\psi}_L(QW_0(t)) = \\ \frac{1}{Z_{\beta L}^2}\int\int e^{-\beta/2H_L}(x,y)Q(x)\frac{e^{-\beta/2H_L}(x+t^{(0)},y)}{e^{-\beta/2H_L}(x,y)}e^{-\beta/2H_L}(x,y)dxdy$$

where $Z^2_{\beta L}$ is the normalization constant

$$Z^2_{\beta L} = \int\int (e^{-\beta/2H_L}(x,y))^2 dxdy = \int e^{-\beta H_L}(x,x)dx.$$

and $x + t^{(0)}$ is the addition of $t$ to $x$ at the component corresponding to the origin of the integer lattice $\mathbf{Z}$.

For $x = (\cdots, x_{-1}, x_0, x_1, \cdots)$ and $y = (\cdots, y_{-1}, y_0, y_1, \cdots)$, we claim that

$$e^{-c(t)} \frac{k^h_\beta(x_0 + t, y_0)}{k^h_\beta(x_0, y_0)} \le \frac{e^{-\beta/2H_L}(x + t^{(0)}, y)}{e^{-\beta/2H_L}(x, y)} \le e^{c(t)} \frac{k^h_\beta(x_0 + t, y_0)}{k^h_\beta(x_0, y_0)} \tag{14}$$

where

$$\begin{aligned} c(t) = &\sup_{x_0} |V(x_0 + t) - V(x_0)| + \sup_{x_0, x_1} |\varphi(x_0 - x_1 + t) - \varphi(x_0 - x_1)| \\ &+ \sup_{x_0, x_{-1}} |\varphi(x_{-1} - x_0 + t) - \varphi(x_{-1} - x_0)|. \end{aligned}$$

$\lim_{t\to 0} c(t) = 0$ due to uniform continuity of $V$ and $\varphi$ and this bound implies regularity.

We now show (14). Note the following tautological equalities holds. For any $n \in \mathbf{N}$,

$$\begin{aligned} k^h_\beta(x_0 + t, y_0) &= \int \cdots \int k^h_{\beta/n}(x_0 + t, z_1) k^h_{\beta/n}(z_1, z_2) \cdots k^h_{\beta/n}(z_n, y_0) dz_1 \cdots dz_n \\ &= \int \cdots \int k^h_{\beta/n}(x_0 + t, z_1 + s_1) k^h_{\beta/n}(z_1 + s_1, z_2 + s_2) \cdots k^h_{\beta/n}(z_n + s_n, y_0) dz_1 \cdots dz_n \end{aligned} \tag{15}$$

for any constants $s_k$. Then, up to a multipliticave factor, $\widetilde{C}_{nt}$,

$$\begin{aligned} &k^h_{\beta/n}(x_0 + t, z_1 + s_1) k^h_{\beta/n}(z_1 + s_1, z_2 + s_2) \cdots k^h_{\beta/n}(z_n + s_n, y_0) = \\ &\widetilde{C}_{nt} \times \exp[-\frac{1}{2\sinh(2\beta/n)} \sum_{k=0}^{n} \{\cosh(2\beta/n)((z_k + s_k)^2 + (z_{k+1} + s_{k+1})^2) \\ &\quad -2(z_k + s_k)(z_{k+1} + s_{k+1})\}] \end{aligned} \tag{16}$$

where we set $z_0 = x$, $z_{n+1} = y$. In the exponent, we can write

$$\begin{aligned} &\sum_{k=0}^{n} \{\cosh(2\beta/n)((z_k + s_k)^2 + (z_{k+1} + s_{k+1})^2) - 2(z_k + s_k)(z_{k+1} + s_{k+1})\} \\ &= \sum_{k=0}^{n} \{\cosh(2\beta/n)(z_k^2 + z_{k+1}^2) - 2z_k z_{k+1}\} \end{aligned}$$

$$+\sum_{k=0}^{n+1} A_{n,k}(s)z_k + \Sigma_n(t, x_0, y_0) \tag{17}$$

where $A_{n,k}(s)$(linear in $s_k$) and $\Sigma_n(t, x_0, y_0)$(quadratic in $s_k$) are terms independent of $z_k$. Now, we choose the constants $s_k$ satisfying the condition $A_{n,k}(s) = 0$ $s_0 = t$ $s_{n+1} = 0$. We do not need the exact form of $A_{n,k}(s)$ and $\Sigma_n(s,t)$ here but what we need are bounds $|s_k| \leq \widetilde{A}|t|$ independent of $n$.
Thus, we obtain

$$k^h_\beta(x_0 + t, y_0) = \exp\left[-\frac{\Sigma_n(t, x_0, y_0)}{2\sinh(2\beta/n)}\right] \times$$
$$\int \cdots \int k^h_{\beta/n}(x_0, z_1)k^h_{\beta/n}(z_1, z_2)\cdots k^h_{\beta/n}(z_n, y_0)dz_1\cdots dz_n \tag{18}$$

and

$$\lim_{n\to\infty} \exp\left[-\frac{\Sigma_n(t, x_0, y_0)}{2\sinh(2\beta/n)}\right] = \frac{k^h_\beta(x_0 + t, y_0)}{k^h_\beta(x_0, y_0)} \tag{19}$$

To show (14), we apply the Trotter–Kato formula again to

$$e^{-\beta H_L^{\text{har}}}(x + t^{(0)}, y) = \lim_{n\to\infty} (e^{-\frac{\beta}{n}H_L^{\text{har}}} e^{-\frac{\beta\Upsilon_L}{n}})^n(x + t^{(0)}, y) \tag{20}$$

We consider now $(e^{-\frac{\beta H_L^{\text{har}}}{n}} e^{-\frac{\beta\Upsilon_L}{n}})^n(x + t^{(0)}, y)$ at each $n$ in (20). The integral kernel of $(e^{-\frac{\beta H_L^{\text{har}}}{n}} e^{-\frac{\beta\Upsilon_L}{n}})^n$ is iteration of integral in which change of the variable $x + t^{(0)}$ affects only to the integral associated to the particle at the origin and its nearest neighbor. In that integral, we denote the variable at the site $-1$ at the lattice by $z_k^{(-1)}$ and that at the site 1 at the lattice by $z_k^{(1)}$. Then, the contribution to the iterated integral from the origin in $(e^{-\frac{\beta H_L^{\text{har}}}{n}} e^{-\frac{\beta\Upsilon_L}{n}})^n(x, y)$ is

$$\begin{aligned}\int \cdots \int & k^h_{\beta/n}(x_0, z_1)\exp\left[-\frac{\beta}{n}(V(z_1) + \varphi(z_1 - z_1^{(-1)}) + \varphi(z_1^{(1)} - z_1))\right] \\ &\times k^h_{\beta/n}(z_1, z_2)\exp\left[-\frac{\beta}{n}(V(z_2) + \varphi(z_2 - z_2^{(-1)}) + \varphi(z_2^{(1)} - z_2))\right] \\ &\cdots k^h_{\beta/n}(z_n, y_0)\exp\left[-\frac{\beta}{n}(V(y_0) + \varphi(y_0 - y_{-1}) + \varphi(y_1 - y_0))\right]dz_1\cdots dz_n\end{aligned} \tag{21}$$

After change of variable $z_k \to z_k + s_k$ as in (15), the corresponding integral for $(e^{-\frac{\beta H_L^h}{n}} e^{-\frac{\beta\Upsilon_L}{n}})^n(x + t^{(0)}, y)$ is

$$\int \cdots \int k^h_{\beta/n}(x_0, z_1) \exp\left[-\frac{\beta}{n}(V(z_1) + \varphi(z_1 - s_1 - z_1^{(-1)}) + \varphi(z_1^{(1)} - z_1 + s_1))\right]$$
$$\times k^h_{\beta/n}(z_1, z_2) \exp\left[-\frac{\beta}{n}(V(z_2 - s_2) + \varphi(z_2 - s_2 - z_2^{(-1)}) + \varphi(z_2^{(1)} - z_2 + s_2))\right]$$
$$\cdots dz_1 \cdots dz_n \times \exp\left[-\frac{\Sigma_n(t, x_0, y_0)}{2\sinh(2\beta/n)}\right] \tag{22}$$

Then,

$$(21) \times e^{-c(\widetilde{A}t)} \exp\left[-\frac{\Sigma_n(t, x_0, y_0)}{2\sinh(2\beta/n)}\right]$$
$$\leq (22) \leq (21) \times e^{c(\widetilde{A}t)} \exp\left[-\frac{\Sigma_n(t, x_0, y_0)}{2\sinh(2\beta/n)}\right] \tag{23}$$

By taking the limit $n \to \infty$ we obtain the bound (14).

Finally, we can show the regularity of $\psi$ by using (14), Proposition 2 and the Lebesgue dominated convergence theorem. □

## 4 Relative Entropy

First, let us recall the definition of quasi-containment. Let $\mathscr{A}$ be a C*-algebra and let $(\mathfrak{H}_1, \pi_1)$ and $(\mathfrak{H}_2, \pi_2)$ be nondegenerate representations of $\mathscr{A}$. The representations $\pi_1$ and $\pi_2$ is quasi-equivalent, if there exists an isomorphism $\gamma : \pi_1(\mathscr{A})'' \mapsto \pi_2(\mathscr{A})''$ such that $\gamma(\pi_1(A)) = \pi_2(A)$ for all $A \in \mathscr{A}$ (see also [6, Definition 2.4.25] and [6, Theorem 2.4.26]). If a subrepresentation of $\pi_1$ is quasi-equivalent to $\pi_2$, then $\pi_1$ is quasi-contain $\pi_2$. The next lemma is essentially due to [2, Lemma 1].

**Lemma 5** *Let $\psi_1$ and $\psi_2$ be regular states on $\mathfrak{R}$ and $(\mathfrak{H}_1, \pi_1, \xi_1)$ and $(\mathfrak{H}_2, \pi_2, \xi_2)$ be the GNS representations associated with $\psi_1$ and $\psi_2$, respectively. If $\pi_1$ does not quasi-contain $\pi_2$, then there exists a sequence of projections $e_m \in \bigcup_{L \in \mathbf{N}} \mathfrak{R}_L$ such that*

$$\lim_m \psi_1(e_m) = 0, \tag{24}$$
$$\lim_m \psi_2(e_m) = a > 0. \tag{25}$$

*Proof* Put $\mathfrak{H} = \mathfrak{H}_1 \oplus \mathfrak{H}_2$, $\pi = \pi_1 \oplus \pi_2$, $\widetilde{\xi}_1 = \xi_1 \oplus 0$, $\widetilde{\xi}_2 = 0 \oplus \xi_2$ and $\mathfrak{M} = \pi(\mathfrak{R})''$. Note that $\pi$ is a regular representation of $\mathfrak{R}$. Let $E_1$ and $E_2$ be the projections from $\mathfrak{H}$ onto $\mathfrak{H}_1$ and $\mathfrak{H}_2$, respectively. By the assumption, there exists a central projection $E \in \mathfrak{M} \cap \mathfrak{M}'$ such that $EE_1 = 0$, $E \leq E_2$. By Kaplansky's density theorem, there exists selfadjoint elements $a_n \in \mathfrak{R}_{L(n)}$ such that

$$\text{st-}\lim_n \pi(a_n) = E,$$

where the st-lim is the strong limit in $\mathfrak{H}$. By the regularity of $\pi$, Lemma 1 and Kaplansky's density theorem, there exists selfadjoint elements $b_m^{(n)} \in \mathscr{K}_{L(n)}$ such that

$$\text{st-}\lim_m \pi(b_m^{(n)}) = \pi(a_n).$$

Note that the spectral projections of $b_m^{(n)}$ are also contained in $\mathfrak{R}_{L(n)}$. Let $e_n$ be the spectral projection of $b_m^{(n)}$ for an interval $[1-\delta, 1+\delta]$, where $\delta \in (0, 1)$ is fixed. Then, $e_n \in \mathscr{K}_{L(n)} \subset \mathfrak{R}_{L(n)}$ and $\text{st-}\lim_n \pi(e_n) = E$.

Thus, $\lim_n \psi_1(e_n) = 0, \quad \lim_n \psi_2(e_n) = a > 0.$ □

Next, let us recall basic properties of the relative entropy of normal states of von Neumann algebras and $*$-algebras. The relative entropy for positive normal linear functionals of a von Neumann algebra was introduced by H. Araki in [3, 4]. Let $\psi_1$ and $\psi_2$ be positive normal linear functionals over a von Neumann algebra $\mathfrak{M}$. In a standard form of our von Neumann algebra $\mathfrak{M}$ acting as a Hilbert space $\mathfrak{H}$, let $\xi_1$ (respectively $\xi_2$) be the vector in the positive cone $\mathfrak{H}$ satisfying $\psi_1(a) = (\xi_1, a\xi_1)$ (resp. $\psi_2(a) = (\xi_2, a\xi_2)$) for $a \in \mathfrak{M}$.

As usual, $S_{\xi_2,\xi_1}$ is the densely defined closed conjugate linear operator $S_{\xi_2,\xi_1}$ satisfying

$$S_{\xi_2,\xi_1} a\xi_1 = a^*\xi_2, \qquad a \in \mathfrak{M},$$

and the relative modular operator $\Delta_{\xi_2,\xi_1}$ is, by definition, the square root of the positive part of in the polar decomposition of $S_{\xi_2,\xi_1}$.

$$\Delta_{\xi_2,\xi_1} = S^*_{\xi_2,\xi_1} S_{\xi_2,\xi_1}, \tag{26}$$

We denote the orthogonal projection to the subspace $\overline{\mathfrak{M}'\xi_1}$ (respectively $\overline{\mathfrak{M}'\xi_2}$ on $\mathfrak{H}$) by $s(\psi_1)$ (respectively $s(\psi_2)$). The relative entropy $S_A(\psi_1, \psi_2)$ is defined by

$$S_A(\psi_1, \psi_2) = \begin{cases} -\int_0^\infty \log(\lambda) d(\xi_1, E(\lambda)\xi_1), & \text{if } (s(\psi_1) \leq s(\psi_2)) \\ \infty & \text{otherwise} \end{cases}. \tag{27}$$

The relative entropy we use later is neither that of finite matrix algebras nor that of a von Neuman algebra but that of the resolvent algebra so that a caution is in order. The one we use there is due to A. Uhlmann who extended the notion of relative entropy to positive linear functionals on a (not necessarily normed) $*$-algebra. (c.f. [26].) We denote the relative entropy defined by A. Uhlmann by $S_U$. $S_U$ coincides with that of H. Araki when the $*$-algebra we concern is a von Neumann algebra as we explain below. (See [14].)

**Lemma 6** ([14, Lemma 3.1]) *Let $\mathscr{A}$ be a unital C*-algebra and $\pi$ be a nondegenerate representation of $\mathscr{A}$ on a Hilbert space. Let $\psi_1$ and $\psi_2$ be positive linear functionals of $\mathscr{A}$ and we denote normal extensions to $\pi(\mathscr{A})''$ by $\widehat{\psi_1}$ and $\widehat{\psi_2}$. Then,*

$$S_U(\psi_1, \psi_2) = S_A(\widehat{\psi_1}, \widehat{\psi_2}). \tag{28}$$

In view of this lemma, by the relative entropy of states $\psi_1$ and $\psi_2$ on a unital C*-algebra, we mean the relative entropy of normal extension of $\widehat{\psi_1}$ and $\widehat{\psi_2}$ if $\psi_1$ and $\psi_2$ are quasi-equivalent. If $\psi_1$ and $\psi_2$ are not quasi-equivalent, the relative entropy is infinite.

**Lemma 7** ([24, Corollary 5.12 (iii)]) *Let $\mathfrak{N} \subset \mathfrak{M}$ be von Neumann algebras and $\psi_1$ and $\psi_2$ be normal states on $\mathfrak{M}$. Assume there exists a norm one projection from $\mathfrak{M}$ to $\mathfrak{N}$. Then*

$$0 \leq S(\psi_1 \restriction_{\mathfrak{N}}, \psi_2 \restriction_{\mathfrak{N}}) \leq S(\psi_1, \psi_2). \tag{29}$$

By the same argument as that in [2], we have the following.

**Lemma 8** *Let $\psi_1$ and $\psi_2$ be regular states on $\mathfrak{R}$. If*

$$\sup_{L \in \mathbf{N}} S(\psi_1 \restriction_{\mathfrak{R}_L}, \psi_2 \restriction_{\mathfrak{R}_L}) \equiv \mu < \infty, \tag{30}$$

*then $\pi_2$ quasi contains $\pi_1$ where $\pi_j$ is the GNS representation of $\mathfrak{R}$ associated with $\psi_j$, $j = 1, 2$.*

*Proof* Assume that $\pi_2$ does not quasi-contain $\pi_1$. By Lemma 5, there exists a sequence of projections $e_n \in \mathfrak{R}_{L(n)}$ such that

$$\lim_n \psi_1(e_n) = a > 0,$$
$$\lim_n \psi_2(e_n) = 0.$$

Then,

$$-\psi_1(e_n) \log \psi_2(e_n) \to \infty.$$

Consider the C*-sub-algebra $\mathscr{B}_n$ of $\mathfrak{R}_{L(n)}$ generated by $e_n$ and $1 - e_n$. Due to Lemma 7,

$$\begin{aligned} S(\psi_1 \restriction_{\mathfrak{R}_L}, \psi_2 \restriction_{\mathfrak{R}_L}) &= S(\widehat{\psi_1} \restriction_{\pi_0(\mathfrak{R}_{L(n)})''}, \widehat{\psi_2} \restriction_{\pi_0(\mathfrak{R}_{L(n)})''}) \\ &\geq S(\widehat{\psi_1} \restriction_{\pi_0(\mathscr{B}_n)''}, \widehat{\psi_2} \restriction_{\pi_0(\mathscr{B}_n)''}) \\ &= \psi_1(e_n) \log \frac{\psi_1(e_n)}{\psi_2(e_n)} + \psi_1(1 - e_n) \log \frac{\psi_1(1 - e_n)}{\psi_2(1 - e_n)}. \end{aligned}$$

The above estimate is contradictory to our assumption. □

We will use the continuity of the relative entropy.

**Lemma 9** ([24, Corollary 5.12 (i)]) *Let $\psi_i$, $\psi$, $\phi_i$ and $\phi$ be normal states on a von Neumann algebra $\mathfrak{M}$. If $\psi_i$ and $\phi_i$ converge to $\psi$ and $\phi$ in $\sigma(\mathfrak{M}_*, \mathfrak{M})$ topology, respectively, then*

$$S(\psi, \phi) \leq \liminf_i S(\psi_i, \phi_i). \tag{31}$$

## 5 Proof of Theorem 3

For regular states, the KMS condition is equivalent to the Arak–Gibbs condition. We consider the perturbation of a regular state and the automorphism $\alpha$. The perturbation of an automorphism and a state on a C*-algebra or a von Neumann algebra is defined in [6, Proposition 5.4.1] and [6, Theorem 5.4.4].

**Proposition 4** *Let* $\phi$ *be a regular* $(\alpha, \beta)$*-KMS state on* $\mathfrak{R}$.
*Put* $W(L) := \pi_0^{-1}(\Phi(\{L, L+1\})) + \pi_0^{-1}(\Phi(\{-L, -L+1\}))$, $L \in \mathbf{N}$. *Then* $\phi$ *satisfies the following condition:*

$$\phi^{\beta W(L)} = \psi_L \otimes \widetilde{\phi} \tag{32}$$

*for all* $L \in \mathbf{N}$, *where* $\widetilde{\phi}$ *is a state over* $\mathfrak{R}_{L^c}$, $\phi^{\beta W(L)}$ *is a perturbed state of* $\phi$ *by* $\beta W(L)$.

*Proof* For positive integers $L < L'$, let $\gamma_t^{L',L}$ be the perturbed automorphism of $\alpha_t^{L'}$ by $\beta W(L)$. Since $H_{L'} - \pi_0(W(L)) = H_{L'\setminus L} + H_L$ and $H_L$ and $H_{L'\setminus L}$ are commute, $\gamma_t^{L',L} = \alpha_t^L \otimes \alpha_t^{L'\setminus L}$. The automorphism $\alpha_t^{L'\setminus L}$ converges strongly to an automorphism $\alpha_t^{L^c}$ on $\mathfrak{R}_{L^c}$ when $L' \to \infty$. Note that the perturbed state $\phi^{\beta W(L)}$ is a $(\gamma, \beta)$-KMS state by construction, where $\gamma_t = \alpha_t^L \otimes \alpha_t^{L^c}$. For $0 < R \in \mathfrak{R}_{L^c}$, we define the state $\phi_R^{\beta W(L)}$ on $\mathfrak{R}_L$ by

$$\phi_R^{\beta W(L)}(Q) = \frac{\phi^{\beta W(L)}(QR)}{\phi^{\beta W(L)}(R)}, \quad Q \in \mathfrak{R}_L.$$

Note that $\phi_R^{\beta W(L)}$ is a regular state by construction and by $\gamma_t = \alpha_t^L \otimes \alpha_t^{L^c}$, $\phi_R^{\beta W(L)}$ is an $(\alpha^L, \beta)$-KMS state. Regular states of our Weyl algebra those of resolvent CCR algebras $\phi_R^{\beta W(L)} = \psi_L$. For all $Q \in \mathfrak{R}_L$ and $0 < R \in \mathfrak{R}_{L^c}$

$$\phi^{\beta W(L)}(QR) = \psi_L(Q)\phi^{\beta W(L)}(R). \tag{33}$$

For any selfadjoint element $R \in \mathfrak{R}_{L^c}$ and any $\varepsilon > 0$, $R + (\|R\| + \varepsilon)1$ is a strictly positive operator. Then we obtain

$$\phi^{\beta W(L)}(Q(R + (\|R\| + \varepsilon)1)) = \psi_L(Q)\phi^{\beta W(L)}(R + (\|R\| + \varepsilon)1).$$

Since for any element $R \in \mathfrak{R}_{L^c}$ can decompose two selfadjoint elements $R_1$ and $R_2$ such that $R = R_1 + iR_2$. By the linearity of $\phi^{\beta W(L)}$, the equation (33) holds for any elements $Q \in \mathfrak{R}_L$ and $R \in \mathfrak{R}_{L^c}$. Thus,

$$\phi^{\beta W(L)} = \psi_L \otimes \phi^{\beta W(L)}|_{\mathfrak{R}_{L^c}}.$$

Put $\widetilde{\phi} = \phi^{\beta W(L)}|_{\mathfrak{R}_{L^c}}$, then we get the claim.

*Remark 1* For $Q \in \mathfrak{R}$, it may not be a linear combination of the form of $A \otimes B$ for $A \in \mathfrak{R}_L$ and $B \in \mathfrak{R}_{L^c}$. However, by Lemma 1 and [8, Theorem 4.2 (v)], for a regular state $\phi$ on $\mathfrak{R}$ and any positive integer $L$, $\pi_\phi(\mathfrak{R}_L) \bigotimes \pi_\phi(\mathfrak{R}_{L^c})$ is a weakly dense sub-algebra in $\pi_\phi(\mathfrak{R})$. For $Q \in \mathfrak{R}$, there exists a positive integer $\{L(n)\}_{n \in \mathbf{N}}$ such that $L \leq L(n)$ for any $n \in \mathbf{N}$ and a sequence $\sum_i a_i^{(n)} R_i^{(n)} \otimes K_i^{(n)}$ such that $a_i^{(n)} \in \mathbf{C}$, $R_i^{(n)} \in \mathfrak{R}_L$, $K_i^{(n)} \in \mathfrak{K}(\Lambda_{L(n)} \backslash \Lambda_L)$ and

$$\pi_\phi(Q) = \text{w-}\lim_n \sum_i a_i^{(n)} \pi_\phi(A_i^{(n)}) \otimes \pi_\phi(K_i^{(n)}).$$

and we can define the product state $\psi_L \otimes \phi^{\beta W(L)}|_{\mathfrak{R}_{L^c}}$ for any $Q \in \mathfrak{R}$.

Finally, we show uniqueness of $(\alpha, \beta)$-KMS state for $\beta > 0$ in Theorem 3. Due to Theorem 4, $\psi$ gives rise to a regular state on $\mathscr{W}$ and hence a regular state of $\mathfrak{R}$.

**Proof of Theorem** 3
First, we show $\psi(Q\alpha_t(R))$ is continuous in $t \in \mathbf{R}$ for any $Q, R \in \mathfrak{R}$. Since $\bigcup_{L \in \mathbf{N}} \mathfrak{R}_L$ is norm dense in $\mathfrak{R}$ and $\mathfrak{R}_L \subset \mathfrak{R}_{L'}$ whenever $L \leq L'$, we show $\psi(Q\alpha_t(R))$ is continuous in $t \in \mathbf{R}$ for any $Q, R \in \mathfrak{R}_L$.

For any positive integer $L$ and any $Q, R \in \mathfrak{R}_L$, $\psi(Q\alpha_t(R))$ is continuous in $t \in \mathbf{R}$. In fact, for any $\varepsilon > 0$, there exists a positive integer $L$ such that $||\alpha_t(R) - \alpha_t^L(R)|| < \frac{\varepsilon}{4}$, $|\psi(QR) - \psi_L(QR)| < \frac{\varepsilon}{4}$ and $|\psi(Q\alpha_t^L(R)) - \psi_L(Q\alpha_t(R))| < \frac{\varepsilon}{4}$ and a $\delta > 0$ such that $|\psi_L(Q\alpha_t^L(R) - QR)| < \frac{\varepsilon}{4}$ for $|t| < \delta$. Then, for $|t| < \delta$

$$\begin{aligned}|\psi(Q\alpha_t(R) - QR)| \leq |\psi(Q\alpha_t(R) - Q\alpha_t^L(R))| + |\psi(Q\alpha_t^L(R)) - \psi_L(Q\alpha_t^L(R))| \\ + |\psi_L(Q\alpha_t^L(R)) - \psi_L(QR)| + |\psi_L(QR) - \psi(QR)| < \varepsilon.\end{aligned}$$

Next, we show that $\psi$ is an $(\alpha, \beta)$-KMS state as specified in Definition 1. Note that the following inequality are valid for any $Q, R \in \mathfrak{R}$:

$$\begin{aligned}|\psi(Q\alpha_t(R)) - \psi_L(Q\alpha_t^L(R))| \leq |\psi(Q\alpha_t(R)) - \psi_L(Q\alpha_t(R))| \\ + |\psi_L(Q\alpha_t(R)) - \psi_L(Q\alpha_t^L(R))|. \quad (34)\end{aligned}$$

Using the integral representation of an analytic function in a strip $I_\beta = \{z \in \mathbf{C} \mid 0 < \text{Im} z < \beta\}$ (c.f. the proof of [25, Theorem 2.2]), $\psi$ is an $(\alpha, \beta)$-KMS state.

Finally, we show the uniqueness of $(\alpha, \beta)$-KMS state. Let $\phi$ be an arbitrary extremal $(\alpha, \beta)$-KMS regular state at $\beta$. Let $(\mathfrak{H}_\psi, \pi_\psi, \Omega_\psi)$ (resp. $(\mathfrak{H}_\phi, \pi_\phi, \Omega_\phi)$) be the GNS representation associated with $\psi$(rest. $\phi$). By $\widehat{\psi}$ and $\widehat{\phi}$, we denote the normal extension to the von Neumann algebras $\pi_\psi(\mathfrak{R})''$ and $\pi_\phi(\mathfrak{R})''$.

Let $\widehat{\phi}_N = \widehat{\phi}^{\beta W(N)}$, $N \in \mathbf{N}$, be the perturbed state of $\widehat{\phi}$ by $\beta W(N)$, where $W(N)$ is defined in Proposition 4. Put $\mathfrak{M} = \pi_\phi(\mathfrak{R})''$ and $\mathfrak{N}_L = \pi_\phi(\mathfrak{R}_L)''$, $L \in \mathbf{N}$. By Lemma 7, for $L \leq N$ we obtain

$$\begin{aligned} 0 &\le S(\widehat{\phi_N}|_{\mathfrak{R}_L}, \widehat{\phi}|_{\mathfrak{R}_L}) \le S(\widehat{\phi_N}, \widehat{\phi}) = \widehat{\phi_N}(\beta W(N)) - \log \widehat{\phi_N}(1) \\ &\le \widehat{\phi_N}(\beta W(N)) - \widehat{\phi}(\beta W((N)) \le 4|\beta| \cdot |\varphi|_\infty. \end{aligned}$$

This follows from Peierls–Bogoliubov inequality:

$$\log \widehat{\phi_N}(1) \ge \log e^{\widehat{\phi}(\beta W(N))} = \widehat{\phi}(\beta W(N)).$$

Due to Lemmas 4 and 6, for $L \le N$

$$\begin{aligned} S(\widehat{\phi_N}|_{\mathfrak{R}_L}, \widehat{\phi}|_{\mathfrak{R}_L}) &= S(\phi_N|_{\mathfrak{R}_L}, \phi|_{\mathfrak{R}_L}) = S(\psi_N|_{\mathfrak{R}_L}, \phi|_{\mathfrak{R}_L}) \\ &= S(\widehat{\psi_N}|_{\pi_0(\mathfrak{R}_L)''}, \widehat{\phi}|_{\pi_0(\mathfrak{R}_L)''}) \le 4|\beta| \cdot |\varphi|_\infty. \end{aligned}$$

Note that $\widehat{\psi_N}|_{\mathfrak{R}_L}$ converge to $\widehat{\psi}|_{\mathfrak{R}_L}$ in $\sigma(\mathscr{B}(\mathfrak{H}_{\Lambda_L})_*, \mathscr{B}(\mathfrak{H}_{\Lambda_L}))$ topology. By Lemma 9, it follows that

$$\begin{aligned} S(\psi|_{\mathfrak{R}_L}, \phi|_{\mathfrak{R}_L}) &= S(\widehat{\psi}|_{\pi_0(\mathfrak{R}_L)''}, \widehat{\phi}|_{\pi_0(\mathfrak{R}_L)''}) \\ &\le \liminf_N S(\widehat{\psi_N}|_{\pi_0(\mathfrak{R}_L)''}, \widehat{\phi}|_{\pi_0(\mathfrak{R}_L)''}) \le 4|\beta| \cdot |\varphi|_\infty. \end{aligned}$$

By Lemma 5 and [2, Lemma 3], we complete our proof. □

## References


1. Albeverio, S., Kondratiev, Y., Kozitsky, Y., Rockner, M.: The Statistical Mechanics of Quantum Lattice Systems: A Path Integral Approach. Ems Tracts in Mathematics. European Mathematical Society, Zürich (2009)
2. Araki, H.: On uniqueness of KMS states of one-dimensional quantum lattice systems. Commun. Math. Phys. **44**(1), 1–7 (1975)
3. Araki, H.: Relative entropy of states of von Neumann algebras. Publ. RIMS, Kyoto Univ. **13**, 173–192 (1977)
4. Araki, H.: Relative entropy for states of von Neumann algebras II. Publ. RIMS, Kyoto Univ. **11**, 809–833 (1976)
5. Bratteli, O., Robinson, D.: Operator Algebras and Quantum Statistical Mechanics I, 2nd edn. Springer, Berlin (1987)
6. Bratteli, O., Robinson, D.: Operator Algebras and Quantum Statistical Mechanics II, 2nd edn. Springer, Berlin (1997)
7. Buchholz, D., Grundling, H.: Algebraic supersymmetry: a case study. Commun. Math. Phys. **272**, 699–750 (2007)
8. Buchholz, D., Grundling, H.: The resolvent algebra: a new approach to canonical quantum systems. J. Funct. Anal. **254**, 2725–2779 (2008)
9. Buchholz, D., Grundling, H.: Lie algebras of derivations and resolvent algebras. Commun. Math. Phys. **320**, 455–467 (2013)
10. Buchholz, D.: The resolvent algebra for oscillating lattice systems: dynamics, ground and equilibrium states. Commun. Math. Phys. **353**, 949–981 (2018)
11. Buchholz, D.: The resolvent algebra of non-relativistic Bose fields: observables, dynamics and states. Commun. Math. Phys. **362**, 691–716 (2018)

12. Fannes, M., Verbeure, A.: On the time evolution automorphisms of the CCR-algebra for quantum mechanics. Commun. Math. Phys. **35**, 257–264 (1974)
13. Hastings, M.B., Koma, T.: Spectral gap and exponential decay of correlations. Commun. Math. Phys. **265**, 781–804 (2006)
14. Hiai, F., Ohya, M., Tsukada, M.: Sufficiency and relative entropy in $*$-algebras with applications in quantum systems. Pac. J. Math. **107**(1), 117–140 (1983)
15. Lieb, E.H., Robinson, D.W.: The finite group velocity of quantum spin systems. Commun. Math. Phys. **28**, 251–257 (1972)
16. Minlos, R., Verbeure, A., Zagrebnov, V.: A quantum crystal model in the light-mass limit: Gibbs states. Rev. Math. Phys. **12**(7), 981–1032 (2000)
17. Nachtergaele, B., Sims, R.: Lieb-Robinson bounds and the exponential clustering theorem. Commun. Math. Phys. **265**, 119–130 (2006)
18. Nachtergaele, B., Sims, R.: Recent progress in quantum spin systems. Markov Process. Relat. Fields **13**, 315–329 (2007)
19. Nachtergaele, B., Sims, R.: Locality estimates for quantum spin systems. arXiv:math-ph/0712.3318v1
20. Nachtergaele, B., Raz, H., Schlein, B., Sims, R.: Lieb-Robinson bounds for harmonic and anharmonic lattice systems. Commun. Math. Phys. **286**, 1073–1098 (2009)
21. Nachtergaele, B., Sims, R.: Lieb-Robinson Bounds in Quantum Many-body Physics. Entropy and the Quantum. Contemporary Mathematics, vol. 529, pp. 141–176. American Mathematical Society, Providence (2010)
22. Nachtergaele, B., Sims, R., Schlein, B., Starr, S., Zagrebnov, V.: On the existence of the dynamics for anharmonic quantum oscillator systems. Rev. Math. Phys. **22**, 207–231 (2010)
23. Nachtergaele, B., Sims, R.: On the dynamics of lattice systems with unbounded on-site terms in the Hamiltonian. arXiv:math-ph/1410.8174
24. Ohya, M., Petz, D.: Quantum Entropy and its Use, Corrected Second Printing. Texts and Monographs in Physics. Springer, Berlin (2004)
25. Powers, R.T., Sakai, S.: Existence of ground states and KMS states for approximately inner dynamics. Commun. Math. Phys. **39**, 273–288 (1975)
26. Uhlmann, A.: Relative entropy and the Wigner-Yanase-Dyson-Lieb concavity in an interpolation theory. Commun. Math. Phys. **54**(1), 21–32 (1977)

# Trotter Product Formula and Linear Evolution Equations on Hilbert Spaces

**Hagen Neidhardt, Artur Stephan and Valentin A. Zagrebnov**

**Abstract** The paper is devoted to evolution equations of the form

$$\frac{\partial}{\partial t}u(t) = -(A + B(t))u(t), \quad t \in \mathcal{I} = [0, T],$$

on separable Hilbert spaces where $A$ is a non-negative self-adjoint operator and $B(\cdot)$ is family of non-negative self-adjoint operators such that $\mathrm{dom}(A^\alpha) \subseteq \mathrm{dom}(B(t))$ for some $\alpha \in [0, 1)$ and the map $A^{-\alpha}B(\cdot)A^{-\alpha}$ is Hölder continuous with the Hölder exponent $\beta \in (0, 1)$. It is shown that the solution operator $U(t, s)$ of the evolution equation can be approximated in the operator norm by a combination of semigroups generated by $A$ and $B(t)$ provided the condition $\beta > 2\alpha - 1$ is satisfied. The convergence rate for the approximation is given by the Hölder exponent $\beta$. The result is proved using the evolution semigroup approach.

## 1 Introduction

A closer look to Kato's work shows that abstract evolution equations and Trotter product formula were topics of high interest for Kato. Already at the beginning of his scientific career, Kato was interested in evolution equations [16, 17]. This interest

On the occasion of the 100th birthday of Tosio Kato

H. Neidhardt · A. Stephan
Weierstrass Institute for Applied Analysis and Stochastics, Mohrenstr. 39, 10117 Berlin, Germany
e-mail: hagen.neidhardt@wias-berlin.de

A. Stephan
e-mail: artur.stephan@wias-berlin.de

V. A. Zagrebnov (✉)
Institut de Mathématiques de Marseille (UMR 7373), Université d'Aix-Marseille, CMI - Technopôle Château-Gombert, 39 rue F. Joliot Curie, 13453 Marseille, France
e-mail: valentin.zagrebnov@univ-amu.fr

© Springer Nature Switzerland AG 2019

T. M. Rassias and V. A. Zagrebnov (eds.), *Analysis and Operator Theory*, Springer Optimization and Its Applications 146, https://doi.org/10.1007/978-3-030-12661-2_13

has lasted a lifetime [18–21, 26, 27, 29]. Another topic of great interest for him was the so-called Trotter product formula [22–24, 28]. Even the paper [23] has inspired further developments in this field [15].

The topic of the present paper is to link evolution equations with the Trotter product formula. To this end, we consider an abstract evolution equation of type

$$\begin{aligned} \frac{\partial u(t)}{\partial t} &= - C(t)u(t), \quad u(s) = x_s, \quad s \in [0,T), \\ C(t) &= A + B(t), \end{aligned} \qquad t \in \mathcal{I} := [0,T], \tag{1.1}$$

on the separable Hilbert space $\mathfrak{H}$. Evolution equations of that type on Hilbert or Banach spaces are widely investigated, cf. [1–4, 7, 29, 31, 32, 43–48, 51–53, 55, 56] or the books [5, 49, 54]. We consider the equation (1.1) under the following assumptions.

**Assumption 1.1**

(S1) The operator $A$ is self-adjoint in the Hilbert space $\mathfrak{H}$ such that $A \geq I$. Let $\{B(t)\}_{t\in\mathcal{I}}$ be a family of non-negative self-adjoint operators in $\mathfrak{H}$ such that the function $(I + B(\cdot))^{-1} : \mathcal{I} \longrightarrow \mathcal{L}(\mathfrak{H})$ is strongly measurable.

(S2) There is an $\alpha \in [0, 1)$ such that for a.e. $t \in \mathcal{I}$ the inclusion $\mathrm{dom}(A^\alpha) \subseteq \mathrm{dom}(B(t))$ holds. Moreover, the function $B(\cdot)A^{-\alpha} : \mathcal{I} \longrightarrow \mathcal{L}(\mathfrak{H})$ is strongly measurable and essentially bounded, i.e.,

$$C_\alpha := \operatorname*{ess\,sup}_{t\in\mathcal{I}} \|B(\cdot)A^{-\alpha}\| < \infty. \tag{1.2}$$

(S3) The map $A^{-\alpha}B(\cdot)A^{-\alpha} : \mathcal{I} \longrightarrow \mathcal{L}(\mathfrak{H})$ is Hölder continuous, i.e, for some $\beta \in (0, 1)$ there is a constant $L_{\alpha,\beta} > 0$ such that the estimate

$$\|A^{-\alpha}(B(t) - B(s))A^{-\alpha}\| \leq L_{\alpha,\beta}|t - s|^\beta, \quad (t,s) \in \mathcal{I} \times \mathcal{I}, \tag{1.3}$$

holds. △

Notice that under the assumption (S2), the operator $C(t)$ is also an invertible non-negative self-adjoint operator for each $t \in \mathcal{I}$. Assumptions of that type were made in [13, 14, 34, 35, 56]. One checks that the assumptions (S1)–(S3) and the additional assumption $\beta > \alpha$ imply the assumptions (I), (VI), and (VII) of [56] for the family $\{C(t)\}_{\in\mathcal{I}}$. Hence, Proposition 3.1 and Theorem 3.2 of [56] yield the existence of a so-called *solution* (or *evolution*) operator for the evolution equation (1.1), i.e., a strongly continuous, uniformly bounded family of bounded operators $\{U(t,s)\}_{(t,s)\in\Delta}$, $\Delta := \{(t,s) \in \mathcal{I} \times \mathcal{I} : 0 \leq s \leq t \leq T\}$, such that the conditions

$$\begin{aligned} U(t,t) &= I, \quad \text{for} \quad t \in \mathcal{I}, \\ U(t,r)U(r,s) &= U(t,s), \quad \text{for} \quad t,r,s \in \mathcal{I} \quad \text{with} \quad s \leq r \leq t, \end{aligned} \tag{1.4}$$

are satisfied and $u(t) = U(t,0)x$ is for every $x \in \mathfrak{H}$ a strict solution of (1.1), see Definition 1.1 of [56]. Because the involved operators are self-adjoint and non-negative one checks that the solution operator consists of contractions.

The aim of the present paper is to analyze the convergence of the following approximation to the solution operator $\{U(t,s)\}_{(t,s)\in\Delta}$. Let

$$s =: t_0 < t_1 < \ldots < t_{n-1} < t_n := t, \quad t_j := s + j\tfrac{t-s}{n}, \tag{1.5}$$

$j = \{0, 1, 2, \ldots, n\}$, $n \in \mathbb{N}$, be a partition of the interval $[s,t]$. Let

$$\begin{aligned} G_j(t,s;n) :=& e^{-\frac{t-s}{n}A} e^{-\frac{t-s}{n}B(t_j)}, \quad j = 0, 1, 2, \ldots, n, \\ V_n(t,s) :=& G_{n-1}(t,s;n) G_{n-1}(t,s;n) \times \cdots \times G_2(t,s;n) G_0(t,s;n), \end{aligned} \tag{1.6}$$

$n \in \mathbb{N}$. The main result in the paper is the following. If the assumptions (S1)–(S3) are satisfied and in addition the condition $\beta > \alpha$ holds, then the solution operator $\{U(t,s)\}_{(t,s)\in\Delta}$ of [56] admits the approximation

$$\operatorname*{ess\,sup}_{(t,s)\in\Delta} \|V_n(t,s) - U(t,s)\| \le \frac{R_\beta}{n^\beta}, \quad n \in \mathbb{N}, \tag{1.7}$$

with some constant $R_\beta > 0$. The result shows that the convergence of the approximation $\{V_n(t,s)\}_{(t,s)\in\Delta}$ is determined by the smoothness of the perturbation $B(\cdot)$.

If the map $A^{-\alpha}B(\cdot)A^{-\alpha} : \mathcal{I} \longrightarrow \mathcal{L}(\mathfrak{H})$ is Lipschitz continuous, then the map is of course Hölder continuous with any exponent $\gamma \in (\alpha, 1)$. Hence from (1.7), it immediately follows that for any $\gamma \in (\alpha, 1)$, there is a constant $R_\gamma$ such that

$$\operatorname*{ess\,sup}_{(t,s)\in\Delta} \|V_n(t,s) - U(t,s)\| \le \frac{R_\gamma}{n^\gamma}, \quad n \in \mathbb{N}. \tag{1.8}$$

In particular, for any $\gamma$ close to one the estimate (1.8) holds.

In [14], the Lipschitz case was considered. It was shown that there is a constant $\Upsilon_0 > 0$ such that the estimate

$$\operatorname*{ess\,sup}_{t\in\mathcal{I}} \|V_n(t,0) - U(t,0)\| \le \Upsilon_0 \frac{\log(n)}{n}, \quad n = 2, 3, \ldots. \tag{1.9}$$

holds. It is obvious that the estimate (1.9) is stronger than

$$\operatorname*{ess\,sup}_{t\in\mathcal{I}} \|V_n(t,0) - U(t,0)\| \le \frac{R_\gamma}{n^\gamma}, \quad n \in \mathbb{N}.$$

(which follows from (1.8)) for any $\gamma$ independent of how close it is to one.

To prove (1.7), we use the so-called evolution semigroup approach which allows not only to verify the estimate (1.7) but also to generalise it. The approach is quite different from the technique used in [14, 56]. We have successfully applied this approach already in [34, 35]. The key idea is to forget about the evolution equation (1.1) and to consider instead of it the operators $\mathcal{K}_0$ and $\mathcal{K}$ on $\mathfrak{K} = L^2(\mathcal{I}, \mathfrak{H})$. The operator $\mathcal{K}_0$ is the generator of the contraction semigroup $\{\mathcal{U}_0(\tau)\}_{\tau\in\mathbb{R}_+}$,

$$(\mathcal{U}_0(\tau)f)(t) := e^{-\tau A}\chi_{\mathcal{I}}(t-\tau)f(t-\tau), \quad f \in L^2(\mathcal{I}, \mathfrak{H}), \tag{1.10}$$

and $\mathcal{K}$ is given by

$$\mathcal{K} = \mathcal{K}_0 + \mathcal{B}, \quad \mathrm{dom}(\mathcal{K}) = \mathrm{dom}(\mathcal{K}_0) \cap \mathrm{dom}(\mathcal{B}),$$

where $\mathcal{B}$ is the multiplication operator induced by the family $\{B(t)\}_{\in\mathcal{I}}$ in $L^2(\mathcal{I}, \mathfrak{H})$ which is self-adjoint and non-negative, for more details see Section 2. It turns out that under the assumptions (S1) and (S2), the operator $\mathcal{K}$ is the generator of a contraction semigroup $\{\mathcal{U}(\tau)\}_{\tau\in\mathbb{R}_+}$ on $L^2(\mathcal{I}, \mathfrak{H})$. For the pair $\{\mathcal{K}_0, \mathcal{B}\}$, we consider the Lie–Trotter product formula. From the original paper of Trotter [50], one gets that

$$\text{s-}\lim_{n\to\infty} \left(e^{-\frac{\tau}{n}\mathcal{K}_0}e^{-\frac{\tau}{n}\mathcal{B}}\right)^n = e^{-\tau\mathcal{K}}, \quad \tau \in \mathbb{R}_+ := [0, \infty), \tag{1.11}$$

holds uniformly in $\tau$ on any bounded interval of $\mathbb{R}_+$. Since $e^{-\tau\mathcal{K}_0} = 0$ and $e^{-\tau\mathcal{K}} = 0$ for $\tau \geq T$ one gets even uniformly in $\tau \in \mathbb{R}_+$.

Previously, it was shown that under certain assumptions, the strong convergence can be improved to operator-norm convergence on Hilbert spaces, see [9, 10, 15, 38, 42] as well as on Banach spaces, see [11]. For an overview, the reader is referred to [37]. To consider the Trotter product formula for evolution, equations is relatively new and was first realized in [34, 35] for Banach spaces.

In the following, we improve the convergence (1.11) to operator-norm convergence. We show that under the assumptions (S1)–(S3) and $\beta > 2\alpha - 1$, there is a constant $R_\beta > 0$ such that

$$\sup_{\tau\in\mathbb{R}_+} \left\| \left(e^{-\frac{\tau}{n}\mathcal{K}_0}e^{-\frac{\tau}{n}\mathcal{B}}\right)^n - e^{-\tau\mathcal{K}} \right\| \leq \frac{R_\beta}{n^\beta}, \quad n \in \mathbb{N}, \tag{1.12}$$

holds.

It turns out that $\mathcal{K}$ is the generator of an evolution semigroup. This means, there is a propagator $\{U(t,s)\}_{(t,s)\in\Delta_0}$, $\Delta_0 := \{(t,s) \in \mathcal{I}_0 \times \mathcal{I}_0 : s \leq t\}$, $\mathcal{I}_0 = (0, T]$, such that the contraction semigroup $\{\mathcal{U}(\tau) = e^{-\tau\mathcal{K}}\}_{\tau\in\mathbb{R}_+}$ admits the representation

$$(\mathcal{U}(\tau)f)(t) = U(t, t-\tau)\chi_{\mathcal{I}}(t-\tau)f(t-\tau), \qquad f \in L^2(\mathcal{I}, \mathfrak{H}). \tag{1.13}$$

We recall that a strongly continuous, uniformly bounded family of bounded operators $\{U(t,s)\}_{(t,s)\in\Delta_0}$ is called a propagator if (1.4) is satisfied for $\mathcal{I}_0$ and $\Delta_0$ instead of $\mathcal{I}$ and $\Delta$, respectively. Roughly speaking, a propagator is a solution operator restricted to $\Delta_0$ where the assumption that $U(t,0)x$ should be a strict solution is dropped. Obviously, the notion of a propagator is weaker then that one of a solution operator. For its existence, one needs only the assumptions (S1) and (S2), see Theorem 4.4 and 4.5 in [34] or Theorem 3.3 [35]. Of course, the propagator coincides with the solution operator of [56] if the assumptions (S1)–(S3) are satisfied and $\beta > \alpha$.

By Proposition 3.8 of [37] and (1.12) we immediately get that under the assumptions (S1)–(S3) and $\beta > 2\alpha - 1$ the estimate

$$\operatorname*{ess\,sup}_{(t,s)\in\Delta_0} \| V_n(t,s) - U(t,s)\| \le \frac{R_\beta}{n^\beta}, \quad n \in \mathbb{N}, \tag{1.14}$$

holds, where the constant $R_\beta$ is that one of (1.12). Notice that the condition $\beta > 2\alpha - 1$ is weaker than $\beta > \alpha$, i.e., if $\beta > \alpha$, then $\beta > 2\alpha - 1$ holds. If $\alpha$ satisfies the condition $\frac{1+\beta}{2} > \alpha > \beta$, then the assumptions (I), (VI), and (VII) of [56] for the family $\{C(t)\}_{\in\mathcal{I}}$ are not valid but nevertheless, we get an approximation of the corresponding propagator $\{U(t,s)\}_{(t,s)\in\Delta_0}$.

The results are stronger than those in [34, 35] for Banach spaces. In [34], a convergence rate $O(n^{-(\beta-\alpha)})$ was found, whereas in [35], the Lipschitz case has been considered and the rate $O(n^{-(1-\alpha)})$ for $\alpha \in (\frac{1}{2}, 1)$ was proved.

It turns out that the result (1.7) can be hardly improved. Indeed in [36], the simple case $\mathfrak{H} := \mathbb{C}$ and $A = 1$ was considered. In that case, the family $\{B(t)\}_{t\in\mathcal{I}}$ reduces to a non-negative bounded measurable function: $b(\cdot) : \mathcal{I} \longrightarrow \mathbb{R}$ which has to be Hölder continuous with exponent $\beta \in (0, 1)$. For that case, it was found in [36] that the convergence rate is $O(n^{-\beta})$ which coincides with (1.7). For the Lipschitz case, it was found $O(n^{-1})$ which suggests that (1.8) and (1.9) might be not optimal.

The paper is organised as follows. In Section 2, we give a short introduction into evolution semigroups. For more details, the reader referred to [33, 34, 39, 40]. The results are proven in Section 3. In Section 3.1, we prove auxiliary results which are necessary to prove the main results of Section 3.2.

**Notation**: Spaces, in particular, Hilbert are denoted by Gothic capital letters like $\mathfrak{H}$, $\mathfrak{K}$, etc. Operators are denoted by Latin or italic capital letters. The Banach space of bounded operators on space is denoted by $\mathcal{L}(\cdot)$, like $\mathcal{L}(\mathfrak{H})$. We set $\mathbb{R}_+ := [0, \infty)$. If a function is called measurable, then it means Lebesgue measurable. The notation "a.e." means that a statement or relation fails at most for a set of Lebesgue measure zero. In the following, we use the notation $\operatorname{ess\,sup}_{(t,s)\in\Delta}$ or $\operatorname{ess\,sup}_{(t,s)\in\Delta_0}$. In that case, the Lebesgue measure of $\mathbb{R}^2$ is meant.

We point out that we call operator $\mathcal{K}$ to be generator of a semigroup $\{e^{-\tau\mathcal{K}}\}_{\tau\in\mathbb{R}_+}$, see e.g., [41], although in [12, 25] it is the operator $-\mathcal{K}$, which is called the generator.

## 2 Evolution Semigroups

Below, we consider the Hilbert space $\mathfrak{K} = L^2(\mathcal{I}, \mathfrak{H})$ consisting of all measurable functions $f(\cdot) : \mathcal{I} \longrightarrow \mathfrak{H}$ such that the norm function $\|f(\cdot)\| : \mathcal{I} \longrightarrow \mathbb{R}_+$ is square integrable. Further, let $D_0$ be the generator of the right-hand sift semigroup on $L^2(\mathcal{I}, \mathfrak{H})$, i.e.,

$$(e^{-\tau D_0} f)(t) = \chi_{\mathcal{I}}(t-\tau) f(t-\tau), \quad t \in \mathcal{I}, \quad \tau \in \mathbb{R}_+, \quad f \in L^2(\mathcal{I}, \mathfrak{H}).$$

Notice that $e^{-\tau D_0} = 0$ for $\tau \geq T$. The generator $D_0$ is given by

$$\begin{aligned} (D_0 f)(t) &:= \frac{\partial}{\partial t} f(t), \\ f \in \mathrm{dom}(D_0) &:= W_0^{1,2}(\mathcal{I}, \mathfrak{H}) = \{f \in W^{1,2}(\mathcal{I}, \mathfrak{H}) : f(0) = 0\}. \end{aligned} \tag{2.1}$$

We remark that $D_0$ is a closure of the maximal symmetric operator and its semigroup is contractive.

Further, we consider the multiplication operator $\mathcal{A}$ in $L^2(\mathcal{I}, \mathfrak{H})$,

$$\begin{aligned} (\mathcal{A} f)(t) &:= A f(t), \\ f \in \mathrm{dom}(\mathcal{A}) &:= \left\{ f \in L^2(\mathcal{I}, \mathfrak{H}) : \begin{array}{l} f(t) \in \mathrm{dom}(A) \quad \text{for a.e. } t \in \mathcal{I} \\ A f(t) \in L^2(\mathcal{I}, \mathfrak{H}). \end{array} \right\} \end{aligned}$$

If (S1) is satisfied, then $\mathcal{A}$ is self-adjoint and $\mathcal{A} \geq I_{L^2(\mathcal{I}, \mathfrak{H})}$. For the resolvent, one has the representation

$$((\mathcal{A} - z)^{-1} f)(t) = (A - z)^{-1} f(t), \; t \in \mathcal{I}_0, \; f \in L^2(\mathcal{I}, \mathfrak{H}), \quad z \in \rho(\mathcal{A}) = \rho(A),$$

and the corresponding semigroup $\{e^{-\tau \mathcal{A}}\}_{\tau \in \mathbb{R}_+}$ is given by

$$(e^{-\tau \mathcal{A}} f)(t) = e^{-\tau A} f(t), \; t \in \mathcal{I}, \; f \in L^2(\mathcal{I}, \mathfrak{H}), \quad \tau \in \mathbb{R}_+. \tag{2.2}$$

Notice that the operators $e^{-\tau D_0}$ and $e^{-\tau \mathcal{A}}$ commute. Let us consider the contraction semigroup

$$\mathcal{U}_0(\tau) := e^{-\tau D_0} e^{-\tau \mathcal{A}}, \quad \tau \in \mathbb{R}_+. \tag{2.3}$$

Obviously, the semigroup $\{\mathcal{U}_0(\tau)\}_{\tau \in \mathbb{R}_+}$ admits the representation (1.13). Due to the maximal $L^2$-regularity of $A$, cf. [6], its generator $\mathcal{K}_0$ is given by

$$\mathcal{K}_0 := D_0 + \mathcal{A}, \quad \mathrm{dom}(\mathcal{K}_0) := \mathrm{dom}(D_0) \cap \mathrm{dom}(\mathcal{A}).$$

Further, we consider the multiplication operator $\mathcal{B}$, defined as

$$\begin{aligned}(\mathcal{B}f)(t) &:= B(t)f(t)\\ f \in \mathrm{dom}(\mathcal{B}) &:= \left\{ f \in L^2(\mathcal{I},\mathfrak{H}) : \begin{array}{c} f(t) \in \mathrm{dom}(B(t)) \quad \text{for a.e. } t \in \mathcal{I} \\ B(t)f(t) \in L^2(\mathcal{I},\mathfrak{H}) \end{array} \right\}.\end{aligned} \tag{2.4}$$

If (S1) is satisfied, then $\mathcal{B}$ is self-adjoint and non-negative. For the resolvent, we have the representation

$$((\mathcal{B}-z)^{-1}f)(t) = (B(t)-z)^{-1}f(t), \quad f \in L^2(\mathcal{I},\mathfrak{H}), \quad z \in \mathbb{C}_{--},$$

for a.e. $t \in \mathcal{I}$. The semigroup $\{e^{-\tau\mathcal{B}}\}_{\tau\in\mathbb{R}_+}$, admits the representation

$$(e^{-\tau\mathcal{B}}f)(t) = e^{-\tau B(t)}f(t), \quad f \in L^2(\mathcal{I},\mathfrak{H}),$$

for a.e. $t \in \mathcal{I}$.

By [34, Proposition 4.4], we get that under the assumptions (S1) and (S2) the operator

$$\mathcal{K} := \mathcal{K}_0 + \mathcal{B}, \quad \mathrm{dom}(\mathcal{K}) := \mathrm{dom}(\mathcal{K}_0) \cap \mathrm{dom}(\mathcal{B}),$$

is a generator of a contraction semigroup on $L^2(\mathcal{I},\mathfrak{H})$. From [34, Proposition 4.5], we obtain that $\mathcal{K}$ is the generator of an evolution semigroup. Because $\mathcal{K}$ is a generator of a contraction semigroup it turns out that the corresponding propagator consists of contractions.

If $\{U(t,s)\}_{(t,s)\in\Delta_0}$ is a propagator, then by virtue of (1.13) it defines a semigroup, which by definition is an evolution semigroup. It turns out that there is a one-to-one correspondence between the set of evolution semigroups on $L^2(\mathcal{I},\mathfrak{H})$ and propagators. It is interesting to note that evolution generators can be characterised quite independent from a propagator, see [33, Theorem 2.8] or [34, Theorem 3.3].

## 3 Results

We start with a general observation concerning the conditions (S1)–(S3).

*Remark 3.1* If the conditions (S1)–(S3) are satisfied for some $\alpha \in [0,1)$, then they are also satisfied for each $\alpha' \in (\alpha, 1]$. Indeed, the condition (S1) is obviously satisfied. To show (S2), we note that $\mathrm{dom}(A^{\alpha'}) \subseteq \mathrm{dom}(A^{\alpha}) \subseteq \mathrm{dom}(B(t))$ for a.e. $t \in \mathcal{I}$. Using the representation

$$B(t)A^{-\alpha'} = B(t)A^{-\alpha}A^{-(\alpha'-\alpha)} \tag{3.1}$$

for a.e. $t \in \mathcal{I}$ we get that the map $B(\cdot)A^{-\alpha'} : \mathcal{I} \longrightarrow \mathcal{L}(\mathfrak{H})$ is strongly measurable. Further, from (3.1)

$$C_{\alpha'} := \operatorname*{ess\,sup}_{t\in\mathcal{I}} \|B(t)A^{-\alpha'}\| \le \operatorname*{ess\,sup}_{t\in\mathcal{I}} \|B(t)A^{-\alpha}\| = C_\alpha < \infty.$$

Moreover we have

$$\|A^{-\alpha'}(B(t) - B(s))A^{-\alpha'}\| \le \|A^{-\alpha}(B(t) - B(s))A^{-\alpha}\| \le L_{\alpha,\beta}|t - s|^{\beta},$$

$(t, s) \in \mathcal{I} \times \mathcal{I}$, which shows that there is a constant $L_{\alpha',\beta} \le L_{\alpha,\beta}$ such that

$$\|A^{-\alpha'}(B(t) - B(s))A^{-\alpha'}\| \le L_{\alpha',\beta}|t - s|^{\beta} \quad (t, s) \in \mathcal{I} \times \mathcal{I}.$$

holds for $(t, s) \in \mathcal{I} \times \mathcal{I}$. △

Since $A$ is self-adjoint and non-negative, one has $\|A^{\gamma} e^{-\tau A}\| \le 1/\tau^{\gamma}$ for any $\tau \in \mathbb{R}_+$ and $\gamma \in [0, 1]$. Then by virtue of (2.2) and of (1.10), (2.3), one gets the estimates

$$\|\mathcal{A}^{\gamma} e^{-\tau\mathcal{A}}\| = \|\overline{e^{-\tau\mathcal{A}}\mathcal{A}^{\gamma}}\| \le \frac{1}{\tau^{\gamma}} \text{ and } \|\mathcal{A}^{\gamma} e^{-\tau\mathcal{K}_0}\| = \|\overline{e^{-\tau\mathcal{K}_0}\mathcal{A}^{\gamma}}\| \le \frac{1}{\tau^{\gamma}}. \quad (3.2)$$

## 3.1 Auxiliary Estimates

In this section, we prove a series of estimates necessary to establish (1.12). The following lemma can be partially derived from [34, Lemma 7.4].

**Lemma 3.2** *Let the assumptions (S1) and (S2) be satisfied. Then for any $\gamma \in [\alpha, 1)$, there is a constant $\Lambda_{\gamma} \ge 1$ such that*

$$\|\mathcal{A}^{\gamma} e^{-\tau\mathcal{K}}\| \le \frac{\Lambda_{\gamma}}{\tau^{\gamma}} \quad \textit{and} \quad \|\overline{e^{-\tau\mathcal{K}}\mathcal{A}^{\gamma}}\| \le \frac{\Lambda_{\gamma}}{\tau^{\gamma}}, \qquad \tau > 0, \quad (3.3)$$

*holds.*

*Proof* The proof of the first estimate follows from Lemma 7.4 of [34] and Remark 3.1. The second estimate can be proved similarly as the first one. One has only to modify the proof of Lemma 7.4 of [34] in a suitable manner and to apply again Remark 3.1. □

*Remark 3.3* Lemma 2.1 of [14] claims that for the Lipschitz case, the solution operator $\{U(t, s)\}_{(t,s)\in\Delta}$ of (1.1) admits the estimates

$$\sup_{(t,s)\in\Delta} (t - s)^{\gamma}\|A^{\gamma}U(t, s)\| < \infty \quad \text{and} \quad \sup_{(t,s)\in\Delta} (t - s)^{\gamma}\|U(t, s)A^{\gamma}\| < \infty$$

for $\gamma \in [0, 1]$. Proposition 2.1 of [36] immediately yields that the corresponding evolution semigroup $\{\mathcal{U}(\tau) = e^{-\tau\mathcal{K}}\}_{\tau\in\mathbb{R}_+}$ satisfies the estimates (3.3) for $\gamma = 1$. △

Now we set

$$\mathcal{T}(\tau) = e^{-\tau\mathcal{K}_0}e^{-\tau\mathcal{B}}, \quad \tau \in \mathbb{R}_+. \quad (3.4)$$

Notice that $\mathcal{T}(\tau) = 0$ for $\tau \ge T$.

**Lemma 3.4** *Let the assumptions (S1) and (S2) be satisfied. Then for any $\gamma \in [\alpha, 1)$, the estimates*

$$\|\mathcal{A}^{-\gamma}(\mathcal{T}(\tau) - \mathcal{U}(\tau))\| \leq 2C_\gamma \tau \quad \text{and} \quad \|(\mathcal{T}(\tau) - \mathcal{U}(\tau))\mathcal{A}^{-\gamma}\| \leq 2C_\gamma \tau, \tag{3.5}$$

*hold for $\tau \geq 0$, where*

$$C_\gamma := \operatorname*{ess\,sup}_{t \in \mathcal{I}} \|B(t)A^{-\gamma}\|. \tag{3.6}$$

*Proof* The proof of the first estimate follows from Lemma 7.6 of [34] and Remark 3.1. The specific constant $2C_\gamma$ is obtained following carefully by the proof of Lemma 7.6 of [34]. The second estimate can be proved by modifying the proof of the first estimate in an obvious manner. □

**Lemma 3.5** *Let the assumptions (S1)–(S3) be satisfied. Then for any $\gamma \in [\alpha, 1)$ and $\beta \in (0, 1)$, there is a constant $Z_{\gamma,\beta} > 0$ such that*

$$\|\mathcal{A}^{-\gamma}(\mathcal{T}(\tau) - \mathcal{U}(\tau))\mathcal{A}^{-\gamma}\| \leq Z_{\gamma,\beta}\tau^{1+\varkappa}, \quad \tau \in \mathbb{R}_+, \tag{3.7}$$

*holds where $\varkappa := \min\{\gamma, \beta\}$.*

*Proof* We use the representation

$$\begin{aligned}\frac{d}{d\sigma}e^{-(\tau-\sigma)\mathcal{K}}e^{-\sigma\mathcal{K}_0}e^{-\sigma\mathcal{B}} =& e^{-(\tau-\sigma)\mathcal{K}}\left\{\mathcal{K}e^{-\sigma\mathcal{K}_0} - e^{-\sigma\mathcal{K}_0}\mathcal{K}_0 - e^{-\sigma\mathcal{K}_0}\mathcal{B}\right\}e^{-\sigma\mathcal{B}}\\ =& e^{-(\tau-\sigma)\mathcal{K}}\left\{\mathcal{B}e^{-\sigma\mathcal{K}_0} - e^{-\sigma\mathcal{K}_0}\mathcal{B}\right\}e^{-\sigma\mathcal{B}}\end{aligned}$$

which yields

$$\begin{aligned}&e^{-(\tau-\sigma)\mathcal{K}}\left\{\mathcal{B}e^{-\sigma\mathcal{K}_0} - e^{-\sigma\mathcal{K}_0}\mathcal{B}\right\}e^{-\sigma\mathcal{B}}\\ &\quad=(e^{-(\tau-\sigma)\mathcal{K}} - e^{-(\tau-\sigma)\mathcal{K}_0})\left\{\mathcal{B}e^{-\sigma\mathcal{K}_0} - e^{-\sigma\mathcal{K}_0}\mathcal{B}\right\}(e^{-\sigma\mathcal{B}} - I)+\\ &\quad\quad e^{-(\tau-\sigma)\mathcal{K}_0}\left\{\mathcal{B}e^{-\sigma\mathcal{K}_0} - e^{-\sigma\mathcal{K}_0}\mathcal{B}\right\}(e^{-\sigma\mathcal{B}} - I)+\\ &\quad\quad (e^{-(\tau-\sigma)\mathcal{K}} - e^{-(\tau-\sigma)\mathcal{K}_0})\left\{\mathcal{B}e^{-\sigma\mathcal{K}_0} - e^{-\sigma\mathcal{K}_0}\mathcal{B}\right\}+\\ &\quad\quad e^{-(\tau-\sigma)\mathcal{K}_0}\left\{\mathcal{B}e^{-\sigma\mathcal{K}_0} - e^{-\sigma\mathcal{K}_0}\mathcal{B}\right\}.\end{aligned}$$

Hence, we obtain the identity

$$\begin{aligned}&\mathcal{A}^{-\gamma}e^{-(\tau-\sigma)\mathcal{K}}\left\{\mathcal{B}e^{-\sigma\mathcal{K}_0} - e^{-\sigma\mathcal{K}_0}\mathcal{B}\right\}e^{-\sigma\mathcal{B}}\mathcal{A}^{-\gamma}\\ &\quad=\mathcal{A}^{-\gamma}(e^{-(\tau-\sigma)\mathcal{K}} - e^{-(\tau-\sigma)\mathcal{K}_0})\left\{\mathcal{B}e^{-\sigma\mathcal{K}_0} - e^{-\sigma\mathcal{K}_0}\mathcal{B}\right\}(e^{-\sigma\mathcal{B}} - I)\mathcal{A}^{-\gamma}+\\ &\quad\quad e^{-(\tau-\sigma)\mathcal{K}_0}\mathcal{A}^{-\gamma}\left\{\mathcal{B}e^{-\sigma\mathcal{K}_0} - e^{-\sigma\mathcal{K}_0}\mathcal{B}\right\}(e^{-\sigma\mathcal{B}} - I)\mathcal{A}^{-\gamma}+\\ &\quad\quad \mathcal{A}^{-\gamma}(e^{-(\tau-\sigma)\mathcal{K}} - e^{-(\tau-\sigma)\mathcal{K}_0})\left\{\mathcal{B}e^{-\sigma\mathcal{K}_0} - e^{-\sigma\mathcal{K}_0}\mathcal{B}\right\}\mathcal{A}^{-\gamma}+\end{aligned}$$

$$e^{-(\tau-\sigma)\mathcal{K}_0}\mathcal{A}^{-\gamma}\Big\{\mathcal{B}e^{-\sigma\mathcal{K}_0} - e^{-\sigma\mathcal{K}_0}\mathcal{B}\Big\}\mathcal{A}^{-\gamma}$$

which leads to the estimate

$$\begin{aligned}
&\left\|\mathcal{A}^{-\gamma}e^{-(\tau-\sigma)\mathcal{K}}\left\{\mathcal{B}e^{-\sigma\mathcal{K}_0} - e^{-\sigma\mathcal{K}_0}\mathcal{B}\right\}e^{-\sigma\mathcal{B}}\mathcal{A}^{-\gamma}\right\| \leq \\
&\quad\left\|\mathcal{A}^{-\gamma}(e^{-(\tau-\sigma)\mathcal{K}} - e^{-(\tau-\sigma)\mathcal{K}_0})\Big\{\mathcal{B}e^{-\sigma\mathcal{K}_0} - e^{-\sigma\mathcal{K}_0}\mathcal{B}\Big\}(e^{-\sigma\mathcal{B}} - I)\mathcal{A}^{-\gamma}\right\| + \qquad (3.8)\\
&\quad\left\|e^{-(\tau-\sigma)\mathcal{K}_0}\mathcal{A}^{-\gamma}\Big\{\mathcal{B}e^{-\sigma\mathcal{K}_0} - e^{-\sigma\mathcal{K}_0}\mathcal{B}\Big\}(e^{-\sigma\mathcal{B}} - I)\mathcal{A}^{-\gamma}\right\| + \qquad (3.9)\\
&\quad\left\|\mathcal{A}^{-\gamma}(e^{-(\tau-\sigma)\mathcal{K}} - e^{-(\tau-\sigma)\mathcal{K}_0})\Big\{\mathcal{B}e^{-\sigma\mathcal{K}_0} - e^{-\sigma\mathcal{K}_0}\mathcal{B}\Big\}\mathcal{A}^{-\gamma}\right\| + \qquad (3.10)\\
&\quad\left\|e^{-(\tau-\sigma)\mathcal{K}_0}\mathcal{A}^{-\gamma}\Big\{\mathcal{B}e^{-\sigma\mathcal{K}_0} - e^{-\sigma\mathcal{K}_0}\mathcal{B}\Big\}\mathcal{A}^{-\gamma}\right\|. \qquad (3.11)
\end{aligned}$$

Note that by (3.2) and (3.6), one gets

$$\begin{aligned}
\|e^{-\sigma\mathcal{K}_0}\mathcal{B}\| &= \|e^{-\sigma\mathcal{K}_0}\mathcal{A}^{\gamma}\overline{\mathcal{A}^{-\gamma}\mathcal{B}}\| \leq \sigma^{-\gamma}\|\overline{\mathcal{A}^{-\gamma}\mathcal{B}}\| = C_\gamma\sigma^{-\gamma},\\
\|\mathcal{B}e^{-\sigma\mathcal{K}_0}\| &= \|\mathcal{B}\mathcal{A}^{-\gamma}\mathcal{A}^{\gamma}e^{-\sigma\mathcal{K}_0}\| \leq \sigma^{-\gamma}\|\mathcal{B}\mathcal{A}^{-\gamma}\| = C_\gamma\sigma^{-\gamma},
\end{aligned} \qquad (3.12)$$

for $\sigma > 0$. Due to (3.12), one estimates (3.8) as

$$\begin{aligned}
&\left\|\mathcal{A}^{-\gamma}(e^{-(\tau-\sigma)\mathcal{K}} - e^{-(\tau-\sigma)\mathcal{K}_0})\Big\{\mathcal{B}e^{-\sigma\mathcal{K}_0} - e^{-\sigma\mathcal{K}_0}\mathcal{B}\Big\}(e^{-\sigma\mathcal{B}} - I)\mathcal{A}^{-\gamma}\right\|\\
&\quad\leq 2\,C_\gamma\sigma^{-\gamma}\left\|\mathcal{A}^{-\gamma}(e^{-(\tau-\sigma)\mathcal{K}} - e^{-(\tau-\sigma)\mathcal{K}_0})\right\|\left\|(e^{-\sigma\mathcal{B}} - I)\mathcal{A}^{-\gamma}\right\|.
\end{aligned}$$

Since the fundamental properties of semigroups and (3.6) yield

$$\left\|(e^{-\sigma\mathcal{B}} - I)\mathcal{A}^{-\gamma}\right\| \leq \|\mathcal{B}\mathcal{A}^{-\gamma}\|\,\sigma \leq C_\gamma\,\sigma, \quad \sigma \in \mathbb{R}_+, \qquad (3.13)$$

and

$$\left\|\mathcal{A}^{-\gamma}(e^{-(\tau-\sigma)\mathcal{K}} - e^{-(\tau-\sigma)\mathcal{K}_0})\right\| \leq C_\gamma(\tau-\sigma), \quad \sigma \in \mathbb{R}_+,$$

we get for (3.8) the estimate

$$\begin{aligned}
&\left\|\mathcal{A}^{-\gamma}(e^{-(\tau-\sigma)\mathcal{K}} - e^{-(\tau-\sigma)\mathcal{K}_0})\Big\{\mathcal{B}e^{-\sigma\mathcal{K}_0} - e^{-\sigma\mathcal{K}_0}\mathcal{B}\Big\}(e^{-\sigma\mathcal{B}} - I)\mathcal{A}^{-\gamma}\right\|\\
&\leq 2\,C_\gamma^3\sigma^{1-\gamma}(\tau-\sigma), \quad 0 \leq \sigma \leq \tau.
\end{aligned} \qquad (3.14)$$

To estimate (3.9), we recall that $\mathcal{A}$ and $\mathcal{K}_0$ commute. Then by (3.6), one gets

$$\begin{aligned}
&\left\|e^{-(\tau-\sigma)\mathcal{K}_0}\mathcal{A}^{-\gamma}\Big\{\mathcal{B}e^{-\sigma\mathcal{K}_0} - e^{-\sigma\mathcal{K}_0}\mathcal{B}\Big\}(e^{-\sigma\mathcal{B}} - I)\mathcal{A}^{-\gamma}\right\|\\
&\quad\leq 2C_\gamma\|(e^{-\sigma\mathcal{B}} - I)\mathcal{A}^{-\gamma}\| \leq 2\,C_\gamma^2\,\sigma, \quad 0 \leq \sigma \leq \tau,
\end{aligned} \qquad (3.15)$$

where (3.13) was used for the last inequality.

To estimate (3.10), we have

$$\begin{aligned}&\left\|\mathcal{A}^{-\gamma}(e^{-(\tau-\sigma)\mathcal{K}}-e^{-(\tau-\sigma)\mathcal{K}_0})\left\{\mathcal{B}e^{-\sigma\mathcal{K}_0}-e^{-\sigma\mathcal{K}_0}\mathcal{B}\right\}\mathcal{A}^{-\gamma}\right\|\\ \le&2\,C_\gamma\|\mathcal{A}^{-\gamma}(e^{-(\tau-\sigma)\mathcal{K}}-e^{-(\tau-\sigma)\mathcal{K}_0})\|\le 2\,C_\gamma^2(\tau-\sigma),\quad 0\le\sigma\le\tau.\end{aligned} \tag{3.16}$$

To estimate (3.11), we use the representation

$$\begin{aligned}&\mathcal{A}^{-\gamma}\left\{\mathcal{B}e^{-\sigma\mathcal{K}_0}-e^{-\sigma\mathcal{K}_0}\mathcal{B}\right\}\mathcal{A}^{-\gamma}\\ &=\mathcal{A}^{-\gamma}\mathcal{B}(e^{-\sigma\mathcal{A}}-I)\mathcal{A}^{-\gamma}e^{-\sigma\mathcal{D}_0}-e^{-\sigma\mathcal{D}_0}\mathcal{A}^{-\gamma}(e^{-\sigma\mathcal{A}}-I)\mathcal{B}\mathcal{A}^{-\gamma}+\\ &\quad\mathcal{A}^{-\gamma}\left\{\mathcal{B}e^{-\sigma\mathcal{D}_0}-e^{-\sigma\mathcal{D}_0}\mathcal{B}\right\}\mathcal{A}^{-\gamma},\end{aligned}$$

that yields

$$\begin{aligned}&\left\|\mathcal{A}^{-\gamma}\left\{\mathcal{B}e^{-\sigma\mathcal{K}_0}-e^{-\sigma\mathcal{K}_0}\mathcal{B}\right\}\mathcal{A}^{-\gamma}\right\|\\ &\le\left\|\mathcal{A}^{-\gamma}\mathcal{B}(e^{-\sigma\mathcal{A}}-I)\mathcal{A}^{-\gamma}\right\|+\left\|\mathcal{A}^{-\gamma}(e^{-\sigma\mathcal{A}}-I)\mathcal{B}\mathcal{A}^{-\gamma}\right\|+\\ &\quad\left\|\mathcal{A}^{-\gamma}\left\{\mathcal{B}e^{-\sigma\mathcal{D}_0}-e^{-\sigma\mathcal{D}_0}\mathcal{B}\right\}\mathcal{A}^{-\gamma}\right\|.\end{aligned}$$

Then by (3.6) and by semigroup properties one gets

$$\left\|\mathcal{A}^{-\gamma}\mathcal{B}(e^{-\sigma\mathcal{A}}-I)\mathcal{A}^{-\gamma}\right\|\le\frac{C_\gamma}{\gamma}\sigma^\gamma,\quad\sigma\in\mathbb{R}_+,$$

and

$$\left\|\mathcal{A}^{-\gamma}(e^{-\sigma\mathcal{A}}-I)\mathcal{B}\mathcal{A}^{-\gamma}\right\|\le\frac{C_\gamma}{\gamma}\sigma^\gamma,\quad\sigma\in\mathbb{R}_+.$$

The last term is obtained by using (S3) (for $\alpha$ substituted by $\gamma$) and the definitions (2.1), (2.4):

$$\left\|\mathcal{A}^{-\gamma}\left\{\mathcal{B}e^{-\sigma\mathcal{D}_0}-e^{-\sigma\mathcal{D}_0}\mathcal{B}\right\}\mathcal{A}^{-\gamma}\right\|\le\sigma^\beta L_{\gamma,\beta},\quad\sigma\in\mathbb{R}_+.$$

Summing up one finds that

$$\left\|\mathcal{A}^{-\gamma}\left\{\mathcal{B}e^{-\sigma\mathcal{K}_0}-e^{-\sigma\mathcal{K}_0}\mathcal{B}\right\}\mathcal{A}^{-\gamma}\right\|\le\frac{2C_\gamma}{\gamma}\sigma^\gamma+L_{\gamma,\beta}\sigma^\beta,\quad\sigma\in\mathbb{R}_+. \tag{3.17}$$

Using the estimates (3.14)–(3.17), we get

$$\left\| \mathcal{A}^{-\gamma} e^{-(\tau-\sigma)\mathcal{K}} \left\{ \mathcal{B} e^{-\sigma \mathcal{K}_0} - e^{-\sigma \mathcal{K}_0} \mathcal{B} \right\} e^{-\sigma \mathcal{B}} \mathcal{A}^{-\gamma} \right\|$$
$$\leq 2C_\gamma^3 \sigma^{1-\gamma}(\tau - \sigma) + 2C_\gamma^2 \sigma + 2C_\gamma^2(\tau - \sigma) + \frac{2C_\gamma}{\gamma}\sigma^\gamma + L_{\gamma,\beta}\sigma^\beta$$
$$= 2C_\gamma^3 \sigma^{1-\gamma}(\tau - \sigma) + 2C_\gamma^2 \tau + \frac{2C_\gamma}{\gamma}\sigma^\gamma + L_{\gamma,\beta}\sigma^\beta,$$

or returning back to its derivative

$$\left\| \mathcal{A}^{-\gamma} \frac{d}{d\sigma} e^{-(\tau-\sigma)\mathcal{K}} e^{-\sigma \mathcal{K}_0} e^{-\sigma \mathcal{B}} \mathcal{A}^{-\gamma} \right\|$$
$$\leq 2C_\gamma^3 \sigma^{1-\gamma}(\tau - \sigma) + 2C_\gamma^2 \tau + \frac{2C_\gamma}{\gamma}\sigma^\gamma + L_{\gamma,\beta}\sigma^\beta, \quad 0 \leq \sigma \leq \tau.$$

Since

$$\mathcal{A}^{-\gamma}(e^{-\tau \mathcal{K}_0} e^{-\tau \mathcal{B}} - e^{-\tau \mathcal{K}})\mathcal{A}^{-\gamma} = \int_0^\tau \mathcal{A}^{-\gamma} \frac{d}{d\sigma} e^{-(\tau-\sigma)\mathcal{K}} e^{-\sigma \mathcal{K}_0} e^{-\sigma \mathcal{B}} \mathcal{A}^{-\gamma} d\sigma,$$

we find the estimate

$$\left\| \mathcal{A}^{-\gamma}(e^{-\tau \mathcal{K}_0} e^{-\tau \mathcal{B}} - e^{-\tau \mathcal{K}})\mathcal{A}^{-\gamma} \right\| \leq \int_0^\tau \left\| \mathcal{A}^{-\gamma} \frac{d}{d\sigma} e^{-(\tau-\sigma)\mathcal{K}} e^{-\sigma \mathcal{K}_0} e^{-\sigma \mathcal{B}} \mathcal{A}^{-\gamma} \right\| d\sigma,$$

which yields the estimate

$$\left\| \mathcal{A}^{-\gamma}(e^{-\tau \mathcal{K}_0} e^{-\tau \mathcal{B}} - e^{-\tau \mathcal{K}})\mathcal{A}^{-\gamma} \right\|$$
$$\leq 2C_\gamma^3 \int_0^\tau \sigma^{1-\gamma}(\tau - \sigma) d\sigma + 2C_\gamma^2 \tau^2 + \frac{2C_\gamma}{(1+\gamma)\gamma}\tau^{1+\gamma} + \frac{L_{\gamma,\beta}}{1+\beta}\tau^{1+\beta}$$

or after integration

$$\left\| \mathcal{A}^{-\gamma}(e^{-\tau \mathcal{K}_0} e^{-\tau \mathcal{B}} - e^{-\tau \mathcal{K}})\mathcal{A}^{-\gamma} \right\|$$
$$\leq \frac{2C_\gamma^3}{(2-\gamma)(3-\gamma)}\tau^{3-\gamma} + 2C_\gamma^2 \tau^2 + \frac{2C_\gamma}{(1+\gamma)\gamma}\tau^{1+\gamma} + \frac{L_{\gamma,\beta}}{1+\beta}\tau^{1+\beta}, \quad \tau \in \mathbb{R}_+.$$

If $\gamma \in [\alpha, 1)$ and $\gamma \leq \beta < 1$, then one gets

$$\left\| \mathcal{A}^{-\gamma}(e^{-\tau \mathcal{K}_0} e^{-\tau \mathcal{B}} - e^{-\tau \mathcal{K}})\mathcal{A}^{-\gamma} \right\| \tag{3.18}$$
$$\leq \left( \frac{2C_\gamma^3}{(2-\gamma)(3-\gamma)}\tau^{2-2\gamma} + 2C_\gamma^2 \tau^{1-\gamma} + \frac{2C_\gamma}{(1+\gamma)\gamma} + \frac{L_{\gamma,\beta}}{1+\beta}\tau^{\beta-\gamma} \right)\tau^{1+\gamma},$$

$\tau \in \mathbb{R}_+$, which immediately yields (3.7).

If $\gamma \in [\alpha, 1)$ and $0 < \beta < \gamma$, then one can rewrite it as

$$\left\| \mathcal{A}^{-\gamma}(e^{-\tau\mathcal{K}_0}e^{-\tau\mathcal{B}} - e^{-\tau\mathcal{K}})\mathcal{A}^{-\gamma} \right\| \tag{3.19}$$
$$\leq \left( \frac{2C_\gamma^3}{(2-\gamma)(3-\gamma)}\tau^{2-\gamma-\beta} + 2C_\gamma^2\tau^{1-\beta} + \frac{2C_\gamma}{(1+\gamma)\gamma}\tau^{\gamma-\beta} + \frac{L_{\gamma,\beta}}{1+\beta} \right)\tau^{1+\beta},$$

$\tau \in \mathbb{R}_+$, which shows (3.7) for this choice of $\gamma$ and $\beta$. □

*Remark 3.6* For $\gamma \in [\alpha, 1)$ and $\gamma \leq \beta < 1$, we find from (3.18) that

$$Z_{\gamma,\beta} := \frac{2C_\gamma^3}{(2-\gamma)(3-\gamma)}T^{2-2\gamma} + 2C_\gamma^2 T^{1-\gamma} + \frac{2C_\gamma}{(1+\gamma)\gamma} + \frac{L_{\gamma,\beta}}{1+\beta}T^{\beta-\gamma} \, . \tag{3.20}$$

For $\gamma \in [\alpha, 1)$ and $0 < \beta < \gamma$, we get from (3.19) that

$$Z_{\gamma,\beta} := \frac{2C_\gamma^3}{(2-\gamma)(3-\gamma)}T^{2-\gamma-\beta} + 2C_\gamma^2 T^{1-\beta} + \frac{2C_\gamma}{(1+\gamma)\gamma}T^{\gamma-\beta} + \frac{L_{\gamma,\beta}}{1+\beta} \, .$$

Here, $C_\gamma := \operatorname{ess\,sup}_{t\in\mathcal{I}} \|\mathcal{B}\mathcal{A}^{-\gamma}\|$, see (3.3), and $L_{\gamma,\beta}$ is the Hölder constant of the function $A^{-\gamma}B(\cdot)A^{-\gamma} : \mathcal{I} \longrightarrow \mathcal{L}(\mathfrak{H})$, see (S3).

**Lemma 3.7** *Let the assumptions (S1) and (S2) be satisfied. Then*

$$\|\mathcal{A}^\gamma(\mathcal{U}(\tau) - \mathcal{T}(\tau))\mathcal{A}^{-\gamma}\| \leq \left(\tfrac{\Lambda_\gamma}{1-\gamma} + 1\right) C_\gamma \tau^{1-\gamma}, \quad \tau \in \mathbb{R}_+, \tag{3.21}$$

*for* $\gamma \in [\alpha, 1)$.

*Proof* We use the representation

$$\mathcal{U}(\tau) - \mathcal{T}(\tau) = e^{-\tau\mathcal{K}} - e^{-\tau\mathcal{K}_0} + e^{-\tau\mathcal{K}_0}(I - e^{-\tau\mathcal{B}})$$

which yields

$$\mathcal{A}^\gamma(\mathcal{U}(\tau) - \mathcal{T}(\tau))\mathcal{A}^{-\gamma} = \mathcal{A}^\gamma(e^{-\tau\mathcal{K}} - e^{-\tau\mathcal{K}_0})\mathcal{A}^{-\gamma} + \mathcal{A}^\gamma e^{-\tau\mathcal{K}_0}(I - e^{-\tau\mathcal{B}})\mathcal{A}^{-\gamma}$$

Using the semigroup property we obtain for the first term the representation:

$$\mathcal{A}^\gamma(e^{-\tau\mathcal{K}} - e^{-\tau\mathcal{K}_0})\mathcal{A}^{-\gamma} = -\int_0^\tau \mathcal{A}^\gamma e^{-(\tau-x)\mathcal{K}}\mathcal{B}\mathcal{A}^{-\gamma}e^{-x\mathcal{K}_0}dx \, .$$

Hence, by (3.3) and (3.6) one gets

$$\|\mathcal{A}^\gamma(e^{-\tau\mathcal{K}} - e^{-\tau\mathcal{K}_0})\mathcal{A}^{-\gamma}\| \le \int_0^\tau \|\mathcal{A}^\gamma e^{-(\tau-x)\mathcal{K}}\|\|\mathcal{B}\mathcal{A}^{-\gamma}\|dx \le$$

$$\Lambda_\gamma C_\gamma \int_0^\tau \frac{1}{(\tau-x)^\gamma}dx = \frac{\Lambda_\gamma C_\gamma}{1-\gamma}\tau^{1-\gamma}\,. \tag{3.22}$$

To estimate the second term we use the inequality

$$\|\mathcal{A}^\gamma e^{-\tau\mathcal{K}_0}(I - e^{-\tau\mathcal{B}})\mathcal{A}^{-\gamma}\| \le \|\mathcal{A}^\gamma e^{-\tau\mathcal{K}_0}\|\|(I - e^{-\tau\mathcal{B}})\mathcal{A}^{-\gamma}\|.$$

Using (3.2) and (3.13) we estimate the second term as

$$\|\mathcal{A}^\gamma e^{-\tau\mathcal{K}_0}(I - e^{-\tau\mathcal{B}})\mathcal{A}^{-\gamma}\| \le C_\gamma\, \tau^{1-\gamma}\,. \tag{3.23}$$

Now the estimates (3.22) and (3.23) yield (3.21). □

**Lemma 3.8** *Let the assumption (S1) be satisfied. If for each $\gamma \in [\alpha, 1)$ there is a constant $M_\gamma > 0$ such that*

$$\|\mathcal{A}^\gamma \mathcal{T}(\tau)^m\| \le \frac{M_\gamma}{(m\tau)^\gamma}, \quad m \in \mathbb{N}, \quad \tau \in \mathbb{R}_+, \tag{3.24}$$

*holds for $\mathcal{T}(\tau)$ defined in (3.4), then*

$$\|\mathcal{A}^\sigma \mathcal{T}(\tau)^m\| \le \frac{M_\gamma^\delta}{(m\tau)^\sigma}, \quad m \in \mathbb{N}, \tag{3.25}$$

*holds for $\sigma \in [0, \gamma]$ and $\delta := \sigma/\gamma$.*

*Proof* If (3.24) is satisfied, then

$$\|\overline{(\mathcal{T}(\tau)^*)^m \mathcal{A}^\gamma}\| \le \frac{M_\gamma}{(m\tau)^\gamma}, \quad m \in \mathbb{N},$$

holds, which is equivalent to

$$\mathcal{A}^\gamma \mathcal{T}(\tau)^m \overline{(\mathcal{T}(\tau)^*)^m \mathcal{A}^\gamma} \le \frac{M_\gamma^2}{(m\tau)^{2\gamma}}, \quad m \in \mathbb{N},$$

or

$$\mathcal{T}(\tau)^m (\mathcal{T}(\tau)^*)^m \le \frac{M_\gamma^2}{(m\tau)^{2\gamma}}\mathcal{A}^{-2\gamma}, \quad m \in \mathbb{N}.$$

Let $\delta \in (0, 1)$. Using the Heinz inequality [8, Theorem X.4.2] we get

$$\Big(\mathcal{T}(\tau)^m (\mathcal{T}(\tau)^*)^m\Big)^\delta \le \frac{M_\gamma^{2\delta}}{(m\tau)^{2\delta\gamma}}\mathcal{A}^{-2\delta\gamma}, \quad m \in \mathbb{N}.$$

Since $\mathcal{T}(\tau)^m(\mathcal{T}(\tau)^*)^m$ is a self-adjoint contraction we get

$$\mathcal{T}(\tau)^m(\mathcal{T}(\tau)^*)^m \leq \left(\mathcal{T}(\tau)^m(\mathcal{T}(\tau)^*)^m\right)^\delta, \quad m \in \mathbb{N},$$

which yields

$$\mathcal{T}(\tau)^m(\mathcal{T}(\tau)^*)^m \leq \frac{M_\gamma^{2\delta}}{(m\tau)^{2\delta\gamma}} \mathcal{A}^{-2\delta\gamma}, \quad m \in \mathbb{N},$$

or

$$\mathcal{A}^{\delta\gamma}\mathcal{T}(\tau)^m\overline{(\mathcal{T}(\tau)^*)^m\mathcal{A}^{\delta\gamma}} \leq \frac{M_\gamma^{2\delta}}{(m\tau)^{2\delta\gamma}}, \quad m \in \mathbb{N}.$$

Therefore, one gets

$$\left\|\overline{(\mathcal{T}(\tau)^*)^m\mathcal{A}^{\delta\gamma}}\right\| \leq \frac{M_\gamma^{\delta}}{(m\tau)^{\delta\gamma}}, \quad m \in \mathbb{N},$$

or

$$\left\|\mathcal{A}^{\delta\gamma}\mathcal{T}(\tau)^m\right\| \leq \frac{M_\gamma^{\delta}}{(m\tau)^{\delta\gamma}}, \quad m \in \mathbb{N}.$$

Setting $\delta = \sigma/\gamma$ we obtain the proof of (3.25). □

**Lemma 3.9** *Let the assumptions (S1) and (S2) be satisfied and let $\gamma \in (\alpha, 1)$. Then there is a constant $M_\gamma > 0$ such that*

$$\|\mathcal{A}^\gamma\mathcal{T}(\tau)^m\| \leq \frac{M_\gamma}{(m\tau)^\gamma}, \quad m = 1, 2, \ldots n, \tag{3.26}$$

*holds for any $T > 0$ if $\tau \in (0, \frac{T}{n})$ and $n \geq n_0$ where $n_0 := \lfloor (2(\frac{\Lambda_\gamma}{1-\gamma} + 1) C_\gamma)^{\frac{1}{1-\gamma}} T \rfloor + 1$ and $\lfloor x \rfloor$ denotes the largest integer smaller than $x$.*

*Proof* Let $M_\gamma > 0$ be a constant which satisfies the inequality

$$5\Lambda_\gamma + 2\left(\frac{\Lambda_\gamma}{1-\gamma} + 1\right) C_\gamma M_\gamma T^{1-\gamma} \frac{1}{n^{1-\gamma}} + 4\Lambda_\gamma C_\gamma M_\gamma^{\frac{\alpha}{\gamma}} B(1-\alpha, 1-\gamma)\, T^{1-\alpha} \leq M_\gamma \tag{3.27}$$

for $n \geq n_0$. Here, constants $\Lambda_\gamma$ and $C_\gamma$ are defined by Lemmas 3.2 and 3.4, respectively, while $B(\cdot, \cdot)$ denotes the Euler Beta-function. (Note that such $M_\gamma > 0$ always exists, see Remark 3.10 below.)

Let $m = 1$. Then by (3.2) and (3.4) we get

$$\|\mathcal{A}^\gamma\mathcal{T}(\tau)\| \leq \|\mathcal{A}^\gamma e^{-\tau\mathcal{K}_0}\| \leq \frac{1}{\tau^\gamma} \leq \frac{\Lambda_\gamma}{\tau^\gamma} \leq \frac{M_\gamma}{\tau^\gamma},$$

for $\tau > 0$ and, in particular, for $\tau \in (0, T/n)$. Hence (3.26) holds for $m = 1$.

Let us assume that (3.26) holds for $l = 1, 2, \dots, m-1$, with $m \le n$, i.e.,

$$\|\mathcal{A}^\gamma \mathcal{T}(\tau)^l\| \le \frac{M_\gamma}{(l\tau)^\gamma}, \quad l = 1, 2, \dots m-1, \tag{3.28}$$

for $\tau \in (0, T/n)$. We are going to show that (3.28) holds for $l = m$. To this aim, we use the representation

$$\mathcal{U}(\tau)^m - \mathcal{T}(\tau)^m = \sum_{k=0}^{m-1} \mathcal{U}(\tau)^{m-1-k}(\mathcal{U}(\tau) - \mathcal{T}(\tau))\mathcal{T}(\tau)^k, \quad m = 2, 3, \dots,$$

which implies

$$\mathcal{T}(\tau)^m = \mathcal{U}(\tau)^m - \sum_{k=0}^{m-1} \mathcal{U}(\tau)^{m-1-k}(\mathcal{U}(\tau) - \mathcal{T}(\tau))\mathcal{T}(\tau)^k, \quad m = 2, 3, \dots.$$

Hence

$$\mathcal{A}^\gamma \mathcal{T}(\tau)^m = \mathcal{A}^\gamma \mathcal{U}(\tau)^m - \sum_{k=0}^{m-1} \mathcal{A}^\gamma \mathcal{U}(\tau)^{m-1-k}(\mathcal{U}(\tau) - \mathcal{T}(\tau))\mathcal{T}(\tau)^k$$

or

$$\begin{aligned}\mathcal{A}^\gamma \mathcal{T}(\tau)^m &= \mathcal{A}^\gamma \mathcal{U}(\tau)^m - \mathcal{A}^\gamma \mathcal{U}(\tau)^{m-1}(\mathcal{U}(\tau) - \mathcal{T}(\tau)) \\ &\quad - \mathcal{A}^\gamma (\mathcal{U}(\tau) - \mathcal{T}(\tau))\mathcal{T}(\tau)^{m-1} - \sum_{k=1}^{m-2} \mathcal{A}^\gamma \mathcal{U}(\tau)^{m-1-k}(\mathcal{U}(\tau) - \mathcal{T}(\tau))\mathcal{T}(\tau)^k.\end{aligned}$$

for $m = 3, 4, \dots$. This yields the inequality

$$\begin{aligned}&\|\mathcal{A}^\gamma \mathcal{T}(\tau)^m\| \le \|\mathcal{A}^\gamma \mathcal{U}(\tau)^m\| + \|\mathcal{A}^\gamma \mathcal{U}(\tau)^{m-1}(\mathcal{U}(\tau) - \mathcal{T}(\tau))\| + \\ &\|\mathcal{A}^\gamma (\mathcal{U}(\tau) - \mathcal{T}(\tau))\mathcal{T}(\tau)^{m-1}\| + \sum_{k=1}^{m-2} \|\mathcal{A}^\gamma \mathcal{U}(\tau)^{m-1-k}(\mathcal{U}(\tau) - \mathcal{T}(\tau))\mathcal{T}(\tau)^k\|\end{aligned} \tag{3.29}$$

for $m = 3, 4, \dots$. From Lemma 3.2, we get the estimates

$$\|\mathcal{A}^\gamma \mathcal{U}(\tau)^m\| \le \frac{\Lambda_\gamma}{(m\tau)^\gamma}, \quad m = 2, 3, \dots,$$

and consequently

$$\|\mathcal{A}^\gamma \mathcal{U}(\tau)^{m-1}(\mathcal{U}(\tau) - \mathcal{T}(\tau))\| \le \frac{2\Lambda_\gamma}{((m-1)\tau)^\gamma} \le \frac{4\Lambda_\gamma}{(m\tau)^\gamma}, \quad m = 2, 3, \dots.$$

Then, summing up estimates for the first two terms in the right-hand side of (3.29), we obtain

$$\|\mathcal{A}^\gamma \mathcal{U}(\tau)^m\| + \|\mathcal{A}^\gamma \mathcal{U}(\tau)^{m-1}(\mathcal{U}(\tau) - \mathcal{T}(\tau))\| \le \frac{5\Lambda_\gamma}{(m\tau)^\gamma}, \quad m = 2, 3, \dots . \tag{3.30}$$

Next, we get for the third term in the right-hand side of (3.29), the estimate

$$\|\mathcal{A}^\gamma (\mathcal{U}(\tau) - \mathcal{T}(\tau))\mathcal{T}(\tau)^{m-1}\| \le \|\mathcal{A}^\gamma (\mathcal{U}(\tau) - \mathcal{T}(\tau))\mathcal{A}^{-\gamma}\| \|\mathcal{A}^\gamma \mathcal{T}(\tau)^{m-1}\|,$$

$m = 2, 3, \dots$. Then using Lemma 3.7 we find that

$$\|\mathcal{A}^\gamma (\mathcal{U}(\tau) - \mathcal{T}(\tau))\mathcal{T}(\tau)^{m-1}\| \le \left(\frac{\Lambda_\gamma}{1-\gamma} + 1\right) C_\gamma \tau^{1-\gamma} \|\mathcal{A}^\gamma \mathcal{T}(\tau)^{m-1}\|, \quad m = 2, 3, \dots .$$

By assumption (3.28) this yields

$$\|\mathcal{A}^\gamma (\mathcal{U}(\tau) - \mathcal{T}(\tau))\mathcal{T}(\tau)^{m-1}\| \le \left(\frac{\Lambda_\gamma}{1-\gamma} + 1\right) M_\gamma C_\gamma \frac{1}{((m-1)\tau)^\gamma} \tau^{1-\gamma}, \quad m = 2, 3, \dots ,$$

for $\tau \in (0, T/n)$, which leads to

$$\|\mathcal{A}^\gamma (\mathcal{U}(\tau) - \mathcal{T}(\tau))\mathcal{T}(\tau)^{m-1}\| \le \left(\frac{\Lambda_\gamma}{1-\gamma} + 1\right) M_\gamma C_\gamma \frac{2}{(m\tau)^\gamma} \tau^{1-\gamma}, \quad m = 2, 3, \dots . \tag{3.31}$$

Finally, one gets for the sum in (3.29)

$$\sum_{k=1}^{m-2} \|\mathcal{A}^\gamma \mathcal{U}(\tau)^{m-1-k}(\mathcal{U}(\tau) - \mathcal{T}(\tau))\mathcal{T}(\tau)^k\|$$

$$\le \sum_{k=1}^{m-2} \|\mathcal{A}^\gamma \mathcal{U}(\tau)^{m-1-k}\| \|(\mathcal{U}(\tau) - \mathcal{T}(\tau))\mathcal{A}^{-\alpha}\| \|\mathcal{A}^\alpha \mathcal{T}(\tau)^k\|, \quad m = 2, 3, \dots .$$

Then by Lemma 3.2 this implies

$$\sum_{k=1}^{m-2} \|\mathcal{A}^\gamma \mathcal{U}(\tau)^{m-1-k}(\mathcal{U}(\tau) - \mathcal{T}(\tau))\mathcal{T}(\tau)^k\|$$

$$\le \Lambda_\gamma \sum_{k=1}^{m-2} \frac{1}{((m-1-k)\tau)^\gamma} \|(\mathcal{U}(\tau) - \mathcal{T}(\tau))\mathcal{A}^{-\alpha}\| \|\mathcal{A}^\alpha \mathcal{T}(\tau)^k\|, \quad m = 2, 3, \dots .$$

Taking into account Lemma 3.4 we get

$$\sum_{k=1}^{m-2} \|\mathcal{A}^{\gamma}\mathcal{U}(\tau)^{m-1-k}(\mathcal{U}(\tau) - \mathcal{T}(\tau))\mathcal{T}(\tau)^k\|$$
$$\leq 2\Lambda_\gamma C_\alpha \sum_{k=1}^{m-2} \frac{\tau}{((m-1-k)\tau)^{\gamma}} \|\mathcal{A}^{\alpha}\mathcal{T}(\tau)^k\|, \quad m = 2, 3, \dots .$$

Finally, using assumption (3.28) and Lemma 3.8 one obtains

$$\sum_{k=1}^{m-2} \|\mathcal{A}^{\gamma}\mathcal{U}(\tau)^{m-1-k}(\mathcal{U}(\tau) - \mathcal{T}(\tau))\mathcal{T}(\tau)^k\|$$
$$\leq 2\Lambda_\gamma C_\alpha M_\gamma^{\frac{\alpha}{\gamma}} \sum_{k=1}^{m-2} \frac{\tau}{((m-1-k)\tau)^{\gamma}} \frac{1}{(k\tau)^{\alpha}}, \quad m = 2, 3, \dots ,$$

or

$$\sum_{k=1}^{m-2} \|\mathcal{A}^{\gamma}\mathcal{U}(\tau)^{m-1-k}(\mathcal{U}(\tau) - \mathcal{T}(\tau))\mathcal{T}(\tau)^k\|$$
$$\leq 2\Lambda_\gamma C_\gamma M_\gamma^{\frac{\alpha}{\gamma}} \left(\sum_{k=1}^{m-2} \frac{1}{(m-1-k)^{\gamma}} \frac{1}{k^{\alpha}}\right) \tau^{1-\gamma-\alpha}, \quad m = 2, 3, \dots ,$$

for $\tau \in (0, T/n)$. Since Lemma 3.11 below yields

$$\sum_{k=1}^{m-2} \frac{1}{(m-1-k)^{\gamma}} \frac{1}{k^{\alpha}} \leq B(1-\alpha, 1-\gamma)(m-1)^{1-\gamma-\alpha}, \quad m = 2, 3, \dots , \tag{3.32}$$

where $B(\cdot, \cdot)$ is the Euler Beta-function, we get

$$\sum_{k=1}^{m-2} \|\mathcal{A}^{\gamma}\mathcal{U}(\tau)^{m-1-k}(\mathcal{U}(\tau) - \mathcal{T}(\tau))\mathcal{T}(\tau)^k\|$$
$$\leq 2\Lambda_\gamma C_\gamma M_\gamma^{\frac{\alpha}{\gamma}} B(1-\alpha, 1-\gamma)\tau^{1-\gamma-\alpha}(m-1)^{1-\gamma-\alpha}, \quad m = 2, 3, \dots ,$$

which in turn leads to

$$\begin{aligned} &\sum_{k=1}^{m-2} \|\mathcal{A}^{\gamma}\mathcal{U}(\tau)^{m-1-k}(\mathcal{U}(\tau) - \mathcal{T}(\tau))\mathcal{T}(\tau)^k\| \\ &\qquad \leq \frac{4\Lambda_\gamma C_\gamma M_\gamma^{\frac{\alpha}{\gamma}} B(1-\alpha, 1-\gamma)}{(m\tau)^{\gamma}} \tau^{1-\alpha} m^{1-\alpha}, \end{aligned} \tag{3.33}$$

for $m = 2, 3, \dots$ and any $\tau \in (0, T/n)$.

Now we take into account (3.29)–(3.31) and (3.33) to conclude that

$$\|\mathcal{A}^{\gamma}\mathcal{T}(\tau)^m\| \leq$$
$$\left\{5\Lambda_\gamma + 2\left(\frac{\Lambda_\gamma}{1-\gamma}+1\right)M_\gamma C_\gamma \tau^{1-\gamma} + 4\Lambda_\gamma C_\gamma M_\gamma^{\frac{\alpha}{\gamma}}\, B(1-\alpha, 1-\gamma)\tau^{1-\alpha}m^{1-\alpha}\right\}\frac{1}{(m\tau)^\gamma},$$

for $m = 2, 3, \ldots$ and $\tau \in (0, T/n)$. Then

$$\|\mathcal{A}^{\gamma}\mathcal{T}(\tau)^m\| \leq$$
$$\left\{5\Lambda_\gamma + 2\left(\frac{\Lambda_\gamma}{1-\gamma}+1\right)M_\gamma C_\gamma T^{1-\gamma}\frac{1}{n^{1-\gamma}} + 4\Lambda_\gamma C_\gamma M_\gamma^{\frac{\alpha}{\gamma}}\, B(1-\alpha, 1-\gamma)\, T^{1-\alpha}\right\}\frac{1}{(m\tau)^\gamma}.$$

From assumption (3.27) we get

$$5\Lambda_\gamma + 2\left(\frac{\Lambda_\gamma}{1-\gamma}+1\right)M_\gamma C_\gamma T^{1-\gamma}\frac{1}{n^{1-\gamma}} + 4\Lambda_\gamma C_\gamma M_\gamma^{\frac{\alpha}{\gamma}}\, B(1-\alpha, 1-\gamma)\, T^{1-\alpha} \leq M_\gamma$$

for $n \geq n_0$, which shows that (3.28) holds for $l = 1, 2, 3, \ldots, n$ and $n \geq n_0$ which proves (3.26). □

*Remark 3.10* One checks that condition (3.27) is always satisfied for sufficiently large $M = M_\gamma$ and $n \geq n_0$. Indeed, after setting

$$c_0 := 5\Lambda_\gamma, \quad c_1 := 2\left(\frac{\Lambda_\gamma}{1-\gamma}+1\right)C_\gamma T^{1-\gamma}, \quad c_2 := 4\Lambda_\gamma C_\gamma B(1-\alpha, 1-\gamma)\, T^{1-\alpha}$$

we get the condition

$$c_0 + \frac{c_1}{n^{1-\gamma}}M + c_2 M^{\frac{\alpha}{\gamma}} \leq M$$

which yields

$$c_0 + c_2 M^{\frac{\alpha}{\gamma}} \leq (1 - \frac{c_1}{n^{1-\gamma}})M$$

or

$$\frac{c_0}{M} + \frac{c_2}{M^{1-\frac{\alpha}{\gamma}}} \leq 1 - \frac{c_1}{n^{1-\gamma}}$$

Since $n > c_1^{\frac{1}{1-\gamma}}$ we have $1 - c_1/n^{1-\gamma} > 0$. The left-hand side tends to zero if $M \to \infty$. Hence, choosing $M$ sufficiently large we guarantee the existence of $M_\gamma$ such that condition (3.27) is satisfied for any $n \geq n_0$. △

It remains only to verify the following statement.

**Lemma 3.11** *Let $\alpha \in [0, 1)$ and $\gamma \in [\alpha, 1)$. Then*

$$\sum_{k=1}^{n-1}\frac{1}{(n-k)^\gamma}\frac{1}{k^\alpha} \leq B(1-\alpha, 1-\gamma)n^{1-\gamma-\alpha}, \quad n \in 2, 3, \ldots.$$

*the estimate holds where $B(\cdot,\cdot)$ is the Euler Beta-function.*

$$B(1-\alpha, 1-\gamma) := \int_0^1 \frac{1}{x^\alpha(1-x)^\gamma}dx$$

*Proof* If $x \in (k-1, k]$, then

$$\frac{1}{k^\alpha} \le \frac{1}{x^\alpha} \quad \text{and} \quad \frac{1}{(n-k)^\gamma} \le \frac{1}{(n-1-x)^\gamma}$$

for $k = 1, 2, \ldots, n-1$. Hence

$$\frac{1}{(n-k)^\gamma k^\alpha} \le \frac{1}{(n-1-x)^\gamma x^\alpha}, \quad x \in (k-1, k].$$

Therefore

$$\frac{1}{(n-k)^\gamma k^\alpha} = \int_{k-1}^k \frac{1}{(n-k)^\gamma k^\alpha}dx \le \int_{k-1}^k \frac{1}{(n-1-x)^\gamma x^\alpha}dx, \quad x \in (k-1, k],$$

or

$$\begin{aligned}\sum_{k=1}^{n-1} \frac{1}{(n-k)^\gamma k^\alpha} &= \sum_{k=1}^{n-1} \int_{k-1}^k \frac{1}{(n-k)^\gamma k^\alpha}dx \le \sum_{k=1}^{n-1} \int_{k-1}^k \frac{1}{(n-1-x)^\gamma x^\alpha}dx \\ &= \int_0^{n-1} \frac{1}{(n-1-x)^\gamma x^\alpha}dx = B(1-\alpha, 1-\gamma)n^{1-\alpha-\gamma}.\end{aligned}$$

□

## 3.2 Main Results

In this section, we collect our main results and their proofs. They are based on preliminaries established in Section 3.1.

**Theorem 3.12** *Let the assumptions (S1)–(S3) be satisfied and let $\beta > 2\alpha - 1$. Then there is a constant $R_\beta > 0$ such that*

$$\sup_{\tau \in \mathbb{R}_+} \|\mathcal{U}(\tau) - \mathcal{T}(\tau/n)^n\| \le \frac{R_\beta}{n^\beta} \tag{3.34}$$

*holds for $n \in \mathbb{N}$ and $\tau \in \mathbb{R}_+$.*

*Proof* Taking into account the representation

$$\mathcal{U}(\tau/n)^n - \mathcal{T}(\tau/n)^n = \sum_{m=0}^{n-1} \mathcal{U}(\tau/n)^{n-m-1}(\mathcal{U}(\tau/n) - \mathcal{T}(\tau/n))\mathcal{T}(\tau/n)^m, \quad n \in \mathbb{N},$$

or, identically,

$$\begin{aligned}
&\mathcal{U}(\tau/n)^n - \mathcal{T}(\tau/n)^n \\
&\quad = \mathcal{U}(\tau/n)^{n-1}(\mathcal{U}(\tau/n) - \mathcal{T}(\tau/n)) + (\mathcal{U}(\tau/n) - \mathcal{T}(\tau/n))\mathcal{T}(\tau/n)^{n-1} + \\
&\quad \sum_{m=1}^{n-2} \mathcal{U}(\tau/n)^{n-m-1}(\mathcal{U}(\tau/n) - \mathcal{T}(\tau/n))\mathcal{T}(\tau/n)^m, \quad n = 3, 4, \ldots,
\end{aligned}$$

we obtain the estimate

$$\begin{aligned}
&\|\mathcal{U}(\tau/n)^n - \mathcal{T}(\tau/n)^n\| \\
&\le \|\mathcal{U}(\tau/n)^{n-1}\mathcal{A}^\gamma\| \|\mathcal{A}^{-\gamma}(\mathcal{U}(\tau/n) - \mathcal{T}(\tau/n))\| \\
&\quad + \|(\mathcal{U}(\tau/n) - \mathcal{T}(\tau/n))\mathcal{A}^{-\gamma}\| \|\mathcal{A}^\gamma \mathcal{T}(\tau/n)^{n-1}\| \qquad (3.35) \\
&\quad + \sum_{m=1}^{n-2} \|\mathcal{U}(\tau/n)^{n-m-1}\mathcal{A}^\gamma\| \|\mathcal{A}^{-\gamma}(\mathcal{U}(\tau/n) - \mathcal{T}(\tau/n))\mathcal{A}^{-\gamma}\| \|\mathcal{A}^\gamma \mathcal{T}(\tau/n)^m\|,
\end{aligned}$$

$n = 3, 4, \ldots$.

Note that using Lemmas 3.2 and 3.4 one gets

$$\|\mathcal{U}(\tau/n)^{n-1}\mathcal{A}^\gamma\| \|\mathcal{A}^{-\gamma}(\mathcal{U}(\tau/n) - \mathcal{T}(\tau/n))\| \le 2 \frac{\Lambda_\gamma C_\gamma}{(\tau(n-1)/n)^\gamma} \frac{\tau}{n},$$

which yields

$$\|\mathcal{U}(\tau/n)^{n-1}\mathcal{A}^\gamma\| \|\mathcal{A}^{-\gamma}(\mathcal{U}(\tau/n) - \mathcal{T}(\tau/n))\| \le 2^{1+\gamma} \Lambda_\gamma C_\gamma T^{1-\gamma} \frac{1}{n}. \qquad (3.36)$$

for $n = 3, 4, \ldots$ and $\tau \in [0, T]$.

Now using Lemmas 3.4 and 3.9 for $m = n - 1$ we find

$$\|(\mathcal{U}(\tau/n) - \mathcal{T}(\tau/n))\mathcal{A}^{-\gamma}\| \|\mathcal{A}^\gamma \mathcal{T}(\tau/n)^{n-1}\| \le 2\, C_\gamma \frac{\tau}{n} \frac{M_\gamma}{(\tau(n-1)/n)^\gamma},$$

for $n \ge n_0$, where $n_0$ is defined in Lemma 3.9 and $\tau \in [0, T]$. Hence,

$$\|(\mathcal{U}(\tau/n) - \mathcal{T}(\tau/n))\mathcal{A}^{-\gamma}\| \|\mathcal{A}^\gamma \mathcal{T}(\tau/n)^{n-1}\| \le 2^{1+\gamma}\, C_\gamma M_\gamma\, T^{1-\gamma} \frac{1}{n}. \qquad (3.37)$$

Taking into account Lemmas 3.2, 3.5, and 3.9 (for $\varkappa = \min\{\gamma, \beta\}$) one gets

$$\sum_{m=1}^{n-2} \|\mathcal{U}(\tau/n)^{n-m-1}\mathcal{A}^{\gamma}\|\|\mathcal{A}^{-\gamma}(\mathcal{U}(\tau/n)-\mathcal{T}(\tau/n))\mathcal{A}^{-\gamma}\|\|\mathcal{A}^{\gamma}\mathcal{T}(\tau/n)^m\|$$
$$\leq \sum_{m=1}^{n-2} \frac{\Lambda_\gamma\, Z_{\gamma,\beta}}{((n-m-1)\,\tau/n)^\gamma}\left(\frac{\tau}{n}\right)^{1+\varkappa}\frac{M_\gamma}{(m\,\tau/n)^\gamma}$$
$$= \frac{\Lambda_\gamma Z_{\gamma,\beta} M_\gamma \tau^{1+\varkappa-2\gamma}}{n^{1+\varkappa-2\gamma}}\sum_{m=1}^{n-2}\frac{1}{(n-m-1)^\gamma}\frac{1}{m^\gamma},$$

for $n > \max\{2, n_0\}$ and $\tau \in [0, T]$. Then by (3.32) we obtain

$$\sum_{m=1}^{n-2} \|\mathcal{U}(\tau/n)^{n-m-1}\mathcal{A}^{\gamma}\|\|\mathcal{A}^{-\gamma}(\mathcal{U}(\tau/n)-\mathcal{T}(\tau/n))\mathcal{A}^{-\gamma}\|\|\mathcal{A}^{\gamma}\mathcal{T}(\tau/n)^m\|$$
$$\leq \frac{\Lambda_\gamma Z_{\gamma,\beta} M_\gamma \tau^{1+\varkappa-2\gamma}}{n^{1+\varkappa-2\gamma}}\, B(1-\gamma, 1-\gamma)\, n^{1-2\gamma},$$

or

$$\sum_{m=1}^{n-2} \|\mathcal{U}(\tau/n)^{n-m-1}\mathcal{A}^{\gamma}\|\|\mathcal{A}^{-\gamma}(\mathcal{U}(\tau/n)-\mathcal{T}(\tau/n))\mathcal{A}^{-\gamma}\|\|\mathcal{A}^{\gamma}\mathcal{T}(\tau/n)^m\| \tag{3.38}$$
$$\leq \Lambda_\gamma Z_{\gamma,\beta} M_\gamma B(1-\gamma, 1-\gamma)\, T^{1+\varkappa-2\gamma}\,\frac{1}{n^\varkappa}.$$

Therefore, by virtue of (3.35)–(3.38) we get for $n > \max\{2, n_0\}$ and $\tau \in [0, T]$ the estimate

$$\|\mathcal{U}(\tau) - \mathcal{T}(\tau/n)^n\| = \|\mathcal{U}(\tau/n)^n - \mathcal{T}(\tau/n)^n\|$$
$$\leq 2^{1+\gamma}\Lambda_\gamma C_\gamma T^{1-\gamma}\frac{1}{n} + 2^{1+\gamma} C_\gamma M_\gamma T^{1-\gamma}\frac{1}{n} + \Lambda_\gamma Z_{\gamma,\beta} M_\gamma B(1-\gamma, 1-\gamma) T^{1+\varkappa-2\gamma}\frac{1}{n^\varkappa}$$
$$\leq \left\{2^{1+\gamma}\Lambda_\gamma C_\gamma T^{1-\gamma} + 2^{1+\gamma} C_\gamma M_\gamma T^{1-\gamma} + \Lambda_\gamma Z_{\gamma,\beta} M_\gamma B(1-\gamma, 1-\gamma) T^{1+\varkappa-2\gamma}\right\}\frac{1}{n^\varkappa}.$$

If $\alpha < \beta < 1$, then we choose $\gamma = \beta$, i.e., $\varkappa = \beta$ and $1 + \varkappa - 2\gamma = 1 - \beta \geq 0$. Setting

$$R'_\beta := 2^{1+\beta}\Lambda_\beta C_\beta T^{1-\beta} + 2^{1+\beta} C_\beta M_\beta T^{1-\beta} + \Lambda_\beta Z_{\beta,\beta} M_\beta B(1-\beta, 1-\beta) T^{1-\beta}$$

one obtains the estimate

$$\|\mathcal{U}(\tau)^n - \mathcal{T}(\tau/n)^n\| \leq \frac{R'_\beta}{n^\beta}, \tag{3.39}$$

for $n > \max\{2, n_0\}$ and $\tau \in [0, T]$.

Now let $0 < \beta \le \alpha$. Since $1 + \beta - 2\alpha > 0$, there exists $\gamma \in (\alpha, 1)$ such that $1 + \beta - 2\gamma \ge 0$. Indeed, there is a $\varepsilon > 0$ verifying $1 + \beta - 2\alpha > 2\varepsilon$. Setting $\gamma = \alpha + \varepsilon$ we get $1 + \beta - 2\gamma > 0$. Notice that $\varkappa = \beta$. Then setting

$$R'_\beta := 2^{1+\gamma} \Lambda_\gamma C_\gamma T^{1-\gamma} + 2^{1+\gamma} C_\gamma M_\gamma T^{1-\gamma} + \Lambda_\gamma Z_{\gamma,\beta} M_\gamma B(1-\gamma, 1-\gamma) T^{1+\beta-2\gamma},$$

we obtain (3.39) for $n > \max\{2, n_0\}$.

Both results immediately imply that there is a constant $R_\gamma$ such that (3.34) holds for $\tau \in [0, T]$ and $n \in \mathbb{N}$. Finally, using $\mathcal{U}(\tau) = 0$ and $\mathcal{T}(\tau/n)^n = 0$ for $\tau \ge T$ we obtain (3.28) for any $\tau \in \mathbb{R}_+$. □

Now we set

$$\widetilde{\mathcal{T}}(\tau) := e^{-\tau\mathcal{B}} e^{-\tau\mathcal{K}_0}, \quad \tau \in \mathbb{R}_+.$$

**Corollary 3.13** *Let the assumptions (S1)–(S3) be satisfied and $\beta > 2\alpha - 1$. Then there exists $\widetilde{R}_\beta > 0$ such that estimate*

$$\sup_{\tau \in \mathbb{R}_+} \|\mathcal{U}(\tau) - \widetilde{\mathcal{T}}(\tau/n)^n\| \le \frac{\widetilde{R}_\beta}{n^\beta} \tag{3.40}$$

*holds for $n \in \mathbb{N}$ and $\tau \in \mathbb{R}_+$.*

*Proof* Notice that

$$\widetilde{\mathcal{T}}(\tau/n)^{n+1} = e^{-\tau\mathcal{B}/n} \mathcal{T}(\tau/n)^n e^{-\tau\mathcal{K}_0/n}, \quad \tau \in \mathbb{R}_+, \quad n \in \mathbb{N}.$$

Hence

$$\begin{aligned}\mathcal{U}((n+1)\tau/n) - \widetilde{\mathcal{T}}(\tau/n)^{n+1} &= e^{-(n+1)\tau\mathcal{K}/n} - e^{-\tau\mathcal{B}/n}\mathcal{T}(\tau/n)^n e^{-\tau\mathcal{K}_0/n}\\ &= e^{-(n+1)\tau\mathcal{K}/n} - e^{-\tau\mathcal{B}/n} e^{-\tau\mathcal{K}} e^{-\tau\mathcal{K}_0/n} + e^{-\tau\mathcal{B}/n}(\mathcal{U}(\tau) - \mathcal{T}(\tau/n)^n) e^{-\tau\mathcal{K}_0/n}\\ &= (I - e^{-\tau\mathcal{B}/n}) e^{-\tau\mathcal{K}} e^{-\tau\mathcal{K}_0/n} + e^{-\tau\mathcal{K}}(e^{-\tau\mathcal{K}/n} - e^{-\tau\mathcal{K}_0/n}) +\\ &\quad e^{-\tau\mathcal{B}/n}(\mathcal{U}(\tau) - \mathcal{T}(\tau/n)^n) e^{-\tau\mathcal{K}_0/n}, \quad \tau \in \mathbb{R}_+, \quad n \in \mathbb{N},\end{aligned}$$

which yields the estimate

$$\begin{aligned}&\|\mathcal{U}((n+1)\tfrac{\tau}{n}) - \widetilde{\mathcal{T}}(\tfrac{\tau}{n})^{n+1}\|\\ &\quad \le \|(I - e^{-\frac{\tau}{n}\mathcal{B}}) e^{-\tau\mathcal{K}}\| + \|e^{-\tau\mathcal{K}}(e^{-\frac{\tau}{n}\mathcal{K}} - e^{-\frac{\tau}{n}\mathcal{K}_0})\| +\\ &\quad \|\mathcal{U}(\tau) - \mathcal{T}(\tfrac{\tau}{n})^n\|, \quad \tau \in \mathbb{R}_+, \quad n \in \mathbb{N}.\end{aligned} \tag{3.41}$$

Obviously, one has

$$\|(I - e^{-\frac{\tau}{n}\mathcal{B}}) e^{-\tau\mathcal{K}}\| \le \|(I - e^{-\frac{\tau}{n}\mathcal{B}}) \mathcal{A}^{-\alpha}\| \|\mathcal{A}^\alpha e^{-\tau\mathcal{K}}\|, \quad \tau \in \mathbb{R}_+, \quad n \in \mathbb{N}.$$

Using

$$(I - e^{-\frac{\tau}{n}\mathcal{B}})\mathcal{A}^{-\alpha} = \int_0^{\frac{\tau}{n}} e^{-\sigma\mathcal{B}}\mathcal{B}\mathcal{A}^{-\alpha}d\sigma, \quad \tau \in \mathbb{R}_+, \quad n \in \mathbb{N},$$

we get the estimate

$$\|(I - e^{-\frac{\tau}{n}\mathcal{B}})\mathcal{A}^{-\alpha}\| \leq C_\alpha \frac{\tau}{n}, \quad \tau \in \mathbb{R}_+, \quad n \in \mathbb{N}.$$

Taking into account condition (S2) and Lemma 3.2 we find

$$\|(I - e^{-\frac{\tau}{n}\mathcal{B}})e^{-\tau\mathcal{K}}\| \leq C_\alpha \Lambda_\alpha \frac{\tau^{1-\alpha}}{n} \leq C_\alpha \Lambda_\alpha T^{1-\alpha}\frac{1}{n}, \quad \tau \in \mathbb{R}_+, \quad n \in \mathbb{N}, \tag{3.42}$$

where we have used that $e^{-\tau\mathcal{K}} = 0$ for $\tau \geq T$.

Further, we have

$$\|e^{-\tau\mathcal{K}}(e^{-\frac{\tau}{n}\mathcal{K}} - e^{-\frac{\tau}{n}\mathcal{K}_0})\| \leq \|e^{-\tau\mathcal{K}}\mathcal{A}^\alpha\| \, \|\mathcal{A}^{-\alpha}(e^{-\frac{\tau}{n}\mathcal{K}} - e^{-\frac{\tau}{n}\mathcal{K}_0})\|,$$

$\tau \in \mathbb{R}_+$, $n \in \mathbb{N}$. Then using

$$\mathcal{A}^{-\alpha}(e^{-\frac{\tau}{n}\mathcal{K}} - e^{-\frac{\tau}{n}\mathcal{K}_0}) = -\int_0^{\frac{\tau}{n}} e^{-\sigma\mathcal{K}_0}\overline{\mathcal{A}^{-\alpha}\mathcal{B}}e^{-(\tau-\sigma)\mathcal{K}}d\sigma,$$

$\tau \in \mathbb{R}_+$, $n \in \mathbb{N}$, we find the estimate

$$\|\mathcal{A}^{-\alpha}(e^{-\frac{\tau}{n}\mathcal{K}} - e^{-\frac{\tau}{n}\mathcal{K}_0})\| \leq C_\alpha \frac{\tau}{n}, \quad \tau \in \mathbb{R}_+, \quad n \in \mathbb{N}.$$

Applying again Lemma 3.2 one gets

$$\|e^{-\tau\mathcal{K}}(e^{-\frac{\tau}{n}\mathcal{K}} - e^{-\frac{\tau}{n}\mathcal{K}_0})\| \leq C_\alpha \Lambda_\alpha T^{1-\alpha}\frac{1}{n}, \quad \tau \in \mathbb{R}_+, \quad n \in \mathbb{N}. \tag{3.43}$$

The insertion of (3.42) and (3.43) into (3.41) yields

$$\|\mathcal{U}((n+1)\tfrac{\tau}{n}) - \widetilde{\mathcal{T}}(\tfrac{\tau}{n})^{n+1}\| \leq 2C_\alpha\Lambda_\alpha\frac{1}{n} + \|\mathcal{U}(\tau) - \mathcal{T}(\tfrac{\tau}{n})^n)\|, \quad \tau \in \mathbb{R}_+, \quad n \in \mathbb{N}.$$

Then by Theorem 3.12 we obtain

$$\|\mathcal{U}((n+1)\tfrac{\tau}{n}) - \widetilde{\mathcal{T}}(\tfrac{\tau}{n})^{n+1}\| \leq 2C_\alpha\Lambda_\alpha\frac{1}{n} + R_\gamma\frac{1}{n^\gamma}, \quad \tau \in \mathbb{R}_+, \quad n \in \mathbb{N}.$$

Therefore, by setting $R'_\gamma := 2C_\alpha\Lambda_\alpha + R_\gamma$ we obtain

$$\|\mathcal{U}((n+1)\tfrac{\tau}{n}) - \widetilde{\mathcal{T}}(\tfrac{\tau}{n})^{n+1}\| \le \frac{R'_\gamma}{n^\gamma}, \quad \tau \in \mathbb{R}_+, \quad n \in \mathbb{N}.$$

which yields

$$\sup_{\tau\in\mathbb{R}_+} \|\mathcal{U}((n+1)\tfrac{\tau}{n}) - \widetilde{\mathcal{T}}(\tfrac{\tau}{n})^{n+1}\| \le \frac{R'_\gamma}{n^\gamma}, \quad \tau \in \mathbb{R}_+, \quad n \in \mathbb{N}.$$

Let $\tau = \tau' n/(n+1)$ for $\tau' \in \mathbb{R}_+$. Then

$$\sup_{\tau\in\mathbb{R}_+} \|\mathcal{U}((n+1)\tfrac{\tau}{n}) - \widetilde{\mathcal{T}}(\tfrac{\tau}{n})^{n+1}\| = \sup_{\tau'\in\mathbb{R}_+} \|\mathcal{U}(\tau') - \widetilde{\mathcal{T}}(\tfrac{\tau'}{n+1})^{n+1}\| \le \frac{R'_\gamma}{n^\gamma},$$

or

$$\sup_{\tau'\in\mathbb{R}_+} \|\mathcal{U}(\tau') - \widetilde{\mathcal{T}}(\tfrac{\tau'}{n+1})^{n+1}\| \le 2^\gamma \frac{R'_\gamma}{(n+1)^\gamma},$$

$\tau \in \mathbb{R}_+$, $n \in \mathbb{N}$. Setting $\widetilde{R}_\gamma := \max\{2, 2^\gamma R'_\gamma\}$ we prove (3.40). □

These results can be immediately extended to propagators. To this end we set

$$\begin{aligned} \widetilde{G}_j(t,s;n) :=& e^{-\frac{t-s}{n}B(t_j)} e^{-\frac{t-s}{n}A}, \quad j = 0,1,2,\ldots,n, \\ \widetilde{V}_n(t,s) :=& \widetilde{G}_n(t,s;n)\widetilde{G}_{n-1}(t,s;n) \times \cdots \times \widetilde{G}_2(t,s;n)\widetilde{G}_1(t,s;n), \end{aligned} \tag{3.44}$$

$t_j := s + j\frac{t-s}{n}$, $j = 0,1,2,\ldots,n$, in analogy to (1.6).

**Theorem 3.14** *Let the assumptions (S1)–(S3) be satisfied. Further, let* $\{U(t,s)\}_{(t,s)\in\Delta_0}$ *be the propagator corresponding to the evolution generator* $\mathcal{K}$ *and let* $\{V_n(t,s)\}_{(t,s)\in\Delta_0}$ *and* $\{\widetilde{V}_n(t,s)\}_{(t,s)\in\Delta_0}$ *be defined by* (1.6) *and* (3.44)*, respectively. If* $\beta > 2\alpha - 1$*, then the estimates*

$$\operatorname*{ess\,sup}_{(t,s)\in\Delta_0} \|U(t,s) - V_n(t,s)\| \le \frac{R_\beta}{n^\beta} \quad \text{and} \quad \operatorname*{ess\,sup}_{(t,s)\in\Delta_0} \|U(t,s) - \widetilde{V}_n(t,s)\| \le \frac{\widetilde{R}_\beta}{n^\beta} \tag{3.45}$$

*hold for* $n \in \mathbb{N}$*, where the constants* $R_\gamma$ *and* $\widetilde{R}_\gamma$ *are those of Theorem 3.12 and Corollary 3.13.*

*Proof* Note that Proposition 2.1 of [36] yields

$$\sup_{\tau\in\mathbb{R}_+} \|\mathcal{U}(\tau) - \mathcal{T}(\tfrac{\tau}{n})^n\| = \operatorname*{ess\,sup}_{(t,s)\in\Delta_0} \|U(t,s) - V_n(t,s)\|, \quad n \in \mathbb{N}.$$

Then applying Theorem 3.12 we prove (3.45).

To proof the second estimate, we use Proposition 3.8 of [37] where the relation

$$\sup_{\tau \in \mathbb{R}_+} \|\mathcal{U}(\tau) - \widetilde{\mathcal{T}}(\tfrac{\tau}{n})^n\| = \operatorname*{ess\,sup}_{(t,s)\in\Delta_0} \|U(t,s) - \widetilde{V}_n(t,s)\|, \quad n \in \mathbb{N}.$$

was shown. Applying Corollary 3.13 we complete the proof. □

# 4 Example

As an example, we consider the diffusion equation perturbed by a time-dependent scalar potential. For this aim, let $\mathfrak{H} = L^2(\Omega)$, where $\Omega \subset \mathbb{R}^3$ is a bounded domain with sufficiently smooth boundary. Domains in higher dimension can be treated analogously. The equation reads as

$$\dot{u}(t) = \Delta u(t) - B(t)u(t), \quad u(s) = u_s \in \mathfrak{H}, \quad t, s \in [0, T], \tag{4.1}$$

where $\Delta$ denotes the Laplace operator in $L^2(\Omega)$ with Dirichlet boundary conditions, i.e., $\Delta : \mathrm{dom}(\Delta) = H^2(\Omega) \cap H_0^1(\Omega) \to L^2(\Omega)$ and $H_0^1(\Omega)$ denotes the subset of functions that vanish at the boundary. Then operator $-\Delta$ is self-adjoint on $\mathfrak{H}$ and positive. For any $\alpha \in (0, 1)$ the fractional power of operator $-\Delta$ is defined on the domain $\mathrm{dom}((-\Delta)^\alpha)$, i.e., $(-\Delta)^\alpha : \mathrm{dom}((-\Delta)^\alpha) \to L^2(\Omega)$. The domain is given by a fractional Sobolev space and for $\alpha > 1/2$, we have $\mathrm{dom}((-\Delta)^\alpha) = H_0^{2\alpha}(\Omega) \subset H^{2\alpha}(\Omega)$ (see [30] for more information).

Moreover, let $B(t)$ denote a time-dependent scalar-valued multiplication operator given by

$$\begin{aligned} (B(t)f)(x) &= V(t,x)f(x), \\ \mathrm{dom}(B(t)) &= \{f \in L^2(\mathcal{I}, \mathfrak{H}) : V(\cdot, x)f(x) \in L^2(\mathcal{I}, \mathfrak{H})\} \end{aligned} \tag{4.2}$$

where $V : \mathcal{I} \times \Omega \to \mathbb{R}$ is measurable. We assume that the potential $V(\cdot,\cdot)$ is real and non-negative. Then $B(t)$ is obviously self-adjoint and non-negative on $\mathfrak{H}$.

**Theorem 4.1** *Let $A$ be the Laplacian operator $-\Delta$ with Dirichlet boundary conditions in $L^2(\Omega)$, see above. Further, let $\{B(t)\}_{i\in\mathcal{I}}$ be the family of multiplication operators defined by* (4.2)*. If $V(\cdot,\cdot) : \mathcal{I} \times \Omega \longrightarrow \mathbb{R}$ is measurable, real, non-negative with regularity $V \in L^\infty(\mathcal{I}, L^{2+\varepsilon}(\Omega)) \cap C^\beta(\mathcal{I}, L^{1+\varepsilon}(\Omega))$ for $\beta \in (0, 1)$ and some $\varepsilon > 0$, then the assumptions (S1)–(S3) are satisfied with $\alpha \in [3/4, 1)$. Moreover, if $\beta > 2\alpha - 1$ then the converging rates of Theorem 3.12, Corollary 3.13 and Theorem 3.14 hold.*

*Proof* Since $\Omega$ is bounded there one has $\inf \sigma(A) > 0$ which does not satisfy $A \geq I$ in general and, hence, assumption (S1) is not satisfied. Nevertheless, $\inf \sigma(A) > 0$ is sufficient to prove the converging results. So we can believe that (S1) is satisfied.

Let $\alpha \geq 3/4$. Using the Sobolev space embeddings, we get that $H^{2\alpha}(\Omega) \subset L^\gamma(\Omega)$ for any $\gamma \in [2, \infty[$. Hence, if $V \in L^\infty(\mathcal{I}, L^{2+\varepsilon}(\Omega))$, we conclude that the function

$[0, T] \ni t \mapsto B(t)(-\Delta)^{-\alpha}$ is essentially operator-norm bounded in $t \in \mathcal{I}$ and thus, (S2) is satisfied. Now, we consider

$$F(t) := (-\Delta)^{-\alpha} B(t)(-\Delta)^{-\alpha} : L^2(\Omega) \to H^{2\alpha}(\Omega) \subset L^2(\Omega).$$

The function $F(\cdot) : \mathcal{I} \to \mathcal{L}(\mathfrak{H})$ is bounded for fixed $t \in [0, T]$ if for any $f, g \in H^{2\alpha}(\Omega)$ the function $\langle f, B(t)g \rangle$ is bounded. This holds since $V(t, \cdot) \in L^{1+\varepsilon}(\Omega)$ and $H^{2\alpha}(\Omega) \subset L^{\gamma}(\Omega)$ for any $\gamma \in [2, \infty[$. Hence we conclude that (S3) is satisfied and the claim is proved. □

Theorem 4.1 provides a convergence rate of an approximation of the solution of (4.1) by the time-ordered product

$$\widetilde{V}_n(t, s) = \prod_{j=1}^{n} e^{-\frac{t-s}{n} V(\frac{jt+(n-j)s}{n}, \cdot)} e^{\frac{t-s}{n} \Delta} \tag{4.3}$$

This looks elaborate, but is indeed simple. There are strategies to compute the semigroup of the Laplace operator for bounded domains and there are also explicit formulas on special domains like disks, etc. The factors $e^{-\tau V(t_j)}$, $j = 1, 2, \ldots, n$ are scalar valued and can be easily computed.

**Acknowledgements** We thank Takashi Ichinose and Hideo Tamura for the explanation of details of the proof of Theorem 1.1 of [14], which makes possible to prove Lemmas 3.8 and 3.9.

## References


1. Acquistapace, P., Terreni, B.: On the abstract Cauchy problem in the case of constant domains. Atti Accad. Naz. Lincei Rend. Cl. Sci. Fis. Mat. Natur. (8) **76**(1), 7–13 (1984)
2. Acquistapace, P., Terreni, B.: Maximal space regularity for abstract linear nonautonomous parabolic equations. J. Funct. Anal. **60**(2), 168–210 (1985)
3. Amann, H.: On abstract parabolic fundamental solutions. J. Math. Soc. Jpn. **39**(1), 93–116 (1987)
4. Amann, H.: Parabolic evolution equations in interpolation and extrapolation spaces. J. Funct. Anal. **78**(2), 233–270 (1988)
5. Amann, H.: Linear and Quasilinear Parabolic Problems. Vol. I. Monographs in Mathematics, vol. 89. Birkhäuser Boston, Inc., Boston (1995). Abstract linear theory
6. Arendt, W., Chill, R., Fornaro, S., Poupaud, C.: $L^p$-maximal regularity for non-autonomous evolution equations. J. Differ. Equ. **237**(1), 1–26 (2007)
7. Arendt, W., Dier, D., Ouhabaz, E.-M.: Invariance of convex sets for non-autonomous evolution equations governed by forms. J. Lond. Math. Soc. (2), **89**(3), 903–916 (2014)
8. Birman, M.S., Solomjak, M.Z.: Spectral Theory of Selfadjoint Operators in Hilbert Space. Mathematics and its Applications (Soviet Series). D. Reidel Publishing Co., Dordrecht (1987)
9. Cachia, V., Neidhardt, H., Zagrebnov, V.A.: Accretive perturbations and error estimates for the Trotter product formula. Integr. Equ. Oper. Theory **39**(4), 396–412 (2001)
10. Cachia, V., Neidhardt, H., Zagrebnov, V.A.: Comments on the Trotter product formula error-bound estimates for nonself-adjoint semigroups. Integr. Equ. Oper. Theory **42**(4), 425–448 (2002)

11. Cachia, V., Zagrebnov, V.A.: Operator-norm approximation of semigroups by quasi-sectorial contractions. J. Funct. Anal. **180**(1), 176–194 (2001)
12. Engel, K.-J., Nagel, R.: One-Parameter Semigroups for Linear Evolution Equations. Graduate Texts in Mathematics, vol. 194. Springer, New York (2000)
13. Fujie, Y., Tanabe, H.: On some parabolic equations of evolution in Hilbert space. Osaka J. Math. **10**, 115–130 (1973)
14. Ichinose, T., Tamura, H.: Error estimate in operator norm of exponential product formulas for propagators of parabolic evolution equations. Osaka J. Math. **35**(4), 751–770 (1998)
15. Ichinose, T., Tamura, Hideo, Tamura, Hiroshi, Zagrebnov, V.A.: Note on the paper: "The norm convergence of the Trotter-Kato product formula with error bound" by T. Ichinose and H. Tamura. Commun. Math. Phys. **221**(3), 499–510 (2001)
16. Kato, T.: Integration of the equation of evolution in a Banach space. J. Math. Soc. Jpn. **5**, 208–234 (1953)
17. Kato, T.: On linear differential equations in Banach spaces. Commun. Pure Appl. Math. **9**, 479–486 (1956)
18. Kato, T.: Abstract evolution equations of parabolic type in Banach and Hilbert spaces. Nagoya Math. J. **19**, 93–125 (1961)
19. Kato, T.: Nonlinear evolution equations in Banach spaces. In: Proceedings of Symposia in Applied Mathematics, vol. XVII, pp. 50–67. American Mathematical Society, Providence (1965)
20. Kato, T.: Linear evolution equations of "hyperbolic" type. J. Fac. Sci. Univ. Tokyo Sect. **I**(17), 241–258 (1970)
21. Kato, T.: Linear evolution equations of "hyperbolic" type. II. J. Math. Soc. Jpn. **25**, 648–666 (1973)
22. Kato, T.: On the Trotter-Lie product formula. Proc. Jpn. Acad. **50**, 694–698 (1974)
23. Kato, T.: Trotter's product formula for an arbitrary pair of self-adjoint contraction semigroups. In: Topics in Functional Analysis. Advances in Mathematics Supplement Studies, vol. 3, pp. 185–195. Academic, New York (1978)
24. Kato, T.: Trotter's product formula for some nonlinear semigroups. In: Nonlinear Evolution Equations (Proceedings of a Symposium Conducted by the Mathematics Research Center, the University of Wisconsin, Madison, 1977). Publication of the Mathematics Research Center, University of Wisconsin, vol. 40, pp. 155–162. Academic, New York (1978)
25. Kato, T.: Perturbation Theory for Linear Operators. Classics in Mathematics. Springer, Berlin (1995)
26. Kato, T.: Linear and quasi-linear equations of evolution of hyperbolic type. In: Hyperbolicity. C.I.M.E. Summer Sch., vol. 72, pp. 125–191. Springer, Heidelberg (2011)
27. Kato, T.: Semi-groups and temporally inhomogenous evolution equations. In: Abstract Differential Equations (Italian). C.I.M.E. Summer School, vol. 29, pp. 1–43. Springer, Heidelberg (2011)
28. Kato, T., Masuda, K.: Trotter's product formula for nonlinear semigroups generated by the subdifferentials of convex functionals. J. Math. Soc. Jpn. **30**(1), 169–178 (1978)
29. Kato, T., Tanabe, H.: On the abstract evolution equation. Osaka Math. J. **14**, 107–133 (1962)
30. Lions, J.L., Magenes, E.: Non-homogeneous Boundary Value Problems and Applications, vol. I. Springer, New York (1972)
31. Lunardi, A.: On the evolution operator for abstract parabolic equations. Israel J. Math. **60**(3), 281–314 (1987)
32. Monniaux, S., Rhandi, A.: Semigroup methods to solve non-autonomous evolution equations. Semigroup Forum **60**(1), 122–134 (2000)
33. Neidhardt, H.: On abstract linear evolution equations. I. Math. Nachr. **103**, 283–298 (1981)
34. Neidhardt, H., Stephan, A., Zagrebnov, V.A.: Convergence rate estimates for Trotter product approximations of solution operators for non-autonomous Cauchy problems. arXiv: 1612.06147 [math.FA], December 2016. (to appear in Publications of RIMS, Kyoto University)
35. Neidhardt, H., Stephan, A., Zagrebnov, V.A.: On convergence rate estimates for approximations of solution operators for linear nonautonomous evolution equations. Nanosyst.: Phys., Chem., Math. **8**, 202–215 (2017)

36. Neidhardt, H., Stephan, A., Zagrebnov, V.A.: Remarks on the operator-norm convergence of the Trotter product formula. Integr. Equ. Oper. Theory **90**(2) (2018)
37. Neidhardt, H., Stephan, A., Zagrebnov, V.A.: Operator-Norm Convergence of the Trotter Product Formula on Hilbert and Banach Spaces: A Short Survey, pp. 229–247. Springer International Publishing, Berlin (2018)
38. Neidhardt, H., Zagrebnov, V.A.: On error estimates for the Trotter-Kato product formula. Lett. Math. Phys. **44**(3), 169–186 (1998)
39. Neidhardt, H., Zagrebnov, V.A.: Linear non-autonomous Cauchy problems and evolution semigroups. Adv. Differ. Equ. **14**(3–4), 289–340 (2009)
40. Nickel, G.: On evolution semigroups and nonautonomous Cauchy problems. Diss. Summ. Math. **1**(1–2), 195–202 (1996)
41. Reed, M., Simon, B.: Methods of Modern Mathematical Physics. II. Fourier Analysis, Self-adjointness. Academic, New York (1975)
42. Rogava, D.L.: On the error of Trotter-type formulas in the case of selfadjoint operators. Funktsional. Anal. i Prilozhen., **27**(3), 84–86 (1993)
43. Tanabe, H.: A class of the equations of evolution in a Banach space. Osaka Math. J. **11**, 121–145 (1959)
44. Tanabe, H.: On the equations of evolution in a Banach space. Osaka Math. J. **12**, 363–376 (1960)
45. Tanabe, H.: Remarks on the equations of evolution in a Banach space. Osaka Math. J. **12**, 145–166 (1960)
46. Tanabe, H.: Evolutional equations of parabolic type. Proc. Jpn. Acad. **37**, 610–613 (1961)
47. Tanabe, H.: On regularity of solutions of abstract differential equations in Banach space. Proc. Jpn. Acad. **43**, 305–307 (1967)
48. Tanabe, H.: On regularity of solutions of abstract differential equations of parabolic type in Banach space. J. Math. Soc. Jpn. **19**, 521–542 (1967)
49. Tanabe, H.: Equations of Evolution. Monographs and Studies in Mathematics, vol. 6. Pitman (Advanced Publishing Program), Boston (1979)
50. Trotter, H.F.: On the product of semi-groups of operators. Proc. Am. Math. Soc. **10**, 545–551 (1959)
51. Yagi, A.: On the abstract linear evolution equations in Banach spaces. J. Math. Soc. Jpn. **28**(2), 290–303 (1976)
52. Yagi, A.: On the abstract evolution equation of parabolic type. Osaka J. Math. **14**(3), 557–568 (1977)
53. Yagi, A.: Fractional powers of operators and evolution equations of parabolic type. Proc. Jpn. Acad. Ser. A Math. Sci. **64**(7), 227–230 (1988)
54. Yagi, A.: Abstract Parabolic Evolution Equations and Their Applications. Springer Monographs in Mathematics. Springer, Berlin (2010)
55. Yagi, A.: Parabolic evolution equations in which the coefficients are the generators of infinitely differentiable semigroups. Funkcial. Ekvac. **32**(1), 107–124 (1989)
56. Yagi, A.: Parabolic evolution equations in which the coefficients are the generators of infinitely differentiable semigroups. II. Funkcial. Ekvac. **33**(1), 139–150 (1990)

# Exact Solutions to Problems with Perturbed Differential and Boundary Operators

**I. N. Parasidis and E. Providas**

*On the occasion of the 100th birthday of Tosio Kato*

**Abstract** We elaborate on a perturbation technique for examining the existence and uniqueness and delivering the solution in closed form of a composite boundary value problem involving the sum of a linear differential and an integral or loaded operator with nonlocal or integral boundary conditions, assuming that the exact solution for the differential operator with conventional boundary conditions is known. We apply this perturbation method to solve partial integro-differential, or loaded differential, equations with nonlocal, or integral, boundary conditions.

## 1 Introduction

Kato has completed and advanced the perturbation theory for linear operators originated by Rayleigh and Schrödinger and developed further by Rellich, Nagy and others [2]. In the introduction of his book [3] it is stated that: 'Perturbations theories are based on the idea of studying a system deviating slightly from a simple ideal system for which the complete solution of the problem under consideration is known; but the problems they treat and the tools they use are quite different'. Recently, an increasing number of papers have appeared using perturbation techniques to study the existence of solutions and other properties of composite problems, termed 'perturbed'

I. N. Parasidis
Department of Electrical Engineering, TEI of Thessaly, 41110 Larissa, Greece
e-mail: paras@teilar.gr

E. Providas (✉)
Department of Mechanical Engineering, TEI of Thessaly, 41110 Larissa, Greece
e-mail: providas@teilar.gr

© Springer Nature Switzerland AG 2019

T. M. Rassias and V. A. Zagrebnov (eds.), *Analysis and Operator Theory*,
Springer Optimization and Its Applications 146,
https://doi.org/10.1007/978-3-030-12661-2_14

problems, by presupposing that the corresponding characteristics for a simpler problem, named 'unperturbed' problem, are known. For example, Dhage [1] investigates the solvability of ordinary nonlinear differential or integral or integro-differential equations involving linear or quadratic perturbations related to whether they can be formulated as the sum or the product of two operators, respectively. Sadybekov and Imanbaev [8] study the spectral properties of boundary value problems encompassing an operator with integral perturbations of its boundary conditions assuming that the corresponding spectral properties of the unperturbed operator are known. In [6, 7] closed form solutions are obtained for the compound problem

$$\begin{aligned} Bu &= Au + Qu = f, \\ D(B) &= \{u : u \in D(A),\ \Gamma u = 0\}, \end{aligned} \tag{1}$$

for all $f \in Y$, where $X, Y, Z$ are Banach spaces, $A : X \to Y$ is a linear maximal differential operator, $\Gamma : X_A \overset{on}{\to} Z$ is a linear bounded boundary operator, $X_A = \big(D(A), ||\cdot||_{X_A}\big)$, $||\cdot||_{X_A}$ denotes a graph norm and $Q$ is a linear or nonlinear integral operator, presuming that the unique solution to the unperturbed problem

$$\begin{aligned} \widehat{A}u &= f, \\ D(\widehat{A}) &= \{u : u \in D(A),\ \Gamma u = 0\}, \end{aligned} \tag{2}$$

where $\widehat{A}$ is a correct restriction of $A$, exists and can be derived in closed form.

In this article, we deal with problems of the type (1) but over and above that with integral perturbations of its boundary conditions. More specifically, we examine solvability conditions and find the exact solution for every $f \in Y$ of the following two boundary value problems:

$$\begin{aligned} Bu &= Au - gF(Au) = f, \\ D(B) &= \{u : u \in D(A),\ \Gamma u = v\Psi(u)\}, \end{aligned} \tag{3}$$

where $F$, $\Psi$ are vectors of linear bounded functionals on $Y$ and $X_A$, respectively, and the vectors $g \in Y^n$, $v \in Z^m$, and

$$\begin{aligned} B_1u &= Au - gG(u) = f, \\ D(B_1) &= \{u : u \in D(A),\ \Gamma u = v\Psi(u) + wF(Au)\}, \end{aligned} \tag{4}$$

where now $G$, $\Psi$ and $F$ are vectors of linear bounded functionals on $X_A$ and $Y$, respectively, and the vectors $g \in Y^n$, $v \in Z^m$, $w \in Z^l$.

The proposed method is easily programmable to any Computer Algebra System and can be handled by scientists and practitioners from different disciplines. We apply this perturbation method to solve partial integro-differential, or loaded differential, equations with nonlocal, or integral, boundary conditions.

The rest of the paper is organized as follows: In Section 2 we give some definitions and preliminary results. In Section 3 we present the perturbation technique for constructing closed form solutions to perturbed boundary value problems. Finally, in Section 4, two applications are included to illustrate the implementation and efficiency of the abstract theory explicated in Section 3.

## 2 Preliminaries

Let $X, Y$ be complex Banach spaces, $A : X \to Y$ an operator and $D(A)$ and $R(A)$ the domain and the range of $A$, respectively. The following definitions are useful in the sequel.

Definition 1: An operator $A_2$ is said to be an *extension* of an operator $A_1$, or $A_1$ is said to be a *restriction* of $A_2$, in symbol $A_1 \subset A_2$, if $D(A_2) \supseteq D(A_1)$ and $A_1 x = A_2 x$ for all $x \in D(A_1)$.
Definition 2: An operator $A : X \to Y$ is called *closed* if for every sequence $x_n$ in $D(A)$ converging to $x_0$ with $Ax_n \to f_0$, it follows that $x_0 \in D(A)$ and $Ax_0 = f_0$.
Definition 3: An operator $A$ is called *maximal* if $R(A) = Y$ and $\ker A \neq \{0\}$.
Definition 4: An operator $\widehat{A} : X \to Y$ is called *correct* if $R(\widehat{A}) = Y$ and the inverse $\widehat{A}^{-1}$ exists and is bounded on $Y$.
Definition 5: An operator $\widehat{A}$ is called a *correct restriction* of the maximal operator $A$ if it is a correct operator and $\widehat{A} \subset A$.
Definition 6: An operator $\widehat{A} : X \to Y$ is called *injective and everywhere solvable* if $R(\widehat{A}) = Y$ and the inverse $\widehat{A}^{-1}$ exists.

Denote by $X^*$ the adjoint space of $X$, i.e. the set of all complex-valued linear and bounded functionals on $X$. Consider the functionals $\Psi_i \in X^*, \ i = 1, \dots, m$ and the vector $\Psi = \text{col}(\Psi_1, \dots, \Psi_m)$ with $\Psi(x) = \text{col}(\Psi_1(x), \dots, \Psi_m(x))$. Let $g = (g_1, \dots, g_n)$ be a vector of $X^n$. Then $\Psi(g)$ is the $m \times n$ matrix whose $i, j$th entry $\Psi_i(g_j)$ is the value of the functional $\Psi_i$ on element $g_j$. Note that $\Psi(gC) = \Psi(g)C$, where $C$ is a $n \times l$ constant matrix. We designate by $0_{ln}$ the $l \times n$ zero matrix, $I_n$ the $n \times n$ identity matrix, $\mathbf{c}$ a constant column vector and $\mathbf{0}$ the zero column vector.

## 3 Main Results

Let now $X, Y, Z$ be complex Banach spaces, $A : X \to Y$ a linear maximal closed operator and $\widehat{A}$ a correct restriction of $A$ defined by

$$\widehat{A}u = Au, \quad D(\widehat{A}) = \{u : u \in D(A), \ \Gamma u = 0\} = \ker \Gamma, \tag{5}$$

where $\Gamma : X_A \overset{on}{\to} Z$ is a linear bounded boundary operator and

$$X_A = \left(D(A), ||\cdot||_{X_A}\right), \quad ||x||_{X_A} = ||x||_X + ||Ax||_Y, \quad \forall x \in D(A). \tag{6}$$

Note that the space $X_A$ is a Banach space if $A$ is a closed operator. Further, let $\widehat{\Gamma}$ be a restriction of $\Gamma$ defined by

$$\widehat{\Gamma}u = \Gamma u, \quad D(\widehat{\Gamma}) = \ker A, \tag{7}$$

where we mention that $\ker A$ is a Banach space in the induced topology of $X$. We prove the next lemma initiated in [5].

**Lemma 1** *Let the operator $\widehat{\Gamma}$ be defined by (7). Then $\widehat{\Gamma} : \ker A \overset{on}{\to} Z$ is bounded and injective and has a bounded inverse $\widehat{\Gamma}^{-1} : Z \overset{on}{\to} \ker A$, i.e. $\widehat{\Gamma}$ is correct.*

*Proof* By assumption the operator $\Gamma$ is bounded from $X_A$ onto $Z$, i.e. there exists a constant $c > 0$ not depending on $x \in X_A$ such that

$$||\Gamma x||_Z \le c||x||_{X_A} = c(||x||_X + ||Ax||_Y), \quad \forall x \in D(A). \tag{8}$$

For every $x_0 \in \ker A$, we have $||\Gamma x_0||_Z = ||\widehat{\Gamma}x_0||_Z \le c||x_0||_X$. Hence, the operator $\widehat{\Gamma}$ is bounded on $\ker A$. Moreover, $\widehat{\Gamma}$ is closed because $\ker A$ is closed. From the condition $\ker \Gamma = D(\widehat{A})$ and the decomposition [4],

$$D(A) = D(\widehat{A}) \oplus \ker A, \tag{9}$$

it follows that $\ker \Gamma \cap \ker A = \{0\}$ and the operator $\widehat{\Gamma}$ is injective. From (9) it is implied that $Z = R(\Gamma) = \Gamma D(A) = \Gamma \ker A = \widehat{\Gamma} \ker A$, thus $Z = R(\widehat{\Gamma})$ and the domain of $\widehat{\Gamma}^{-1}$ is the whole of $Z$. Since $\widehat{\Gamma}^{-1}$ is closed, because $\widehat{\Gamma}$ is closed, $D(\widehat{\Gamma}^{-1}) = Z$ and $Z$ is a Banach space, then by the Closed-Graph Theorem the operator $\widehat{\Gamma}^{-1}$ is bounded. The lemma has been proved.

We first study perturbed boundary value problems as in (3) and prove the following theorem for examining the existence and uniqueness of their solutions and obtaining them in closed form.

**Theorem 1** *Let $X, Y, Z$ be complex Banach spaces, $A : X \to Y$ a linear maximal closed operator and $\widehat{A}$ a correct restriction of $A$ defined by (5). Let $X_A$ as in (6), $\Gamma : X_A \overset{on}{\to} Z$ be a bounded boundary operator and $\widehat{\Gamma}$ a restriction of $\Gamma$ defined by (7). Consider the functionals $F_i \in Y^*$, $i = 1, \dots, n$, $\Psi_j \in X_A^*$, $j = 1, \dots, m$, and the vectors $F = col(F_1, \dots, F_n)$, $\Psi = col(\Psi_1, \dots, \Psi_m)$ and suppose that the components of vectors $g = (g_1, \dots, g_n) \in Y^n$ and $v = (v_1, \dots, v_m) \in Z^m$ are linearly independent. Then:*

*(i) The operator $B : X \to Y$ defined by*

$$\begin{aligned} Bu &= Au - gF(Au) = f, \quad f \in Y, \\ D(B) &= \{u : u \in D(A), \ \Gamma u = v\Psi(u)\}, \end{aligned} \tag{10}$$

*is injective if and only if*

$$\det W = \det\,[I_n - F(g)] \neq 0, \quad \text{and}$$
$$\det V = \det\left[I_m - \Psi(\widehat{\Gamma}^{-1}v)\right] \neq 0, \tag{11}$$

*where* $I_n$, $I_m$ *designate respectively the* $n \times n$ *and* $m \times m$ *identity matrices.*

(ii) *If* (*i*) *is true then* $B$ *is correct and for all* $f \in Y$ *the unique solution of (10) is obtained as follows*

$$u = B^{-1}f = \widehat{A}^{-1}f + \left(\widehat{A}^{-1}g + \widehat{\Gamma}^{-1}vV^{-1}\Psi(\widehat{A}^{-1}g)\right)W^{-1}F(f)$$
$$+\,\widehat{\Gamma}^{-1}vV^{-1}\Psi(\widehat{A}^{-1}f). \tag{12}$$

*Proof* (i) Let $\det W \neq 0$, $\det V \neq 0$ and $u \in \ker B$. Then from (10) we have in succession

$$Bu = Au - gF(Au) = 0,$$
$$F(Au - gF(Au)) = F(0),$$
$$[I_n - F(g)]\,F(Au) = \mathbf{0}. \tag{13}$$

Since $\det W = \det\,[I_n - F(g)] \neq 0$, it follows from (13) that $F(Au) = \mathbf{0}$ and therefore $Bu = Au = 0$ and consequently $u \in \ker A$. Then by means of (7) and (10), we have

$$\Gamma u = \widehat{\Gamma}u = v\Psi(u),$$
$$\widehat{\Gamma}\left(u - \widehat{\Gamma}^{-1}v\Psi(u)\right) = 0. \tag{14}$$

By Lemma (1) and acting with $\Psi$, we get

$$u - \widehat{\Gamma}^{-1}v\Psi(u) = 0,$$
$$\Psi\left(u - \widehat{\Gamma}^{-1}v\Psi(u)\right) = \Psi(0),$$
$$\left[I_m - \Psi(\widehat{\Gamma}^{-1}v)\right]\Psi(u) = \mathbf{0}. \tag{15}$$

Because $\det V = \det\,[I_m - \Psi(\widehat{\Gamma}^{-1}v)] \neq 0$, it follows from (15) that $\Psi(u) = \mathbf{0}$ and therefore $u = 0$. That is $\ker B = \{0\}$ and hence $B$ is an injective operator.

Conversely: Let $\det V = 0$. Then there exists a vector $\mathbf{c} = \operatorname{col}(c_1, \ldots, c_m) \neq \mathbf{0}$ such that $V\mathbf{c} = 0$. Take the element $u_0 = \widehat{\Gamma}^{-1}v\mathbf{c}$ and note that $u_0 \in \ker A \subset D(A)$ and $u_0 \neq 0$, otherwise $\mathbf{c} = \mathbf{0}$ since $v_1, \ldots, v_m$ are linearly independent. Moreover, $u_0 \in D(B)$ since $\Gamma(u_0) = v\mathbf{c}$, $\Psi(u_0) = \Psi(\widehat{\Gamma}^{-1}v)\mathbf{c}$ and

$$\Gamma(u_0) - v\Psi(u_0) = v\mathbf{c} - v\Psi(\widehat{\Gamma}^{-1}v)\mathbf{c} = v[I_m - \Psi(\widehat{\Gamma}^{-1}v)]\mathbf{c} = vV\mathbf{c} = 0.$$

Then from (10) it is implied that $u_0 \in \ker B$ and hence $\ker B \neq \{0\}$ meaning that $B$ is not injective. Let now $\det V \neq 0$, but $\det W = 0$. Then there exists a vector of constants $\mathbf{c} = \mathrm{col}(c_1, \ldots, c_n) \neq \mathbf{0}$ such that $W\mathbf{c} = \mathbf{0}$. Note that $g\mathbf{c} \neq 0$, because the vector components $g_1, \ldots, g_n$ are linearly independent, and that the element $u_0 = (\widehat{A}^{-1}g + \widehat{\Gamma}^{-1}vV^{-1}\Psi(\widehat{A}^{-1}g))\mathbf{c} \neq \mathbf{0}$, otherwise $g = \mathbf{0}$. Moreover, $u_0 \in D(B)$ since

$$\begin{aligned}\Gamma(u_0) - v\Psi(u_0) &= vV^{-1}\Psi(\widehat{A}^{-1}g)\mathbf{c} - v\Psi(\widehat{A}^{-1}g)\mathbf{c} - v\Psi(\widehat{\Gamma}^{-1}v)V^{-1}\Psi(\widehat{A}^{-1}g)\mathbf{c}\\ &= v[I_m - \Psi(\widehat{\Gamma}^{-1}v)]V^{-1}\Psi(\widehat{A}^{-1}g)\mathbf{c} - v\Psi(\widehat{A}^{-1}g)\mathbf{c}\\ &= \left(v\Psi(\widehat{A}^{-1}g) - v\Psi(\widehat{A}^{-1}g)\right)\mathbf{c} = 0,\end{aligned} \tag{16}$$

and

$$\begin{aligned}Bu_0 &= Au_0 - gF(Au_0)\\ &= g\mathbf{c} - gF(g)\mathbf{c}\\ &= g[I_n - F(g)]\mathbf{c}\\ &= gW\mathbf{c} = 0.\end{aligned} \tag{17}$$

Thus $u_0 \in \ker B$, $\ker B \neq \{0\}$ and $B$ is not injective. The statement (i) holds.
(ii) Let $\det W \neq 0$ and $\det V \neq 0$. By applying the vector $F$ on (10), we obtain

$$\begin{aligned}[I_n - F(g)]\,F(Au) &= F(f),\\ F(Au) &= W^{-1}F(f).\end{aligned} \tag{18}$$

By discerning that for every $v \in Z^m$ the element $\widehat{\Gamma}^{-1}v\Psi(u) \in \ker A$ and $\widehat{\Gamma} \subset \Gamma$, the problem (10) may be written as

$$\begin{aligned}Bu &= A\left(u - \widehat{\Gamma}^{-1}v\Psi(u)\right) - gF(Au) = f, \quad f \in Y,\\ D(B) &= \{u : u \in D(A),\ \Gamma(u - \widehat{\Gamma}^{-1}v\Psi(u)) = 0\}.\end{aligned} \tag{19}$$

Since (5) and $\widehat{A} \subset A$, we have $u - \widehat{\Gamma}^{-1}v\Psi(u) \in D(\widehat{A})$ and

$$Bu = \widehat{A}\left(u - \widehat{\Gamma}^{-1}v\Psi(u)\right) - gF(Au) = f, \quad u \in D(B), \quad f \in Y. \tag{20}$$

Utilizing the correctness of $\widehat{A}$ and thus applying the inverse $\widehat{A}^{-1}$ on both sides of (20), we get

$$u - \widehat{\Gamma}^{-1}v\Psi(u) = \widehat{A}^{-1}gF(Au) + \widehat{A}^{-1}f. \tag{21}$$

Then acting by the vector $\Psi$ and using (18), we obtain

$$\Psi(u) - \Psi(\widehat{\Gamma}^{-1}v)\Psi(u) = \Psi(\widehat{A}^{-1}g)F(Au) + \Psi(\widehat{A}^{-1}f),$$
$$\left[I_m - \Psi(\widehat{\Gamma}^{-1}v)\right]\Psi(u) = \Psi(\widehat{A}^{-1}g)W^{-1}F(f) + \Psi(\widehat{A}^{-1}f),$$
$$\Psi(u) = V^{-1}\left(\Psi(\widehat{A}^{-1}g)W^{-1}F(f) + \Psi(\widehat{A}^{-1}f)\right). \tag{22}$$

Substituting into (21), we have

$$u = B^{-1}f = \widehat{A}^{-1}f + \widehat{A}^{-1}gW^{-1}F(f)$$
$$+\widehat{\Gamma}^{-1}vV^{-1}\left(\Psi(\widehat{A}^{-1}g)W^{-1}F(f) + \Psi(\widehat{A}^{-1}f)\right), \tag{23}$$

and hence the unique solution (12) to the problem (10). Because $f$ in (12) is arbitrary, it is implied that $R(B) = Y$. Moreover, because the operators $\widehat{A}^{-1}$, $\widehat{\Gamma}^{-1}$ and the functionals $F_1, \ldots, F_n$, $\Psi_1, \ldots, \Psi_m$ are bounded, it is derived that $B^{-1}$ is bounded. That is, the operator $B$ is correct if and only if (11) holds and the unique solution of (10) is given by (12). The theorem is proved. □

*Remark 1* In the case where $\Gamma u = \mathrm{col}(\Gamma_1 u, \ldots, \Gamma_k u)$, with the boundary operators $\Gamma_i : X_A \overset{\text{on}}{\to} Z$, $i = 1, \ldots, k$, the elements of $v = (v_1, \ldots, v_m)$ are column vectors $v_j = \mathrm{col}(v_{1j}, \ldots, v_{kj})$, $j = 1, \ldots, m$, and in matrix form we have

$$\begin{pmatrix} \Gamma_1 u \\ \vdots \\ \Gamma_k u \end{pmatrix} = \begin{pmatrix} v_{11} & \cdots & v_{1m} \\ \vdots & \cdots & \vdots \\ v_{k1} & \cdots & v_{km} \end{pmatrix} \begin{pmatrix} \Psi_1(u) \\ \vdots \\ \Psi_m(u) \end{pmatrix}. \tag{24}$$

Boundary value problems of the type in (4) contain drastic perturbations of their boundary conditions and therefore they are more difficult to solve. The theorem that follows elaborates in the construction of their exact unique solutions.

**Theorem 2** *Let $X, Y, Z$ be complex Banach spaces, $A : X \to Y$ be a linear maximal closed operator and $\widehat{A}$ a correct restriction of $A$ defined by (5). Further, let $X_A$ as in (6), $\Gamma : X_A \overset{on}{\to} Z$ be a linear bounded boundary operator and $\widehat{\Gamma}$ a restriction of $\Gamma$ defined by (7). Consider the functionals $G_i \in X_A^*$, $i = 1, \ldots, n$, $\Psi_j \in X_A^*$, $j = 1, \ldots, m$, $F_k \in Y^*$, $k = 1, \ldots, l$, and the vectors $G = \mathrm{col}(G_1, \ldots, G_n)$, $\Psi = \mathrm{col}(\Psi_1, \ldots, \Psi_m)$, $F = \mathrm{col}(F_1, \ldots, F_l)$. Suppose that the components of the vectors $g = (g_1, \ldots, g_n) \in Y^n$ and $v = (v_1, \ldots, v_m) \in Z^m$ are linearly independent and that $w = (w_1, \ldots, w_l) \in Z^l$. Then:*

*(i) The operator $B_1 : X \to Y$ defined by*

$$B_1 u = Au - gG(u) = f,$$
$$D(B_1) = \{u : u \in D(A),\ \Gamma u = v\Psi(u) + wF(Au)\}, \tag{25}$$

*is injective if and only if*

$$\det L_1 = \det \begin{pmatrix} \Psi(\widehat{\Gamma}^{-1}v) - I_m & \Psi(\widehat{A}^{-1}g) & \Psi(\widehat{\Gamma}^{-1}w) \\ G(\widehat{\Gamma}^{-1}v) & G(\widehat{A}^{-1}g) - I_n & G(\widehat{\Gamma}^{-1}w) \\ 0_{lm} & F(g) & -I_l \end{pmatrix} \neq 0. \tag{26}$$

*(ii) If (i) holds then B is correct and for every $f \in Y$ the unique solution of (25) is given by*

$$u = B_1^{-1} f = \widehat{A}^{-1} f - \left(\widehat{\Gamma}^{-1}v,\ \widehat{A}^{-1}g,\ \widehat{\Gamma}^{-1}w\right) L_1^{-1} \begin{pmatrix} \Psi(\widehat{A}^{-1}f) \\ G(\widehat{A}^{-1}f) \\ F(f) \end{pmatrix}. \tag{27}$$

*Proof* (i) Let $\det L_1 \neq 0$ and $u \in \ker B_1$. Then

$$B_1 u = Au - gG(u) = 0, \tag{28}$$

and sequentially

$$\begin{aligned} \Gamma u &= v\Psi(u) + wF(Au), \\ \Gamma\left(u - \widehat{\Gamma}^{-1}v\Psi(u) - \widehat{\Gamma}^{-1}wF(Au)\right) &= 0, \end{aligned} \tag{29}$$

by Lemma 1 and observing that $\widehat{\Gamma}^{-1}v\Psi(u),\ \widehat{\Gamma}^{-1}wF(Au) \in \ker A$. From (5) and (29) follows that $u - \widehat{\Gamma}^{-1}v\Psi(u) - \widehat{\Gamma}^{-1}wF(Au) \in D(\widehat{A})$. Hence,

$$\begin{aligned} B_1 u &= Au - gG(u) \\ &= A\left(u - \widehat{\Gamma}^{-1}v\Psi(u) - \widehat{\Gamma}^{-1}wF(Au)\right) - gG(u) \\ &= \widehat{A}\left(u - \widehat{\Gamma}^{-1}v\Psi(u) - \widehat{\Gamma}^{-1}wF(Au) - \widehat{A}^{-1}gG(u)\right) = 0, \end{aligned} \tag{30}$$

and as a result

$$u = \widehat{\Gamma}^{-1}v\Psi(u) + \widehat{\Gamma}^{-1}wF(Au) + \widehat{A}^{-1}gG(u). \tag{31}$$

Acting by the vectors $\Psi$, $G$ on (31) and by the vector $F$ on (28), we obtain

$$\begin{aligned} [I_m - \Psi(\widehat{\Gamma}^{-1}v)]\Psi(u) - \Psi(A^{-1}g)G(u) - \Psi(\widehat{\Gamma}^{-1}w)F(Au) &= 0, \\ -G(\widehat{\Gamma}^{-1}v)\Psi(u) + [I_n - G(\widehat{A}^{-1}g)]G(u) - G(\widehat{\Gamma}^{-1}w)F(Au) &= 0, \\ F(g)G(u) - F(Au) &= 0, \end{aligned}$$

or in matrix form

$$\begin{pmatrix} \Psi(\widehat{\Gamma}^{-1}v) - I_m & \Psi(\widehat{A}^{-1}g) & \Psi(\widehat{\Gamma}^{-1}w) \\ G(\widehat{\Gamma}^{-1}v) & G(\widehat{A}^{-1}g) - I_n & G(\widehat{\Gamma}^{-1}w) \\ 0_{lm} & F(g) & -I_l \end{pmatrix} \begin{pmatrix} \Psi(u) \\ G(u) \\ F(Au) \end{pmatrix} = L_1 \begin{pmatrix} \Psi(u) \\ G(u) \\ F(Au) \end{pmatrix} = \mathbf{0}. \tag{32}$$

Since by assumption $\det L_1 \neq 0$, it is implied that $\Psi(u) = \mathbf{0}$, $G(u) = \mathbf{0}$ and $F(Au) = \mathbf{0}$, and as a consequence from (28) and (29) follows that $B_1 u = Au = 0$ and $\Gamma u = 0$, respectively. Thus $u \in D(\widehat{A})$ and $u = 0$. Therefore $\ker B_1 = \{0\}$ and $B_1$ is an injective operator.

Conversely: Let

$$\det L_1 = \det \begin{pmatrix} \Psi(\widehat{\Gamma}^{-1}v) - I_m & \Psi(\widehat{A}^{-1}g) & \Psi(\widehat{\Gamma}^{-1}w) \\ G(\widehat{\Gamma}^{-1}v) & G(\widehat{A}^{-1}g) - I_n & G(\widehat{\Gamma}^{-1}w) \\ 0_{lm} & F(g) & -I_l \end{pmatrix} = 0. \tag{33}$$

Multiplying the third column of the matrix $L_1$ from the right by $F(g)$ and then adding to the second column, we get

$$\begin{aligned} \det L_1 &= \det \begin{pmatrix} \Psi(\widehat{\Gamma}^{-1}v) - I_m & \Psi(\widehat{A}^{-1}g) + \Psi(\widehat{\Gamma}^{-1}w)F(g) & \Psi(\widehat{\Gamma}^{-1}w) \\ G(\widehat{\Gamma}^{-1}v) & G(\widehat{A}^{-1}g) - I_n + G(\widehat{\Gamma}^{-1}w)F(g) & G(\widehat{\Gamma}^{-1}w) \\ 0_{lm} & 0_{ln} & -I_l \end{pmatrix} \\ &= (-1)^l \det \begin{pmatrix} \Psi(\widehat{\Gamma}^{-1}v) - I_m & \Psi(\widehat{A}^{-1}g) + \Psi(\widehat{\Gamma}^{-1}w)F(g) \\ G(\widehat{\Gamma}^{-1}v) & G(\widehat{A}^{-1}g) - I_n + G(\widehat{\Gamma}^{-1}w)F(g) \end{pmatrix} \\ &= (-1)^l \det L = 0. \end{aligned} \tag{34}$$

Then $\det L = 0$ and hence there exists a nonzero vector $\mathbf{c} = \mathrm{col}(\mathbf{c}_1,\ \mathbf{c}_2)$, with $\mathbf{c}_1 = \mathrm{col}(c_{11}, \ldots, c_{1m})$ and $\mathbf{c}_2 = col(c_{21}, \ldots, c_{2n})$, such that $L\mathbf{c} = 0$ or

$$\begin{pmatrix} \Psi(\widehat{\Gamma}^{-1}v) - I_m & \Psi(\widehat{A}^{-1}g) + \Psi(\widehat{\Gamma}^{-1}w)F(g) \\ G(\widehat{\Gamma}^{-1}v) & G(\widehat{A}^{-1}g) - I_n + G(\widehat{\Gamma}^{-1}w)F(g) \end{pmatrix} \begin{pmatrix} \mathbf{c}_1 \\ \mathbf{c}_2 \end{pmatrix} = \mathbf{0}. \tag{35}$$

Take the element $u_0 = \widehat{\Gamma}^{-1}v\mathbf{c}_1 + (\widehat{A}^{-1}g + \widehat{\Gamma}^{-1}wF(g))\mathbf{c}_2$ and notice that $u_0 \neq 0$; otherwise since $\widehat{A}^{-1}g \in D(\widehat{A})$, $\widehat{\Gamma}^{-1}v$, $\widehat{\Gamma}^{-1}wF(g) \in \ker A$ and $D(\widehat{A}) \cap \ker A = \{0\}$, we have $\widehat{A}^{-1}g\mathbf{c}_2 = 0$ and $\widehat{\Gamma}^{-1}v\mathbf{c}_1 + \widehat{\Gamma}^{-1}wF(g)\mathbf{c}_2 = 0$ from where it is implied that $\mathbf{c}_2 = \mathbf{0}$, because the components of the vector $g$ are linearly independent, and as a consequence $\widehat{\Gamma}^{-1}v\mathbf{c}_1 = 0$ meaning that $\mathbf{c}_1 = 0$, since $v_1, \ldots, v_m$ are linearly independent, and therefore $\mathbf{c} = \mathrm{col}(\mathbf{c}_1, \mathbf{c}_2) = \mathbf{0}$, which contradicts the hypothesis that $\mathbf{c} \neq 0$. By substituting $u_0$ into (25), taking into account that $\widehat{\Gamma} \subset \Gamma$, $\ker \Gamma = D(\widehat{A})$, $\widehat{A} \subset A$ and using (35), we obtain

$$\begin{aligned} \Gamma u_0 - v\Psi(u_0) - wF(Au_0) &= v\mathbf{c}_1 + wF(g)\mathbf{c}_2 - v\Psi(\widehat{\Gamma}^{-1}v)\mathbf{c}_1 \\ &\quad - v\Psi\left(\widehat{A}^{-1}g + \widehat{\Gamma}^{-1}wF(g)\right)\mathbf{c}_2 - wF(g)\mathbf{c}_2 \\ &= -v\left(\Psi(\widehat{\Gamma}^{-1}v) - I_m\right)\mathbf{c}_1 \\ &\quad - v\left(\Psi(\widehat{A}^{-1}g) + \Psi(\widehat{\Gamma}^{-1}w)F(g)\right)\mathbf{c}_2 \\ &= 0. \end{aligned} \tag{36}$$

Thus $u_0 \in D(B_1)$. Analogously, we have

$$\begin{aligned} B_1 u_0 &= A u_0 - g G(u_0) \\ &= g \mathbf{c}_2 - g G(\widehat{\Gamma}^{-1} v)\mathbf{c}_1 - g G\left(\widehat{A}^{-1} g + \widehat{\Gamma}^{-1} w F(g)\right)\mathbf{c}_2 \\ &= -g\left(G(\widehat{\Gamma}^{-1} v)\right)\mathbf{c}_1 - g\left(G(\widehat{A}^{-1} g) - I_n + G(\widehat{\Gamma}^{-1} w) F(g)\right)\mathbf{c}_2 \\ &= 0. \end{aligned} \tag{37}$$

Hence $u_0 \in \ker B_1$. That is $\ker B_1 \neq \{0\}$ and $B_1$ is not injective. The statement (i) holds true.

(ii) Let $\det L_1 \neq 0$. From (25), working in like manner as above, we get

$$\begin{aligned} B_1 u &= A u - g G(u) \\ &= \widehat{A}[u - \widehat{\Gamma}^{-1} v \Psi(u) - \widehat{\Gamma}^{-1} w F(Au) - \widehat{A}^{-1} g G(u)] = f. \end{aligned} \tag{38}$$

Multiplying by $\widehat{A}^{-1}$ and then solving for $u$, we have

$$u = \widehat{\Gamma}^{-1} v \Psi(u) + \widehat{\Gamma}^{-1} w F(Au) + \widehat{A}^{-1} g G(u) + \widehat{A}^{-1} f. \tag{39}$$

Acting by the vectors $\Psi$, $G$ on (39) and by the vector $F$ on (25), we acquire

$$\begin{aligned} \left(I_m - \Psi(\widehat{\Gamma}^{-1} v)\right)\Psi(u) - \Psi(A^{-1} g) G(u) - \Psi(\widehat{\Gamma}^{-1} w) F(Au) &= \Psi(A^{-1} f), \\ - G(\widehat{\Gamma}^{-1} v)\Psi(u) + \left(I_n - G(\widehat{A}^{-1} g)\right) G(u) - G(\widehat{\Gamma}^{-1} w) F(Au) &= G(A^{-1} f), \\ - F(g) G(u) + F(Au) &= F(f). \end{aligned}$$

or in matrix form

$$L_1 \begin{pmatrix} \Psi(u) \\ G(u) \\ F(Au) \end{pmatrix} = - \begin{pmatrix} \Psi(\widehat{A}^{-1} f) \\ G(\widehat{A}^{-1} f) \\ F(f) \end{pmatrix}, \tag{40}$$

which, since $\det L_1 \neq 0$, yields

$$\begin{pmatrix} \Psi(u) \\ G(u) \\ F(Au) \end{pmatrix} = -L_1^{-1} \begin{pmatrix} \Psi(\widehat{A}^{-1} f) \\ G(\widehat{A}^{-1} f) \\ F(f) \end{pmatrix}. \tag{41}$$

Writing (39) in the matrix form

$$u = \widehat{A}^{-1} f + \left(\widehat{\Gamma}^{-1} v,\ \widehat{A}^{-1} g,\ \widehat{\Gamma}^{-1} w\right) \begin{pmatrix} \Psi(u) \\ G(u) \\ F(Au) \end{pmatrix}, \tag{42}$$

and substituting (41) into (42), we obtain (27). Notice that (27) holds for every $f \in Y$ and therefore $R(B_1) = Y$. Moreover, the operators $\widehat{A}^{-1}$, $\widehat{\Gamma}^{-1}$ and the functionals $G_1, \ldots, G_n$, $\Psi_1, \ldots, \Psi_m$ and $F_1, \ldots, F_l$ are bounded and therefore $B_1^{-1}$ is bounded too. Hence, the operator $B_1$ is correct if and only if (26) holds and the unique solution to the problem (25) is provided by (27). The proof is completed. □

*Remark 2* When $\Gamma u = \mathrm{col}(\Gamma_1 u, \ldots, \Gamma_k u)$, where $\Gamma_i : X_A \overset{\text{on}}{\to} Z,\ i = 1, \ldots, k$, the elements of vectors $v = (v_1, \ldots, v_m)$ and $w = (w_1, \ldots, w_l)$ are column vectors $v_j = \mathrm{col}(v_{1j}, \ldots, v_{kj}),\ j = 1, \ldots, m$ and $w_j = \mathrm{col}(w_{1j}, \ldots, w_{kj}),\ j = 1, \ldots, l$, respectively, and in matrix form we write

$$\begin{pmatrix} \Gamma_1 u \\ \vdots \\ \Gamma_k u \end{pmatrix} = \begin{pmatrix} v_{11} & \cdots & v_{1m} \\ \vdots & \cdots & \vdots \\ v_{k1} & \cdots & v_{km} \end{pmatrix} \begin{pmatrix} \psi_1(u) \\ \vdots \\ \psi_m(u) \end{pmatrix} + \begin{pmatrix} w_{11} & \cdots & w_{1l} \\ \vdots & \cdots & \vdots \\ w_{k1} & \cdots & w_{kl} \end{pmatrix} \begin{pmatrix} F_1(Au) \\ \vdots \\ F_l(Au) \end{pmatrix}. \tag{43}$$

The analysis above and the results presented so far demand the operator $A$ be closed and the operator $\Gamma$ be linear and bounded. These two requirements are necessary for the correctness of the operator $\widehat{\Gamma}$ and hence the operators $B$ and $B_1$. However, for most applications it suffices $\widehat{A},\ \widehat{\Gamma}$ to be injective and everywhere solvable. Thus, we can allow $A : X \to Y$ be linear maximal, generally not closed operator, and $\widehat{A}$ an injective and everywhere solvable restriction of $A$ defined by

$$\widehat{A}u = Au, \quad D(\widehat{A}) = \{u : u \in D(A),\ \Gamma u = 0\}, \tag{44}$$

where the linear boundary operator

$$\Gamma : \bar{X}_A \overset{on}{\to} Z, \quad \bar{X}_A = \left(D(A), ||\cdot||_{\bar{X}_A}\right), \tag{45}$$

and $\bar{X}_A$ is a Banach subspace of $X$. Note that for any maximal closed operator $A$, we can take $\bar{X}_A = X_A$ with the graph norm $||u||_{X_A} = ||u||_X + ||Au||_Y$. In the special case wherein $A = A_2 A_1,\ A_1 : X \to S,\ A_2 : S \to Y$, where $S$ is a Banach space, we can take the norm $||u||_{\bar{X}_A} = ||u||_X + ||A_1 u||_S + ||A_2 A_1 u||_Y$, which is useful if the functionals contain derivatives of the unknown function.

Accordingly, Lemma 1 and Theorems 1, 2 can be made less formal by stating the following corresponding Lemma and two Theorems.

**Lemma 2** *Let the operator $\Gamma$ be defined by (45) and $\widehat{\Gamma}$ be a restriction of $\Gamma$ to* $\ker A$. *Then $\widehat{\Gamma} : \ker A \overset{on}{\to} Z$ is injective and $\widehat{\Gamma}^{-1} : Z \overset{on}{\to} \ker A$, i.e. $\widehat{\Gamma}$ is injective and everywhere solvable.*

**Theorem 3** *Let $X, Y, Z$ be complex Banach spaces, $A : X \to Y$ a linear maximal, generally not closed operator, and $\widehat{A}$ is an injective and everywhere solvable restriction of $A$ defined by (44). Moreover, let $\bar{X}_A$ and the operator $\Gamma$ as in (45) and $\widehat{\Gamma}$ as in Lemma 2. Consider the functionals $F_i \in Y^*,\ i = 1, \ldots, n,\ \Psi_j \in X_A^*,\ j = 1, \ldots, m$, and the vectors $F = col(F_1, \ldots, F_n)$, $\Psi = col(\Psi_1, \ldots, \Psi_m)$ and suppose that the components of vectors $g = (g_1, \ldots, g_n) \in Y^n$ and $v = (v_1, \ldots, v_m) \in Z^m$ are linearly independent. Then:*

*(i) The operator $B : X \to Y$ defined by*

$$Bu = Au - gF(Au) = f, \quad f \in Y,$$
$$D(B) = \{u : u \in D(A),\ \Gamma u = v\Psi(u)\}, \tag{46}$$

*is injective and everywhere solvable if and only if (11) holds.*

*(ii) If (i) is true then for all $f \in Y$ the unique solution of (46) is given by (12).*

*(iii) If in addition to (ii), the operators $\widehat{A}^{-1}$ and $\widehat{\Gamma}^{-1}$ are bounded on the whole Y and Z, respectively, then $\widehat{A}$ and $\widehat{\Gamma}$ are correct and hence B is correct whereas for all $f \in Y$ the unique solution of ( 46) is provided by (12).*

**Theorem 4** *Let $X, Y, Z$ be complex Banach spaces, $A : X \to Y$ a linear maximal, generally not closed operator, and $\widehat{A}$ is an injective and everywhere solvable restriction of A defined by (44). Moreover, let $\bar{X}_A$ and the operator $\Gamma$ as in (45) and $\widehat{\Gamma}$ as in Lemma 2. Consider the functionals $G_i \in X_A^*$, $i = 1, \ldots, n$, $\Psi_j \in X_A^*$, $j = 1, \ldots, m$, $F_k \in Y^*$, $k = 1, \ldots, l$, and the vectors $G = \mathrm{col}(G_1, \ldots, G_n)$, $\Psi = \mathrm{col}(\Psi_1, \ldots, \Psi_m)$, $F = \mathrm{col}(F_1, \ldots, F_l)$. Suppose that the components of the vectors $g = (g_1, \ldots, g_n) \in Y^n$ and $v = (v_1, \ldots, v_m) \in Z^m$ are linearly independent and that $w = (w_1, \ldots, w_l) \in Z^l$. Then:*

*(i) The operator $B_1 : X \to Y$ defined by*

$$B_1 u = Au - gG(u) = f,$$
$$D(B_1) = \{u : u \in D(A),\ \Gamma u = v\Psi(u) + wF(Au)\}, \tag{47}$$

*is injective and everywhere solvable if and only if (26) is true.*

*(ii) If (i) holds then for every $f \in Y$ the unique solution of (47) is given by (27).*

*(iii) If in addition to (ii), the operators $\widehat{A}^{-1}$ and $\widehat{\Gamma}^{-1}$ are bounded on the whole Y and Z, respectively, then $\widehat{A}$ and $\widehat{\Gamma}$ are correct and hence $B_1$ is correct, while for all $f \in Y$ the unique solution of (47) is given by (27).*

The proofs of Lemma 2 and Theorems 3, 4 are analogous to proofs of their counterparts Lemma 1 and Theorems 1, 2, respectively.

## 4 Application to Perturbed Partial Differential Problems

We apply the method developed in Section 3 to boundary value problems for perturbed partial differential equations with nonlocal or integral boundary conditions.

**Problem 1** We consider first the following boundary value problem for a partial integro-differential equation of Fredholm type with nonlocal perturbations of its boundary conditions, viz.

$$u_{xy} - (x+y)\int_0^1\int_0^1 y u_{xy}(x,y)dxdy = 3x^2 - \tfrac{1}{2}(x+y), \qquad 0 \le x, y \le 1,$$
$$u_x(x,0) = xu(1,1),$$
$$u(0,y) = \tfrac{2}{15}(5y+4)u\left(\tfrac{1}{2},\tfrac{1}{2}\right). \tag{48}$$

Let $\bar{\Omega} = [0,1] \times [0,1]$, the maximal operator $A : C(\bar{\Omega}) \to C(\bar{\Omega})$ be defined as

$$Au = u_{xy}, \quad D(A) = \{u : u \in C(\bar{\Omega}) : u_x, u_{xy} \in C(\bar{\Omega})\}, \tag{49}$$

and

$$F(Au) = \int_0^1\int_0^1 y u_{xy}(x,y)dxdy,$$
$$g = x + y,$$
$$f = 3x^2 - \frac{1}{2}(x+y). \tag{50}$$

In addition,

$$\Gamma u(x,y) = \begin{pmatrix} u_x(x,0) \\ u(0,y) \end{pmatrix} = \begin{pmatrix} x \\ 0 \end{pmatrix} u(1,1) + \begin{pmatrix} 0 \\ \frac{2}{15}(5y+4) \end{pmatrix} u(\frac{1}{2},\frac{1}{2}), \tag{51}$$

where $X_A = \left(D(A), ||\cdot||_{X_A}\right)$, $||u||_{X_A} = ||u||_C + ||u_{xy}||_C$, $Z = R(\Gamma) = C[0,1] \otimes C[0,1]$ and for every $h = \mathrm{col}(h_1(x),\ h_2(y)) \in Z$ the norm $||h||_Z = ||h_1(x)||_C + ||h_2(y)||_C$. From (51), we get

$$v = (v_1, v_2) = \begin{pmatrix} x & 0 \\ 0 & \frac{2}{15}(5y+4) \end{pmatrix},$$
$$\Psi(u) - \begin{pmatrix} \Psi_1(u) \\ \Psi_2(u) \end{pmatrix} = \begin{pmatrix} u(1,1) \\ u(\frac{1}{2},\frac{1}{2}) \end{pmatrix}. \tag{52}$$

The boundary value problem (48) can be now formulated as in (46) where the operator $B : C(\bar{\Omega}) \to C(\bar{\Omega})$.

The unperturbed problem is defined by

$$\widehat{A}u = u_{xy}(x,y) = f, \quad D(\widehat{A}) = \{u : u \in D(A),\ \Gamma u = 0\}, \tag{53}$$

where $\widehat{A} \subset A$ is correct, and its unique solution is given by the formula

$$u = \widehat{A}^{-1}f = \int_0^x\int_0^y f(t,s)dsdt, \quad \text{for all } f \in C(\bar{\Omega}). \tag{54}$$

The boundary operator $\widehat{\Gamma} \subset \Gamma$ by Lemma 2 is injective and everywhere solvable and for every $h = \mathrm{col}(h_1(x),\ h_2(y)) \in Z$ the problem

$$\widehat{\Gamma}u = h = \begin{pmatrix} h_1(x) \\ h_2(y) \end{pmatrix}, \tag{55}$$

which is equivalent to the problem

$$u_{xy} = 0, \quad u_x(x,0) = h_1(x), \quad u(0,y) = h_2(y), \tag{56}$$

admits the unique solution

$$u = \widehat{\Gamma}^{-1}h = \int_0^x h_1(s)ds + h_2(y). \tag{57}$$

Therefore,

$$\widehat{\Gamma}^{-1}v = (\widehat{\Gamma}^{-1}v_1,\ \widehat{\Gamma}^{-1}v_2) = \left(\frac{x^2}{2},\ \frac{2}{15}(5y+4)\right). \tag{58}$$

Moreover, the operator $\widehat{\Gamma}^{-1}$ is bounded on $Z$, since for every $h \in Z$ from (57) we have $||\widehat{\Gamma}^{-1}h||_C \leq ||h_1(x)||_C + ||h_2(y)||_C = ||h||_Z$, and so $\widehat{\Gamma}$ is correct. The functionals $F,\ \Psi_1,\ \Psi_2$ are bounded too. We compute

$$F(g) = \int_0^1\int_0^1 yg(x,y)dxdy = \frac{7}{12},$$

$$\Psi(\widehat{\Gamma}^{-1}v) = \begin{pmatrix} \Psi_1(\widehat{\Gamma}^{-1}v_1) & \Psi_1(\widehat{\Gamma}^{-1}v_2) \\ \Psi_2(\widehat{\Gamma}^{-1}v_1) & \Psi_2(\widehat{\Gamma}^{-1}v_2) \end{pmatrix} = \begin{pmatrix} \frac{1}{2} & \frac{18}{15} \\ \frac{1}{8} & \frac{13}{15} \end{pmatrix}, \tag{59}$$

and construct the matrices

$$W = [I_1 - F(g)] = \frac{5}{12},$$

$$V = \left[I_2 - \Psi(\widehat{\Gamma}^{-1}v)\right] = \begin{pmatrix} \frac{1}{2} & -\frac{18}{15} \\ -\frac{1}{8} & \frac{2}{15} \end{pmatrix}. \tag{60}$$

Since $\det W = \frac{5}{12} \neq 0$, $\det V = \frac{17}{60} \neq 0$, it follows from Theorem 3, case (iii), that the problem (48) is correct and the solution is given by (12). By first computing

$$\widehat{A}^{-1}f = \int_0^x\int_0^y f(t,s)dsdt = \frac{1}{4}xy(4x^2 - x - y),$$

$$\widehat{A}^{-1}g = \int_0^x\int_0^y g(t,s)dsdt = \frac{1}{2}yx(x+y),$$

$$F(f) = \int_0^1\int_0^1 yf(x,y)dxdy = \frac{5}{24},$$

$$\Psi(\widehat{A}^{-1}f) = \begin{pmatrix} \Psi_1(\widehat{A}^{-1}f) \\ \Psi_2(\widehat{A}^{-1}f) \end{pmatrix} = \begin{pmatrix} \frac{1}{2} \\ 0 \end{pmatrix},$$
$$\Psi(\widehat{A}^{-1}g) = \begin{pmatrix} \Psi_1(\widehat{A}^{-1}g) \\ \Psi_2(\widehat{A}^{-1}g) \end{pmatrix} = \begin{pmatrix} 1 \\ \frac{1}{8} \end{pmatrix}, \tag{61}$$

and then substituting into (12) we get the solution

$$u(x, y) = \frac{1}{4}(4x^3y - 5x^2 - 5y - 4). \tag{62}$$

**Problem 2** As a second example we contemplate a partial differential equation perturbed with a load, namely

$$-\Delta u(x) - g(x)u(x_0) = f(x), \quad x \in \Omega, \tag{63}$$

subject to an integral perturbation of the boundary condition

$$u(x)|_{\partial\Omega} = v(t)\int_\Omega u(x)dx - w(t)\int_\Omega \phi(x)\Delta u(x)dx, \quad t \in \partial\Omega, \tag{64}$$

where $\Omega = \{x : x \in \mathrm{R}^3,\ |x| < 1\}$, $\partial\Omega = \{x : x \in \mathrm{R}^3,\ |x| = 1\}$, $x_0 \in \Omega$ and $f(x)$, $\phi(x) \in L_2(\Omega)$ are given. Examine for which functions $g(x),\ v(t),\ w(t)$ the problem is everywhere solvable and find its unique solution.

We put

$$\begin{aligned} Au &= -\Delta u, \quad D(A) = \bar{X}_A = \mathrm{W}_2^2(\Omega), \\ \widehat{A}u &= Au, \quad D(\widehat{A}) = \{u : u \in D(A),\ u|_{\partial\Omega} = 0\}, \\ \Gamma u &= u|_{\partial\Omega} = u(t), \quad t \in \partial\Omega, \\ \widehat{\Gamma}u &= \Gamma u, \quad D(\widehat{\Gamma}) = \{u : u \in D(A),\ \Delta u = 0\}, \end{aligned}$$
$$G(u) = u(x_0), \quad \Psi(u) = \int_\Omega u(x)dx, \quad F(Au) = -\int_\Omega \phi(x)\Delta u(x)dx. \tag{65}$$

Note that $\Gamma u(x)$ is the trace of $u(x)$ on the sphere $|x| = 1$ and by the embedding theorem for each $u(x) \in \mathrm{W}_2^2(\Omega)$ follows that $\Gamma u(x) = u(t) \in \mathrm{W}_2^{3/2}(\partial\Omega)$. Then $Z = R(\Gamma) = \mathrm{W}_2^{3/2}(\partial\Omega)$. Thus, according to Theorem 4, $g(x) \in L_2(\Omega)$ and $v(t),\ w(t) \in Z$. The problem (63), (64) can be now expressed as in (47) where $B : L_2(\Omega) \to L_2(\Omega)$.

The unperturbed problem

$$\begin{aligned} -\Delta u(x) = f(x), \quad u(x) \in \mathrm{W}_2^2(\Omega), \quad f(x) \in L_2(\Omega), \\ u(x)|_{\partial\Omega} = 0, \end{aligned} \tag{66}$$

is the well known Dirichlet problem for the Poisson equation and it is known that the operator $\widehat{A}$, corresponding to (66) is injective and everywhere solvable and the unique solution of (66) is given by

$$u(x) = \widehat{A}^{-1} f(x) = \int_{\Omega} \mathscr{G}(x, y) f(y) dy, \quad \forall f \in L_2(\Omega), \tag{67}$$

where $\mathscr{G}(x, y)$ is the Green's function. The problem $\widehat{\Gamma} u = h(t)$, $h(t) \in Z$ is equivalent to

$$\Delta u(x) = 0, \quad u(x)|_{\partial\Omega} = h(t), \quad u(x) \in W_2^2(\Omega), \quad h(t) \in Z, \tag{68}$$

which is the Dirichlet problem for the Laplace equation and it is known that the operator $\widehat{\Gamma}$, corresponding to (68) is injective and everywhere solvable and the unique solution of (68) is given by

$$u(x) = \widehat{\Gamma}^{-1} h(t) = \int_{\partial\Omega} \mathscr{P}(x, t) h(t) dt, \quad \forall\, h(t) \in Z, \tag{69}$$

where $\mathscr{P}(x, t)$ is the Poisson's kernel. Notice that $F \in L_2^*(\Omega)$ and $\Psi,\ G \in C^*(\bar{\Omega})$. Because of the embedding $W_2^2(\Omega) \subset C(\bar{\Omega})$, we can suppose that for each $u(x) \in W_2^2(\Omega)$ holds $u(x) \in C(\bar{\Omega})$ and there exist the embedding $[C(\bar{\Omega})]^* \subset [W_2^2(\Omega)]^*$. Then $\Psi,\ G \in [W_2^2(\Omega)]^* = X_A^*$. Now Theorem 4 can be applied. For any $g(x) \in L_2(\Omega)$ and $v(t),\ w(t) \in Z$, we find

$$\begin{aligned}
\widehat{A}^{-1} g &= \int_{\Omega} \mathscr{G}(x, y) g(y) dy, \\
\widehat{\Gamma}^{-1} v &= \int_{\partial\Omega} \mathscr{P}(x, t) v(t) dt, \\
\widehat{\Gamma}^{-1} w &= \int_{\partial\Omega} \mathscr{P}(x, t) w(t) dt,
\end{aligned} \tag{70}$$

and then

$$\begin{aligned}
G(\widehat{A}^{-1} g) &= \int_{\Omega} \mathscr{G}(x_0, y) g(y) dy, \\
G(\widehat{\Gamma}^{-1} v) &= \int_{\partial\Omega} \mathscr{P}(x_0, t) v(t) dt, \\
G(\widehat{\Gamma}^{-1} w) &= \int_{\partial\Omega} \mathscr{P}(x_0, t) w(t) dt, \\
\Psi(\widehat{A}^{-1} g) &= \int_{\Omega} \int_{\Omega} \mathscr{G}(x, y) g(y) dy dx, \\
\Psi(\widehat{\Gamma}^{-1} v) &= \int_{\Omega} \int_{\partial\Omega} \mathscr{P}(x, t) v(t) dt dx,
\end{aligned}$$

$$\Psi(\widehat{\Gamma}^{-1}w) = \int_{\Omega}\int_{\partial\Omega} \mathscr{P}(x,t)w(t)dtdx,$$
$$F(g) = \int_{\Omega} \phi(x)g(x)dx. \tag{71}$$

Substitution into (26) yields det $L_1$. If and only if det $L_1 \neq 0$, then the problem (63), (64) has a unique solution which can be delivered in closed form by (27).

## References


1. Dhage, B.C.: Quadratic perturbations of periodic boundary value problems of second ordinary differential equations. Differ. Equ. Appl. **2**(4), 465–486 (2010)
2. Kato, T.: On the perturbation theory of closed linear operators. J. Math. Soc. Jpn. **4**(3–4), 323–337 (1952)
3. Kato, T.: Perturbation Theory for Linear Operators, Springer, Berlin (1995). https://doi.org/10.1007/978-3-642-66282-9
4. Kokebaev, B.K., Otelbaev, M., Shynybekov, A.N.: About restrictions and extensions of operators. D.A.N. SSSR **271**, 1307–1310 (1983) [Russian]
5. Oinarov, R.O., Parasidis, I.N.: Correct extensions of operators with finite defect in Banach spaces. Izv. Akad. Kaz. SSR **5**, 42–46 (1988) [Russian]
6. Parasidis, I.N., Providas, E.: Extension operator method for the exact solution of integro-differential equations. In: Pardalos, P.M., Rassias T.M. (eds.) Contributions in Mathematics and Engineering, pp. 473–496. Springer, Berlin (2016). https://doi.org/10.1007/978-3-319-31317-7
7. Parasidis, I.N., Providas, E.: On the exact solution of nonlinear integro-differential equations: In: Rassias, T.M. (ed.) Applications of Nonlinear Analysis, pp. 591–609. Springer, Cham (2018).https://doi.org/10.1007/978-3-319-89815-5
8. Sadybekov, M.A., Imanbaev, N.S.: A regular differential operator with perturbed boundary condition. Math. Notes **101**(5), 878–887 (2017). https://doi.org/10.1134/S0001434617050133

# On a Few Equivalent Statements of a Hilbert-Type Integral Inequality in the Whole Plane with the Hurwitz Zeta Function

**Themistocles M. Rassias and Bicheng Yang**

*On the occasion of the 100th birthday of Tosio Kato*

**Abstract** In the present paper, we prove some equivalent statements of a Hilbert-type integral inequality in the whole plane with intermediate variables. In our theorems, the constant factor is associated to the Hurwitz zeta function and we prove that it is the best possible. We also derive various special cases and applications.

**Mathematics Subject Classification** 26D15 · 47A07

## 1 Introduction

If

$$0 < \int_0^\infty f^2(x)dx < \infty \text{ and } 0 < \int_0^\infty g^2(y)dy < \infty .$$

then the following well-known Hilbert integral inequality (see [1]) holds true:

T. M. Rassias (✉)
Department of Mathematics, National Technical University of Athens,
Zografou Campus, 157 80 Athens, Greece
e-mail: trassias@math.ntua.gr

B. Yang
Department of Mathematics, Guangdong University of Education, Guangzhou, Guangdong
510303, People's Republic of China
e-mail: bcyang@gdei.edu.cn; bcyang818@163.com

© Springer Nature Switzerland AG 2019

T. M. Rassias and V. A. Zagrebnov (eds.), *Analysis and Operator Theory*,
Springer Optimization and Its Applications 146,
https://doi.org/10.1007/978-3-030-12661-2_15

$$\int_0^\infty \int_0^\infty \frac{f(x)g(y)}{x+y} dxdy < \pi \left( \int_0^\infty f^2(x)dx \int_0^\infty g^2(y)dy \right)^{\frac{1}{2}}, \tag{1}$$

where the constant factor $\pi$ is the best possible.

In 1925, by introducing one pair of conjugate exponents $(p, q)$, Hardy [2] extended (1) as follows:

For $p > 1$, $\frac{1}{p} + \frac{1}{q} = 1$, $f(x), g(y) \geq 0$,

$$0 < \int_0^\infty f^p(x)dx < \infty \text{ and } 0 < \int_0^\infty g^q(y)dy < \infty,$$

it holds that:

$$\int_0^\infty \int_0^\infty \frac{f(x)g(y)}{x+y} dxdy < \frac{\pi}{\sin(\pi/p)} \left( \int_0^\infty f^p(x)dx \right)^{\frac{1}{p}} \left( \int_0^\infty g^q(y)dy \right)^{\frac{1}{q}}, \tag{2}$$

where the constant factor $\frac{\pi}{\sin(\pi/p)}$ is the best possible.

Subsequently, in 1934, Hardy et al. extended (2) with the kernel $k_1(x, y)$, where $k_1(x, y)$ is a general nonnegative homogeneous function of degree $-1$ (cf. [3], Theorem 319). Simultaneously, the following Hilbert-type integral inequality with a non-homogeneous kernel is proved:

If $p > 1$, $\frac{1}{p} + \frac{1}{q} = 1$, $h(u) > 0$,

$$\phi(\sigma) = \int_0^\infty h(u)u^{\sigma-1}du \in \mathbf{R}_+,$$

then

$$\int_0^\infty \int_0^\infty h(xy)f(x)g(y)dxdy < \phi(\frac{1}{p}) \left( \int_0^\infty x^{p-2} f^p(x)dx \right)^{\frac{1}{p}} \left( \int_0^\infty g^q(y)dy \right)^{\frac{1}{q}}, \tag{3}$$

where the constant factor $\phi(\frac{1}{p})$ is the best possible (cf. [3], Theorem 350).

In 1998, by introducing an independent parameter $\lambda > 0$, Yang proved an extension of (1) with the kernel

$$\frac{1}{(x+y)^\lambda},$$

(cf. [4, 5]).

In 2004, by introducing another pair of conjugate exponents

$$(r, s)\ (r > 1, \frac{1}{r} + \frac{1}{s} = 1),$$

Yang [6] obtained an extension of (2) with the kernel

$$\frac{1}{x^\lambda + y^\lambda}\ (\lambda > 0).$$

One year later, Yang et al. [7] extended also (2) and the result of [4] with the kernel

$$\frac{1}{(x + y)^\lambda}\ (\lambda > 0).$$

Krnić et al. [8–16] obtained some extensions of (1), (2) and (3). Kato, in various works [17–21], proved various other very interesting types of inequalities and operators.

In 2009, Yang proved the following extension of (2) (cf. [22, 23]):

If $\lambda_1 + \lambda_2 = \lambda \in \mathbf{R} = (-\infty, \infty)$, $k_\lambda(x, y)$ is a nonnegative homogeneous function of degree $-\lambda$, satisfying

$$k_\lambda(ux, uy) = u^{-\lambda} k_\lambda(x, y)(u, x, y > 0),$$

$$k(\lambda_1) = \int_0^\infty k_\lambda(u, 1) u^{\lambda_1 - 1} du \in \mathbf{R}_+ = (0, \infty),$$

then for $p > 1, \frac{1}{p} + \frac{1}{q} = 1$, we have

$$\int_0^\infty \int_0^\infty k_\lambda(x, y) f(x) g(y) dx dy$$
$$< k(\lambda_1) \left( \int_0^\infty x^{p(1-\lambda_1)-1} f^p(x) dx \right)^{\frac{1}{p}} \left( \int_0^\infty y^{q(1-\lambda_2)-1} g^q(y) dy \right)^{\frac{1}{q}}, \tag{4}$$

where the constant factor $k(\lambda_1)$ is the best possible; for $0 < p < 1, \frac{1}{p} + \frac{1}{q} = 1$, the reverse of (4) is satisfied.

The following extension of (3) was given:

For $p > 1, \frac{1}{p} + \frac{1}{q} = 1$, we have

$$\int_0^\infty \int_0^\infty h(xy) f(x) g(y) dx dy$$
$$< \phi(\sigma) \left( \int_0^\infty x^{p(1-\sigma)-1} f^p(x) dx \right)^{\frac{1}{p}} \left( \int_0^\infty y^{q(1-\sigma)-1} g^q(y) dy \right)^{\frac{1}{q}}, \tag{5}$$

where the constant factor $\phi(\sigma)$ is the best possible; for $0 < p < 1, \frac{1}{p} + \frac{1}{q} = 1$, the reverse of (5) is satisfied (cf. [24]).

In 2007, Yang [25] proved the following Hilbert-type integral inequality in the whole plane with the exponent function as the intermediate variable:

$$\int_{-\infty}^{\infty}\int_{-\infty}^{\infty}\frac{f(x)g(y)}{(1+e^{x+y})^{\lambda}}dxdy$$
$$< B(\frac{\lambda}{2},\frac{\lambda}{2})\left(\int_{-\infty}^{\infty}e^{-\lambda x}f^2(x)dx\int_{-\infty}^{\infty}e^{-\lambda y}g^2(y)dy\right)^{\frac{1}{2}}, \tag{6}$$

where the constant factor $B(\frac{\lambda}{2}, \frac{\lambda}{2})$ ($\lambda > 0$, $B(u, v)$ stands for the beta function) is proved to be the best possible.

He et al. [15, 26–36] proved some new Hilbert-type integral inequalities in the whole plane with the best possible constant factors.

Some inequalities equivalent to (4) and (5) have been considered in [23]. In 2013, Yang [24] also studied the equivalency between (4) and (5). In 2017, Hong [37] obtained an equivalent condition between (4) and some parameters. Several authors continue to investigate this topic for other types of integral inequalities and operators (cf. [38–43]).

In the present paper, we prove some equivalent statements of a Hilbert-type integral inequality in the whole plane with intermediate variables (see Theorem 1). In our theorems, the constant factor is associated to the Hurwitz zeta function and we prove that it is the best possible (see Theorem 2). We also consider some particular cases in part of Theorem 2 and Remark 1, operator expressions in Theorems 3 and 4, as well as equivalent reverses with the best possible constant factor in Theorem 5.

## 2 An Example and a Few Lemmas

*Example 1* For $\gamma > -1$, $\rho, \mu, \sigma > 0$, $\mu + \sigma = \lambda$, setting

$$h(u) := \frac{|\ln \rho u^{\lambda}|^{\gamma}}{1 + \rho u^{\lambda}} (u > 0),$$

we obtain

$$\begin{aligned} k_{\lambda}^{(\gamma)}(\sigma) &:= \int_0^{\infty} h(u)u^{\sigma-1}du = \int_0^{\infty}\frac{|\ln \rho u^{\lambda}|^{\gamma}}{1+\rho u^{\lambda}}u^{\sigma-1}du \\ &= \frac{1}{\lambda\rho^{\sigma/\lambda}}\int_0^{\infty}\frac{|\ln v|^{\gamma}}{1+v}v^{\frac{\sigma}{\lambda}-1}dv \ (v = \rho u^{\lambda}) \\ &= \frac{1}{\lambda\rho^{\sigma/\lambda}}\int_0^{1}\frac{(-\ln v)^{\gamma}}{1+v}(v^{\frac{\sigma}{\lambda}-1} + v^{\frac{\mu}{\lambda}-1})dv \end{aligned}$$

$$= \frac{1}{\lambda\rho^{\sigma/\lambda}} \int_0^1 (-\ln v)^\gamma \sum_{k=0}^{\infty} (-1)^k v^k (v^{\frac{\sigma}{\lambda}-1} + v^{\frac{\mu}{\lambda}-1}) dv$$

$$= \frac{1}{\lambda\rho^{\sigma/\lambda}} \int_0^1 (-\ln v)^\gamma \sum_{k=0}^{\infty} (v^{2k} - v^{2k+1})(v^{\frac{\sigma}{\lambda}-1} + v^{\frac{\mu}{\lambda}-1}) dv.$$

By the Lebesgue term by term theorem (cf. [44]), it follows that

$$\begin{aligned} k_\lambda^{(\gamma)}(\sigma) &= \frac{1}{\lambda\rho^{\sigma/\lambda}} \sum_{k=0}^{\infty} \int_0^1 (-\ln v)^\gamma (v^{2k+\frac{\sigma}{\lambda}-1} + v^{2k+\frac{\mu}{\lambda}}) dv \\ &= \frac{1}{\lambda\rho^{\sigma/\lambda}} \sum_{k=0}^{\infty} (-1)^k \int_0^1 (-\ln v)^\gamma (v^{k+\frac{\sigma}{\lambda}-1} + v^{k+\frac{\mu}{\lambda}-1}) dv \\ &= \frac{1}{\lambda\rho^{\sigma/\lambda}} \int_0^\infty e^{-t} t^\gamma dt \left[ \sum_{k=0}^{\infty} \frac{(-1)^k}{(k+\frac{\sigma}{\lambda})^{\gamma+1}} + \sum_{k=0}^{\infty} \frac{(-1)^k}{(k+\frac{\mu}{\lambda})^{\gamma+1}} \right] \\ &= \frac{\Gamma(\gamma+1)}{\lambda\rho^{\sigma/\lambda}} \left( \xi(\gamma+1, \frac{\sigma}{\lambda}) + \xi(\gamma+1, \frac{\mu}{\lambda}) \right) \in \mathbf{R}_+, \end{aligned} \quad (7)$$

where

$$\Gamma(s) := \int_0^\infty e^{-v} v^{s-1} dv \ (\mathrm{Re}\, s > 0)$$

stands for the gamma function and

$$\xi(s, a) := \sum_{k=0}^{\infty} \frac{(-1)^k}{(k+a)^s} \ (\mathrm{Re}\, s > 0; a > 0)$$

(cf. [45]).

In particular:

(1) for $\gamma = 0$, setting $v = \rho u^\lambda$, we have

$$\begin{aligned} k_\lambda^{(0)}(\sigma) &= \int_0^\infty \frac{u^{\sigma-1} du}{1+\rho u^\lambda} = \frac{1}{\lambda\rho^{\sigma/\lambda}} \int_0^\infty \frac{v^{\frac{\sigma}{\lambda}-1}}{1+v} dv \\ &= \frac{\pi}{\lambda\rho^{\sigma/\lambda} \sin(\frac{\pi\sigma}{\lambda})}; \end{aligned} \quad (8)$$

(2) for $\gamma > 0$, we have

$$\begin{aligned} &\sum_{k=0}^{\infty} \frac{(-1)^k}{(k+\frac{\sigma}{\lambda})^{\gamma+1}} \\ &= \sum_{k=0}^{\infty} \frac{1}{(k+\frac{\sigma}{\lambda})^{\gamma+1}} - 2\sum_{k=0}^{\infty} \frac{1}{(2k+1+\frac{\sigma}{\lambda})^{\gamma+1}} \end{aligned}$$

$$= \sum_{k=0}^{\infty} \frac{1}{(k+\frac{\sigma}{\lambda})^{\gamma+1}} - \frac{1}{2^{\gamma}} \sum_{k=0}^{\infty} \frac{1}{(k+\frac{\lambda+\sigma}{2\lambda})^{\gamma+1}}$$
$$= \zeta(\gamma+1, \frac{\sigma}{\lambda}) - \frac{1}{2^{\gamma}} \zeta(\gamma+1, \frac{\lambda+\sigma}{2\lambda}),$$

where

$$\zeta(s,a) := \sum_{k=0}^{\infty} \frac{(-1)^k}{(k+a)^s} \ (\mathrm{Re}\, s > 1; 0 < a \leq 1)$$

is the Hurwitz zeta function (cf. [45]). Then it follows that

$$k_{\lambda}^{(\gamma)}(\sigma) = \frac{\Gamma(\gamma+1)}{\lambda \rho^{\sigma/\lambda}} \left( \zeta(\gamma+1, \frac{\sigma}{\lambda}) - \frac{1}{2^{\gamma}} \zeta(\gamma+1, \frac{\lambda+\sigma}{2\lambda}) + \zeta(\gamma+1, \frac{\mu}{\lambda}) - \frac{1}{2^{\gamma}} \zeta(\gamma+1, \frac{\lambda+\mu}{2\lambda}) \right). \tag{9}$$

For $\delta \in \{-1, 1\}$, $\alpha, \beta \in (-1, 1)$, we set

$$x_{\alpha} := |x| + \alpha x,\ y_{\beta} := |y| + \beta y \ \ (x, y \in \mathbf{R}),$$
$$E_{\delta} := \{t \in \mathbf{R};\ |t|^{\delta} \geq 1\},\ E_{-\delta} = \{t \in \mathbf{R};\ |t|^{\delta} \leq 1\}.$$

**Lemma 1** *For $c > 0$, $\eta = \alpha, \beta$, we have*

$$\int_{E_{\delta}} t_{\eta}^{-c\delta-1} dt = \frac{1}{c} \left[ \frac{1}{(1+\eta)^{c\delta+1}} + \frac{1}{(1-\eta)^{c\delta+1}} \right], \tag{10}$$
$$\int_{E_{-\delta}} t_{\eta}^{c\delta-1} dt = \frac{1}{c} \left[ \frac{1}{(1+\eta)^{-c\delta+1}} + \frac{1}{(1-\eta)^{-c\delta+1}} \right]; \tag{11}$$

*for $c \leq 0$, it follows that*

$$\int_{E_{\delta}} t_{\eta}^{-c\delta-1} dt = \int_{E_{-\delta}} t_{\eta}^{c\delta-1} dt = \infty.$$

*Proof* Setting

$$E_{\delta}^{+} := \{t \in \mathbf{R}_{+}; t^{\delta} \geq 1\},\ \ E_{\delta}^{-} := \{-t \in \mathbf{R}_{+};\ (-t)^{\delta} \geq 1\},$$

we have that $E_{\delta} = E_{\delta}^{+} \cup E_{\delta}^{-}$ and

$$\int_{E_{\delta}} t_{\eta}^{-c\delta-1} dt = \int_{E_{\delta}^{+}} [(1+\eta)t]^{-c\delta-1} dt + \int_{E_{\delta}^{-}} [(1-\eta)(-t)]^{-c\delta-1} dt$$

$$= \left[\frac{1}{(1+\eta)^{c\delta+1}} + \frac{1}{(1-\eta)^{c\delta+1}}\right] \int_{E_\delta^+} t^{-c\delta-1} dt.$$

Setting $u = t^\delta$ (or $t = u^{\frac{1}{\delta}}$), we derive that

$$\int_{E_\delta^+} t^{-c\delta-1} dt = \frac{1}{|\delta|} \int_1^\infty u^{\frac{1}{\delta}(-c\delta-1)} u^{\frac{1}{\delta}-1} du = \int_1^\infty u^{-c-1} du.$$

Hence, for $c > 0$, (10) follows and for $c \le 0$, it holds

$$\int_{E_\delta} t_\eta^{-c\delta-1} dt = \infty.$$

Since

$$\int_{E_{-\delta}} t_\eta^{c\delta-1} dt = \int_{E_{(-\delta)}} t_\eta^{-c(-\delta)-1} dt,$$

in view of (10), for $c > 0$, (11) follows, and for $c \le 0$, we get

$$\int_{E_{-\delta}} t_\eta^{c\delta-1} dt = \infty.$$

This completes the proof of the lemma. □

In the sequel, we assume that

$$p > 0\,(\neq 1),\ \frac{1}{p} + \frac{1}{q} = 1,\ \delta \in \{-1, 1\},\ \alpha,\ \beta \in (-1, 1),\ \gamma > -1,$$

$$\rho,\ \mu,\ \sigma > 0,\ \mu + \sigma = \lambda,\ \sigma_1 \in \mathbf{R}_+,$$

$k_\lambda^{(\gamma)}(\sigma)$ is as in (7), and

$$K_{\alpha,\beta}^{(\gamma)}(\sigma) := \frac{2k_\lambda^{(\gamma)}(\sigma)}{(1-\alpha^2)^{1/q}(1-\beta^2)^{1/p}}. \tag{12}$$

For $p > 1$, $n \in \mathbf{N} = \{1, 2, \cdots\}$, $E_{-1} = [-1, 1]$, $x \in E_\delta$, we define the following expressions:

$$I^{(-)}(x) := \int_{-1}^0 \frac{|\ln \rho(x_\alpha^\delta y_\beta)^\lambda|^\gamma}{1+\rho(x_\alpha^\delta y_\beta)^\lambda} y_\beta^{\sigma+\frac{1}{qn}-1} dy,$$

$$I^{(+)}(x) := \int_0^1 \frac{|\ln \rho(x_\alpha^\delta y_\beta)^\lambda|^\gamma}{1+\rho(x_\alpha^\delta y_\beta)^\lambda} y_\beta^{\sigma+\frac{1}{qn}-1} dy,$$

$$I(x) := I^{(-)}(x) + I^{(+)}(x) = \int_{E_{-1}} \frac{|\ln \rho(x_\alpha^\delta y_\beta)^\lambda|^\gamma}{1+\rho(x_\alpha^\delta y_\beta)^\lambda} y_\beta^{\sigma+\frac{1}{qn}-1} dy.$$

Since $y_\beta = (sgn(y) + \beta)y$, where

$$sgn(y) := \begin{cases} -1, y < 0 \\ 0, y = 0 \\ 1, y > 0 \end{cases},$$

it follows that

$$x_\alpha^\delta = (1 + \alpha \cdot sgn(x))^\delta |x|^\delta \geq \min_{\delta \in \{-,,1\}} (1 \pm |\alpha|)^\delta \ (x \in E_\delta),$$

$1 - |\alpha| \leq (1 + |\alpha|)^{-1} \leq 1 + |\alpha| \leq (1 - |\alpha|)^{-1}$, and

$$(1 \pm \beta)x_\alpha^\delta \geq m_{\alpha,\beta} := (1 - |\beta|)(1 - |\alpha|) > 0 \ (x \in E_\delta). \tag{13}$$

For fixed $x \in E_\delta$, setting $u = x_\alpha^\delta y_\beta$, we obtain that

$$\begin{aligned} I^{(-)}(x) &= \frac{x_\alpha^{-\delta(\sigma + \frac{1}{qn})}}{1 - \beta} \int_0^{(1-\beta)x_\alpha^\delta} \frac{|\ln \rho u^\lambda|^\gamma}{1 + \rho u^\lambda} u^{\sigma + \frac{1}{qn} - 1} du \\ &\geq \frac{x_\alpha^{-\delta(\sigma + \frac{1}{qn})}}{1 - \beta} \int_0^{m_{\alpha,\beta}} \frac{|\ln \rho u^\lambda|^\gamma}{1 + \rho u^\lambda} u^{\sigma + \frac{1}{qn} - 1} du, \end{aligned}$$

$$\begin{aligned} I^{(+)}(x) &= \frac{x_\alpha^{-\delta(\sigma + \frac{1}{qn})}}{1 + \beta} \int_0^{(1+\beta)x_\alpha^\delta} \frac{|\ln \rho u^\lambda|^\gamma}{1 + \rho u^\lambda} u^{\sigma + \frac{1}{qn} - 1} du \\ &\geq \frac{x_\alpha^{-\delta(\sigma + \frac{1}{qn})}}{1 + \beta} \int_0^{m_{\alpha,\beta}} \frac{|\ln \rho u^\lambda|^\gamma}{1 + \rho u^\lambda} u^{\sigma + \frac{1}{qn} - 1} du, \\ I(x) &= x_\alpha^{-\delta(\sigma + \frac{1}{qn})} \left[ \frac{1}{1 - \beta} \int_0^{(1-\beta)x_\alpha^\delta} \frac{|\ln \rho u^\lambda|^\gamma}{1 + \rho u^\lambda} u^{\sigma + \frac{1}{qn} - 1} du \right. \\ &\quad \left. + \frac{1}{1 + \beta} \int_0^{(1+\beta)x_\alpha^\delta} \frac{|\ln \rho u^\lambda|^\gamma}{1 + \rho u^\lambda} u^{\sigma + \frac{1}{qn} - 1} du \right] \\ &\geq \frac{2x_\alpha^{-\delta(\sigma + \frac{1}{qn})}}{1 - \beta^2} \int_0^{m_{\alpha,\beta}} \frac{|\ln \rho u^\lambda|^\gamma}{1 + \rho u^\lambda} u^{\sigma + \frac{1}{qn} - 1} du. \end{aligned} \tag{14}$$

For $p > 1$, $n \in \mathbf{N}$, $x \in F_\delta$, we define the following expressions:

$$\begin{aligned} J^{(-)}(x) &:= \int_{-\infty}^{-1} \frac{|\ln \rho(x_\alpha^\delta y_\beta)^\lambda|^\gamma}{|1 - \rho(x_\alpha^\delta y_\beta)^\lambda|} y_\beta^{\sigma - \frac{1}{qn} - 1} dy, \\ J^{(+)}(x) &:= \int_1^{\infty} \frac{|\ln \rho(x_\alpha^\delta y_\beta)^\lambda|^\gamma}{|1 - \rho(x_\alpha^\delta y_\beta)^\lambda|} y_\beta^{\sigma - \frac{1}{qn} - 1} dy, \end{aligned}$$

$$J(x) := J^{(-)}(x) + J^{(+)}(x)$$
$$= \int_{E_1} \frac{|\ln \rho(x_\alpha^\delta y_\beta)^\lambda|^\gamma}{|1-\rho(x_\alpha^\delta y_\beta)^\lambda|} y_\beta^{\sigma-\frac{1}{qn}-1} dy.$$

Since for $x \in E_{-\delta}$,

$$x_\alpha^\delta = (1+\alpha \cdot sgn(x))^\delta |x|^\delta$$
$$\leq \max_{\delta \in \{-1,1\}} \{(1 \pm |\alpha|)^\delta\} = (1-|\alpha|)^{-1},$$

we have

$$M_{\alpha,\beta} := (1+|\beta|)(1-|\alpha|)^{-1} \geq (1 \pm \beta) x_\alpha^\delta \ (x \in E_{-\delta}). \tag{15}$$

For fixed $x \in E_{-\delta}$, setting $u = x_\alpha^\delta y_\beta$, we derive that

$$J^{(-)}(x) = \frac{x_\alpha^{-\delta(\sigma-\frac{1}{qn})}}{1-\beta} \int_{(1-\beta)x_\alpha^\delta}^\infty \frac{|\ln \rho u^\lambda|^\gamma}{1+\rho u^\lambda} u^{\sigma-\frac{1}{qn}-1} du$$
$$\geq \frac{x_\alpha^{-\delta(\sigma-\frac{1}{qn})}}{1-\beta} \int_{M_{\alpha,\beta}}^\infty \frac{|\ln \rho u^\lambda|^\gamma}{1+\rho u^\lambda} u^{\sigma-\frac{1}{qn}-1} du,$$

$$J^{(+)}(x) = \frac{x_\alpha^{-\delta(\sigma-\frac{1}{qn})}}{1+\beta} \int_{(1+\beta)x_\alpha^\delta}^\infty \frac{|\ln \rho u^\lambda|^\gamma}{1+\rho u^\lambda} u^{\sigma-\frac{1}{qn}-1} du$$
$$\geq \frac{x_\alpha^{-\delta(\sigma-\frac{1}{qn})}}{1+\beta} \int_{M_{\alpha,\beta}}^\infty \frac{|\ln \rho u^\lambda|^\gamma}{1+\rho u^\lambda} u^{\sigma-\frac{1}{qn}-1} du,$$

$$J(x) = x_\alpha^{-\delta(\sigma-\frac{1}{qn})} \left[ \frac{1}{1-\beta} \int_{(1-\beta)x_\alpha^\delta}^\infty \frac{|\ln \rho u^\lambda|^\gamma}{1+\rho u^\lambda} u^{\sigma-\frac{1}{qn}-1} du \right.$$
$$\left. + \frac{1}{1+\beta} \int_{(1+\beta)x_\alpha^\delta}^\infty \frac{|\ln \rho u^\lambda|^\gamma}{1+\rho u^\lambda} u^{\sigma-\frac{1}{qn}-1} du \right]$$
$$\geq \frac{2x_\alpha^{-\delta(\sigma-\frac{1}{qn})}}{1-\beta^2} \int_{M_{\alpha,\beta}}^\infty \frac{|\ln \rho u^\lambda|^\gamma}{1+\rho u^\lambda} u^{\sigma-\frac{1}{qn}-1} du. \tag{16}$$

In view of (14) and (16), the following lemma follows:

**Lemma 2** *For $p > 1$, the following inequalities hold true:*

$$I_1 := \int_{E_\delta} I(x) x_\alpha^{\delta(\sigma_1-\frac{1}{pn})-1} dx$$

$$\geq \frac{2}{1-\beta^2}\int_{E_\delta} x_\alpha^{-\delta(\sigma-\sigma_1+\frac{1}{n})-1}dx\int_0^{m_{\alpha,\beta}}\frac{|\ln\rho u^\lambda|^\gamma}{1+\rho u^\lambda}u^{\sigma+\frac{1}{qn}-1}du, \tag{17}$$

$$\begin{aligned} J_1 &:= \int_{F_\delta} J(x)x_\alpha^{\delta(\sigma_1+\frac{1}{pn})-1}dx \\ &\geq \frac{2}{1-\beta^2}\int_{E_{-\delta}} x_\alpha^{\delta(\sigma_1-\sigma+\frac{1}{n})-1}dx\int_{M_{\alpha,\beta}}^{\infty}\frac{|\ln\rho u^\lambda|^\gamma}{1+\rho u^\lambda}u^{\sigma-\frac{1}{qn}-1}du. \end{aligned} \tag{18}$$

**Lemma 3** *If $p>1$, there exists a constant $M$, such that for any nonnegative measurable functions $f(x)$ and $g(y)$ in $\mathbf{R}$, satisfying*

$$\int_{-\infty}^{\infty} x_\alpha^{p(1-\delta\sigma_1)-1}f^p(x)dx<\infty, \text{ and } \int_{-\infty}^{\infty} y_\beta^{q(1-\sigma)-1}g^q(y)dy<\infty,$$

*the following inequality*

$$\begin{aligned} I &:= \int_{-\infty}^{\infty}\int_{-\infty}^{\infty}\frac{|\ln\rho(x_\alpha^\delta y_\beta)^\lambda|^\gamma}{1+\rho(x_\alpha^\delta y_\beta)^\lambda}f(x)g(y)dxdy \\ &\leq M\left[\int_{-\infty}^{\infty} x_\alpha^{p(1-\delta\sigma_1)-1}f^p(x)dx\right]^{\frac{1}{p}}\left[\int_{-\infty}^{\infty} y_\beta^{q(1-\sigma)-1}g^q(y)dy\right]^{\frac{1}{q}} \end{aligned} \tag{19}$$

*holds true, then we have $\sigma_1=\sigma$.*

*Proof* If $\sigma_1>\sigma$, then for $n\geq\frac{1}{\sigma_1-\sigma}$ $(n\in\mathbf{N})$, we set the following functions:

$$f_n(x):=\begin{cases} x_\alpha^{\delta(\sigma_1-\frac{1}{pn})-1}, x\in E_\delta \\ 0, x\in\mathbf{R}\backslash E_\delta \end{cases},$$

$$g_n(y):=\begin{cases} y_\beta^{\sigma+\frac{1}{qn}-1}, y\in E_{-1} \\ 0, y\in\mathbf{R}\backslash E_{-1} \end{cases}.$$

By (10) and (11), we obtain that

$$\begin{aligned} \widetilde{J}_1 &:= \left[\int_{-\infty}^{\infty} x_\alpha^{p(1-\delta\sigma_1)-1}f_n^p(x)dx\right]^{\frac{1}{p}}\left[\int_{-\infty}^{\infty} y_\beta^{q(1-\sigma)-1}g_n^q(y)dy\right]^{\frac{1}{q}} \\ &= \left(\int_{E_\delta} x_\alpha^{-\frac{\delta}{n}-1}dx\right)^{\frac{1}{p}}\left(\int_{E_{-1}} y_\beta^{\frac{1}{n}-1}dy\right)^{\frac{1}{q}} \\ &= n\left[\frac{1}{(1+\alpha)^{\frac{\delta}{n}+1}}+\frac{1}{(1-\alpha)^{\frac{\delta}{n}+1}}\right]^{\frac{1}{p}} \end{aligned}$$

$$\times\left[\frac{1}{(1+\beta)^{-\frac{1}{n}+1}}+\frac{1}{(1-\beta)^{-\frac{1}{n}+1}}\right]^{\frac{1}{q}}<\infty.$$

By (17) and (19) (for $f=f_n, g=g_n$), we have

$$\begin{aligned}
&\frac{2}{1-\beta^2}\int_{E_\delta} x_\alpha^{-\delta(\sigma-\sigma_1+\frac{1}{n})-1}dx\int_0^{m_{\alpha,\beta}}\frac{|\ln\rho u^\lambda|^\gamma}{1+\rho u^\lambda}u^{\sigma+\frac{1}{qn}-1}du\\
&\leq I_1=\int_{-\infty}^{\infty}\int_{-\infty}^{\infty}\frac{|\ln\rho(x_\alpha^\delta y_\beta)^\lambda|^\gamma}{1+\rho(x_\alpha^\delta y_\beta)^\lambda}f_n(x)g_n(y)dxdy\\
&\leq M\widetilde{J}_1<\infty.
\end{aligned}$$

Since for any $n\geq\frac{1}{\sigma_1-\sigma}$ ($n\in\mathbf{N}$), we have $\sigma-\sigma_1+\frac{1}{n}\leq 0$, by Lemma 1, it follows that

$$\int_{E_\delta}x_\alpha^{-\delta(\sigma-\sigma_1+\frac{1}{n})-1}dx=\infty.$$

In view of

$$\int_0^{m_{\alpha,\beta}}\frac{|\ln\rho u^\lambda|^\gamma}{1+\rho u^\lambda}u^{\sigma+\frac{1}{qn}-1}du>0,$$

we deduce that

$$\infty\leq M\widetilde{J}_1<\infty,$$

which is a contradiction.

If $\sigma>\sigma_1$, then for $n\geq\frac{1}{\sigma-\sigma_1}$ ($n\in\mathbf{N}$), we set the functions

$$\widetilde{f}_n(x):=\begin{cases}x_\alpha^{\delta(\sigma_1+\frac{1}{pn})-1}, x\in E_{-\delta}\\ 0, x\in\mathbf{R}\backslash E_{-\delta}\end{cases},$$

$$\widetilde{g}_n(y):=\begin{cases}y_\beta^{\sigma-\frac{1}{qn}-1}, y\in E_1\\ 0, y\in\mathbf{R}\backslash E_1\end{cases},$$

and thus by (10) and (11), we obtain that

$$\begin{aligned}
\widetilde{J}_2&:=\left[\int_{-\infty}^{\infty}x_\alpha^{p(1-\delta\sigma_1)-1}\widetilde{f}_n^p(x)dx\right]^{\frac{1}{p}}\left[\int_{-\infty}^{\infty}y_\beta^{q(1-\sigma)-1}\widetilde{g}_n^q(y)dy\right]^{\frac{1}{q}}\\
&=\left(\int_{E_{-\delta}}x_\alpha^{\frac{\delta}{n}-1}dx\right)^{\frac{1}{p}}\left(\int_{E_1}y_\beta^{-\frac{1}{n}-1}dy\right)^{\frac{1}{q}}\\
&=n\left[\frac{1}{(1+\alpha)^{-\frac{\delta}{n}+1}}+\frac{1}{(1-\alpha)^{-\frac{\delta}{n}+1}}\right]^{\frac{1}{p}}
\end{aligned}$$

$$\times\left[\frac{1}{(1+\beta)^{\frac{1}{n}+1}}+\frac{1}{(1-\beta)^{\frac{1}{n}+1}}\right]^{\frac{1}{q}}<\infty.$$

In view of (18) and (19) (for $f=\widetilde{f}_n, g=\widetilde{g}_n$), we have

$$\begin{aligned}&\frac{2}{1-\beta^2}\int_{E_{-\delta}} x_\alpha^{\delta(\sigma_1-\sigma+\frac{1}{n})-1}dx\int_{M_{\alpha,\beta}}^{\infty}\frac{|\ln\rho u^\lambda|^\gamma}{1+\rho u^\lambda}u^{\sigma-\frac{1}{qn}-1}du\\&\leq J_1=\int_{-\infty}^{\infty}\int_{-\infty}^{\infty}\frac{|\ln\rho(x_\alpha^\delta y_\beta)^\lambda|^\gamma}{1+\rho(x_\alpha^\delta y_\beta)^\lambda}\widetilde{f}_n(x)\widetilde{g}_n(y)dxdy\\&\leq M\widetilde{J}_2<\infty.\end{aligned}$$

Since for $n\geq\frac{1}{\sigma-\sigma_1}$ ($n\in\mathbf{N}$), we have $\sigma_1-\sigma+\frac{1}{n}\leq 0$, by Lemma 1, it follows that

$$\int_{E_{-\delta}} x_\alpha^{\delta(\sigma_1-\sigma+\frac{1}{n})-1}dx=\infty.$$

In view of

$$\int_{M_{\alpha,\beta}}^{\infty}\frac{|\ln\rho u^\lambda|^\gamma}{1+\rho u^\lambda}u^{\sigma-\frac{1}{qn}-1}du>0,$$

we have

$$\infty\leq M\widetilde{J}_2<\infty,$$

which is a contradiction.

Hence, we conclude that $\sigma_1=\sigma$.

This completes the proof of the lemma.

For $\sigma_1=\sigma$, the following lemma also holds true:

**Lemma 4** *If $p>1$, there exists a constant $M$, such that for any nonnegative measurable functions $f(x)$ and $g(y)$ in $\mathbf{R}$, satisfying*

$$\int_{-\infty}^{\infty}x_\alpha^{p(1-\delta\sigma)-1}f^p(x)dx<\infty,\ and\ \int_{-\infty}^{\infty}y_\beta^{q(1-\sigma)-1}g^q(y)dy<\infty,$$

*the following inequality*

$$\begin{aligned}&\int_{-\infty}^{\infty}\int_{-\infty}^{\infty}\frac{|\ln\rho(x_\alpha^\delta y_\beta)^\lambda|^\gamma}{1+\rho(x_\alpha^\delta y_\beta)^\lambda}f(x)g(y)dxdy\\&\leq M\left[\int_{-\infty}^{\infty}x_\alpha^{p(1-\delta\sigma)-1}f^p(x)dx\right]^{\frac{1}{p}}\left[\int_{-\infty}^{\infty}y_\beta^{q(1-\sigma)-1}g^q(y)dy\right]^{\frac{1}{q}}\end{aligned}\tag{20}$$

*holds true, then we have $K_{\alpha,\beta}^{(\gamma)}(\sigma)\leq M$.*

*Proof* By (14), we have

$$I_1 = \int_{E_\delta} I(x) x_\alpha^{\delta(\sigma - \frac{1}{pn}) - 1} dx = \frac{L^{(-)} + L^{(+)}}{1 - \beta} + \frac{K^{(-)} + K^{(+)}}{1 + \beta}, \tag{21}$$

where

$$L^{(+)} := \int_{E_\delta^+} x_\alpha^{-\frac{\delta}{n} - 1} \int_0^{(1-\beta)x_\alpha^\delta} \frac{|\ln \rho u^\lambda|^\gamma}{1 + \rho u^\lambda} u^{\sigma + \frac{1}{qn} - 1} du dx,$$

$$L^{(-)} := \int_{E_\delta^-} x_\alpha^{-\frac{\delta}{n} - 1} \int_0^{(1-\beta)x_\alpha^\delta} \frac{|\ln \rho u^\lambda|^\gamma}{1 + \rho u^\lambda} u^{\sigma + \frac{1}{qn} - 1} du dx,$$

$$K^{(+)} := \int_{E_\delta^+} x_\alpha^{-\frac{\delta}{n} - 1} \int_0^{(1+\beta)x_\alpha^\delta} \frac{|\ln \rho u^\lambda|^\gamma}{1 + \rho u^\lambda} u^{\sigma + \frac{1}{qn} - 1} du dx,$$

$$K^{(-)} := \int_{E_\delta^-} x_\alpha^{-\frac{\delta}{n} - 1} \int_0^{(1+\beta)x_\alpha^\delta} \frac{|\ln \rho u^\lambda|^\gamma}{1 + \rho u^\lambda} u^{\sigma + \frac{1}{qn} - 1} du dx.$$

By Fubini's theorem (cf. [44]), we obtain that

$$\begin{aligned}
L^{(+)} &= (1+\alpha)^{-\frac{\delta}{n} - 1} \int_{E_\delta^+} x^{-\frac{\delta}{n} - 1} \int_0^{(1-\beta)(1+\alpha)^\delta x^\delta} \frac{|\ln \rho u^\lambda|^\gamma}{1 + \rho u^\lambda} u^{\sigma + \frac{1}{qn} - 1} du dx \\
&= (1+\alpha)^{-\frac{\delta}{n} - 1} \int_1^\infty y^{-\frac{1}{n} - 1} \int_0^{(1-\beta)(1+\alpha)^\delta y} \frac{|\ln \rho u^\lambda|^\gamma}{1 + \rho u^\lambda} u^{\sigma + \frac{1}{qn} - 1} du dy \\
&= (1+\alpha)^{-\frac{\delta}{n} - 1} \left[ \int_1^\infty y^{-\frac{1}{n} - 1} dy \int_0^{(1-\beta)(1+\alpha)^\delta} \frac{|\ln \rho u^\lambda|^\gamma}{1 + \rho u^\lambda} u^{\sigma + \frac{1}{qn} - 1} du \right. \\
&\quad \left. + \int_1^\infty y^{-\frac{1}{n} - 1} \int_{(1-\beta)(1+\alpha)^\delta}^{(1-\beta)(1+\alpha)^\delta y} \frac{|\ln \rho u^\lambda|^\gamma}{1 + \rho u^\lambda} u^{\sigma + \frac{1}{qn} - 1} du dy \right] \\
&= (1+\alpha)^{-\frac{\delta}{n} - 1} \left\{ n \int_0^{(1-\beta)(1+\alpha)^\delta} \frac{|\ln \rho u^\lambda|^\gamma}{1 + \rho u^\lambda} u^{\sigma + \frac{1}{qn} - 1} du \right. \\
&\quad \left. + \int_{(1-\beta)(1+\alpha)^\delta}^\infty \left[ \int_{\frac{u}{(1-\beta)(1+\alpha)^\delta}}^\infty y^{-\frac{1}{n} - 1} dy \right] \frac{|\ln \rho u^\lambda|^\gamma}{1 + \rho u^\lambda} u^{\sigma + \frac{1}{qn} - 1} du \right\} \\
&= \frac{n}{(1+\alpha)^{\frac{\delta}{n} + 1}} \left[ \int_0^{(1-\beta)(1+\alpha)^\delta} \frac{|\ln \rho u^\lambda|^\gamma}{1 + \rho u^\lambda} u^{\sigma + \frac{1}{qn} - 1} du \right. \\
&\quad \left. + (1-\beta)^{\frac{1}{n}} (1+\alpha)^{\frac{\delta}{n}} \int_{(1-\beta)(1+\alpha)^\delta}^\infty \frac{|\ln \rho u^\lambda|^\gamma}{1 + \rho u^\lambda} u^{\sigma - \frac{1}{pn} - 1} du \right],
\end{aligned}$$

$$
\begin{aligned}
L^{(-)} &= (1-\alpha)^{-\frac{\delta}{n}-1}\int_{E_\delta^-}(-x)^{-\frac{\delta}{n}-1}\int_0^{(1-\beta)(1-\alpha)^\delta(-x)^\delta}\frac{|\ln\rho u^\lambda|^\gamma}{1+\rho u^\lambda}u^{\sigma+\frac{1}{qn}-1}dudx\\
&= (1-\alpha)^{-\frac{\delta}{n}-1}\int_1^\infty y^{-\frac{1}{n}-1}\int_0^{(1-\beta)(1-\alpha)^\delta y}\frac{|\ln\rho u^\lambda|^\gamma}{1+\rho u^\lambda}u^{\sigma+\frac{1}{qn}-1}dudy\\
&= (1-\alpha)^{-\frac{\delta}{n}-1}\int_1^\infty y^{-\frac{1}{n}-1}dy\int_0^{(1-\beta)(1-\alpha)^\delta}\frac{|\ln\rho u^\lambda|^\gamma}{1+\rho u^\lambda}u^{\sigma+\frac{1}{qn}-1}du\\
&\quad+(1-\alpha)^{-\frac{\delta}{n}-1}\int_1^\infty y^{-\frac{1}{n}-1}\int_{(1-\beta)(1-\alpha)^\delta}^{(1-\beta)(1-\alpha)^\delta y}\frac{|\ln\rho u^\lambda|^\gamma}{1+\rho u^\lambda}u^{\sigma+\frac{1}{qn}-1}dudy
\end{aligned}
$$

$$
\begin{aligned}
K^{(+)} &= (1+\alpha)^{-\frac{\delta}{n}-1}\int_{E_\delta^+}x^{-\frac{\delta}{n}-1}\int_0^{(1+\beta)(1+\alpha)^\delta x^\delta}\frac{|\ln\rho u^\lambda|^\gamma}{1+\rho u^\lambda}u^{\sigma+\frac{1}{qn}-1}dudx\\
&= (1+\alpha)^{-\frac{\delta}{n}-1}\int_1^\infty y^{-\frac{1}{n}-1}\int_0^{(1+\beta)(1+\alpha)^\delta y}\frac{|\ln\rho u^\lambda|^\gamma}{1+\rho u^\lambda}u^{\sigma+\frac{1}{qn}-1}dudy\\
&= (1+\alpha)^{-\frac{\delta}{n}-1}\left[\int_1^\infty y^{-\frac{1}{n}-1}dy\int_0^{(1+\beta)(1+\alpha)^\delta}\frac{|\ln\rho u^\lambda|^\gamma}{1+\rho u^\lambda}u^{\sigma+\frac{1}{qn}-1}du\right.\\
&\quad\left.+\int_1^\infty y^{-\frac{1}{n}-1}\int_{(1+\beta)(1+\alpha)^\delta}^{(1+\beta)(1+\alpha)^\delta y}\frac{|\ln\rho u^\lambda|^\gamma}{1+\rho u^\lambda}u^{\sigma+\frac{1}{qn}-1}dudy\right]\\
&= \frac{n}{(1+\alpha)^{\frac{\delta}{n}+1}}\left[\int_0^{(1+\beta)(1+\alpha)^\delta}\frac{|\ln\rho u^\lambda|^\gamma}{1+\rho u^\lambda}u^{\sigma+\frac{1}{qn}-1}du\right.\\
&\quad\left.+(1+\beta)^{\frac{1}{n}}(1+\alpha)^{\frac{\delta}{n}}\int_{(1+\beta)(1+\alpha)^\delta}^\infty\frac{|\ln\rho u^\lambda|^\gamma}{1+\rho u^\lambda}u^{\sigma-\frac{1}{pn}-1}du\right],
\end{aligned}
$$

$$
\begin{aligned}
K^{(-)} &= (1-\alpha)^{-\frac{\delta}{n}-1}\int_{E_\delta^-}(-x)^{-\frac{\delta}{n}-1}\int_0^{(1+\beta)(1-\alpha)^\delta(-x)^\delta}\frac{|\ln\rho u^\lambda|^\gamma}{1+\rho u^\lambda}u^{\sigma+\frac{1}{qn}-1}dudx\\
&= (1-\alpha)^{-\frac{\delta}{n}-1}\int_1^\infty y^{-\frac{1}{n}-1}\int_0^{(1+\beta)(1-\alpha)^\delta y}\frac{|\ln\rho u^\lambda|^\gamma}{1+\rho u^\lambda}u^{\sigma+\frac{1}{qn}-1}dudy\\
&= (1-\alpha)^{-\frac{\delta}{n}-1}\left[\int_1^\infty y^{-\frac{1}{n}-1}dy\int_0^{(1+\beta)(1-\alpha)^\delta}\frac{|\ln\rho u^\lambda|^\gamma}{1+\rho u^\lambda}u^{\sigma+\frac{1}{qn}-1}du\right.\\
&\quad\left.+\int_1^\infty y^{-\frac{1}{n}-1}\int_{(1+\beta)(1-\alpha)^\delta}^{(1+\beta)(1-\alpha)^\delta y}\frac{|\ln\rho u^\lambda|^\gamma}{1+\rho u^\lambda}u^{\sigma+\frac{1}{qn}-1}dudy\right]\\
&= \frac{n}{(1-\alpha)^{\frac{\delta}{n}+1}}\left[\int_0^{(1+\beta)(1-\alpha)^\delta}\frac{|\ln\rho u^\lambda|^\gamma}{1+\rho u^\lambda}u^{\sigma+\frac{1}{qn}-1}du\right.\\
&\quad\left.+(1+\beta)^{\frac{1}{n}}(1-\alpha)^{\frac{\delta}{n}}\int_{(1+\beta)(1-\alpha)^\delta}^\infty\frac{|\ln\rho u^\lambda|^\gamma}{1+\rho u^\lambda}u^{\sigma-\frac{1}{pn}-1}du\right].
\end{aligned}
$$

By (19) (for $f = f_n, g = g_n$), we have

$$\frac{1}{n}I_1 = \frac{1}{n}\left(\frac{L^{(-)} + L^{(+)}}{1-\beta} + \frac{K^{(-)} + K^{(+)}}{1+\beta}\right) \le \frac{1}{n}M\widetilde{J}_1.$$

For $n \to \infty$, by Fatou's lemma (cf. [44]) and the present results, we obtain that

$$\frac{2}{1-\beta^2} \cdot \frac{2k_\lambda^{(\gamma)}(\sigma)}{1-\alpha^2} \le M\left(\frac{2}{1-\alpha^2}\right)^{\frac{1}{p}}\left(\frac{2}{1-\beta^2}\right)^{\frac{1}{q}},$$

namely,

$$K_{\alpha,\beta}^{(\gamma)}(\sigma) = \frac{2k_\lambda^{(\gamma)}(\sigma)}{(1-\alpha^2)^{1/q}(1-\beta^2)^{1/p}} \le M.$$

This completes the proof of the lemma. □

**Lemma 5** *We define the following weight functions:*

$$\omega_\delta(\sigma_1, y) := y_\beta^\sigma \int_{-\infty}^{\infty} \frac{|\ln \rho(x_\alpha^\delta y_\beta)^\lambda|^\gamma}{1+\rho(x_\alpha^\delta y_\beta)^\lambda} x_\alpha^{\delta\sigma_1 - 1} dx \ (y \in \mathbf{R}), \tag{22}$$

$$\varpi_\delta(\sigma, x) := x_\alpha^{\delta\sigma_1} \int_{-\infty}^{\infty} \frac{|\ln \rho(x_\alpha^\delta y_\beta)^\lambda|^\gamma}{1+\rho(x_\alpha^\delta y_\beta)^\lambda} y_\beta^{\sigma - 1} dy \ (x \in \mathbf{R}). \tag{23}$$

*Then we have*

$$\omega_\delta(\sigma_1, y) = \frac{2k_\lambda^{(\gamma)}(\sigma_1)}{1-\alpha^2} y_\beta^{\sigma - \sigma_1},$$

$$\varpi_\delta(\sigma, x) = \frac{2k_\lambda^{(\gamma)}(\sigma)}{1-\beta^2} x_\alpha^{\delta(\sigma_1 - \sigma)} \ (x, y \in \mathbf{R}\backslash\{0\}). \tag{24}$$

*Proof* For fixed $y \in \mathbf{R}\backslash\{0\}$, setting $u = x_\alpha^\delta y_\beta$, we find

$$\omega_\delta(\sigma_1, y) = y_\beta^\sigma \int_{-\infty}^{0} \frac{|\ln \rho(x_\alpha^\delta y_\beta)^\lambda|^\gamma}{1+\rho(x_\alpha^\delta y_\beta)^\lambda} x_\alpha^{\delta\sigma_1 - 1} dx$$

$$+ y_\beta^\sigma \int_{0}^{\infty} \frac{|\ln \rho(x_\alpha^\delta y_\beta)^\lambda|^\gamma}{1+\rho(x_\alpha^\delta y_\beta)^\lambda} x_\alpha^{\delta\sigma_1 - 1} dx$$

$$= \frac{2y_\beta^{\sigma-\sigma_1}}{1-\alpha^2} \int_0^\infty \frac{|\ln \rho u^\lambda|^\gamma}{1+\rho u^\lambda} u^{\sigma_1 - 1} du = \frac{2k_\lambda^{(\gamma)}(\sigma_1)}{1-\alpha^2} y_\beta^{\sigma-\sigma_1};$$

for fixed $x \in \mathbf{R}\backslash\{0\}$, *setting* $u = x_\alpha^\delta y_\beta$, it follows that

$$\varpi_\delta(\sigma,x)=x_\alpha^{\delta\sigma_1}\int_{-\infty}^{0}\frac{|\ln\rho(x_\alpha^\delta y_\beta)^\lambda|^\gamma}{1+\rho(x_\alpha^\delta y_\beta)^\lambda}y_\beta^{\sigma-1}dy$$
$$+x_\alpha^{\delta\sigma_1}\int_{0}^{\infty}\frac{|\ln\rho(x_\alpha^\delta y_\beta)^\lambda|^\gamma}{1+\rho(x_\alpha^\delta y_\beta)^\lambda}y_\beta^{\sigma-1}dy$$
$$=\frac{2x_\alpha^{\delta(\sigma_1-\sigma)}}{1-\beta^2}\int_{0}^{\infty}\frac{|\ln\rho u^\lambda|^\gamma}{1+\rho u^\lambda}u^{\sigma-1}du=\frac{2k_\lambda^{(\gamma)}(\sigma)}{1-\beta^2}x_\alpha^{\delta(\sigma_1-\sigma)}.$$

Hence, we have (24).

This completes the proof of the lemma. □

## 3 Main Results

**Theorem 1** *If $p>1$, $M$ is a constant, then the following statements (i), (ii) and (iii) are equivalent:*

(i) For any $f(x)\geq 0$, satisfying

$$\int_{-\infty}^{\infty}x_\alpha^{p(1-\delta\sigma_1)-1}f^p(x)dx<\infty,$$

the following inequality holds true:

$$J:=\left[\int_{-\infty}^{\infty}y_\beta^{p\sigma-1}\left(\int_{-\infty}^{\infty}\frac{|\ln\rho(x_\alpha^\delta y_\beta)^\lambda|^\gamma}{1+\rho(x_\alpha^\delta y_\beta)^\lambda}f(x)dx\right)^p dy\right]^{\frac{1}{p}}$$
$$\leq M\left[\int_{-\infty}^{\infty}x_\alpha^{p(1-\delta\sigma_1)-1}f^p(x)dx\right]^{\frac{1}{p}}. \tag{25}$$

(ii) For any $f(x), g(y)\geq 0$, satisfying

$$\int_{-\infty}^{\infty}x_\alpha^{p(1-\delta\sigma_1)-1}f^p(x)dx<\infty,\ and\ \int_{-\infty}^{\infty}y_\beta^{q(1-\sigma)-1}g^q(y)dy<\infty,$$

the following inequality holds true:

$$I=\int_{-\infty}^{\infty}\int_{-\infty}^{\infty}\frac{|\ln\rho(x_\alpha^\delta y_\beta)^\lambda|^\gamma}{1+\rho(x_\alpha^\delta y_\beta)^\lambda}f(x)g(y)dxdy$$
$$\leq M\left[\int_{-\infty}^{\infty}x_\alpha^{p(1-\delta\sigma_1)-1}f^p(x)dx\right]^{\frac{1}{p}}\left[\int_{-\infty}^{\infty}y_\beta^{q(1-\sigma)-1}g^q(y)dy\right]^{\frac{1}{q}}. \tag{26}$$

(iii) $\sigma_1=\sigma$, and $K_{\alpha,\beta}^{(\gamma)}(\sigma)\leq M$.

*Proof* $(i) \Rightarrow (ii)$ By Hölder's inequality (cf. [46]), we have

$$I = \int_{-\infty}^{\infty} \left( y_\beta^{\sigma - \frac{1}{p}} \int_{-\infty}^{\infty} \frac{|\ln \rho(x_\alpha^\delta y_\beta)^\lambda|^\gamma}{1+\rho(x_\alpha^\delta y_\beta)^\lambda} f(x)dx \right) \left( y_\beta^{-\sigma + \frac{1}{p}} g(y) \right) dy$$
$$\leq J \left[ \int_{-\infty}^{\infty} y_\beta^{q(1-\sigma)-1} g^q(y)dy \right]^{\frac{1}{q}}. \tag{27}$$

Then by (25), we obtain (26).

$(ii) \Rightarrow (iii)$ By Lemma 1, we have $\sigma_1 = \sigma$. Then by Lemma 2, we derive that

$$K_{\alpha,\beta}^{(\gamma)}(\sigma) \leq M.$$

$(iii) \Rightarrow (i)$ For $\sigma_1 = \sigma$, by Hölder's inequality with weight (see [46]) and (22), we have

$$\left( \int_{-\infty}^{\infty} \frac{|\ln \rho(x_\alpha^\delta y_\beta)^\lambda|^\gamma}{1+\rho(x_\alpha^\delta y_\beta)^\lambda} f(x)dx \right)^p$$
$$= \left\{ \int_{-\infty}^{\infty} \frac{|\ln \rho(x_\alpha^\delta y_\beta)^\lambda|^\gamma}{1+\rho(x_\alpha^\delta y_\beta)^\lambda} \left[ \frac{y_\beta^{(\sigma-1)/p} f(x)}{x_\alpha^{(\delta\sigma-1)/q}} \right] \left[ \frac{x_\alpha^{(\delta\sigma-1)/q}}{y_\beta^{(\sigma-1)/p}} \right] dx \right\}^p$$
$$\leq \int_{-\infty}^{\infty} \frac{|\ln \rho(x_\alpha^\delta y_\beta)^\lambda|^\gamma}{1+\rho(x_\alpha^\delta y_\beta)^\lambda} \frac{y_\beta^{\sigma-1}}{x_\alpha^{(\delta\sigma-1)p/q}} f^p(x)dx$$
$$\times \left[ \int_{-\infty}^{\infty} \frac{|\ln \rho(x_\alpha^\delta y_\beta)^\lambda|^\gamma}{1+\rho(x_\alpha^\delta y_\beta)^\lambda} \frac{x^{\delta\sigma-1}}{y_\beta^{(\sigma-1)q/p}} dx \right]^{p/q}$$
$$= \left[ \omega_\delta(\sigma, y) y_\beta^{q(1-\sigma)-1} \right]^{p-1} \int_{-\infty}^{\infty} \frac{|\ln \rho(x_\alpha^\delta y_\beta)^\lambda|^\gamma}{1+\rho(x_\alpha^\delta y_\beta)^\lambda} \frac{y_\beta^{\sigma-1} f^p(x)}{x_\alpha^{(\delta\sigma-1)p/q}} dx$$
$$= \left( \frac{2k_\lambda^{(\gamma)}(\sigma)}{1-\alpha^2} \right)^{p-1} y_\beta^{-p\sigma+1} \int_{-\infty}^{\infty} \frac{|\ln \rho(x_\alpha^\delta y_\beta)^\lambda|^\gamma}{1+\rho(x_\alpha^\delta y_\beta)^\lambda} \frac{y_\beta^{\sigma-1} f^p(x)}{x_\alpha^{(\delta\sigma-1)p/q}} dx. \tag{28}$$

By Fubini's theorem, (28) and (23), we have

$$J \leq \left( \frac{2k_\lambda^{(\gamma)}(\sigma)}{1-\alpha^2} \right)^{\frac{1}{q}} \left[ \int_{-\infty}^{\infty} \int_{-\infty}^{\infty} \frac{|\ln \rho(x_\alpha^\delta y_\beta)^\lambda|^\gamma}{1+\rho(x_\alpha^\delta y_\beta)^\lambda} \frac{y_\beta^{\sigma-1} f^p(x)}{x_\alpha^{(\delta\sigma-1)p/q}} dxdy \right]^{\frac{1}{p}}$$
$$= \left( \frac{2k_\lambda^{(\gamma)}(\sigma)}{1-\alpha^2} \right)^{\frac{1}{q}}$$
$$\times \left\{ \int_{-\infty}^{\infty} \left[ \int_{-\infty}^{\infty} \frac{|\ln \rho(x_\alpha^\delta y_\beta)^\lambda|^\gamma}{1+\rho(x_\alpha^\delta y_\beta)^\lambda} \frac{y_\beta^{\sigma-1}}{x_\alpha^{(\delta\sigma-1)p/q}} dy \right] f^p(x)dx \right\}^{\frac{1}{p}}$$

$$= \left( \frac{2k_{\lambda}^{(\gamma)}(\sigma)}{1-\alpha^2} \right)^{\frac{1}{q}} \left[ \int_{-\infty}^{\infty} \varpi_{\delta}(\sigma, x) x_{\delta}^{p(1-\delta\sigma)-1} f^p(x) dx \right]^{\frac{1}{p}}$$

$$= K_{\alpha,\beta}^{(\gamma)}(\sigma) \left[ \int_{-\infty}^{\infty} x_{\delta}^{p(1-\delta\sigma)-1} f^p(x) dx \right]^{\frac{1}{p}}.$$

For $K_{\alpha,\beta}^{(\gamma)}(\sigma) \leq M$, we have (25) (for $\sigma_1 = \sigma$).

Therefore, Statements (i), (ii) and (iii) are equivalent.

This completes the proof of the theorem. □

For $\sigma_1 = \sigma$, we have the following theorem:

**Theorem 2** *If $p > 1$, $M$ is a constant, then the following statements (i), (ii) and (iii) are equivalent:*

(i) For any $f(x) \geq 0$, satisfying

$$0 < \int_{-\infty}^{\infty} x_{\alpha}^{p(1-\delta\sigma)-1} f^p(x) dx < \infty,$$

we have the following inequality:

$$\left[ \int_{-\infty}^{\infty} y_{\beta}^{p\sigma-1} \left( \int_{-\infty}^{\infty} \frac{|\ln \rho(x_{\alpha}^{\delta} y_{\beta})^{\lambda}|^{\gamma}}{1+\rho(x_{\alpha}^{\delta} y_{\beta})^{\lambda}} f(x) dx \right)^p dy \right]^{\frac{1}{p}}$$

$$< M \left[ \int_{-\infty}^{\infty} x_{\alpha}^{p(1-\delta\sigma)-1} f^p(x) dx \right]^{\frac{1}{p}}. \tag{29}$$

(ii) For any $f(x), g(y) \geq 0$, satisfying

$$0 < \int_{-\infty}^{\infty} x_{\alpha}^{p(1-\delta\sigma)-1} f^p(x) dx < \infty \text{ and } 0 < \int_{-\infty}^{\infty} y_{\beta}^{q(1-\sigma)-1} g^q(y) dy < \infty,$$

we have the following inequality:

$$\int_{-\infty}^{\infty} \int_{-\infty}^{\infty} \frac{|\ln \rho(x_{\alpha}^{\delta} y_{\beta})^{\lambda}|^{\gamma}}{1+\rho(x_{\alpha}^{\delta} y_{\beta})^{\lambda}} f(x) g(y) dx dy$$

$$< M \left[ \int_{-\infty}^{\infty} x_{\alpha}^{p(1-\delta\sigma)-1} f^p(x) dx \right]^{\frac{1}{p}} \left[ \int_{-\infty}^{\infty} y_{\beta}^{q(1-\sigma)-1} g^q(y) dy \right]^{\frac{1}{q}}. \tag{30}$$

(iii) $K_{\alpha,\beta}^{(\gamma)}(\sigma) \leq M$.

Moreover, if statement (iii) holds true, then the constant factor $M = K_{\alpha,\beta}^{(\gamma)}(\sigma)$ in (29) and (30) is the best possible.

In particular:

(1) for $\delta = 1$, we have the following equivalent inequalities with the nonhomogeneous kernel:

$$\left[\int_{-\infty}^{\infty} y_{\beta}^{p\sigma-1}\left(\int_{-\infty}^{\infty}\frac{|\ln\rho(x_{\alpha}y_{\beta})^{\lambda}|^{\gamma}}{1+\rho(x_{\alpha}y_{\beta})^{\lambda}}f(x)dx\right)^{p}dy\right]^{\frac{1}{p}} < K_{\alpha,\beta}^{(\gamma)}(\sigma)\left[\int_{-\infty}^{\infty} x_{\alpha}^{p(1-\sigma)-1}f^{p}(x)dx\right]^{\frac{1}{p}}, \tag{31}$$

$$\int_{-\infty}^{\infty}\int_{-\infty}^{\infty}\frac{|\ln\rho(x_{\alpha}y_{\beta})^{\lambda}|^{\gamma}}{1+\rho(x_{\alpha}y_{\beta})^{\lambda}}f(x)g(y)dxdy < K_{\alpha,\beta}^{(\gamma)}(\sigma)\left[\int_{-\infty}^{\infty} x_{\alpha}^{p(1-\sigma)-1}f^{p}(x)dx\right]^{\frac{1}{p}}\left[\int_{-\infty}^{\infty} y_{\beta}^{q(1-\sigma)-1}g^{q}(y)dy\right]^{\frac{1}{q}}, \tag{32}$$

where $K_{\alpha,\beta}^{(\gamma)}(\sigma)$ is the best possible constant factor.

For $\sigma = \frac{1}{p} < \lambda$, we have

$$\left[\int_{-\infty}^{\infty}\left(\int_{-\infty}^{\infty}\frac{|\ln\rho(x_{\alpha}y_{\beta})^{\lambda}|^{\gamma}}{1+\rho(x_{\alpha}y_{\beta})^{\lambda}}f(x)dx\right)^{p}dy\right]^{\frac{1}{p}} < K_{\alpha,\beta}^{(\gamma)}(\frac{1}{p})\left[\int_{-\infty}^{\infty} x_{\alpha}^{p-2}f^{p}(x)dx\right]^{\frac{1}{p}}, \tag{33}$$

$$\int_{-\infty}^{\infty}\int_{-\infty}^{\infty}\frac{|\ln\rho(x_{\alpha}y_{\beta})^{\lambda}|^{\gamma}}{1+\rho(x_{\alpha}y_{\beta})^{\lambda}}f(x)g(y)dxdy < K_{\alpha,\beta}^{(\gamma)}(\frac{1}{p})\left[\int_{-\infty}^{\infty} x_{\alpha}^{p-2}f^{p}(x)dx\right]^{\frac{1}{p}}\left[\int_{-\infty}^{\infty} g^{q}(y)dy\right]^{\frac{1}{q}}. \tag{34}$$

(2) for $\delta = -1$, replacing $x^{\lambda}f(x)$ by $f(x)$, we have

$$0 < \int_{-\infty}^{\infty} x_{\alpha}^{p(1-\mu)-1}f^{p}(x)dx < \infty,$$

and the following equivalent inequalities with the homogeneous kernel of degree $-\lambda$:

$$\left[\int_{-\infty}^{\infty} y_{\beta}^{p\sigma-1}\left(\int_{-\infty}^{\infty}\frac{|\ln\rho(y_{\beta}/x_{\alpha})^{\lambda}|^{\gamma}}{x_{\alpha}^{\lambda}+\rho y_{\beta}^{\lambda}}f(x)dx\right)^{p}dy\right]^{\frac{1}{p}}$$

$$< K_{\alpha,\beta}^{(\gamma)}(\sigma)\left[\int_{-\infty}^{\infty} x_{\alpha}^{p(1-\mu)-1} f^{p}(x)dx\right]^{\frac{1}{p}}, \tag{35}$$

$$\int_{-\infty}^{\infty}\int_{-\infty}^{\infty} \frac{|\ln \rho(y_{\beta}/x_{\alpha})^{\lambda}|^{\gamma}}{x_{\alpha}^{\lambda}+\rho y_{\beta}^{\lambda}} f(x)g(y)dxdy$$
$$< K_{\alpha,\beta}^{(\gamma)}(\sigma)\left[\int_{-\infty}^{\infty} x_{\alpha}^{p(1-\mu)-1} f^{p}(x)dx\right]^{\frac{1}{p}}\left[\int_{-\infty}^{\infty} y_{\beta}^{q(1-\sigma)-1} g^{q}(y)dy\right]^{\frac{1}{q}}, \tag{36}$$

where $K_{\alpha,\beta}^{(\gamma)}(\sigma)$ is the best possible constant factor.

For $\lambda = 1, \sigma = \frac{1}{p}, \mu = \frac{1}{q}$, we have

$$\left[\int_{-\infty}^{\infty}\left(\int_{-\infty}^{\infty} \frac{|\ln \rho(y_{\beta}/x_{\alpha})|^{\gamma}}{x_{\alpha}+\rho y_{\beta}} f(x)dx\right)^{p} dy\right]^{\frac{1}{p}}$$
$$< 2\Gamma(\gamma+1)\frac{\xi(\gamma+1,\frac{1}{p})+\xi(\gamma+1,\frac{1}{q})}{\rho^{1/p}(1-\alpha^{2})^{1/q}(1-\beta^{2})^{1/p}}\left(\int_{-\infty}^{\infty} f^{p}(x)dx\right)^{\frac{1}{p}}, \tag{37}$$

$$\int_{-\infty}^{\infty}\int_{-\infty}^{\infty} \frac{|\ln \rho(y_{\beta}/x_{\alpha})|^{\gamma}}{x_{\alpha}+\rho y_{\beta}} f(x)g(y)dxdy$$
$$< 2\Gamma(\gamma+1)\frac{\xi(\gamma+1,\frac{1}{p})+\xi(\gamma+1,\frac{1}{q})}{\rho^{1/p}(1-\alpha^{2})^{1/q}(1-\beta^{2})^{1/p}}$$
$$\times\left(\int_{-\infty}^{\infty} f^{p}(x)dx\right)^{\frac{1}{p}}\left(\int_{-\infty}^{\infty} g^{q}(y)dy\right)^{\frac{1}{q}}. \tag{38}$$

*Proof* For $\sigma_1 = \sigma$ and the assumption of (i), if (28) assumes the form of equality for some $y \in \mathbf{R}\backslash\{0\}$, then (see [46]), there exist constants $A$ and $B$, such that they are not all zero, and

$$A\frac{y_{\beta}^{\sigma-1}}{x_{\alpha}^{(\delta\sigma-1)p/q}} f^{p}(x) = B\frac{x^{\delta\sigma-1}}{y_{\beta}^{(\sigma-1)q/p}} \text{ a.e. in } \mathbf{R}.$$

Let us suppose that $A \neq 0$ (otherwise $B = A = 0$). Then, it follows that

$$x_{\delta}^{p(1-\delta\sigma)-1} f^{p}(x) = y_{\beta}^{q(1-\sigma)}\frac{B}{Ax_{\alpha}} \text{ a.e. in } \mathbf{R}.$$

Since

$$\int_{-\infty}^{\infty} x_{\alpha}^{-1}dx = \infty,$$

this contradicts the fact that

$$0 < \int_{-\infty}^{\infty} x_{\alpha}^{p(1-\delta\sigma)-1} f^{p}(x)dx < \infty.$$

Hence, (28) assumes the form of strict inequality, and so does (25). Hence, (29) and (30) are valid.

In view of Theorem 1, we can still conclude that the statements (i), (ii) and (iii) in Theorem 2 are equivalent.

When the statement (iii) holds true, namely, $K_{\alpha,\beta}^{(\gamma)}(\sigma) \leq M$, if there exists a constant $M \leq K_{\alpha,\beta}^{(\gamma)}(\sigma)$, such that (30) is valid, then we have $M = K_{\alpha,\beta}^{(\gamma)}(\sigma)$, and we can conclude that the constant factor $M = K_{\alpha,\beta}^{(\gamma)}(\sigma)$ in (30) is the best possible.

The constant factor $M = K_{\alpha,\beta}^{(\gamma)}(\sigma)$ in (29) is still the best possible. Otherwise, by (27) (for $\sigma_1 = \sigma$), we would reach a contradiction that the constant factor $M = K_{\alpha,\beta}^{(\gamma)}(\sigma)$ in (30) is not the best possible.

This completes the proof of the theorem. □

## 4 Operator Expressions and a Remark

(1) For $p > 1$, we set the following functions:

$$\varphi(x) := x_{\alpha}^{p(1-\delta\sigma)-1} (x \in \mathbf{R}), \quad \text{and} \quad \psi(y) := y_{\beta}^{q(1-\sigma)-1},$$

wherefrom,

$$\psi^{1-p}(y) = y_{\beta}^{p\sigma-1} \; (y \in \mathbf{R}).$$

We define the following real normed linear spaces:

$$L_{p,\varphi}(\mathbf{R}) := \left\{ f : ||f||_{p,\varphi} := \left( \int_{-\infty}^{\infty} \varphi(x)|f(x)|^{p} dx \right)^{\frac{1}{p}} < \infty \right\},$$

$$L_{q,\psi}(\mathbf{R}) = \left\{ g : ||g||_{q,\psi} = \left( \int_{-\infty}^{\infty} \psi(y)|g(y)|^{q} dy \right)^{\frac{1}{q}} < \infty \right\},$$

$$L_{p,\psi^{1-p}}(\mathbf{R}) = \left\{ h : ||h||_{p,\psi^{1-p}} = \left( \int_{-\infty}^{\infty} \psi^{1-p}(y)|h(y)|^{p} dy \right)^{\frac{1}{p}} < \infty \right\}.$$

In view of Theorem 2, for $f \in L_{p,\varphi}(\mathbf{R})$, setting

$$h_1(y) := \int_{-\infty}^{\infty} \frac{|\ln \rho(x_\alpha^\delta y_\beta)^\lambda|^\gamma}{1+\rho(x_\alpha^\delta y_\beta)^\lambda} f(x)dx \ (y \in \mathbf{R}),$$

by (31), we have

$$||h_1||_{p,\psi^{1-p}} = \left[\int_{-\infty}^{\infty} \psi^{1-p}(y)h_1^p(y)dy\right]^{\frac{1}{p}} \leq M||f||_{p,\varphi} < \infty. \tag{39}$$

**Definition 1** We define a Hilbert-type integral operator with the nonhomogeneous kernel $T_1 : L_{p,\varphi}(\mathbf{R}) \to L_{p,\psi^{1-p}}(\mathbf{R})$ as follows:

For any $f \in L_{p,\varphi}(\mathbf{R})$, there exists a unique representation $T_1 f = h_1 \in L_{p,\psi^{1-p}}(\mathbf{R})$, satisfying $T_1 f(y) = h_1(y)$, for any $y \in \mathbf{R}$.

In view of (39), it follows that

$$||T_1 f||_{p,\psi^{1-p}} = ||h_1||_{p,\psi^{1-p}} \leq M||f||_{p,\varphi},$$

and then the operator $T_1$ is bounded satisfying

$$||T_1|| = \sup_{f(\neq\theta)\in L_{p,\varphi}(\mathbf{R})} \frac{||T_1 f||_{p,\psi^{1-p}}}{||f||_{p,\varphi}} \leq M.$$

If we define the formal inner product of $T_1 f$ and $g$ as follows:

$$(T_1 f, g) := \int_{-\infty}^{\infty} \left(\int_{-\infty}^{\infty} \frac{|\ln \rho(x_\alpha^\delta y_\beta)^\lambda|^\gamma}{1+\rho(x_\alpha^\delta y_\beta)^\lambda} f(x)dx\right) g(y)dy,$$

then we can express Theorem 2 as follows:

**Theorem 3** *If $p > 1$, $M$ is a constant, then the following statements (i), (ii) and (iii) are equivalent:*

(i) For any $f(x) \geq 0$, $f \in L_{p,\varphi}(\mathbf{R})$, $||f||_{p,\varphi} > 0$, we have the following inequality:

$$||T_1 f||_{p,\psi^{1-p}} < M||f||_{p,\varphi}. \tag{40}$$

(ii) For any $f(x), g(y) \geq 0$, $f \in L_{p,\varphi}(\mathbf{R})$, $g \in L_{q,\psi}(\mathbf{R})$, $||f||_{p,\varphi}, ||g||_{q,\psi} > 0$, we have the following inequality:

$$(T_1 f, g) < M||f||_{p,\varphi}||g||_{q,\psi}. \tag{41}$$

(iii) $K_{\alpha,\beta}^{(\gamma)}(\sigma) \leq M$.

Moreover, if the statement (iii) holds true, then the constant factor $M = K_{\alpha,\beta}^{(\gamma)}(\sigma)$ in (40) and (41) is the best possible, namely,

$$||T_1|| = K_{\alpha,\beta}^{(\gamma)}(\sigma).$$

(2) For $p > 1$, we also set the following function:

$$\phi(x) := x_\alpha^{p(1-\mu)-1} \ (x \in \mathbf{R})$$

as well as the real normed linear spaces:

$$L_{p,\phi}(\mathbf{R}) := \left\{ f : ||f||_{p,\phi} := \left( \int_{-\infty}^{\infty} \phi(x)|f(x)|^p dx \right)^{\frac{1}{p}} < \infty \right\}.$$

In view of (35), for $f \in L_{p,\phi}(\mathbf{R})$, setting

$$h_2(y) := \int_{-\infty}^{\infty} \frac{|\ln \rho(y_\beta/x_\alpha)^\lambda|^\gamma}{x_\alpha^\lambda + \rho y_\beta^\lambda} f(x)dx \ (y \in \mathbf{R}),$$

we have

$$||h_2||_{p,\psi^{1-p}} = \left[ \int_{-\infty}^{\infty} \psi^{1-p}(y) h_2^p(y) dy \right]^{\frac{1}{p}} \leq M||f||_{p,\phi} < \infty. \tag{42}$$

**Definition 2** Let us define a Hilbert-type integral operator with the homogeneous kernel $T_2 : L_{p,\phi}(\mathbf{R}) \to L_{p,\psi^{1-p}}(\mathbf{R})$ as follows:

For any $f \in L_{p,\phi}(\mathbf{R})$, there exists a unique representation $T_2 f = h_2 \in L_{p,\psi^{1-p}}$ $(\mathbf{R})$, satisfying $T_2 f(y) = h_2(y)$, for any $y \in \mathbf{R}$.

In view of (42), it follows that

$$||T_2 f||_{p,\psi^{1-p}} = ||h_2||_{p,\psi^{1-p}} \leq M||f||_{p,\phi},$$

and then the operator $T_2$ is bounded satisfying

$$||T_2|| = \sup_{f(\neq\theta)\in L_{p,\phi}(\mathbf{R})} \frac{||T_2 f||_{p,\psi^{1-p}}}{||f||_{p,\phi}} \leq M.$$

If we define the formal inner product of $T_2 f$ and $g$ as follows:

$$(T_2 f, g) := \int_{-\infty}^{\infty} \left( \int_{-\infty}^{\infty} \frac{|\ln \rho(y_\beta/x_\alpha)^\lambda|^\gamma}{x_\alpha^\lambda + \rho y_\beta^\lambda} f(x)dx \right) g(y)dy,$$

then we can express Theorem 2 (for $\delta = -1$, replacing $x^\lambda f(x)$ by $f(x)$) as follows:

**Theorem 4** *If $p > 1$, $M$ is a constant, then the following statements (i), (ii) and (iii) are equivalent:*

(i) For any $f(x) \geq 0, f \in L_{p,\phi}(\mathbf{R})$, $||f||_{p,\phi} > 0$, we have the following inequality:

$$||T_2 f||_{p,\psi^{1-p}} < M||f||_{p,\phi}. \tag{43}$$

(ii) For any $f(x), g(y) \geq 0$, $f \in L_{p,\phi}(\mathbf{R})$, $g \in L_{q,\psi}(\mathbf{R})$, $||f||_{p,\phi}, ||g||_{q,\psi} > 0$, we have the following inequality:

$$(T_2 f, g) < M||f||_{p,\phi}||g||_{q,\psi}. \tag{44}$$

(iii) $K_{\alpha,\beta}^{(\gamma)}(\sigma) \leq M$.

Moreover, if the statement (iii) holds true, then the constant factor $M = K_{\alpha,\beta}^{(\gamma)}(\sigma)$ in (43) and (44) is the best possible, namely,

$$||T_2|| = ||T_1|| = K_{\alpha,\beta}^{(\gamma)}(\sigma).$$

*Remark 1* (1) In particular, for $\alpha = \beta = 0$ in (31) and (32), we have the following equivalent inequalities:

$$\left[\int_{-\infty}^{\infty} |y|^{p\sigma-1}\left(\int_{-\infty}^{\infty} \frac{|\ln \rho|xy|^{\lambda}|^{\gamma}}{1+\rho|xy|^{\lambda}} f(x)dx\right)^p dy\right]^{\frac{1}{p}}$$
$$< 2k_{\lambda}^{(\gamma)}(\sigma)\left[\int_{-\infty}^{\infty} |x|^{p(1-\sigma)-1} f^p(x)dx\right]^{\frac{1}{p}}, \tag{45}$$

$$\int_{-\infty}^{\infty}\int_{-\infty}^{\infty} \frac{|\ln \rho|xy|^{\lambda}|^{\gamma}}{1+\rho|xy|^{\lambda}} f(x)g(y)dxdy$$
$$< 2k_{\lambda}^{(\gamma)}(\sigma)\left[\int_{-\infty}^{\infty} |x|^{p(1-\sigma)-1} f^p(x)dx\right]^{\frac{1}{p}}\left[\int_{-\infty}^{\infty} |y|^{q(1-\sigma)-1} g^q(y)dy\right]^{\frac{1}{q}}, \tag{46}$$

where $2k_{\lambda}^{(\gamma)}(\sigma)$ is the best possible constant factor.

If $f(-x) = f(x)$, $g(-y) = g(y)$ $(x, y \in \mathbf{R}_+)$, then we have the following equivalent inequalities:

$$\left[\int_0^{\infty} y^{p\sigma-1}\left(\int_0^{\infty} \frac{|\ln \rho(xy)^{\lambda}|^{\gamma}}{1+\rho(xy)^{\lambda}} f(x)dx\right)^p dy\right]^{\frac{1}{p}}$$
$$< k_{\lambda}^{(\gamma)}(\sigma)\left[\int_0^{\infty} x^{p(1-\sigma)-1} f^p(x)dx\right]^{\frac{1}{p}}, \tag{47}$$

$$\int_0^{\infty}\int_0^{\infty} \frac{|\ln \rho(xy)^{\lambda}|^{\gamma}}{1+\rho(xy)^{\lambda}} f(x)g(y)dxdy$$
$$< k_{\lambda}^{(\gamma)}(\sigma)\left[\int_0^{\infty} x^{p(1-\sigma)-1} f^p(x)dx\right]^{\frac{1}{p}}\left[\int_0^{\infty} y^{q(1-\sigma)-1} g^q(y)dy\right]^{\frac{1}{q}}, \tag{48}$$

where $k_\lambda^{(\gamma)}(\sigma)$ is the best possible constant factor.

(2) For $\alpha = \beta = 0$ in (35) and (36), we have the following equivalent inequalities:

$$\left[\int_{-\infty}^{\infty} |y|^{p\sigma-1}\left(\int_{-\infty}^{\infty} \frac{|\ln \rho|y/x|^\lambda|^\gamma}{|x|^\lambda + \rho|y|^\lambda} f(x)dx\right)^p dy\right]^{\frac{1}{p}} < 2k_\lambda^{(\gamma)}(\sigma)\left[\int_{-\infty}^{\infty} |x|^{p(1-\mu)-1} f^p(x)dx\right]^{\frac{1}{p}}, \tag{49}$$

$$\int_{-\infty}^{\infty}\int_{-\infty}^{\infty} \frac{|\ln \rho|y/x|^\lambda|^\gamma}{|x|^\lambda + \rho|y|^\lambda} f(x)g(y)dxdy < 2k_\lambda^{(\gamma)}(\sigma)\left[\int_{-\infty}^{\infty} |x|^{p(1-\mu)-1} f^p(x)dx\right]^{\frac{1}{p}}\left[\int_{-\infty}^{\infty} |y|^{q(1-\sigma)-1} g^q(y)dy\right]^{\frac{1}{q}}, \tag{50}$$

where, $2k_\lambda^{(\gamma)}(\sigma)$ is the best possible constant factor.

If $f(-x) = f(x), g(-y) = g(y)$ $(x, y \in \mathbf{R}_+)$, then we have the following equivalent inequalities:

$$\left[\int_0^{\infty} y^{p\sigma-1}\left(\int_0^{\infty} \frac{|\ln \rho(y/x)^\lambda|^\gamma}{x^\lambda + \rho y^\lambda} f(x)dx\right)^p dy\right]^{\frac{1}{p}} < k_\lambda^{(\gamma)}(\sigma)\left[\int_0^{\infty} x^{p(1-\mu)-1} f^p(x)dx\right]^{\frac{1}{p}}, \tag{51}$$

$$\int_0^{\infty}\int_0^{\infty} \frac{|\ln \rho(y/x)^\lambda|^\gamma}{x^\lambda + \rho y^\lambda} f(x)g(y)dxdy < k_\lambda^{(\gamma)}(\sigma)\left[\int_0^{\infty} x^{p(1-\mu)-1} f^p(x)dx\right]^{\frac{1}{p}}\left[\int_0^{\infty} y^{q(1-\sigma)-1} g^q(y)dy\right]^{\frac{1}{q}}, \tag{52}$$

where $k_\lambda^{(\gamma)}(\sigma)$ is the best possible constant factor.

## 5 Some Equivalent Reverses

**Lemma 6** *If* $0 < p < 1$, *then the following statements (i), (ii) and (iii) are true and equivalent:*

(i) For any $f(x) \geq 0$, satisfying

$$\int_{-\infty}^{\infty} x_\alpha^{p(1-\delta\sigma)-1} f^p(x)dx < \infty,$$

we have the following inequality:

$$J = \left[\int_{-\infty}^{\infty} y_\beta^{p\sigma-1}\left(\int_{-\infty}^{\infty} \frac{|\ln \rho(x_\alpha^\delta y_\beta)^\lambda|^\gamma}{1+\rho(x_\alpha^\delta y_\beta)^\lambda} f(x)dx\right)^p dy\right]^{\frac{1}{p}}$$
$$\geq K_{\alpha,\beta}^{(\gamma)}(\sigma)\left[\int_{-\infty}^{\infty} x_\alpha^{p(1-\delta\sigma)-1} f^p(x)dx\right]^{\frac{1}{p}}. \tag{53}$$

(ii) For any $g(y) \geq 0$, satisfying

$$\int_{-\infty}^{\infty} y_\beta^{q(1-\sigma)-1} g^q(y)dy < \infty,$$

we have the following inequality:

$$L := \left[\int_{-\infty}^{\infty} x^{q\delta\sigma-1}\left(\int_{-\infty}^{\infty} \frac{|\ln \rho(x_\alpha^\delta y_\beta)^\lambda|^\gamma}{1+\rho(x_\alpha^\delta y_\beta)^\lambda} g(y)dy\right)^q dx\right]^{\frac{1}{q}}$$
$$\geq K_{\alpha,\beta}^{(\gamma)}(\sigma)\left[\int_{-\infty}^{\infty} y_\beta^{q(1-\sigma)-1} g^q(y)dy\right]^{\frac{1}{q}}. \tag{54}$$

(iii) For any $f(x), g(y) \geq 0$, satisfying

$$\int_{-\infty}^{\infty} x_\alpha^{p(1-\delta\sigma)-1} f^p(x)dx < \infty, \text{ and } \int_{-\infty}^{\infty} y_\beta^{q(1-\sigma)-1} g^q(y)dy < \infty,$$

we have the following inequality:

$$I = \int_{-\infty}^{\infty}\int_{-\infty}^{\infty} \frac{|\ln \rho(x_\alpha^\delta y_\beta)^\lambda|^\gamma}{1+\rho(x_\alpha^\delta y_\beta)^\lambda} f(x)g(y)dxdy$$
$$\geq K_{\alpha,\beta}^{(\gamma)}(\sigma)\left[\int_{-\infty}^{\infty} x_\alpha^{p(1-\delta\sigma)-1} f^p(x)dx\right]^{\frac{1}{p}}\left[\int_{-\infty}^{\infty} y_\beta^{q(1-\sigma)-1} g^q(y)dy\right]^{\frac{1}{q}}. \tag{55}$$

Moreover, the constant factor $K_{\alpha,\beta}^{(\gamma)}(\sigma)$ in (53), (54) and (55) is the best possible.

*Proof* By the reverse Hölder inequality (cf. [46]), we deduce the reverse of (28). Then we can similarly obtain (53).

$(i) \Rightarrow (iii)$ By the reverse Hölder inequality, we have

$$I = \int_{-\infty}^{\infty}\left(y_\beta^{\sigma-\frac{1}{p}}\int_{-\infty}^{\infty} \frac{|\ln \rho(x_\alpha^\delta y_\beta)^\lambda|^\gamma}{1+\rho(x_\alpha^\delta y_\beta)^\lambda} f(x)dx\right)\left(y_\beta^{-\sigma+\frac{1}{p}} g(y)\right)dy$$

$$\geq J\left[\int_{-\infty}^{\infty} y_{\beta}^{q(1-\sigma)-1} g^{q}(y)dy\right]^{\frac{1}{q}}. \tag{56}$$

Then by (53), we have (55).
$(ii) \Rightarrow (iii)$ By the reverse Hölder inequality (cf. [46]), we have

$$I = \int_{-\infty}^{\infty}\left(x_{\alpha}^{-\delta\sigma+\frac{1}{q}} f(x)\right)\left(x_{\alpha}^{\delta\sigma-\frac{1}{q}}\int_{-\infty}^{\infty}\frac{|\ln\rho(x_{\alpha}^{\delta}y_{\beta})^{\lambda}|^{\gamma}}{1+\rho(x_{\alpha}^{\delta}y_{\beta})^{\lambda}} g(y)dy\right)dx$$
$$\geq \left[\int_{-\infty}^{\infty} x_{\alpha}^{p(1-\delta\sigma)-1} f^{p}(x)dx\right]^{\frac{1}{p}} L. \tag{57}$$

Then by (54), we have (55).
$(iii) \Rightarrow (i)$ We set

$$g(y) := y_{\beta}^{p\sigma-1}\left(\int_{-\infty}^{\infty}\frac{|\ln\rho(x_{\alpha}^{\delta}y_{\beta})^{\lambda}|^{\gamma}}{1+\rho(x_{\alpha}^{\delta}y_{\beta})^{\lambda}} f(x)dx\right)^{p-1} \quad (y\in\mathbf{R}).$$

If $J = 0$, then (53) assumes the form of equality; if $J = \infty$, then (53) remains true. Suppose that $0 < J < \infty$. By (55), we obtain that

$$\infty > \int_{-\infty}^{\infty} y_{\beta}^{q(1-\sigma)-1} g^{q}(y)dy = J^{p} = I$$
$$\geq K_{\alpha,\beta}^{(\gamma)}(\sigma)\left[\int_{-\infty}^{\infty} x_{\alpha}^{p(1-\delta\sigma)-1} f^{p}(x)dx\right]^{\frac{1}{p}}\left[\int_{-\infty}^{\infty} y_{\beta}^{q(1-\sigma)-1} g^{q}(y)dy\right]^{\frac{1}{q}} > 0,$$

$$\left[\int_{-\infty}^{\infty} y_{\beta}^{q(1-\sigma)-1} g^{q}(y)dy\right]^{\frac{1}{p}} = J \geq K_{\alpha,\beta}^{(\gamma)}(\sigma)\left[\int_{-\infty}^{\infty} x_{\alpha}^{p(1-\delta\sigma)-1} f^{p}(x)dx\right]^{\frac{1}{p}},$$

namely, (53) follows.
$(iii) \Rightarrow (ii)$ Similarly, setting

$$f(x) := x^{q\delta\sigma-1}\left(\int_{-\infty}^{\infty}\frac{|\ln\rho(x_{\alpha}^{\delta}y_{\beta})^{\lambda}|^{\gamma}}{1+\rho(x_{\alpha}^{\delta}y_{\beta})^{\lambda}} g(y)dy\right)^{q-1} \quad (x\in\mathbf{R}),$$

by (55), we obtain (54).

Since (53) is true, then the statements (i), (ii) and (iii) are true and equivalent.

For $n \geq \frac{1}{p\sigma}$ $(n\in\mathbf{N})$, we set the following functions:

$$f_{n}(x) = \begin{cases} x_{\alpha}^{\delta(\sigma-\frac{1}{pn})-1}, x\in E_{\delta} \\ 0, x\in\mathbf{R}\backslash E_{\delta} \end{cases},$$

$$g_n(y) = \begin{cases} y_\beta^{\sigma+\frac{1}{qn}-1}, y \in E_{-1} \\ 0, y \in \mathbf{R} \backslash E_{-1} \end{cases}.$$

If there exists a constant $M(\geq K_{\alpha,\beta}^{(\gamma)}(\sigma))$, such that (55) is valid when replacing $K_{\alpha,\beta}^{(\gamma)}(\sigma)$ by $M$, then for $f = f_n, g = g_n$, we have

$$\begin{aligned} I_n &:= \int_{-\infty}^{\infty}\int_{-\infty}^{\infty} \frac{|\ln \rho(x_\alpha^\delta y_\beta)^\lambda|^\gamma}{1+\rho(x_\alpha^\delta y_\beta)^\lambda} f_n(x)g_n(y)dxdy \\ &\geq M\left[\int_{-\infty}^{\infty} x_\alpha^{p(1-\delta\sigma)-1} f_n^p(x)dx\right]^{\frac{1}{p}} \left[\int_{-\infty}^{\infty} y_\beta^{q(1-\sigma)-1} g_n^q(y)dy\right]^{\frac{1}{q}}. \end{aligned}$$

We obtain

$$\begin{aligned} J_n &:= \left[\int_{-\infty}^{\infty} x_\alpha^{p(1-\delta\sigma)-1} f_n^p(x)dx\right]^{\frac{1}{p}} \left[\int_{-\infty}^{\infty} y_\beta^{q(1-\sigma)-1} g_n^q(y)dy\right]^{\frac{1}{q}} \\ &= n\left[\frac{1}{(1+\alpha)^{\frac{\delta}{n}+1}} + \frac{1}{(1-\alpha)^{\frac{\delta}{n}+1}}\right]^{\frac{1}{p}} \\ &\times \left[\frac{1}{(1+\beta)^{-\frac{1}{n}+1}} + \frac{1}{(1-\beta)^{-\frac{1}{n}+1}}\right]^{\frac{1}{q}}, \end{aligned}$$

$$\begin{aligned} \frac{1}{n}I_n &= \frac{1}{n}\int_{E_{-1}}\left[y_\beta^{(\sigma-\frac{1}{pn})}\int_{E_\delta} \frac{|\ln \rho(x_\alpha^\delta y_\beta)^\lambda|^\gamma}{1+\rho(x_\alpha^\delta y_\beta)^\lambda} x_\alpha^{\delta(\sigma-\frac{1}{pn})-1}dx\right] y_\beta^{\frac{1}{n}-1}dy \\ &\geq \frac{1}{n}\int_{E_{-1}} \omega_\delta(\sigma-\frac{1}{pn}, y) y_\beta^{\frac{1}{n}-1}dy \\ &= \frac{2k_\lambda^{(\gamma)}(\sigma-\frac{1}{pn})}{1-\alpha^2}\left[\frac{1}{(1+\beta)^{-\frac{1}{n}+1}} + \frac{1}{(1-\beta)^{-\frac{1}{n}+1}}\right] \\ &\geq \frac{1}{n}MJ_n. \end{aligned} \tag{58}$$

Setting $\sigma_0 \in (0, \sigma), n \geq \frac{1}{p(\sigma-\sigma_0)}$, it follows that

$$\begin{aligned} k_\lambda^{(\gamma)}(\sigma-\frac{1}{pn}) &= \int_0^1 h(u)u^{\sigma-\frac{1}{pn}-1}du + \int_1^\infty h(u)u^{\sigma-\frac{1}{pn}-1}du \\ &\leq \int_0^1 h(u)u^{\sigma_0-1}du + \int_1^\infty h(u)u^{\sigma-1}du \\ &\leq k_\lambda^{(\gamma)}(\sigma_0) + k_\lambda^{(\gamma)}(\sigma) < \infty. \end{aligned}$$

By the Lebesgue control convergence theorem (cf. [44]), we have

$$k_{\lambda}^{(\gamma)}(\sigma-\frac{1}{pn}) \to k_{\lambda}^{(\gamma)}(\sigma)\ (n \to \infty).$$

Setting $n \to \infty$ in (58), we have

$$\frac{2k_{\lambda}^{(\gamma)}(\sigma)}{1-\alpha^2}\frac{2}{1-\beta^2} \geq M\left(\frac{2}{1-\alpha^2}\right)^{\frac{1}{p}}\left(\frac{2}{1-\beta^2}\right)^{\frac{1}{q}},$$

namely

$$K_{\alpha,\beta}^{(\gamma)}(\sigma) = \frac{2k_{\lambda}^{(\gamma)}(\sigma)}{(1-\alpha^2)^{1/q}(1-\beta^2)^{1/p}} \geq M.$$

Hence, $M = K_{\alpha,\beta}^{(\gamma)}(\sigma)$ is the best possible constant factor of (55). The constant factor in (53), (54) is still the best possible. Otherwise, by (56), (57) we would reach the contradiction that the constant factor in (55) is not the best possible.

This completes the proof of the lemma. □

**Lemma 7** *If* $0 < p < 1$, *there exists a constant* $M$, *such that for any nonnegative measurable functions* $f(x)$ *and* $g(y)$ *in* $\mathbf{R}$, *satisfying*

$$\int_{-\infty}^{\infty} x_{\alpha}^{p(1-\delta\sigma)-1} f^p(x)dx < \infty, \ and \int_{-\infty}^{\infty} y_{\beta}^{q(1-\sigma)-1} g^q(y)dy < \infty,$$

*the following inequality*

$$\begin{aligned} I &= \int_{-\infty}^{\infty}\int_{-\infty}^{\infty} \frac{|\ln \rho(x_{\alpha}^{\delta} y_{\beta})^{\lambda}|^{\gamma}}{1+\rho(x_{\alpha}^{\delta} y_{\beta})^{\lambda}} f(x)g(y)dxdy \\ &\geq M\left[\int_{-\infty}^{\infty} x_{\alpha}^{p(1-\delta\sigma_1)-1} f^p(x)dx\right]^{\frac{1}{p}}\left[\int_{-\infty}^{\infty} y_{\beta}^{q(1-\sigma)-1} g^q(y)dy\right]^{\frac{1}{q}} \end{aligned} \tag{59}$$

*holds true, then we have* $\sigma_1 = \sigma$, *and* $M \leq K_{\alpha,\beta}^{(\gamma)}(\sigma)$.

*Proof* By the reverse Hölder inequality and (24), we have

$$\begin{aligned} I &= \int_{-\infty}^{\infty}\int_{-\infty}^{\infty} \frac{|\ln \rho(x_{\alpha}^{\delta} y_{\beta})^{\lambda}|^{\gamma}}{1+\rho(x_{\alpha}^{\delta} y_{\beta})^{\lambda}} \left[\frac{y_{\beta}^{(\sigma-1)/p} f(x)}{x_{\alpha}^{(\delta\sigma_1-1)/q}}\right]\left[\frac{x_{\alpha}^{(\delta\sigma_1-1)/q} g(y)}{y_{\beta}^{(\sigma-1)/p}}\right] dxdy \\ &\geq \left[\int_{-\infty}^{\infty} \varpi_{\delta}(\sigma,x) x_{\alpha}^{p(1-\delta\sigma_1)-1} f^p(x)dx\right]^{\frac{1}{p}}\left[\int_{-\infty}^{\infty} \omega(\sigma_1,y) y_{\beta}^{q(1-\sigma)-1} g^q(y)dy\right]^{\frac{1}{q}} \\ &= 2\left(\frac{k_{\lambda}^{(\gamma)}(\sigma)}{1-\beta^2}\right)^{\frac{1}{p}}\left(\frac{k_{\lambda}^{(\gamma)}(\sigma_1)}{1-\alpha^2}\right)^{\frac{1}{q}} \end{aligned}$$

$$\times \left[ \int_{-\infty}^{\infty} x_{\alpha}^{\delta(\sigma_1-\sigma)} x_{\alpha}^{p(1-\delta\sigma_1)-1} f^p(x)dx \right]^{\frac{1}{p}} \left[ \int_{-\infty}^{\infty} y_{\beta}^{\sigma-\sigma_1} y_{\beta}^{q(1-\sigma)-1} g^q(y)dy \right]^{\frac{1}{q}}.$$

In view of (59), we obtain that

$$x_{\alpha}^{\delta(\sigma_1-\sigma)} = \text{Constant and } y_{\beta}^{\sigma-\sigma_1} = \text{Constant},$$

namely, $\sigma_1 = \sigma$. Then we obtain that

$$2\left(\frac{k_{\lambda}^{(\gamma)}(\sigma)}{1-\beta^2}\right)^{\frac{1}{p}} \left(\frac{k_{\lambda}^{(\gamma)}(\sigma_1)}{1-\alpha^2}\right)^{\frac{1}{q}} = K_{\alpha,\beta}^{(\gamma)}(\sigma).$$

By Lemma 6, since the constant factor $K_{\alpha,\beta}^{(\gamma)}(\sigma)$ in (55) is the best possible, it follows that $M \le K_{\alpha,\beta}^{(\gamma)}(\sigma)$.

This completes the proof of the lemma. □

In view of Lemmas 6 and 7, we obtain the following theorem:

**Theorem 5** *If* $0 < p < 1$, $M$ *is a constant, then the following statements (i), (ii), (iii) and (iv) are equivalent:*

(i) For any $f(x) \ge 0$, satisfying

$$0 < \int_{-\infty}^{\infty} x_{\alpha}^{p(1-\delta\sigma)-1} f^p(x)dx < \infty,$$

we have the following inequality:

$$\left[ \int_{-\infty}^{\infty} y_{\beta}^{p\sigma-1} \left( \int_{-\infty}^{\infty} \frac{|\ln \rho(x_{\alpha}^{\delta} y_{\beta})^{\lambda}|^{\gamma}}{1+\rho(x_{\alpha}^{\delta} y_{\beta})^{\lambda}} f(x)dx \right)^p dy \right]^{\frac{1}{p}}$$
$$> M \left[ \int_{-\infty}^{\infty} x_{\alpha}^{p(1-\delta\sigma_1)-1} f^p(x)dx \right]^{\frac{1}{p}}. \tag{60}$$

(ii) For any $g(y) \ge 0$, satisfying

$$0 < \int_{-\infty}^{\infty} y_{\beta}^{q(1-\sigma)-1} g^q(y)dy < \infty,$$

we have the following inequality:

$$\left[ \int_{-\infty}^{\infty} x^{q\delta\sigma_1-1} \left( \int_{-\infty}^{\infty} \frac{|\ln \rho(x_{\alpha}^{\delta} y_{\beta})^{\lambda}|^{\gamma}}{1+\rho(x_{\alpha}^{\delta} y_{\beta})^{\lambda}} g(y)dy \right)^q dx \right]^{\frac{1}{q}}$$
$$> M \left[ \int_{-\infty}^{\infty} y_{\beta}^{q(1-\sigma)-1} g^q(y)dy \right]^{\frac{1}{q}}. \tag{61}$$

(iii) For any $f(x), g(y) \geq 0$, satisfying

$$0 < \int_{-\infty}^{\infty} x_{\alpha}^{p(1-\delta\sigma_1)-1} f^p(x)dx < \infty, \; and\, 0 < \int_{-\infty}^{\infty} y_{\beta}^{q(1-\sigma)-1} g^q(y)dy < \infty,$$

we have the following inequality:

$$\int_{-\infty}^{\infty}\int_{-\infty}^{\infty} \frac{|\ln \rho(x_{\alpha}^{\delta} y_{\beta})^{\lambda}|^{\gamma}}{1+\rho(x_{\alpha}^{\delta} y_{\beta})^{\lambda}} f(x)g(y)dxdy$$
$$> M\left[\int_{-\infty}^{\infty} x_{\alpha}^{p(1-\delta\sigma_1)-1} f^p(x)dx\right]^{\frac{1}{p}} \left[\int_{-\infty}^{\infty} y_{\beta}^{q(1-\sigma)-1} g^q(y)dy\right]^{\frac{1}{q}}. \quad (62)$$

(iv) $\sigma_1 = \sigma$, and $M \leq K_{\alpha,\beta}^{(\gamma)}(\sigma)$.

Moreover, if (iv) is satisfied, then the constant factor $M = K_{\alpha,\beta}^{(\gamma)}(\sigma)$ in (60), (61) and (62) is the best possible.

In particular, for $\delta = -1$, replacing $x^{\lambda} f(x)$ by $f(x)$, the following Statements (i), (ii), (iii) and (iv) are equivalent:

(i) For any $f(x) \geq 0$, satisfying

$$0 < \int_{-\infty}^{\infty} x_{\alpha}^{p(1-\mu_1)-1} f^p(x)dx < \infty,$$

we have the following inequality:

$$\left[\int_{-\infty}^{\infty} y_{\beta}^{p\sigma-1}\left(\int_{-\infty}^{\infty} \frac{|\ln \rho(y_{\beta}/x_{\alpha})^{\lambda}|^{\gamma}}{x_{\alpha}^{\lambda}+\rho y_{\beta}^{\lambda}} f(x)dx\right)^{p} dy\right]^{\frac{1}{p}}$$
$$> M\left[\int_{-\infty}^{\infty} x_{\alpha}^{p(1-\mu_1)-1} f^p(x)dx\right]^{\frac{1}{p}}. \quad (63)$$

(ii) For any $g(y) \geq 0$, satisfying

$$0 < \int_{-\infty}^{\infty} y_{\beta}^{q(1-\sigma)-1} g^q(y)dy < \infty,$$

we have the following inequality:

$$\left[\int_{-\infty}^{\infty} x^{q\mu_1-1}\left(\int_{-\infty}^{\infty} \frac{|\ln \rho(y_{\beta}/x_{\alpha})^{\lambda}|^{\gamma}}{x_{\alpha}^{\lambda}+\rho y_{\beta}^{\lambda}} g(y)dy\right)^{q} dx\right]^{\frac{1}{q}}$$
$$> M\left[\int_{-\infty}^{\infty} y_{\beta}^{q(1-\sigma)-1} g^q(y)dy\right]^{\frac{1}{q}}. \quad (64)$$

(iii) For any $f(x), g(y) \geq 0$, satisfying

$$0 < \int_{-\infty}^{\infty} x_{\alpha}^{p(1-\mu_1)-1} f^p(x)dx < \infty, \; and\, 0 < \int_{-\infty}^{\infty} y_{\beta}^{q(1-\sigma)-1} g^q(y)dy < \infty,$$

we have the following inequality:

$$\int_{-\infty}^{\infty}\int_{-\infty}^{\infty} \frac{|\ln \rho(y_\beta/x_\alpha)^\lambda|^\gamma}{x_\alpha^\lambda + \rho y_\beta^\lambda} f(x)g(y)dxdy$$
$$> M\left[\int_{-\infty}^{\infty} x_{\alpha}^{p(1-\mu_1)-1} f^p(x)dx\right]^{\frac{1}{p}} \left[\int_{-\infty}^{\infty} y_{\beta}^{q(1-\sigma)-1} g^q(y)dy\right]^{\frac{1}{q}}. \quad (65)$$

(iv) $\mu_1 = \mu(= \lambda - \sigma)$, and $M \leq K_{\alpha,\beta}^{(\gamma)}(\sigma)$.

Moreover, if (iv) follows, then the constant factor $M = K_{\alpha,\beta}^{(\gamma)}(\sigma)$ in (63), (64) and (65) is the best possible.

*Proof* Similarly to prove the strict inequalities in Theorem 2, we obtain the strict inequalities (60), (61) and (62). Similarly as we worked in the proof of Lemma 6, we can show that the statements (i), (ii) and (iii) are equivalent, and $(iv) \Rightarrow (iii)$. By Lemma 7, we get $(iii) \Rightarrow (iv)$. Hence, the statements (i), (ii), (iii) and (iv) are equivalent, and the constant factor $M = K_{\alpha,\beta}^{(\gamma)}(\sigma)$ in (60), (61) and (62) is the best possible.

This completes the proof of the theorem. □

**Acknowledgements** This work is supported by the National Natural Science Foundation (No. 617721-40), and Science and Technology Planning Project Item of Guangzhou City (No. 201707010229). We are grateful for their support.

## References

1. Schur, I.: Bernerkungen sur Theorie der beschrankten Billnearformen mit unendlich vielen Veranderlichen. J. Math. **140**, 1–28 (1911)
2. Hardy, G.H.: Note on a theorem of Hilbert concerning series of positive terms. Proc. Lond. Math. Soc. **23**(2) (1925). Records of Proc. xlv-xlvi
3. Hardy, G.H., Littlewood, J.E., Pólya, G.: Inequalities. Cambridge University Press, Cambridge (1934)
4. Yang, B.C.: On Hilbert's integral inequality. J. Math. Anal. Appl. **220**, 778–785 (1998)
5. Yang, B.C.: A note on Hilbert's integral inequality. Chin. Q. J. Math. **13**(4), 83–86 (1998)
6. Yang, B.C.: On an extension of Hilbert's integral inequality with some parameters. Aust. J. Math. Anal. Appl. **1**(1), 1–8 (2004). Art. 11
7. Yang, B.C., Brnetić, I., Krnić, M., Pečarić, J.E.: Generalization of Hilbert and Hardy-Hilbert integral inequalities. Math. Inequal. Appl. **8**(2), 259–272 (2005)
8. Krnić, M., Pečarić, J.E.: Hilbert's inequalities and their reverses. Publ. Math. Debrecen **67**(3–4), 315–331 (2005)

9. Hong, Y.: On Hardy-Hilbert integral inequalities with some parameters. J. Inequal. Pure Appl. Math. **6**(4), 1–10 (2005). Art. 92
10. Arpad, B., Choonghong, O.: Best constant for certain multi linear integral operator. J. Inequal. Appl. 28582 (2006)
11. Li, Y.J., He, B.: On inequalities of Hilbert's type. Bull. Aust. Math. Soc. **76**(1), 1–13 (2007)
12. Zhong, W.Y., Yang, B.C.: On multiple Hardy-Hilbert's integral inequality with kernel. J. Inequal. Appl. **2007**, 17 pp. (2007). Art. ID 27962
13. Xu, J.S.: Hardy-Hilbert's Inequalities with two parameters. Adv. Math. **36**(2), 63–76 (2007)
14. Rassias, M.Th., Yang, B.C.: On a multidimensional half - discrete Hilbert - type inequality related to the hyperbolic cotangent function. Appl. Math. Comput. **242**, 800–813 (2014)
15. Rassias, M.Th., Yang, B.C.: A Hilbert - type integral inequality in the whole plane related to the hyper geometric function and the beta function. J. Math. Anal. Appl. **428**(2), 1286–1308 (2015)
16. Rassias, M.Th., Yang, B.C.: On a Hardy-Hilbert-type inequality with a general homogeneous kernel. Int. J. Nonlinear Anal. Appl. **7**(1), 249–269 (2016)
17. Kato, T.: Perturbation Theory for Linear Operators. Classics in Mathematics. Springer, Berlin (1995)
18. Kato, T.: Variation of discrete spectra. Commun. Math. Phys. **111**(3), 501–504 (1987)
19. Kato, T., Satake, I.: An algebraic theory of Landau-Kolmogorov inequalities. Tohoku Math. J. **33**(3), 421–428 (1981)
20. Kato, T.: On an inequality of Hardy, Littlewood, and Pólya. Adv. Math. **7**, 217–218 (1971)
21. Kato, T.: Demicontinuity, hemicontinuity and monotonicity, II. Bull. Am. Math. Soc. **73**, 886–889 (1967)
22. Yang, B.C.: The Norm of Operator and Hilbert-Type Inequalities. Science Press, Beijing (2009)
23. Yang, B.C.: Hilbert-Type Integral Inequalities. Bentham Science Publishers Ltd., The United Emirates (2009)
24. Yang, B.C.: On Hilbert-type integral inequalities and their operator expressions. J. Guangaong Univ. Edu. **33**(5), 1–17 (2013)
25. Yang, B.C.: A new Hilbert-type integral inequality. Soochow J. Math. **33**(4), 849–859 (2007)
26. He, B., Yang, B.C.: On a Hilbert-type integral inequality with the homogeneous kernel of 0-degree and the hypergeometric function. Math. Pract. Theory **40**(18), 105–211 (2010)
27. Yang, B.C.: A new Hilbert-type integral inequality with some parameters. J. Jilin Univ. (Science Edition) **46**(6), 1085–1090 (2008)
28. Zeng, Z., Xie, Z.T.: On a new Hilbert-type integral inequality with the homogeneous kernel of degree 0 and the integral in whole plane. J. Inequal. Appl. **2010**, 9 pp. (2010). Article ID 256796
29. Wang, A.Z., Yang, B.C.: A new Hilbert-type integral inequality in whole plane with the non-homogeneous kernel. J. Inequal. Appl. **2011**, 123 (2011)
30. Xin, D.M., Yang, B.C.: A Hilbert-type integral inequality in whole plane with the homogeneous kernel of degree -2. J. Inequal. Appl. **2011**, 11 pp. (2011). Article ID 401428
31. He, B., Yang, B.C.: On an inequality concerning a non-homogeneous kernel and the hypergeometric function. Tamsul Oxf. J. Inf. Math. Sci. **27**(1), 75–88 (2011)
32. Xie, Z.T., Zeng, Z., Sun, Y.F.: A new Hilbert-type inequality with the homogeneous kernel of degree -2. Adv. Appl. Math. Sci. **12**(7), 391–401 (2013)
33. Huang, Q.L., Wu, S.H., Yang, B.C.: Parameterized Hilbert-type integral inequalities in the whole plane. Sci. World J. **2014**, 8 pp. (2014). Article ID 169061
34. Zheng, Z., Raja Rama Gandhi, K., Xie, Z.T.: A new Hilbert-type inequality with the homogeneous kernel of degree -2 and with the integral. Bull. Math. Sci. Appl. **3**(1), 11–20 (2014)
35. Huang, X.Y., Cao, J.F., He, B., Yang, B.C.: Hilbert-type and Hardy-type integral inequalities with operator expressions and the best constants in the whole plane. J. Inequal. Appl. **2015**, 129 (2015)
36. Gu, Z.H., Yang, B.C.: A Hilbert-type integral inequality in the whole plane with a non-homogeneous kernel and a few parameters. J. Inequal. Appl. **2015**, 314 (2015)

37. Hong, Y.: On the structure character of Hilbert's type integral inequality with homogeneous kernal and applications. J. Jilin Univ. (Science Edition) **55**(2), 189–194 (2017)
38. Rassias, M.Th., Yang, B.C.: Equivalent properties of a Hilbert-type integral inequality with the best constant factor related the Hurwitz zeta function. Ann. Funct. Anal. **9**(2), 282–295 (2018)
39. Hong, Y., Huang, Q.L., Yang, B.C., Liao, J.Q.: The necessary and sufficient conditions for the existence of a kind of Hilbert-type multiple integral inequality with the non-homogeneous kernel and its applications. J. Inequal. Appl. **2017**, 316 (2017)
40. Yang, B.C., Chen, Q.: Equivalent conditions of existence of a class of reverse Hardy-type integral inequalities with nonhomogeneous kernel. J. Jilin Univ. (Science Edition) **55**(4), 804–808 (2017)
41. Yang, B.C.: Equivalent conditions of the existence of Hardy-type and Yang-Hilbert-type integral inequalities with nonhomogeneous kernel. J. Guangdong Univ. Edu. **37**(3), 5–10 (2017)
42. Yang, B.C.: On some equivalent conditions related to the bounded property of Yang-Hilbert-type operator. J. Guangdong Univ. Edu. **37**(5), 5–11 (2017)
43. Yang, Z.M., Yang, B.C.: Equivalent conditions of the existence of the reverse Hardy-type integral inequalities with the nonhomogeneous kernel. J. Guangdong Univ. Edu. **37**(5), 28–32 (2017)
44. Kuang, J.C.: Real and Functional Analysis (Continuation) (second volume). Higher Education Press, Beijing (2015)
45. Wang, J.Q., Guo, D.R.: Introduction to Special Functions. Science Press, Beijing (1979)
46. Kuang, J.C.: Applied Inequalities. Shangdong Science and Technology Press, Jinan (2004)

# Tosio Kato's Work on Non-relativistic Quantum Mechanics: A Brief Report

**Barry Simon**

**Abstract** We present the highpoints of Tosio Kato's work on the mathematics behind Non-relativistic Quantum Mechanics.

## 1 Introduction

In 2017, we celebrated the 100th anniversary of the birth of Tosio Kato (August 25, 1917–October 2, 1999), the founding father of the theory of Schrödinger operators. There was a centennial held in Tokyo in his memory and honor in September. I decided to write a review article on his work in non-relativistic quantum mechanics (NRQM), which, as well see, was only part of his opus. I originally guessed it would be about 80 pages but it turned out to be more than 210! It will appear in Bull. Math. Sci. [83]. Evans Harrell, the Editor of the IAMP newsletter asked if I could produce a Reader's Digest version and that is what this is. I was asked by the Editors of the current volume to also supply that version for this book and I am doing so with the permission of the Editors of Bull. Math. Sci. and the IAMP newsletter. Since the longer article has an over 600-item bibliography, there were no references in the newsletter version, but I decided to include a limited bibliography in the current version. That is the main difference from the IAMP newsletter version. The rest of this introduction will say a little about Kato's life, while the next will summarize some major themes in his work. I will then describe in some depth (but less detail than in my Bull. Math. Sci. article) five topics that were among the most important of Kato's contributions to NRQM.

B. Simon (✉)
IBM Professor of Mathematics and Theoretical Physics, Emeritus; California Institute of Technology, Pasadena, CA 91125, USA
e-mail: bsimon@caltech.edu

© Springer Nature Switzerland AG 2019

T. M. Rassias and V. A. Zagrebnov (eds.), *Analysis and Operator Theory*, Springer Optimization and Its Applications 146,
https://doi.org/10.1007/978-3-030-12661-2_16

Kato's most significant paper [27] was on self-adjointness of atomic Hamiltonians and was published in 1951 in Trans. A.M.S. (see Section 3). I note that he was 34 when it was published (it was submitted a few years earlier as we'll discuss in Section 3). Before it, his most important work was his thesis, awarded in 1951 and published [23, 24, 26] in 1949–51. One might be surprised at his age when this work was published but not if one understands the impact of the war. Kato got his BS from the University of Tokyo in 1941, a year in which he published two (not mathematical) papers in theoretical physics. But during the war, he was evacuated to the countryside. We were at a conference together one evening and Kato described rather harrowing experiences in the camp he was assigned to, especially an evacuation of the camp down a steep wet hill. He contracted TB in the camp. In his acceptance for the Wiener Prize [1], Kato says that his work on essential self-adjointness and on perturbation theory was essentially complete by the end of the war.

Kato at Berkeley

In 1946, Kato returned to the University of Tokyo as an Assistant (a position common for students progressing toward their degrees) in physics, was appointed Assistant Professor of Physics in 1951 and Full Professor in 1958. I've sometimes wondered what his colleagues in physics made of him. He was perhaps influenced by the distinguished Japanese algebraic geometer, Kunihiko Kodaira (1915–1997) 2 years his senior and a 1954 Fields medalist. Kodaira got a BS in physics after his BA in mathematics and was given a joint appointment in 1944, so there was clearly some sympathy toward pure mathematics in the physics department.

Beginning in 1954, Kato started visiting the United States. This bland statement masks some drama. In 1954, Kato was invited to visit Berkeley for a year, I presume arranged by F. Wolf. Of course, Kato needed a visa and it is likely it would have been denied due to his history of TB. Fortunately, just at the time (and only for a period of about a year), the scientific attaché at the US embassy in Tokyo was Otto Laporte (1902–1971) on leave from a professorship in Physics at the University of Michigan. Charles Dolph (1919–1994), a mathematician at Michigan, learned of the problem and contacted Laporte who intervened to get Kato a visa. Dolph once told me that he thought his most important contribution to American mathematics was his helping to allow Kato to come to the US. In 1987, in honor of Kato's 70th birthday, there was a special issue [14] of the Journal of Mathematical Analysis and

Applications and the issue was jointly dedicated to Laporte (he passed away in 1971) and Kato and edited by Dolph and Kato's student Jim Howland.

František Wolf (1904–1989) was important in bringing Kato to Berkeley both as a visitor and later, a regular faculty member. He was a Czech mathematician who had a junior position at Charles University in Prague. Wolf had spent time in Cambridge and did some significant work on trigonometric series under the influence of Littlewood. When the Germans invaded Czechoslovakia in March 1938, he was able to get an invitation to Mittag-Leffler. He got permission from the Germans for a 3-week visa but stayed in Sweden! He was then able to get an instructorship at Macalester College in Minnesota. He made what turned out to be a fateful decision in terms of later developments. Because travel across the Atlantic was difficult, he took the trans-Siberian railroad across the Soviet Union and then through Japan and across the Pacific to the US. This was mid-1941 before the US entered the war and made travel across the Pacific difficult.

Wolf stopped in Berkeley to talk with G. C. Evans (known for his work on potential theory), who was then department chair. Evans knew of Wolf's work and offered him a position on the spot!! After the year he promised to Macalester, Wolf returned to Berkeley and worked his way up the ranks. In 1952, Wolf [91] extended Sz.-Nagy's work [62] to the Banach space case. At about the same time Nagy himself did similar work [63] and in so did Kato [28]. While Wolf and Kato didn't know of each other's work, Wolf learned of Kato's work and that led to his invitation for Kato to visit Berkeley.

Kato as a student

During the mid-1950s, Kato spent close to 3 years visiting US institutions, mainly Berkeley, but also the Courant Institute, American University, National Bureau of Standards, and Caltech. In 1962, he accepted a professorship in Mathematics from Berkeley where he spent the rest of his career and remained after his retirement. One should not underestimate the courage it takes for a 45-year old to move to a very different culture because of a scientific opportunity. That said, I'm told that when he retired and some of his students urged him to live in Japan, he said he liked the weather in Northern California too much to consider it. The reader can consult the Mathematics Genealogy Project (for a list of Kato's students (24 listed there, 3 from Tokyo and 21 from Berkeley; the best known are Ikebe and Kuroda from Tokyo and Balslev and Howland from Berkeley). Notices A.M.S. **47** (2000), 650–657 is a memorial article with lots of reminiscences.

The pictures here are all from the estate of Mizue Kato, Tosio's wife who passed away in 2011. Her will gave control of the pictures to H. Fujita, M. Ishiguro, and S. T. Kuroda. I thank them for permission to use the pictures and H. Okamoto for providing digital versions.

## 2 Overview

While this review will cover a huge array of work, it is important to realize it is only a fraction, albeit a substantial fraction, of Kato's opus. I'd classify his work into four broad areas, NRQM, nonlinear PDEs, linear semigroup theory, and miscellaneous contributions to functional analysis. We will not give references to all this work. The reader can get an (almost) complete bibliography from MathSciNet or, for papers up to 1987, the dedication of the special issue of JMAA on the occasion of Kato's 70th birthday has a bibliography.

Around 1980, one can detect a clear shift in Kato's interest. Before 1980, the bulk of his papers are on NRQM with a sprinkling in the other three areas while after 1980, the bulk are on nonlinear equations with a sprinkling in the other areas including NRQM. Kato's nonlinear work includes looking at the Euler, Navier–Stokes, KdV, and nonlinear Schrödinger equations. He was a pioneer in existence results—we note that his famous 1951 paper can be viewed as a result on the existence of solutions for the time-dependent linear Schrödinger equation! It is almost that when NRQM became too crowded with workers drawn by his work, he moved to a new area which took some time to become popular.

The basic results on generators of semigroups on Banach spaces date back to the early 1950s going under the name Feller-Miyadera-Phillips and Hille–Yosida theorems (with a later 1961 paper of Lumer–Phillips). A basic book with references to this work is by Pazy [65]. This is a subject that Kato returned to often, especially in the 1960s. Pazy lists 19 papers by Kato on the subject. Perhaps the most important of these results are the Trotter–Kato theorems and the definition of fractional powers for generators of (not necessarily self-adjoint) contraction semigroups.

Given this work on the theory on a Banach space, it is interesting to see a quote that his friend Cordes attributes to Kato: "there is no decent Banach space, except Hilbert space." It is likely Kato had in mind the spectral theorem and the theory of quadratic forms of operators, a subject where he made important contributions, especially the monotone convergence theorems for forms.

The fourth area is a catchall for a variety of results that don't fit into the other bins. Among these results is an improvement of the celebrated Calderón–Vaillancourt bounds on pseudodifferential operators. Also, Kato [40] proved the absolute value for operators is not Lipschitz continuous even restricted to the self-adjoint operators but for any pair of bounded, even non-self-adjoint, operators one has that

$$\| |S| - |T| \| \leq \frac{2}{\pi} \|S - T\| \left( 2 + \log \frac{\|S\| + \|T\|}{\|S - T\|} \right) \tag{1}$$

He was also a key player [29] in variants of Loewner's result [60] that the square root is a monotone function on positive operators on a Hilbert space, a result rediscovered by Heinz [17]. It and its variants have been called the Heinz–Loewner or Heinz–Kato inequality. Kato returned several times to this subject, most notably [34] finding a version of the Heinz–Loewner inequality (with an extra constant depending on s) for maximal accretive operators on a Hilbert space.

One can get a feel for Kato's impact by considering the number of theorems, theories, and inequalities with his name on them. Here are some: Kato's theorem (which usually refers to his result on self-adjointness of atomic Hamiltonians), the Kato–Rellich theorem (which Rellich had first), the Kato–Rosenblum theorem and the Kato–Birman theory (where Kato had the most significant results although, as we'll see, Rosenblum should get more credit than he does), the Kato projection lemma and Kato dynamics (used in the adiabatic theorem), the Putnam–Kato theorem, the Trotter–Kato theorem, the Kato cusp condition, Kato smoothness theory, the Kato class of potentials, and Kato–Kuroda eigenfunction expansions. To me, *Kato's inequality* refers to the self-adjointness technique discussed in Section 5, but the term has also been used for the Hardy-like inequality with best constant for $r^{-1}$ in three dimensions, for a result on hyponormal operators that follows from Kato smoothness theory (there is a book with a section called "Kato's inequality" on it) and for the above-mentioned variant of the Heinz–Loewner inequality for maximal accretive operators. There are also Heinz–Kato, Ponce–Kato, and Kato–Temple inequalities. Erhard Seiler and I [72] proved that if $f, g \in L^p(\mathbb{R}^\nu)$, $p \geq 2$, then $f(X)g(-i\nabla)$ is in the trace ideal $\mathcal{I}_p$. At the time, Kato and I had correspondence about the issue and about some results for $p < 2$. In Reed–Simon Volume 3, we mentioned that Kato had this result independently. Although Kato never published anything on the subject, in recent times, it has come to be called the Kato–Seiler–Simon inequality.

Of course, when discussing the impact of Kato's work, one must emphasize the importance of his book *Perturbation Theory for Linear Operators* [42] which has been a bible for several generations of mathematicians. One of its virtues is its comprehensive nature. Percy Deift told me that Peter Lax told him that Friedrichs remarked on the book: "Oh, its easy to write a book when you put everything in it!"

We will not discuss every piece of work that Kato did in NRQM—for example, he wrote several papers on variational bounds on scattering phase shifts whose lasting impact was limited.

Kato was a key figure in eigenvalue perturbation theory, the subject of his thesis, which was codified in his book [42]. For the regular case, the basic results were by Rellich [69] and later Sz.-Nagy [62] but refined by Kato. He was the pioneer in the asymptotic case, originally using his extension of Temple's inequality but later operator techniques. The full article has 5 sections on eigenvalue perturbation theory, but I'be chosen other areas for this shorter version.

Ever since the work of von Neumann about 1930, it has been clear that self-adjointness of quantum Hamiltonians is crucial. In this regard, Kato has been a, indeed, the key figure. His contributions include

1. Self-adjointness of atomic Hamiltonians (see Section 3).

2. His work with Ikebe [20] on $V(x) > -cx^2 - d$, an area where Weinholtz was the pioneer.
3. Kato's inequality, including his theorem that if $V \in L^2_{loc}(\mathbb{R}^\nu)$ and $V \geq 0$, then $-\Delta + V$ is essentially self-adjoint on $C_0^\infty(\mathbb{R}^\nu)$ (see Section 5).
4. He was a pioneer on the use of quadratic forms including his work on perturbations (KLMN theorem) and monotone convergence for forms.

Scattering and Spectral Theory was a major thread in Kato's work from 1955 to 1980, so much so that when he gave a plenary talk at the 1970 ICM [38], it was on this subject. One can identify four areas:

1. Trace class scattering. (see Section 6)
2. Kato Smoothness. (see Section 7)
3. Eigenfunction expansions (Kato–Kuroda theory [37, 49, 50]).
4. Jensen–Kato [22] on low energy and large times.

Three isolated but important contributions:

1. His 1948 paper on the adiabatic theorem. (Section 4)
2. His ultimate Trotter product formula [45]: if $A$ and $B$ are arbitrary positive self-adjoint operators, if $C$ is the self-adjoint form sum on $\overline{Q(A) \cap Q(B)} \equiv \mathcal{K}$, and if $P$ is the projection onto $\mathcal{K}$, then

$$\text{s} - \lim_{n\to\infty} \left[e^{-tA/n} e^{-tB/n}\right]^n = e^{-tC} P$$

3. His 1957 paper [30] on properties of eigenfunctions of general Schrödinger operators and especially of Coulomb Hamiltonians. For the Coulomb case, he proved that eigenfunctions were globally Lipschitz but not generally $C^1$; instead he found the celebrated Kato cusp condition—at coincidence points, radial averages have well defined jumps (like $e^{-|x|}$ for the hydrogen ground state). Trial function calculations for atoms and molecules use the Kato cusp condition making this paper his most quoted one on Google Scholar (with over 1,700 citations!)

## 3 Foundations of Atomic Physics

Ever since von Neumann's work [87] around 1930, it has been clear that a fundamental mathematical problem in quantum theory, indeed the fundamental question in atomic physics, is the self-adjointness of atomic Hamiltonians so Kato's 1951 paper [27] is a pathbreaking contribution of great significance. He considered *N–body Hamiltonians* on $L^2(\mathbb{R}^{\nu N})$ of the formal form

$$H = -\sum_{j=1}^{N} \frac{1}{2m_j} \Delta_j + \sum_{i<j} V_{ij}(x_i - x_j) \tag{1}$$

where $x \in \mathbb{R}^{\nu N}$ is written $\boldsymbol{x} = (x_1, \dots, x_N)$ with $x_j \in \mathbb{R}^\nu$, $\Delta_j$ is the $\nu$-dimensional Laplacian in $x_j$ and each $V_{ij}$ is a real-valued function on $\mathbb{R}^\nu$. In 1951, Kato considered only the physically relevant case $\nu = 3$.

If there are $N + k$ particles in the limit where the masses of particles $N + 1, \dots, N + k$ are infinite, one considers an operator like $H$ but adds terms

$$\sum_{j=1}^{N} V_j(x_j), \qquad V_j(x) = \sum_{\ell=N+1}^{N+k} V_{j\ell}(x - x_\ell) \tag{2}$$

where $x_{N+1}, \dots, x_{N+k}$ are fixed points in $\mathbb{R}^\nu$.

More generally, one wants to consider (as Kato did) Hamiltonians with the center of mass removed. We note that the self-adjointness results on the Hamiltonians of the form (1) easily imply results on Hamiltonians (on $L^2(\mathbb{R}^{(N-1)\nu})$) with the center of mass motion removed. Of especial interest is the Hamiltonian of the form (2) with $N = 1$, i.e.,

$$H = -\Delta + W(x) \tag{3}$$

on $L^2(\mathbb{R}^\nu)$ which we'll call *reduced two-body Hamiltonians* (since, except for a factor of $(2\mu)^{-1}$ in front of $-\Delta$, it is the two-body Hamiltonian with the center of mass removed).

Kato's big 1951 result was (I will use "esa" as shorthand for essentially self-adjoint and "esa$\nu$" for esa on $C_0^\infty(\mathbb{R}^\nu)$).

**Theorem 1** (Kato's Theorem, First Form [27]) *Let* $\nu = 3$. *Let each* $V_{ij}$ *in* (1) *lie in* $L^2(\mathbb{R}^3) + L^\infty(\mathbb{R}^3)$. *Then, the Hamiltonian of* (1) *is self-adjoint on* $D(H) = D(-\Delta)$ *and esa–*$(3N)$.

**Remarks**

1. The same results holds with the terms in (2) added so long as each $V_j$ lies in $L^2(\mathbb{R}^3) + L^\infty(\mathbb{R}^3)$.
2. Kato didn't assume that $V \in L^2(\mathbb{R}^3) + L^\infty(\mathbb{R}^3)$ but rather the stronger hypothesis that for some $R < \infty$, one has that $\int_{|x|<R} |V(x)|^2 d^3x < \infty$ and $\sup_{|x|\geq R} |V(x)| < \infty$, but his proof extends to $L^2(\mathbb{R}^3) + L^\infty(\mathbb{R}^3)$.
3. Kato didn't state that $C_0^\infty(\mathbb{R}^{3N})$ is a core but rather that $\psi$'s of the form $P(x)e^{-\frac{1}{2}x^2}$ with $P$ a polynomial in the coordinates of $x$ is a core (He included the $\frac{1}{2}$ so the set was invariant under Fourier transform.) His result is now usually stated in terms of $C_0^\infty$.

If $v(x) = 1/|x|$ on $\mathbb{R}^3$, then $v \in L^2(\mathbb{R}^3) + L^\infty(\mathbb{R}^3)$, so Theorem 1 has the important Corollary, which includes the Hamiltonians of atoms and molecules:

**Theorem 2** (Kato's Theorem, Second Form [27]) *The Hamiltonian, $H$, of* (1) *with* $\nu = 3$ *and each*

$$V_{ij}(x) = \frac{z_{ij}}{|x|} \tag{4}$$

*and the Hamiltonian with terms of the form* (2) *where*

$$V_j(x) = \sum_{\ell=N+1}^{N+k} \frac{z_{j\ell}}{|x - x_\ell|} \tag{5}$$

*are self-adjoint on* $D(-\Delta)$ *and esa-*$3N$.

**Remark** This result assures that the time-dependent Schrödinger equation $\dot{\psi}_t = -iH\psi_t$ has solutions (since self-adjointness means that $e^{-itH}$ exists as a unitary operator). The analogous problem for Coulomb Newton's equation (i.e., solvability for a.e. initial condition) is open for $N \geq 5$!

As Kato remarks in his Wiener prize acceptance, "the proof turned out to be rather easy." It has three steps:

(1) The Kato–Rellich theorem [82, Theorem 7.1.14] which reduces the proof to showing that each $V_{ij}$ is relatively bounded for Laplacian on $\mathbb{R}^3$ with relative bound 0.
(2) A proof that any function in $L^2(\mathbb{R}^3) + L^\infty(\mathbb{R}^3)$, as an operator on $L^2(\mathbb{R}^3)$, is $-\Delta$-bounded with relative bound 0. This relies on the simple Sobolev estimate [81, Section 6.3] that on $\mathbb{R}^3$ any function in $D(-\Delta)$ is bounded. Alternatively, for the atomic case, Hardy's inequality [81, Theorem 4.1.3] (that $(4r^2)^{-1} \leq -\Delta$) implies this fact.
(3) A piece of simple kinematics that says that the two-body estimate in step 2 extends to one for $v_{ij}(x_i - x_j)$ as an operator on $L^2(\mathbb{R}^{3N})$.

Kato states in the paper that he had found the results by 1944. Kato originally submitted the paper to Physical Review. Physical Review transferred the manuscript to the Transactions of the AMS, where it eventually appeared. They had trouble finding a referee and, in the process, the manuscript was lost (a serious problem in pre-Xerox days!). Eventually, von Neumann got involved and helped get the paper accepted. I've always thought that given how important he knew the paper was, von Neumann should have suggested Annals of Mathematics and used his influence to get it published there. The receipt date of October 15, 1948 on the version published in the Transactions shows a long lag compared to the other papers in the same issue of the Transactions which have receipt dates of Dec. 1949 through June, 1950. Recently, after Kato's widow died and left his papers to some mathematicians (see the end of Section 1) and some fascinating correspondence of Kato with Kemble and von Neumann came to light. There are plans by Kato's students to publish an edited version of these letters.

It is a puzzle why it took so long for this theorem to be found. One factor may have been von Neumann's attitude. Bargmann told me of a conversation several young mathematicians had with von Neumann around 1948 in which von Neumann told them that self-adjointness for atomic Hamiltonians was an impossibly hard problem and that even for the Hydrogen atom, the problem was difficult and open. This is a little strange since, using spherical symmetry, Hydrogen can be reduced to a direct sum of one-dimensional problems. For such ODEs, there is a powerful limit point–limit circle method [82, Theorem 7.4.12] named after Weyl and Titchmarsh (although it was Stone, in his 1932 book, who first made it explicit). Using this, it is easy to see (there is one subtlety for $\ell = 0$ since the operator then is limit point at 0) that the Hydrogen Hamiltonian is self-adjoint and this appears at least as early as Rellich [70] in 1944. Of course, this method doesn't work for multielectron atoms. In any event, it is possible that von Neumann's attitude may have discouraged some from working on the problem.

Still, it is surprising that neither Friedrichs nor Rellich found this result. Rellich used Hardy's inequality in his perturbation theory papers [69] in a closely related context. Namely, he used Hardy's inequality to show that $r^{-1} \leq 4\varepsilon(-\Delta) + \frac{1}{4}\varepsilon^{-1}$ to note the semi-boundedness of the Hydrogen Hamiltonian. Since Rellich certainly knew the Kato–Rellich theorem, it appears that he knew steps 1 and 2.

In a sense, it is pointless to speculate why Rellich didn't find Theorem 2, but it is difficult to resist. It is possible that he never considered the problem of esa of atomic Hamiltonians, settling for a presumption that using the Friedrichs extension suffices (as Kato suggests in his Wiener prize acceptance) but I think that unlikely. It is possible that he thought about the problem but dismissed it as too difficult and never thought hard about it. Perhaps the most likely explanation involves Step 3: once you understand it, it is trivial, but until you conceive that it might be true, it might elude you.

Later authors considered the analogs for dimension $\nu \neq 3$—Stummel [84] was a big hero here. For our discussion in Section 5, we'll need the following: We call $p$, *$\nu$-canonical* if $p = 2$ for $\nu \leq 3$, $p > 2$ if $\nu = 4$ and $p = \nu/2$ if $p \geq 5$. The optimal $L^p$ extension of Theorem 1 is (see [82, Theorem 7.1.19] for a proof):

**Theorem 3** *Let $p$ be $\nu$-canonical. Then $V \in L^p(\mathbb{R}^\nu) + L^\infty(\mathbb{R}^\nu)$ is $-\Delta$-bounded with relative bound zero. If $\nu \geq 5$, $V \in L^p_w(\mathbb{R}^\nu) + L^\infty(\mathbb{R}^\nu)$ is $-\Delta$-bounded on $L^2(\mathbb{R}^\nu)$.*

It is known that this is optimal since when $\nu > 5$ limit point, limit circle methods show that for $C$ sufficiently large $-\Delta - Cr^{-2}$ is not essentially self-adjoint on $C_0^\infty$ and $r^{-2}$ lies in all $L^p + L^\infty$ with $p < \nu/2$. The longer paper includes further discussion of these higher dimensional analogs and also discusses Kato's application in his 1951 paper to the Coulomb-Dirac-Hamiltonian.

## 4 The Adiabatic Theorem

In 1950, Kato published a paper [25] in a physics journal (denoted as based on a presentation in 1948) on the quantum adiabatic theorem. It is his only paper on the subject but has strongly impacted virtually all the huge literature on the subject and related subjects ever since (there are more Google Scholar citations of this paper than of the one on self-adjointness of atomic Hamiltonians). We will begin by describing his theorem and its proof which introduced what he called *adiabatic dynamics* and

I'll call the *Kato dynamics*. We'll see that the Kato dynamics defines a notion of parallel transport on the natural vector bundle over the manifold of all $k$-dimensional subspaces of a Hilbert space, $\mathcal{H}$, and so a connection. This connection is called the Berry connection and its holonomy is the Berry phase (when $k = 1$). All this Berry stuff was certainly not even hinted at in Kato's work but it is implicit in the framework. Then I'll say something about the history before Kato.

The adiabatic theorem considers a family of time-dependent Hamiltonians, $H(s)$, $0 \le s \le 1$ and imagines changing them slowly, i.e., looking at $H(s/T)$, $0 \le s \le T$ for $T$ very large. Thus, we look for $\tilde{U}_T(s)$ solving

$$\frac{d}{ds}\tilde{U}_T(s) = -iH(s/T)\tilde{U}_T(s), \ 0 \le s \le T; \qquad \tilde{U}_T(0) = \mathbf{1} \tag{1}$$

Letting $U_T(s) = \tilde{U}_T(sT)$, $0 \le s \le 1$, we see that $U_T(s)$, $0 \le s \le 1$ solves

$$\frac{d}{ds}U_T(s) = -iTH(s)U_T(s), \ 0 \le s \le 1; \qquad U_T(0) = \mathbf{1} \tag{2}$$

Here is Kato's adiabatic theorem

**Theorem 4** (Kato [25]) *Let $H(s)$ be a $C^2$ family of bounded self-adjoint operators on a (complex, separable) Hilbert space, $\mathcal{H}$. Suppose there is a $C^2$ function, $\lambda(s)$, so that for all $s$, $\lambda(s)$ is an isolated point in the spectrum of $H(s)$ and so that*

$$\alpha \equiv \inf_{0 \le s \le 1} dist(\lambda(s), \sigma(H(s)) \setminus \{\lambda(s)\}) > 0 \tag{3}$$

*Let $P(s)$ be the projection onto the eigenspace for $\lambda(s)$ as an eigenvalue of $H(s)$. Then*

$$\lim_{T \to \infty} (1 - P(s))U_T(s)P(0) = 0 \tag{4}$$

*uniformly in $s$ in* [0, 1].

**Remarks**

1. Thus if $\varphi_0 \in \operatorname{ran} P(0)$, this says that when $T$ is large, $U_T(s)\varphi_0$ is close to lying in $\operatorname{ran} P(s)$. That is as $T \to \infty$, the solution gets very close to the "curve" $\{\operatorname{ran} P(s)\}_{0 \le s \le 1}$.
2. If there is an eigenvalue of constant multiplicity near $\lambda(0)$ for $s$ small, it follows from the contour representation of $P(s)$ that $P(s)$ and $\lambda(s)$ are $C^2$.
3. Kato made no explicit assumptions on regularity in $s$ saying "Our proof given below is rather formal and not faultless from the mathematical point of view. Of course, it is possible to retain mathematical rigor by a detailed argument based on clearly defined assumptions, but it would take us too far into an unnecessary complication and obscure the essentials of the problem." It is hard to imagine the Kato of 1960 using such language! In any event, the proof requires that $P(s)$ be $C^2$.

4. As we'll see, the size estimate for (4) is $O(1/T)$.

Kato's wonderful realization is that there is an explicit dynamics, $W(s)$ for which (4) is exact, i.e.,

$$(1 - P(s))W(s)P(0) = 0 \tag{5}$$

He not only constructs it but proves the theorem by showing that (this formula only holds in case $\lambda(s) \equiv 0$; see (10) below)

$$\lim_{T\to\infty} [U_T(s) - W(s)]P(0) = 0 \tag{6}$$

The $W(s)$ that Kato constructs, he called the *adiabatic dynamics*. It is sometimes called Kato's adiabatic dynamics. We call it the *Kato dynamics*. Here is the basic result:

**Theorem 5** (Kato dynamics [25]) *Let $W(s)$ solve*

$$\frac{d}{ds}W(s) = iA(s)W(s), 0 \leq s \leq 1; \qquad W(0) = \mathbf{1} \tag{7}$$

$$iA(s) \equiv [P'(s), P(s)] \tag{8}$$

*Then $W(s)$ is unitary and obeys*

$$W(s)P(0)W(s)^{-1} = P(s) \tag{9}$$

The proof is not hard (see the longer paper for details). Using $P(s)^2 = P(s)$ and its derivative, one shows that $W(s)^{-1}P(s)W(s)$ has zero derivative.

The proof of Theorem 4 depends on proving that

$$\|U_T(s)P(0) - e^{-iT\int_0^s \lambda(s)\,ds}W(s)P(0)\| = O(1/T) \tag{10}$$

Equation (10) says a lot more than (4). Equation (4) says that as $T \to \infty$, $U_T(s)$ maps ran $P(0)$ to ran $P(s)$. Equation (4) actually tells you what the precise limiting map is! One fancy pants way of describing this is as follows. Fix $k \geq 1$ in $\mathbb{Z}$. Let $\mathcal{M}$ be the manifold of all $k$-dimensional subspaces of some Hilbert space, $\mathcal{H}$. We want $\dim(\mathcal{H}) \geq k$, but it could be finite. Or $\mathcal{M}$ might be a smooth submanifold of the set of all such subspaces. For each $\omega \in \mathcal{M}$, we have the projection $P(\omega)$. There is a natural vector bundle of $k$-dimensional spaces over $\mathcal{M}$, namely, we associate to $\omega \in \mathcal{M}$, the space ran $P(\omega)$. If $k = 1$, we get a complex line bundle.

The Kato dynamics, $W(s)$, tells you how to "parallel transport" a vector $v \in$ ran $P(\gamma(0))$ along a curve $\gamma(s)$; $0 \leq s \leq 1$ in $\mathcal{M}$. In the language of differential geometry, it defines a connection and such a connection has a holonomy and a curvature. In less fancy terms, consider the case $k = 1$. Suppose $\gamma$ is a closed curve. Then $W(1)$ is a unitary map of ran $P(0)$ to itself, so multiplication by $e^{i\Gamma_B(\gamma)}$. Returning to $U_T$, it says that the phase change over a closed curve isn't what one might naively expect, namely $\exp(-i\int_0^T \lambda(s/T)\,ds) = \exp(-iT\int_0^1 \lambda(s)\,ds)$. There is an additional term, $\exp(i\Gamma_B)$. This is the *Berry phase* discovered by Berry [4] in 1983 (it was discovered in 1956 by Pancharatnam [64], but then forgotten). I realized [80] that this was just the holonomy of a natural bundle connection and that, moreover, this bundle and connection is precisely the one whose Chern integers are the TKN$^2$ integers of Thouless et al. [85] (as discussed by Avron–Seiler–Simon [3]). Thouless got a recent physics Nobel Prize in part for the discovery of the TKN$^2$ integers. The holonomy, i.e. Berry's phase, is an integral of the Kato connection $[P, dP]$. As usual, this line integral over a closed curve is the integral of its differential $[dP, dP]$ over a bounding surface. This quantity is the curvature of the bundle and has come to be called the *Berry curvature* (even though Berry did not use the differential geometric language). Naively $[dP, dP]$ would seem to be zero but it is shorthand for the two-form

$$\sum_{i\neq j}\left[\frac{\partial P}{\partial s_i}, \frac{\partial P}{\partial s_j}\right] ds_i \wedge ds_j \tag{11}$$

This formula of Avron–Seiler–Simon for the Berry curvature is a direct descendant of formulae in Kato's paper, although, of course, he did not consider the questions that lead to Berry's phase.

Finally, a short excursion into the history of adiabatic theorems. "Adiabatic" first entered into physics as a term in thermodynamics meaning a process with no heat exchange. In 1916, Ehrenfest [15] discussed the "adiabatic principle" in classical mechanics. The basic example is the realization (earlier than Ehrenfest) that while the energy of a harmonic oscillator is not conserved under time dependent change of the underlying parameters, the action (energy divided by frequency) is fixed in the limit that the parameters are slowly changed (the reader should figure out what Kato's adiabatic theorem says about a harmonic oscillator with slowly varying frequency). Interestingly enough, many adiabatic processes in the thermodynamic sense are quite rapid, so the Ehrenfest use has, at best, a very weak connection to the initial meaning of the term!

Ehrenfest used these ideas by asserting that in old quantum theory, the natural quantum numbers were precisely these adiabatic invariants. Once new quantum mechanics was discovered, Born and Fock [11] in 1928 discussed what they called the quantum adiabatic theorem, essentially Theorem 4 for simple eigenvalues with a complete set of (normalizable) eigenfunctions. It was 20 years before Kato found his wonderful extension (and then more than 30 years before Berry made the next breakthrough). The longer article has a discussion on the considerable further mathematical literature of the quantum adiabatic theorem.

## 5 Kato's Inequality

This section will discuss a self-adjointness method that appeared in a 1972 paper [39] of Kato based on a remarkable distributional inequality. Its consequences are a subject to which Kato returned often with at least seven additional papers [12, 41, 43, 44, 46–48]. It is also his work that most intersected my own—I [73] motivated his initial paper and it, in turn, motivated several of my later papers [74–79].

To explain the background, recall that in Section 3, we defined $p$ to be $\nu$-canonical ($\nu$ is dimension) if $p = 2$ for $\nu \leq 3$, $p > 2$ for $\nu = 4$ and $p = \nu/2$ for $\nu \geq 5$. For now, we focus on $\nu \geq 5$ so that $p = \nu/2$. As we saw, if $V \in L^p(\mathbb{R}^\nu) + L^\infty(\mathbb{R}^\nu)$, then $-\Delta + V$ is esa-$\nu$. The example $V(x) = -C|x|^{-2}$ for $C$ sufficiently large shows that $p = \nu/2$ is sharp. That is, for any $2 \leq q \leq \nu/2$, there is a $V \in L^q(\mathbb{R}^\nu) + L^\infty(\mathbb{R}^\nu)$, so that $-\Delta + V$ is defined on but not esa on $C_0^\infty(\mathbb{R}^\nu)$.

In these counterexamples, though, $V$ is negative. It was known since the late 1950s that while the negative part of $V$ requires some global hypothesis for esa-$\nu$, the positive part does not (e.g., $-\Delta - x^4$ is not esa-$\nu$ while $-\Delta + x^4$ is esa-$\nu$). But when I started looking at these issues around 1970, there was presumption that for local singularities, there was no difference between the positive and negative parts. In retrospect, this shouldn't have been the belief! After all, limit point–limit circle methods show that if $V(x) = |x|^{-\alpha}$ with $\alpha < \nu/2$ (to make $V \in L^2_{loc}$ so that $-\Delta + V$ is defined on $C_0^\infty(\mathbb{R}^\nu)$) then $-\Delta + V$ is esa-$\nu$ although, if $\alpha > 2$, $-\Delta - V$ is not. (Limit point–limit circle methods apply for $-\Delta + V$ for any $\alpha$ if we look at $C_0^\infty(\mathbb{R}^\nu \setminus \{0\})$ but then only when $\alpha < \nu/2$, we can extend the conclusion to $C_0^\infty(\mathbb{R}^\nu)$.) This example shows that the conventional wisdom was faulty but people didn't think about separate local conditions on

$$V_+(x) \equiv \max(V(x), 0); \qquad V_-(x) = \max(-V(x), 0) \tag{1}$$

Kato's result shattered the then conventional wisdom:

**Theorem 6** (Kato [39]) *If* $V \geq 0$ *and* $V \in L^2_{loc}(\mathbb{R}^\nu)$, *then* $-\Delta + V$ *is esa-*$\nu$.

Kato's result was actually a conjecture that I made on the basis of a slightly weaker result that I had proven:

**Theorem 7** (Simon [73]) *If* $V \geq 0$ *and* $V \in L^2(\mathbb{R}^\nu, e^{-cx^2} d^\nu x)$ *for some* $c > 0$, *then* $-\Delta + V$ *is esa-*$\nu$.

Of course, this covers pretty wild growth at infinity but Theorem 6 is the definitive result since one needs that $V \in L^2_{loc}(\mathbb{R}^\nu)$ for $-\Delta + V$ to be defined on all functions in $C_0^\infty(\mathbb{R}^\nu)$. I found Theorem 7 because I was also working at the time in constructive quantum field theory which was then studying the simplest interacting field models $\varphi_2^4$ and $P(\varphi)_2$ (the subscript 2 means two space–time dimensions). I was able to use results in the theory of hypercontractive semigroups that seemed very different from what Kato used although connections were later found as well as a semigroup proof of Theorem 6. In my preprint proving Theorem 7, I conjectured Theorem 6 and

Kato's response was sent within a few weeks of my mailing him my preprint. The longer paper describes both my proof and these later developements but I want to focus here on Kato's arguments. Kato proved:

**Theorem 8** (Kato's inequality [39]) *Let $u \in L^1_{loc}(\mathbb{R}^\nu)$ be such that its distributional Laplacian, $\Delta u$ is also in $L^1_{loc}(\mathbb{R}^\nu)$. Define*

$$\operatorname{sgn}(u)(x) = \begin{cases} \overline{u(x)}/|u(x)|, & \text{if } u(x) \neq 0 \\ 0, & \text{if } u(x) = 0 \end{cases} \tag{2}$$

*(so $u \operatorname{sgn}(u) = |u|$). Then as distributions*

$$\Delta|u| \geq \operatorname{Re}\left[\operatorname{sgn}(u)\Delta u\right] \tag{3}$$

**Remarks**

1. What we call $\operatorname{sgn}(u)$, Kato calls $\operatorname{sgn}(\bar{u})$.
2. We should pause to emphasize what a surprise this was. Kato was a long-established master of operator theory. He was 55 years old. Seemingly from left field, he pulled a distributional inequality out of his hat. It is true, like other analysts, that he'd been introduced to distributional ideas in the study of PDEs, but no one had ever used them in this way. Truly a remarkable discovery.

The proof is not hard. By replacing $u$ by $u * h_n$ with $h_n$ a smooth approximate identity and taking limits (using $\operatorname{sgn}(u * h_n)(x) \to \operatorname{sgn}(u)(x)$ for a.e. $x$ and using a suitable dominated convergence theorem), we can suppose that $u$ is a $C^\infty$ function. In that case, for $\varepsilon > 0$, let $u_\varepsilon = (\bar{u}u + \varepsilon^2)^{1/2}$. From $u_\varepsilon^2 = \bar{u}u + \varepsilon^2$, we get that

$$2u_\varepsilon \vec{\nabla} u_\varepsilon = 2\operatorname{Re}(\bar{u}\vec{\nabla} u) \tag{4}$$

which implies (since $|\bar{u}| \leq u_\varepsilon$) that

$$|\vec{\nabla} u_\varepsilon| \leq |\vec{\nabla} u| \tag{5}$$

Applying $\frac{1}{2}\vec{\nabla} \cdot$ to (4), we get that

$$u_\varepsilon \Delta u_\varepsilon + |\vec{\nabla} u_\varepsilon|^2 = \operatorname{Re}(\bar{u}\Delta(u)) + |\vec{\nabla} u|^2 \tag{6}$$

Using (5) and letting $\operatorname{sgn}_\varepsilon(u) = \bar{u}/u_\varepsilon$, we get that

$$\Delta u_\varepsilon \geq \operatorname{Re}(\operatorname{sgn}_\varepsilon(u)\Delta u) \tag{7}$$

Taking $\varepsilon \downarrow 0$ yields (3).

Once we have (3), here is Kato's proof [39] of Theorem 6. Consider $T$, the operator closure of $-\Delta + V$ on $C_0^\infty(\mathbb{R}^\nu)$. $T \geq 0$, so, by a simple argument, it suffices to show that $\operatorname{ran}(T + \mathbf{1}) = \mathscr{H}$ or equivalently, that $T^*u = -u \Rightarrow u = 0$. So suppose that $u \in L^2(\mathbb{R}^\nu)$ and that

$$T^*u = -u \tag{8}$$

Since $T^*$ is defined via distributions, (3) implies that

$$\Delta u = (V + 1)u \tag{9}$$

Since $u$ and $V + 1$ are both in $L^2_{loc}$, we conclude that $\Delta u \in L^1_{loc}$ so by Kato's inequality

$$\Delta|u| \geq (\operatorname{sgn}(u))(V + 1)u = |u|(V + 1) \geq |u| \tag{10}$$

Convolution with nonnegative functions preserves positivity of distributions, so for any non-negative $h \in C_0^\infty(\mathbb{R}^\nu)$, we have that

$$\Delta(h * |u|) = h * \Delta|u| \geq h * |u| \tag{11}$$

Since $u \in L^2$, $h * |u|$ is a $C^\infty$ function with classical Laplacian in $L^2$, so $h * |u| \in D(-\Delta)$. $(-\Delta + 1)^{-1}$ has a positive integral kernel, so (11)$\Rightarrow (-\Delta + 1)(h * |u|) \leq 0 \Rightarrow h * |u| \leq 0 \Rightarrow h * |u| = 0$. Taking $h_n$ to be an approximate identity, we have that $h_n * u \to u$ in $L^2$, so $u = 0$ completing the proof.

At first sight, Kato's proof seems to have nothing to do with the semigroup ideas used in the proof of Theorem 7 and the proof of Theorem 6 that I found using semigroup methods. But in trying to understand Kato's work, I found the following abstract result:

**Theorem 9** (Simon [77]) *Let $A$ be a positive self-adjoint operator on $L^2(M, d\mu)$ for a $\sigma$-finite, separable measure space $(M, \Sigma, d\mu)$. Then the following are equivalent:*

*(a) ($e^{-tA}$ is positivity preserving)*

$$\forall u \in L^2,\ u \geq 0, t \geq 0 \Rightarrow e^{-tA}u \geq 0$$

*(b) (Beurling–Deny criterion) $u \in Q(A) \Rightarrow |u| \in Q(A)$ and*

$$q_A(|u|) \leq q_A(u) \tag{12}$$

*(c) (Abstract Kato Inequality) $u \in D(A) \Rightarrow |u| \in Q(A)$ and for all $\varphi \in Q(A)$ with $\varphi \geq 0$, one has that*

$$\langle A^{1/2}\varphi, A^{1/2}|u|\rangle \geq \operatorname{Re}\langle \varphi, \operatorname{sgn}(u)Au\rangle \tag{13}$$

In his original paper [39], Kato proved more than (3). He showed that

$$\Delta|u| \geq \operatorname{Re}\left[\operatorname{sgn}(u)(\vec{\nabla} - i\vec{a})^2 u\right] \tag{14}$$

In his initial paper, he required that $\overrightarrow{a}$ to be $C^1(\mathbb{R}^\nu)$ and he then followed his arguments to get Theorem 6 with $-\Delta + V$ replaced by $-(\nabla - ia)^2 + V$ when $a \in C^1(\mathbb{R}^\nu)$, $V \in L^2_{loc}(\mathbb{R}^\nu)$, $V \geq 0$. But there was a more important consequence of (14) than a self-adjointness result. I noted [76] that (14) implies, by approximating $|u|$ by positive $\varphi \in C_0^\infty(\mathbb{R}^\nu)$, that

$$\langle |u|, \Delta |u| \rangle \geq \langle u, D^2 u \rangle$$

which implies that

$$\langle u, (-D^2 + V)u \rangle \geq \langle |u|, (-\Delta + V)|u| \rangle \tag{15}$$

This, in turn, implies that turning on any, even non-constant, magnetic field always increases the ground-state energy (for spinless bosons), something I called *universal diamagnetism*.

If one thinks of this as a zero temperature result, it is natural to expect a finite temperature result (that is, for, say, finite matrices, one has that $\lim_{\beta\to\infty} -\beta^{-1}\mathrm{Tr}(e^{-\beta A}) = \inf \sigma(A)$ which in statistical mechanical terms is saying that as the temperature goes to zero, the free energy approaches a ground-state energy).

$$\mathrm{Tr}(e^{-tH(a,V)}) \leq \mathrm{Tr}(e^{-tH(a=0,V)}) \tag{16}$$

where

$$H(a, V) = -(\nabla - ia)^2 + V \tag{17}$$

This suggested to me the inequality

$$|e^{-tH(a,V)}\varphi| \leq e^{-tH(a=0,V)}|\varphi| \tag{18}$$

I mentioned this conjecture at a brown bag lunch seminar when I was in Princeton. Ed Nelson remarked that formally, it followed from the Feynman–Kac–Ito formula for semigroups in magnetic fields, which says that adding a magnetic field with gauge, $\overrightarrow{a}$, adds a factor $\exp(i\int \overrightarrow{a}(\omega(s)) \cdot d\omega)$ to the Feynman–Kac formula (the integral is an Ito stochastic integral). Equation (18) is immediate from $|\exp(i\int \overrightarrow{a}(\omega(s)) \cdot d\omega)| = 1$ and the positivity of the rest of the Feynman–Kac integrand. Some have called (18) the Nelson–Simon inequality but the name I gave it, namely *diamagnetic inequality*, has stuck.

The issue with Nelson's proof is that at the time, the Feynman–Kac–Ito was only known for smooth $a$'s. One can obtain the Feynman–Kac–Ito for more general $a$'s by independently proving a suitable core result. After successive improvements by me and then Kato, I proved that

**Theorem 10** (Simon [78]) *Equation* (18) *holds for* $V \geq 0$, $V \in L^1_{loc}(\mathbb{R}^\nu)$ *and* $\overrightarrow{a} \in L^2_{loc}$.

The optimal self-adjointness result ($V \geq 0$, $V \in L^2_{loc}$, $a \in L^4_{loc}$, $\operatorname{div} a \in L^2_{loc}$) was proven by Leinfelder–Simader [59].

As with Theorem 9, there is an abstract two-operator Kato inequality result (originally conjectured in Simon), which was proven by Hess–Schader-Uhlenbrock [19] and Simon [79]. For more details as well as a discussion of the Kato class which Kato introduced in his original Kato inequality note [39], see the longer paper.

## 6 Kato–Rosenblum and Kato–Birman

The Kato group, late 1950s.
S.T. Kuroda (standing), T. Kato, T. Ikebe, H. Fujita, Y. Nakata

Starting with Rutherford's 1911 discovery of the atomic nucleus, scattering has been a central tool in fundamental physics, so it isn't surprising that one of the first papers in the new quantum theory was by Born [10] in 1926 on scattering. While scattering is at a deep level a time-dependent phenomenon, Born used eigenfunctions and time-independent ideas. In the early 1940s, the theoretical physics community first considered time-dependent approaches to scattering. Wheeler [90] and Heisenberg [18] defined the $S$-matrix and Møller [61] introduced wave operators as limits (with no precision as to what kind of limit).

It was Friedrichs in a prescient 1948 paper [16], who first considered the invariance of the absolutely continuous spectrum under sufficiently regular perturbations. Friedrichs was Rellich's slightly older contemporary. Both were students of Courant at Göttingen in the late 1920s (in 1925 and 1929 respectively). By 1948, Friedrichs was a professor at Courant's Institute at NYU. He looked at several simple models and for one with purely a.c. spectrum, he could prove the perturbed models were unitarily equivalent to unperturbed models. While Friedrichs neither quoted Møller nor ever wrote down the explicit formulae

$$\Omega^{\pm}(H, H_0) = \mathrm{s} - \lim_{t\to\mp\infty} e^{itH} e^{-itH_0} \tag{1}$$

(the strange $\pm$ versus $\mp$ convention that we use is universal in the theoretical physics community and uncommon among mathematicians and is not the convention that Kato used), he did prove something equivalent to showing that the limit $\Omega^+$ existed and was the unitary he constructed with time-independent methods.

Motivated in part by Friedrichs, in 1957, Kato published two papers [31, 32] that set out the basics of the theory we will discuss in this section. In the first, he had the important idea of defining

$$\Omega^{\pm}(A, B) = \text{s} - \lim_{t \to \mp\infty} e^{itA} e^{-itB} P_{ac}(B) \tag{2}$$

where $P_{ac}(B)$ is the projection onto $\mathcal{H}_{ac}(B)$, the set of all $\varphi \in \mathcal{H}$ for which the spectral measure of $B$ and $\varphi$ is absolutely continuous with respect to Lebesgue measure. If these strong limits exist, we say that the wave operators $\Omega^{\pm}(A, B)$ exist.

By replacing $t$ by $t + s$, one sees that if $\Omega^{\pm}(A, B)$ exist then $e^{isA}\Omega^{\pm} = \Omega^{\pm} e^{isB}$. Since $\Omega^{\pm}$ are unitary maps, $U^{\pm}$, of $\mathcal{H}_{ac}(B)$ to their ranges, we see that $U^{\pm} B \restriction \mathcal{H}_{ac}(B)(U^{\pm})^{-1} = A \restriction \operatorname{ran} \Omega^{\pm}$. In particular, $\operatorname{ran} \Omega^{\pm}$ are invariant subspaces for $A$ and lie in $\mathcal{H}_{ac}(A)$. It is thus natural to define: $\Omega^{\pm}(A, B)$ are said to be *complete* if

$$\operatorname{ran} \Omega^{+}(A, B) = \operatorname{ran} \Omega^{-}(A, B) = \mathcal{H}_{ac}(A) \tag{3}$$

In the first of the 1957 papers, Kato proved the following.

**Theorem 11** (Kato [31]) *Let $\Omega^{\pm}(A, B)$ exist. Then they are complete if and only if $\Omega^{\pm}(B, A)$ exist.*

The proof is almost trivial. It depends on noting that

$$\psi = \lim_{t \to \infty} e^{iAt} e^{-itB} \varphi \iff \varphi = \lim_{t \to \infty} e^{itB} e^{-itA} \psi \tag{4}$$

since $\|\psi - e^{iAt} e^{-itB} \varphi\| = \|e^{itB} e^{-itA} \psi - \varphi\|$. That said, it is a critical realization because it reduces a completeness result to an existence theorem. In particular, it implies that symmetric conditions which imply existence also imply completeness. We'll say more about this below.

To show the importance of this idea, motivated by it in 1977, Deift and Simon [13] proved that completeness of multichannel scattering for $N$-body scattering was equivalent to the existence of geometrically define "inverse" wave operators. All proofs of asymptotic completeness for $N$-body systems prove it by showing the existence of these Deift–Simon wave operators in support of Kato's Theorem 11.

One consequence of Theorem 11 is that a symmetric condition for existence implies completeness also. Using this idea, in his first 1957 paper, Kato proved

**Theorem 12** (Kato [31]) *Let $H_0$ be a self-adjoint operator and $V$ a (bounded) self-adjoint finite rank operator. Then, $H = H_0 + V$ is a self-adjoint operator and the wave operators $\Omega^{\pm}(H, H_0)$ exist and are complete.*

Later in 1957, Kato proved

**Theorem 13** (Kato–Rosenblum Theorem [32, 71]) *The conclusions of Theorem 12 remain true if $V$ is a (bounded) trace class operator.*

In a sense this theorem is optimal. It is a result of Weyl [89]–von Neumann [88] that if $A$ is a self-adjoint operator, one can find a Hilbert–Schmidt operator, C, so that $B = A + C$ has only pure point spectrum. Kato's student, S. T. Kuroda [52], shortly after Kato proved Theorem 13, extended this result of Weyl–von Neumann to any trace ideal strictly bigger than trace class. So within trace ideal perturbations, one cannot do better than Theorem 13.

The name given to this theorem comes from the fact that before Kato proved Theorem 13, Rosenblum [71] proved a special case that motivated Kato: namely, if $A$ and $B$ have purely a.c. spectrum and $A - B$ is trace class, then $\Omega^{\pm}(A, B)$ exist and are unitary (so complete).

S. Kuroda, T. Ikebe, H. Fujita recently

I'd always assumed that Rosenblum's paper [71] was a rapid reaction to Kato's finite rank paper [31] which, in turn, motivated Kato's trace class paper [32]. But I recently learned that this assumption is not correct. Rosenblum was a graduate student of Wolf at Berkeley, who submitted his thesis in March 1955. It contained his trace class result under some additional technical hypotheses; a Dec. 1955 Berkeley technical report had the result as eventually published without the extra technical assumption. Rosenblum submitted a paper to the American Journal of Mathematics which took a long time refereeing it before rejecting it. In April 1956, Rosenblum submitted a revised paper to the Pacific Journal in which it eventually appeared (this version dropped the technical condition; I've no idea what the original journal submission had).

Kato's finite rank paper was submitted to J. Math. Soc. Japan on March 15, 1957 and was published in the issue dated April 1957(!). The full trace class result was submitted to Proc. Japan Acad. on May 15, 1957. Kato's first paper quotes an abstract of a talk Rosenblum gave to an A.M.S. meeting, but I don't think that abstract contained many details. This finite rank paper has a note added in proof thanking Rosenblum for sending the technical report to Kato, quoting its main result and saying that Kato had found the full trace class results ("Details will be published elsewhere."). That second paper used some technical ideas from Rosenblum's paper.

I've heard that Rosenblum always felt that he'd not received sufficient credit for his trace class paper. There is some justice to this. The realization that trace class is the natural class is important. As I've discussed, trace class is maximal in a certain sense. Kato was at Berkeley in 1954 when Rosenblum was a student (albeit some time before his thesis was completed) and Kato was in contact with Wolf. However, there is no indication that Kato knew anything about Rosenblum's work until shortly before he wrote up his finite rank paper when he became aware of Rosenblum's abstract. My surmise is that both, motivated by Friedrichs, independently became interested in scattering.

The longer paper describes further developments and some applications of the trace class theory, some of them due to Kato himself. To me, the heroes of this later work are Kuroda [53], Pearson [66] and especially Birman [5–9].

## 7 Kato Smoothness

In this final section, we discuss the theory of Kato smoothness which is based primarily on two papers of Kato published in 1966 [35] (when Kato was 49) and 1968 [36]. The first is the basic one with four important results: the equivalence of many conditions giving the definition, the connection to spectral analysis, the implications for existence and completeness of wave operators and, finally, a perturbation result. The second paper concerns the Putnam–Kato theorem on positive commutators.

To me, the 1951 self-adjointness paper is Kato's most significant work (with the adiabatic theorem paper a close second), Kato's inequality his deepest and the subject of this section his most beautiful. One of the things that is so beautiful is that there isn't just a relation between the time-independent and time-dependent objects—there is an equivalence! Part of the equivalence depends on the simple formula that holds when $H$ is a self-adjoint operator on a (complex) Hilbert space $\mathcal{H}$ (where $R(\mu) = (H - \mu)^{-1}$)

$$\int_0^\infty e^{-\varepsilon t} e^{it\lambda} e^{-itH} \varphi \, dt = -iR(\lambda + i\varepsilon)\varphi \tag{1}$$

for any $\varphi \in \mathcal{H}$ because $\int_0^\infty e^{-\varepsilon t} e^{i(\lambda - x)t} \, dt = -i(x - \lambda - i\varepsilon)^{-1}$.

Here, is the set of equivalent definitions:

**Theorem 14** (Kato (1966) [35]) *Let $H$ be a self-adjoint operator and $A$ a closed operator. The following are all equal:*

$$\sup_{\substack{\|\varphi\|=1 \\ \varepsilon>0}} \frac{1}{4\pi^2} \int_{-\infty}^\infty \left( \|AR(\lambda + i\varepsilon)\varphi\|^2 + \|AR(\lambda - i\varepsilon)\varphi\|^2 \right) d\lambda \tag{2}$$

$$\sup_{\|\varphi\|=1} \frac{1}{2\pi} \int_{-\infty}^{\infty} \|Ae^{-itH}\varphi\|^2\, dt \tag{3}$$

$$\sup_{\substack{\|\varphi\|=1,\, \varphi\in D(A^*) \\ -\infty<a<b<\infty}} \frac{\|P_{(a,b)}(H)A^*\varphi\|^2}{b-a} \tag{4}$$

$$\sup_{\substack{\mu\notin\mathbb{R},\, \varphi\in D(A^*) \\ \|\varphi\|=1}} \frac{1}{2\pi}|\langle A^*\varphi, [R(\mu) - R(\bar{\mu})]A^*\varphi\rangle| \tag{5}$$

$$\sup_{\substack{\mu\notin\mathbb{R},\, \varphi\in D(A^*) \\ \|\varphi\|=1}} \frac{1}{\pi}\|R(\mu)A^*\varphi\|^2\, |Im\mu| \tag{6}$$

*In particular, if one is finite (resp. infinite), then all are.*

**Remarks**

1. In (2)/(3), we set $\|A\psi\| = \infty$ if $\psi \notin D(A)$, so, for example, to say that (3) is finite implies that for each $\varphi$, we have that $e^{-itH}\varphi \in D(A)$ for Lebesgue a.e. $t \in \mathbb{R}$.
2. If one and so all of the above quantities are finite we say that $A$ is $H$*-smooth.* The common value of these quantities is called $\|A\|_H^2$.
3. The proof is not hard. For example, If the integral in (3) has a factor of $e^{-2\varepsilon t}$ put inside it, the equality of the integrals in (2) and (3) follows from (1) and the Plancherel theorem. By monotone convergence, the sup of the time integral with the $e^{-2\varepsilon t}$ factor is the integral without that factor.

Smoothness has an immediate consequence for the spectral type of $H$:

**Theorem 15** (Kato [35]) *Let $H$ be a self-adjoint operator and let $A$ be $H$-smooth. Then $ran(A^*) \subset \mathcal{H}_{ac}(H)$. In particular, if $\ker(A) = \{0\}$, then $H$ has purely a.c. spectrum.*

The proof is very easy. If $d\nu$ is the $H$-spectral measure for $A^*\varphi$, then (4) says that

$$\nu(I) \le \|A\|_H \|\varphi\|^2 |I| \tag{7}$$

(where $|\cdot|$ is Lebesgue measure) for open intervals, $I$. By taking unions and using outer regularity, (7) holds for all sets, so $\nu$ is absolutely continuous.

Smoothness also implies existence and completeness of wave operators.

**Theorem 16** (Kato [35]) *Let $H, H_0$ be two self-adjoint operators. Let $A, B$ be closed operators so that $A$ is $H$-smooth and $B$ is $H_0$-smooth and so that*

$$H - H_0 = A^*B \tag{8}$$

*Then $\Omega^\pm(H, H_0)$ exist and are complete.*

**Remark** The proof is again easy (indeed, one of the beauties of Kato smoothness theory is how much one gets with simple proofs). The key is to use

$$
\begin{aligned}
|\langle\psi, (W(t)-W(s))\varphi\rangle| &= \left|\int_s^t \langle Ae^{-iuH}\psi, Be^{-iuH_0}\varphi\rangle\, du\right| \\
&\le \left(\int_{-\infty}^{\infty} \|Ae^{-iuH}\psi\|^2\, du\right)^{1/2} \left(\int_{-s}^{t} \|Be^{-iuH_0}\varphi\|^2\, du\right)^{1/2} \\
&\le \sqrt{2\pi}\,\|A\|_H\|\psi\| \left(\int_{-s}^{t} \|Be^{-iuH_0}\varphi\|^2\, du\right)^{1/2}
\end{aligned}
$$

which implies that

$$
\|(W(t)-W(s))\varphi\| \le \sqrt{2\pi}\,\|A\|_H \left(\int_{-s}^{t} \|Be^{-iuH_0}\varphi\|^2\, du\right)^{1/2} \tag{9}
$$

We say that a closed operator, $A$ is $H$*-supersmooth* if and only if

$$
\|A\|_{H,SS}^2 \equiv \sup_{z\in\mathbb{C}\setminus\mathbb{R}} \|A(H-z)^{-1}A^*\| < \infty \tag{10}
$$

The notion is in Kato's basic 1966 paper [35] and the name is from a 1989 paper of Kato–Yajima [51]. The name hasn't stuck but I like it, so I'll use it. The fourth important result in Kato's 1966 paper is

**Theorem 17** (Kato [35]) *Let $H_0$ be a self-adjoint operator. Let $A$ be $H_0$-supersmooth and $C$ a bounded self-adjoint operator so that*

$$
\alpha \equiv \|C\|\|A\|_{H_0,SS}^2 < 1 \tag{11}
$$

*Let $B = A^*CA$. Then $B$ is relatively form bounded with relative form bound at most $\alpha$. If $H = H_0 + B$, then $A$ is also $H$-supersmooth with*

$$
\|A\|_{H,SS} \le \|A\|_{H_0,SS}(1-\alpha)^{-1/2} \tag{12}
$$

*In particular, $\Omega^\pm(H, H_0)$ exist and are complete.*

**Remark** Once again, the proofs are simple. The key is a formal geometric series:

$$
\begin{aligned}
A(H-z)^{-1}A^* &= A(H_0-z)^{-1}A^* \\
&+ \sum_{j=0}^{\infty}(-1)^{j+1}A(H_0-z)^{-1}A^*\left[CA(H_0-z)^{-1}A^*\right]^j CA(H_0-z)^{-1}A^*
\end{aligned} \tag{13}
$$

In his original paper [35], Kato proved for $\nu \ge 3$, any $V \in L^{\frac{1}{2}\nu-\varepsilon} \cap L^{\frac{1}{2}\nu+\varepsilon}$ is supersmooth and Kato–Yajima [51] improved that to $L^{\nu/2}$. Iorio–O'Carroll [21] used supersmooth ideas to get a weak coupling result for $N$-body systems. If one has an

$N$-body system on $\mathbb{R}^\nu$ with two-body potentials in $L^{\nu/2}$ and if $\nu > 2$, then for small coupling, $H$ is purely a.c. with spectrum $[0, \infty)$.

In his 1968 paper [36], Kato used smoothness ideas to prove that if $A$ and $B$ are bounded self-adjoint operators so that $i[A, B] > 0$, then $A$ and $B$ have purely a.c. spectrum (a result proven the year before by very different methods by Putnam [67]). If $A$ is the generator of dilations and $B$ an $N$-body potential with repulsive potentials one has positivity of the commutator but neither operator is bounded so Kato's argument doesn't apply. In a series of papers, Lavine [54–58] was able to use smoothness ideas to prove suitably $N$-body systems with repulsive potentials have purely a.c. spectrum and, with sufficiently fast decay, have complete wave operators. To do this, he introduced the important notion of local smoothness [56]. For details, the reader can consult the longer paper or, better, the relevant section of Reed–Simon [68, Section XIII.8].

The longer paper also discusses a paper of Vakulenko [86] that should be better known. For two-body systems, he proved that if $|V(x)| \leq C(1 + |x|)^{-1-\varepsilon} \equiv g(x)^2$, then $g$ is $-\Delta + V$ locally smooth on $[0, \infty)$ which recovers the results of Agmon–Kato–Kuroda [2, 37] on completeness of wave operators and absence of singular continuous spectrum in this case and also the result of Kato [33] on the absence of positive eigenvalues. The longer paper also has references to work of Yafaev on long range [93] and on $N$-body systems [92] that uses smoothness as one tool in the analysis of such problems.

## References


1. 1980 Wiener and Steele prizes awarded. Not. A.M.S **27**, 528–533 (1980)
2. Agmon, S.: Spectral properties of Schrödinger operators and scattering theory. Ann. Scuola Norm. Sup. Pisa Cl. Sci. **2**, 151–218 (1975)
3. Avron, J., Seiler, R., Simon, B.: Homotopy and Quantization in Condensed Matter Physics. Phys. Rev. Lett. **51**, 51–53 (1983)
4. Berry, M.V.: Quantal phase factors accompanying adiabatic changes. Proc. Roy. Soc. A **392**, 45–57 (1984)
5. Birman, M.Š.: Perturbations of the continuous spectrum of a singular elliptic operator by varying the boundary and the boundary conditions, Vestnik Leningrad. Univ. **17**, 22–55 (1962); Eng Trans.: Am. Math. Soc. Transl. Ser. 2, **225**, Suslina, T., Yafaev, D. (eds.), pp. 19–53. American Mathematical Society, Providence (2008)
6. Birman, M.Š.: A criterion for existence of wave operators. Izv. Akad. Nauk. SSSR Ser. Mat. **27** 883–906 (1963); Eng. Trans.: A.M.S. Transl. **54**, 91–117 (1966)
7. Birman, M.Š.: A local criterion for the existence of wave operators. Izv. Akad. Nauk SSSR Ser. Mat. **32**, 914–942 (1968); Eng. Trans.: Math. USSR–Izv. **2**, 879–906 (1968)
8. Birman, M.Š.: A test of the existence of complete wave operators in scattering theory for a pair of spaces. Spectral Theory. Wave Process. Problems of Mathematical Physics, vol. 4, pp. 22–26. Springer, Berlin (1970)
9. Birman, M.Š., Krein, M.G.: On the theory of wave operators and scattering operators. Dokl. Akad. Nauk. SSSR **144**, 475–478; Eng. Trans: Sov. Math. Dokl. **3**, 740–744 (1962)
10. Born, M.: Quantenmechanik der Stossvorgänge. Zeit. für Physik **38**, 803–827 (1926)
11. Born, M., Fock, V.A.: Beweis des Adiabatensatzes. Zeit. für Physik **A51**, 165–180 (1928)
12. Brézis, H., Kato, T.: Remarks on the Schrödinger operator with singular complex potentials. J. Math. Pures Appl. **58**, 137–151 (1979)

13. Deift, P., Simon, B.: A time-dependent approach to the completeness of multiparticle quantum systems. Commun. Pure Appl. Math. **30**, 573–583 (1977)
14. Dolph, C., Howland, J.: Dedication of special issue in honor of Otto Laporte and Tosio Kato. JMAA **127**, 299–311 (1987)
15. Ehrenfest, P.: Adiabatische Invarianten und Quantentheorie. Ann. d. Phys. **51**, 327–352 (1916)
16. Friedrichs, K.O.: On the perturbation of continuous spectra. Commun. Pure Appl. Math. **1**, 361–406 (1948)
17. Heinz, E.: Beiträge zur Störungstheorie der Spektralzerlegung. Math. Ann. **123**, 415–438 (1951)
18. Heisenberg, W.: Die beobachtbaren Grössen in der Theorie der Elementarteilchen, I, II. Zeit. für Physik **120**(513–538), 673–702 (1943)
19. Hess, H., Schrader, R., Uhlenbrock, D.A.: Domination of semigroups and generalization of Kato's inequality. Duke Math. J. **44**, 893–904 (1977)
20. Ikebe, T., Kato, T.: Uniqueness of the self-adjoint extensions of singular elliptic differential operators. Arch. Ration. Mech. Anal. **9**, 77–92 (1962)
21. Iorio, R.J., O'Carroll, M.: Asymptotic completeness for multi-particle Schroedinger Hamiltonians with weak potentials. Commun. Math. Phys. **27**, 137–145 (1972)
22. Jensen, A., Kato, T.: Spectral properties of Schrödinger operators and time-decay of the wave functions. Duke Math. J. **46**, 583–611 (1979)
23. Kato, T.: On the convergence of the perturbation method, I. Progr. Theor. Phys. **4**, 514–523 (1949)
24. Kato, T.: On the convergence of the perturbation method, II. Progr. Theor. Phys. **5**(95–101), 207–212 (1950)
25. Kato, T.: On the adiabatic theorem for quantum mechanics. J. Phys. Soc. Jpn. **5**, 435–439 (1950)
26. Kato, T.: On the convergence of the perturbation method. J. Fac. Sci. Univ. Tokyo **6**, 145–226 (1951)
27. Kato, T.: Fundamental properties of Hamiltonian operators of Schrödinger type. Trans. A.M.S. **70**, 195–211 (1951)
28. Kato, T.: On the perturbation theory of closed linear operators. J. Math. Soc. Jpn. **4**, 323–337 (1952)
29. Kato, T.: Notes on some inequalities for linear operators. Math. Ann. **125**, 208–212 (1952)
30. Kato, T.: On the eigenfunctions of many particle systems in quantum mechanics. Commun. Pure Appl. Math. **10**, 151–177 (1957)
31. Kato, T.: On finite-dimensional perturbations of self-adjoint operators. J. Math. Soc. Jpn. **9**, 239–249 (1957)
32. Kato, T.: Perturbation of continuous spectra by trace class operators. Proc. Jpn. Acad. **33**, 260–264 (1957)
33. Kato, T.: Growth properties of solutions of the reduced wave equation with a variable coefficient. Commun. Pure Appl. Math. **12**, 403–425 (1959)
34. Kato, T.: A generalization of the Heinz inequality. Proc. Jpn. Acad. Ser. A. Math. Sci. **6**, 305–308 (1961)
35. Kato, T.: Wave operators and similarity for some non-selfadjoint operators. Math. Ann. **162**, 258–279 (1966)
36. Kato, T.: Smooth operators and commutators. Studia Mathematica **31**, 535–546 (1968)
37. Kato, T.: Some results on potential scattering. In: Proceedings of the International Conference on Functional Analysis and Related Topics, Tokyo, 1969, University of Tokyo Press, Tokyo, pp. 206–215 (1970)
38. Kato, T.: Scattering theory and perturbation of continuous spectra. Actes du Congrès International des Mathématiciens **1**, 135–140 (1971)
39. Kato, T.: Schrödinger operators with singular potentials. Israel J. Math. **13**, 135–148 (1972)
40. Kato, T.: Continuity of the map $S \mapsto |S|$ for linear operators. Proc. Jpn. Acad. **49**, 157–160 (1973)

41. Kato, T.: A second look at the essential self-adjointness of the Schrödinger operators. In: Enz, C., Mehra, J. (eds.) Physical Reality and Mathematical Description, pp. 193–201. Reidel, Dordrecht (1974)
42. Kato, T.: Perturbation Theory for Linear Operators. Grundlehren der Mathematischen Wissenschaften, Band, vol. 132, 2nd edn. Springer, Berlin (1976) (1st edn. 1966; there is also a revised and corrected 1980 printing)
43. Kato, T.: Remarks on Schrödinger operators with vector potentials. Integr. Equ. Oper. Theory **1**, 103–113 (1978)
44. Kato, T.: On some Schrödinger operators with a singular complex potential. Ann. Scuola Norm. Super. Pisa, Cl. Sci. **IV. 5**, 105–114 (1978)
45. Kato, T.: Trotter's product formula for an arbitrary pair of self–adjoint contraction semigroups. In: Gohnerg, I., Kac, M. (eds.) Topics in Functional Analysis, Essays dedicated to M. G. Krein. Advances in Mathematics, Supplimentary Studies, vol. 3, pp. 185–195. Academic, New York (1978)
46. Kato, T.: Remarks on the selfadjointness and related problems for differential operators. In: Knowles, I., Lewis, R. (eds.) Spectral Theory of Differential Operators (Proceedings Conference, Birmingham, USA 1981), pp. 253–266. North-Holland, Amsterdam (1981)
47. Kato, T.: Nonselfadjoint Schrödinger operators with singular first-order coefficients. Proc. Roy. Soc. Edinb. Sect. A **96**, 323–329 (1984)
48. Kato, T.: $L^p$-theory of Schrödinger operators with a singular potential. In: Nagel, R., Schlotterbeck, U., Wolff, M. (eds.) Aspects of Positivity in Functional Analysis, pp. 63–78. North-Holland, Amsterdam (1986)
49. Kato, T., Kuroda, S.T.: Theory of simple scattering and eigenfunction expansions. In: Browder, F. (ed.) Functional Analysis and Related Field, pp. 99–131. Springer, Berlin (1970)
50. Kato, T., Kuroda, S.T.: The abstract theory of scattering. Rocky Mt. J. Math. **1**, 127–171 (1971)
51. Kato, T., Yajima, K.: Some examples of smooth operators and the associated smoothing effect. Rev. Math. Phys. **1**, 481–496 (1989)
52. Kuroda, S.T.: On a theorem of Weyl-von Neumann. Proc. Jpn. Acad. **34**, 11–15 (1958)
53. Kuroda, S.T.: On the existence and the unitarity property of the scattering operator. Nuovo Cimento **12**, 431–454 (1959)
54. Lavine, R.: Absolute continuity of Hamiltonian operators with repulsive potentials. Proc. A.M.S. **22**, 55–60 (1969)
55. Lavine, R.: Commutators and scattering theory, I. Repulsive interactions. Commun. Math. Phys. **20**, 301–323 (1971)
56. Lavine, R.: Commutators and scattering theory, II. A class of one-body problems. Indiana Univ. Math. J. **21**, 643–656 (1972)
57. Lavine, R.: Completeness of the wave operators in the repulsive N-body problem. J. Math. Phys. **14**, 376–379 (1973)
58. Lavine, R.: Absolute continuity of positive spectrum for Schrödinger operators with long range potentials. J. Func. Anal. **12**, 30–54 (1973)
59. Leinfelder, H., Simader, C.: Schrödinger operators with singular magnetic vector potentials. Math. Zeit. **176**, 1–19 (1981)
60. Löwner, K.: Über monotone Matrixfunktionen. Math. Zeit. **38**, 177–216 (1934)
61. Møller, C.: General properties of the characteristic matrix in the theory of elementary particles, I. Danske. Vid. Selsk. Mat.-Fys. Medd. **23**, 1–48 (1945)
62. Nagy, B.Sz.: Perturbations des transformations autoadjointes dans l'espace de Hilbert. Commentarii Math. Helv. **19**, 347–366 (1947)
63. Nagy, BSz: Perturbations des transformations linéaires fermées. Acta Sci. Math. Szeged **14**, 125–137 (1951)
64. Pancharatnam, S.: Generalized theory of interference, and its applications. Part I. Coherent pencils. Proc. Indian Acad. Sci. **A44**, 247–262 (1956)
65. Pazy, A.: Semigroups of Linear Operators and Applications to Partial Differential Equations. Springer, New York (1983)
66. Pearson, D.B.: A generalization of Birman's trace theorem. J. Funct. Anal. **28**, 182–186 (1978)

67. Putnam, C.R.: Commutation Properties of Hilbert Space Operators and Related Topics. Springer, New York (1967)
68. Reed, M., Simon, B.: Methods of Modern Mathematical Physics, IV: Analysis of Operators. Academic Press, New York (1978)
69. Rellich, F.: Störungstheorie der Spektralzerlegung. I–V. Math. Ann. **113**, 600–619, 677–685 (1937); **116**, 555–70 (1939); **117**, 356–382 (1940); **118**, 462–484 (1942)
70. Rellich, F.: Die zulässigen Randbedingungen bei den singulären Eigenwertproblemen der mathematischen Physik. Gewöhnliche Differentialgleichungen zweiter Ordnung. Math. Zeit. **49**, 702–723 (1944)
71. Rosenblum, M.: Perturbation of the continuous spectrum and unitary equivalence. Pac. J. Math. **7**, 997–1010 (1957)
72. Seiler, E., Simon, B.: Bounds in the Yukawa quantum field theory: upper bound on the pressure, Hamiltonian bound and linear lower bound. Commun. Math. Phys **45**, 99–114 (1975)
73. Simon, B.: Essential self-adjointness of Schrödinger operators with positive potentials. Math. Ann. **201**, 211–220 (1973)
74. Simon, B.: Essential self-adjointness of Schrödinger operators with singular potentials. Arch. Ration. Mech. Anal. **52**, 44–48 (1973)
75. Simon, B.: Schrödinger operators with singular magnetic vector potentials. Math. Zeit. **131**, 361–370 (1973)
76. Simon, B.: Universal diamagnetism of spinless Bose systems. Phys. Rev. Lett. **36**, 1083–1084 (1976)
77. Simon, B.: An abstract Kato's inequality for generators of positivity preserving semigroups. Indiana Univ. Math. J. **26**, 1067–1073 (1977)
78. Simon, B.: Maximal and minimal Schrödinger forms. J. Oper. Theory **1**, 37–47 (1979)
79. Simon, B.: Kato's inequality and the comparison of semigroups. J. Funct. Anal. **32**, 97–101 (1979)
80. Simon, B.: Holonomy, the Quantum Adiabatic Theorem and Berry's Phase. Phys. Rev. Lett. **51**, 2167–2170 (1983)
81. Simon, B.: A Comprehensive Course in Analysis, Part 3: Harmonic Analysis. American Mathematical Society, Providence (2015)
82. Simon, B.: A Comprehensive Course in Analysis, Part 4: Operator Theory. American Mathematical Society, Providence (2015)
83. Simon, B.: Tosio Kato's work on non–relativistic quantum mechanics. Bull. Math. Sci., to appear
84. Stummel, F.: Singuläre elliptische Differentialoperatoren in Hilbertschen Räumen. Math. Ann. **132**, 150–176 (1956)
85. Thouless, D.J., Kohmoto, M., Nightingale, M.P., den Nijs, M.: Quantized Hall Conductance in a Two-Dimensional Periodic Potential. Phys. Rev. Lett. **49**, 405–408 (1982)
86. Vakulenko, A.F.: A variant of commutator estimates in spectral theory. Kraev. Zadachi Mat. Fiz. i Smezhn. Vopr. Teor. Funktsiĭ **19**, 29–36 (1987); translated in. J. Sov. Math. **49**, 1136–1139 (1990)
87. von Neumann, J.: Mathematical Foundations of Quantum Mechanics. Princeton University Press, Princeton (1996); German original: Mathematische Grundlagen der Quantenmechanik (1932); first English translation (1955)
88. von Neumann, J.: Charakterisierung des Spektrums eines Integraloperators. Actualités Sci. Industr. **229**, 3–20 (1935)
89. Weyl, H.: Über beschränkte quadratische Formen, deren Differenz vollstetig ist. Palermo Rend. **27**, 373–392, 402 (1909)
90. Wheeler, J.A.: On the Mathematical Description of Light Nuclei by the Method of Resonating Group Structure. Phys. Rev. **52**, 1107–1122 (1937)
91. Wolf, F.: Analytic perturbation of operators in Banach spaces. Math. Ann. **124**, 317–333 (1952)
92. Yafaev, D.R.: Radiation conditions and scattering theory for N-particle Hamiltonians. Commun. Math. Phys. **154**, 523–554 (1993)
93. Yafaev, D.: The scattering amplitude for the Schrödinger equation with a long-range potential. Commun. Math. Phys. **191**, 183–218 (1998)

# The Factor Type of Conservative KMS-Weights on Graph $C^*$-Algebras

**Klaus Thomsen**

**Abstract** We determine the factor generated by the GNS-representation defined by a KMS-weight for a generalized gauge action on a simple graph $C^*$-algebra, when the corresponding measure on the path space of the graph is conservative for the shift.

## 1 Introduction

In certain quantum statistical models, the observables are represented by the elements of a $C^*$-algebra and the time-evolution by a one-parameter group of automorphisms on the $C^*$-algebra, [1]. In such models, the equilibrium states are characterized by the KMS-condition [2], and much work has been devoted to the study of these KMS-states. As a result, there are now large classes of $C^*$-algebras and one-parameter groups for which the structure of the equilibrium states is completely understood. An extremal KMS-state, and more generally also an extremal KMS-weight, gives rise in a canonical way to a representation of the $C^*$-algebra which generates a von Neumann algebra factor. Some of the papers on KMS-states have included a determination of this factor for the extremal KMS-states. This is the case in [3–18].[1] The factor types give a natural way to distinguish between KMS-states very similar to the type classification of non-singular transformations in ergodic theory. The purpose of this paper is to determine the factor types for a class of KMS-weights and KMS-states that arise from generalized gauge actions on simple graph $C^*$-algebras. In [19], the author studied such KMS-weights and among other things it was shown that they can be naturally divided into three classes depending on the properties of the measure they induce on the unit space of the groupoid underlying the $C^*$-algebra. For the KMS-weights, we consider here the measure is concentrated on the space of infinite paths in the graph and it is conservative with respect to the action of the shift.

[1]Despite some effort to make this a complete list, presumably it is not.

K. Thomsen (✉)
Department of Mathematics, University of Aarhus, Aarhus, Denmark
e-mail: matkt@math.au.dk

© Springer Nature Switzerland AG 2019

T. M. Rassias and V. A. Zagrebnov (eds.), *Analysis and Operator Theory*, Springer Optimization and Its Applications 146,
https://doi.org/10.1007/978-3-030-12661-2_17

We call them, therefore, conservative KMS-weights. We determine their factor type when the graph $C^*$-algebra is simple by calculating the $\Gamma$-invariant of Connes for the associated factors. As shown in [19], there is a bijective correspondence between the rays of KMS-weights and the KMS-states on a corner defined by a vertex in the graph, at least for the cases we consider in this paper, and we show, therefore, first that the $\Gamma$-invariant for the factor defined by an extremal KMS-weight is the same as the $\Gamma$-invariant for the factor obtained from the corresponding KMS-state on the corner. This is not surprising, but it is helpful because it shows that our calculation of the $\Gamma$-invariant covers two cases.

## 2 The Setting

### 2.1 *Factors from KMS-Weights*

Let $A$ be a $C^*$-algebra and $A_+$ the convex cone of positive elements in $A$. A *weight* on $A$ is a map $\psi : A_+ \to [0, \infty]$ with the properties that $\psi(a + b) = \psi(a) + \psi(b)$ and $\psi(\lambda a) = \lambda\psi(a)$ for all $a, b \in A_+$ and all $\lambda \in \mathbb{R}$, $\lambda > 0$. By definition, $\psi$ is *densely defined* when $\{a \in A_+ : \psi(a) < \infty\}$ is dense in $A_+$ and *lower semicontinuous* when $\{a \in A_+ : \psi(a) \leq \alpha\}$ is closed for all $\alpha \geq 0$. We will use [20–22] as our source of information on weights, and as in [22] we say that a weight is *proper* when it is nonzero, densely defined, and lower semicontinuous.

Let $\psi$ be a proper weight on $A$. Set $\mathcal{N}_\psi = \{a \in A : \psi(a^*a) < \infty\}$ and note that

$$\mathcal{N}_\psi^* \mathcal{N}_\psi = \text{Span}\left\{a^*b : a, b \in \mathcal{N}_\psi\right\}$$

is a dense $*$-subalgebra of $A$, and that there is a unique well-defined linear map $\mathcal{N}_\psi^* \mathcal{N}_\psi \to \mathbb{C}$ which extends $\psi : \mathcal{N}_\psi^* \mathcal{N}_\psi \cap A_+ \to [0, \infty)$. We denote also this densely defined linear map by $\psi$.

Let $\alpha : \mathbb{R} \to \text{Aut}A$ be a point-wise norm-continuous one-parameter group of automorphisms on $A$. Let $\beta \in \mathbb{R}$. Following [23], we say that a proper weight $\psi$ on $A$ is a *$\beta$-KMS weight* for $\alpha$ when

(i) $\psi \circ \alpha_t = \psi$ for all $t \in \mathbb{R}$, and
(ii) for every pair $a, b \in \mathcal{N}_\psi \cap \mathcal{N}_\psi^*$ there is a continuous and bounded function $F$ defined on the closed strip $D_\beta$ in $\mathbb{C}$ consisting of the numbers $z \in \mathbb{C}$ whose imaginary part lies between 0 and $\beta$, and is holomorphic in the interior of the strip and satisfies that

$$F(t) = \psi(a\alpha_t(b)), \ F(t + i\beta) = \psi(\alpha_t(b)a)$$

for all $t \in \mathbb{R}$.[2]

[2]Note that we apply the definition from [23] for the action $\alpha_{-t}$ in order to use the same sign convention as in [1], for example.

A $\beta$-KMS-weight $\psi$ with the property that

$$\sup\{\psi(a) : 0 \leq a \leq 1\} = 1$$

will be called a *$\beta$-KMS-state*. The following is Theorem 2.4 in [19].

**Theorem 1** *Let $\alpha : \mathbb{R} \to \mathrm{Aut}A$ be a point-wise norm-continuous one-parameter group of automorphisms on a $C^*$-algebra A. Let p be a projection in the fixed point algebra of $\alpha$ such that p is full in A. For all $\beta \in \mathbb{R}$, the map*

$$\psi \mapsto \psi(p)^{-1}\psi|_{pAp}$$

*is a bijection between the set of rays of $\beta$-KMS-weights for $\alpha$ and the $\beta$-KMS-states for the restriction of $\alpha$ to $pAp$.*

Given a proper weight $\psi$ on a $C^*$-algebra $A$, there is a GNS-type construction consisting of a Hilbert space $H_\psi$, a linear map $\Lambda_\psi : \mathcal{N}_\psi \to H_\psi$ with dense range and a nondegenerate representation $\pi_\psi$ of $A$ on $H_\psi$ such that

- $\psi(b^*a) = \langle \Lambda_\psi(a), \Lambda_\psi(b)\rangle, \ a, b \in \mathcal{N}_\psi$, and
- $\pi_\psi(a)\Lambda_\psi(b) = \Lambda_\psi(ab), \ a \in A, \ b \in \mathcal{N}_\psi$,

cf. [20–22]. A $\beta$-KMS-weight $\psi$ on $A$ is *extremal* when the only $\beta$-KMS-weights $\varphi$ on $A$ with the property that $\varphi(a) \leq \psi(a)$ for all $a \in A_+$ are scalar multiples of $\psi$, viz. $\varphi = s\psi$ for some $s > 0$.

The following is Lemma 4.9 in [14].

**Lemma 1** *Let A be a separable $C^*$-algebra and $\alpha$ a point-wise norm-continuous one-parameter group of automorphisms on A. Let $\psi$ be an extremal $\beta$-KMS-weight for $\alpha$. Then $\pi_\psi(A)''$ is a factor.*

It is shown in Section 2.2 of [21] that a $\beta$-KMS-weight $\psi$ extends to a normal semi-finite faithful weight $\tilde{\psi}$ on $\pi_\psi(A)''$ such that $\psi = \tilde{\psi} \circ \pi_\psi$, and that the modular group on $\pi_\psi(A)''$ corresponding to $\tilde{\psi}$ is the one-parameter group $\theta$ on $\pi_\psi(A)''$ given by

$$\theta_t = \tilde{\alpha}_{-\beta t} ,$$

where $\tilde{\alpha}$ is the $\sigma$-weakly continuous extension of $\alpha$ defined such that $\tilde{\alpha}_t \circ \pi_\psi = \pi_\psi \circ \alpha_t$. By construction $\tilde{\alpha}_t = \mathrm{Ad}\, U_t$, where $U_t \in B(H_\psi)$ is defined by

$$U_t\Lambda_\psi(a) = \Lambda_\psi(\alpha_t(a)) .$$

In the setting of Theorem 1, let $(\pi_\varphi, H_\varphi, \xi_\varphi)$ be the GNS-representation of the state $\varphi$ on $pAp$ defined such that

$$\varphi(x) = \psi(p)^{-1}\psi(x) .$$

The modular automorphism group $\theta'$ on $\pi_\varphi(pAp)''$ corresponding to the vector state defined by $\xi_\varphi$ is given by

$$\theta'_t\left(\pi_\varphi(pap)\right) = \pi_\varphi\left(\alpha_{-\beta t}(pap)\right) = \pi_\varphi\left(p\alpha_{-\beta t}(a)p\right) .$$

**Lemma 2** *In the setting of Theorem 1 there is an $*$-isomorphism*

$$\pi_\varphi(pAp)'' \simeq \pi_\psi(p)\pi_\psi(A)''\pi_\psi(p) \tag{1}$$

*of von Neumann algebras which is equivariant with respect to $\theta$ and $\theta'$.*

*Proof* Let $q \in B(H_\psi)$ be the orthogonal projection on $\overline{\Lambda_\psi(pAp)}$ and define a unitary $W : H_\varphi \to qH_\psi$ such that

$$W\pi_\varphi(x)\xi_\varphi = \psi(p)^{-\frac{1}{2}}\Lambda_\psi(x)$$

for $x \in pAp$. Conjugation by $W$ gives an isomorphism $\pi_\varphi(pAp)'' \simeq q\pi_\psi(A)''q$. Since $\left(qH_\psi, \pi_\psi, \frac{\Lambda_\psi(p)}{\sqrt{\psi(p)}}\right)$ is the GNS-triple of $\varphi$ it follows from Corollary 5.3.9 in [1] that $\Lambda_\psi(p)$ is separating for $\pi_\psi(p)\pi_\psi(A)''\pi_\psi(p)$. Since $q$ commutes with $\pi_\psi(p)\pi_\psi(A)''\pi_\psi(p)$ and $qH_\psi$ contains $\Lambda_\psi(p)$, the map $m \mapsto mq$ is an isomorphism

$$\pi_\psi(p)\pi_\psi(A)''\pi_\psi(p) \to q\pi_\psi(A)''q .$$

We obtain then the isomorphism (1) as the composition of two isomorphisms, both of which are equivariant.

When $M$ is a $\sigma$-finite von Neumann algebra factor every normal faithful semi-finite weight on $M$ comes together with a modular automorphism group $\theta = (\theta_t)_{t\in\mathbb{R}}$ and the Connes-invariant $\Gamma(M)$ is the intersection

$$\Gamma(M) = \bigcap_q \mathrm{Sp}(qMq) ,$$

where we take the intersection over all projections $q \in M$ fixed by $\theta$, and $\mathrm{Sp}(qMq)$ denotes the Arveson spectrum of the restriction of $\theta$ to $qMq$. In more detail, $\mathrm{Sp}(qMq)$ is defined as follows. For $f \in L^1(\mathbb{R})$ define a linear map $\theta_f : qMq \to qMq$ such that

$$\theta_f(a) = \int_{\mathbb{R}} f(t)\theta_t(a)\, dt .$$

Then

$$\mathrm{Sp}(qMq) = \bigcap\left\{Z(f) : f \in L^1(\mathbb{R}),\ \theta_f(qMq) = \{0\}\right\} ,$$

where

$$Z(f) = \left\{ r \in \mathbb{R} : \int_{\mathbb{R}} e^{itr} f(t)\, dt = 0 \right\} .$$

See [24]. In particular, when $\psi$ is an extremal $\beta$-KMS-weight on $A$ we can calculate $\Gamma\left(\pi_\psi(A)''\right)$ by using the automorphism group $\theta_t = \tilde{\alpha}_{-\beta t}$ and we get the following immediate corollary to Lemma 2.

**Corollary 1** *In the setting of Lemma 2,*

$$\Gamma\left(\pi_\varphi(pAp)''\right) = \Gamma\left(\pi_\psi(A)''\right) .$$

## 2.2 Generalized Gauge Actions on Graph C*-Algebras

Let $\mathscr{G}$ be a countable directed graph with vertex set $\mathscr{G}_V$ and arrow set $\mathscr{G}_{Ar}$. For an arrow $a \in \mathscr{G}_{Ar}$ we denote by $s(a) \in \mathscr{G}_V$ its source and by $r(a) \in \mathscr{G}_V$ its range. A vertex $v$ which does not emit any arrow is a *sink*, while a vertex $v$ which emits infinitely many arrows is called an *infinite emitter*. The set of sinks and infinite emitters in $\mathscr{G}$ is denoted by $V_\infty$. An *infinite path* in $\mathscr{G}$ is an element $p \in (\mathscr{G}_{Ar})^{\mathbb{N}}$ such that $r(p_i) = s(p_{i+1})$ for all $i$. A finite path $\mu = a_1 a_2 \cdots a_n = (a_i)_{i=1}^n \in (\mathscr{G}_{Ar})^n$ is defined similarly. The number of edges in $\mu$ is its *length* and we denote it by $|\mu|$. A vertex $v \in \mathscr{G}_V$ will be considered as a finite path of length 0. We let $P(\mathscr{G})$ denote the (possibly empty) set of infinite paths in $\mathscr{G}$ and $P_f(\mathscr{G})$ the set of finite paths in $\mathscr{G}$. The set $P(\mathscr{G})$ is a complete metric space when the metric is given by

$$d(p,q) = \sum_{i=1}^{\infty} 2^{-i} \delta(p_i, q_i) , \tag{2}$$

where $\delta(a,a) = 0$ and $\delta(a,b) = 1$ when $a \neq b$. We extend the source map to $P(\mathscr{G})$ such that $s(p) = s(p_1)$ when $p = (p_i)_{i=1}^\infty \in P(\mathscr{G})$, and the range and source maps to $P_f(\mathscr{G})$ such that $s(\mu) = s(a_1)$ and $r(\mu) = r(a_{|\mu|})$ when $|\mu| \geq 1$, and $s(v) = r(v) = v$ when $v \in \mathscr{G}_V$. Associated to the finite path $\mu \in P_f(\mathscr{G})$ is the cylinder set

$$Z(\mu) = \left\{ (p_i)_{i=1}^\infty \in P(\mathscr{G}) : \ p_j = a_j, \ j = 1, 2, \cdots, |\mu| \right\}$$

which is an open and closed set in $P(\mathscr{G})$. In particular, when $\mu$ has length 0 and hence is just a vertex $v$,

$$Z(v) = \{ p \in P(\mathscr{G}) : \ s(p) = v \} .$$

The $C^*$-algebra $C^*(\mathscr{G})$ of the graph $\mathscr{G}$ was introduced in this generality in [25] as the universal $C^*$-algebra generated by a collection $S_a$, $a \in \mathscr{G}_{Ar}$, of partial isometries and a collection $P_v$, $v \in \mathscr{G}_V$, of mutually orthogonal projections subject to the conditions that

(1) $S_a^* S_a = P_{r(a)}$, $\forall a \in \mathscr{G}_{Ar}$,
(2) $S_a S_a^* \leq P_{s(a)}$, $\forall a \in \mathscr{G}_{Ar}$,
(3) $\sum_{a \in F} S_a S_a^* \leq P_v$ for every finite subset $F \subseteq s^{-1}(v)$ and all $v \in \mathscr{G}_V$, and
(4) $P_v = \sum_{a \in s^{-1}(v)} S_a S_a^*$, $\forall v \in \mathscr{G}_V \backslash V_\infty$.

For a finite path $\mu = (a_i)_{i=1}^{|\mu|} \in P_f(\mathscr{G})$, we set

$$S_\mu = S_{a_1} S_{a_2} S_{a_3} \cdots S_{a_{|\mu|}} .$$

The elements $S_\mu S_\nu^*$, $\mu, \nu \in P_f(\mathscr{G})$, span a dense $*$-subalgebra $\mathscr{A}$ in $C^*(\mathscr{G})$. The projections

$$P_\mu = S_\mu S_\mu^*$$

will play an important role in the following.

**Lemma 3** *$C^*(\mathscr{G})$ is a nuclear $C^*$-algebra and $\pi(C^*(\mathscr{G}))''$ is a hyperfinite von Neumann algebra for all nondegenerate representations $\pi$ of $C^*(\mathscr{G})$.*

*Proof* The nuclearity of $C^*(\mathscr{G})$ follows from [26] when $\mathscr{G}$ is row-finite in the sense that $\#s^{-1}(v) < \infty$ for all $v \in \mathscr{G}_V$, and the general case follows then from [27]. The second statement is a well-known consequence of the first and goes back to [28] and [29].

We describe next the necessary and sufficient conditions which $\mathscr{G}$ must satisfy for $C^*(\mathscr{G})$ to be simple. These conditions were identified by Szymanski in [30]. A *loop* in $\mathscr{G}$ is a finite path $\mu \in P_f(\mathscr{G})$ of positive length such that $r(\mu) = s(\mu)$. We will say that a loop $\mu$ *has an exit* then $\#s^{-1}(v) \geq 2$ for at least one vertex $v$ in $\mu$. A subset $H \subseteq \mathscr{G}_V$ is *hereditary* when $a \in \mathscr{G}_{Ar}$, $s(a) \in H \Rightarrow r(a) \in H$, and *saturated* when

$$v \in \mathscr{G}_V \backslash V_\infty,\ r(s^{-1}(v)) \subseteq H \ \Rightarrow \ v \in H .$$

In the following we say that $\mathscr{G}$ is *cofinal* when the only nonempty subset of $\mathscr{G}_V$, which is both hereditary and saturated is $\mathscr{G}_V$ itself.

**Theorem 2** (Theorem 12 in [30].) *$C^*(\mathscr{G})$ is simple if and only if $\mathscr{G}$ is cofinal and every loop in $\mathscr{G}$ has an exit.*

A function $F : \mathscr{G}_{Ar} \to \mathbb{R}$ will be called a *potential* on $\mathscr{G}$. Using it we can define a point-wise norm-continuous one-parameter group $\alpha^F = \left(\alpha_t^F\right)_{t \in \mathbb{R}}$ on $C^*(\mathscr{G})$ such that

$$\alpha_t^F(S_a) = e^{iF(a)t} S_a$$

for all $a \in \mathscr{G}_{Ar}$ and

$$\alpha_t^F(P_v) = P_v$$

for all $v \in \mathscr{G}_V$. An action of this sort is called a *generalized gauge action*; the *gauge action* itself being the one-parameter group corresponding to the constant function $F = 1$. To describe the KMS-weights for a generalized gauge action, extend $F$ to a map $F : P_f(\mathscr{G}) \to \mathbb{R}$ such that $F(v) = 0$ when $v \in \mathscr{G}_V$, and

$$F(\mu) = \sum_{i=1}^{n} F(a_i)$$

when $\mu = (a_i)_{i=1}^n \in (\mathscr{G}_{Ar})^n$. For $\beta \in \mathbb{R}$, define the matrix $A(\beta) = \left(A(\beta)_{v,w}\right)_{v,w \in \mathscr{G}_V}$ over $\mathscr{G}_V$ such that

$$A(\beta)_{v,w} = \sum_a e^{-\beta F(a)}$$

where we sum over arrows $a \in \mathscr{G}_{Ar}$ with $s(a) = v$ and $r(a) = w$. As in [19] we say that a nonzero nonnegative vector $\psi = (\psi_v)_{v \in \mathscr{G}_V}$ is *almost $A(\beta)$-harmonic* when

- $\sum_{w \in \mathscr{G}_V} A(\beta)_{v,w} \psi_w \leq \psi_v, \ \forall v \in \mathscr{G}_V$, and
- $\sum_{w \in \mathscr{G}_V} A(\beta)_{v,w} \psi_w = \psi_v, \ \forall v \in \mathscr{G}_V \backslash V_\infty$.

The almost $A(\beta)$-harmonic vectors $\psi$ for which

- $\sum_{w \in \mathscr{G}_V} A(\beta)_{v,w} \psi_w = \psi_v, \ \forall v \in \mathscr{G}_V$,

will be called *$A(\beta)$-harmonic*. In particular, when $\mathscr{G}$ is row-finite without sinks an almost $A(\beta)$-harmonic vector is automatically $A(\beta)$-harmonic. An almost $A(\beta)$-harmonic vector which is not $A(\beta)$-harmonic will be said to be a *proper almost $A(\beta)$-harmonic vector*. It was shown in [19] that an almost $A(\beta)$-harmonic vector $\varphi$ gives rise to a $\beta$-KMS- weight $W_\varphi$ characterized by the properties that $S_\mu^* \in \mathscr{N}_{W_\varphi}$ and

$$W_\varphi\left(S_\mu S_\nu^*\right) = \begin{cases} 0, & \mu \neq \nu \\ e^{-\beta F(\mu)} \varphi_{r(\mu)}, & \mu = \nu \end{cases}$$

when $\mu, \nu \in P_f(\mathscr{G})$, and that all *gauge invariant* KMS-weights arise like this. Generally there can be KMS-weights that are not gauge invariant, but as shown in Proposition 5.6 of [31] all KMS-weights are gauge invariant when $C^*(\mathscr{G})$ is simple. Therefore, in the case that concerns us here, all KMS-weights arise from almost $A(\beta)$-harmonic vectors. Borrowing terminology from harmonic analysis we say that an almost $A(\beta)$-harmonic vector $\varphi$ is *minimal* when it only dominates multiples of itself, i.e., when every almost $A(\beta)$-harmonic vector $\varphi'$ with the property that $\varphi'_v \leq \varphi_v$ for all $v \in \mathscr{G}_V$ is of the form $\varphi' = \lambda\varphi$ for some $\lambda > 0$. Thus, the minimal almost $A(\beta)$-harmonic vectors are those for which the corresponding $\beta$-KMS-weight $W_\varphi$ is extremal.

The minimal almost $A(\beta)$-harmonic vectors can be subdivided into various ways. Here we shall consider three fundamental classes. The first class consists of the minimal proper almost $A(\beta)$-harmonic vectors. The other two classes consist of the $A(\beta)$-harmonic vectors and are distinguished by the properties of the measures they define on $P(\mathscr{G})$. To explain this observe that by Lemma 3.7 in [19] an $A(\beta)$-harmonic vector $\varphi$ defines a Borel measure $m_\varphi$ on $P(\mathscr{G})$ such that

$$m_\varphi(Z(\mu)) = e^{-\beta F(\mu)}\varphi_{r(\mu)} \tag{3}$$

for all $\mu \in P_f(\mathscr{G})$. The Borel measures $m$ on $P(\mathscr{G})$ that arise from $A(\beta)$-harmonic vectors in this way are characterized by the two properties that

- $m(Z(v)) < \infty$ for all $v \in \mathscr{G}_V$, and
- $m(\sigma(B \cap Z(a))) = e^{\beta F(a)} m(B \cap Z(a))$ for every edge $a \in \mathscr{G}_{Ar}$ and every Borel subset $B$ of $P(\mathscr{G})$.

Here $\sigma$ denotes the shift on $P(\mathscr{G})$, defined such that $\sigma(p)_i = p_{i+1}$. Nonzero Borel measures on $P(\mathscr{G})$ with these two properties are called $e^{\beta F}$-conformal, [32]. They are the measures that were called harmonic $\beta$-KMS measures in [19]. Let $\mathscr{G}$ be a cofinal graph. As in [14, 19] we say that a vertex $v \in \mathscr{G}_V$ is *non-wandering* when there is a finite path $\mu \in P_f(\mathscr{G})$ of positive length such that $v = s(\mu) = r(\mu)$. When the set $NW_{\mathscr{G}}$ of non-wandering vertexes is not empty it is a hereditary subset of $\mathscr{G}_V$, and together with the arrows

$$NW_{Ar} = \{a \in \mathscr{G}_{Ar} : \ s(a) \in NW_{\mathscr{G}}\}$$

they constitute a strongly connected subgraph of $\mathscr{G}$ which we also denote by $NW_{\mathscr{G}}$, cf. Proposition 4.3 in [19]. Set

$$P(\mathscr{G})_{rec} = \bigcap_{a \in NW_{Ar}} \{p \in P(\mathscr{G}) : \ p_i = a \text{ for infinitely many } i\}$$

and

$$P(\mathscr{G})_{wan} = \bigcap_{v \in \mathscr{G}_V} \{p \in P(\mathscr{G}) : \ \#\{i \in \mathbb{N} : \ s(p_i) = v\} < \infty\} \ .$$

We say that an $A(\beta)$-harmonic vector $\varphi$ is *conservative* when $m_\varphi$ is concentrated on $P(\mathscr{G})_{rec}$ and that $\varphi$ is *dissipative* when $m_\varphi$ is concentrated on $P(\mathscr{G})_{wan}$. When $NW_{\mathscr{G}}$ is empty, $P(\mathscr{G})_{rec} = \emptyset$ and $P(\mathscr{G})_{wan} = P(\mathscr{G})$, and hence all $A(\beta)$-harmonic vectors are dissipative. To see that we have introduced a genuine dichotomy we need the following. For strongly connected graphs it is Theorem 4.10 in [19].

**Theorem 3** *Let $\mathscr{G}$ be a cofinal digraph such that $NW_{\mathscr{G}} \neq \emptyset$. Every $e^{\beta F}$-conformal measure $m$ is concentrated either on $P(\mathscr{G})_{rec}$ or on $P(\mathscr{G})_{wan}$, and*

- *$m$ is concentrated on $P(\mathscr{G})_{rec}$ if and only if*

$$\sum_{n=0}^{\infty} A(\beta)^n_{v,v} = \infty$$

*for one, and hence all $v \in NW_{\mathscr{G}}$, and*

- *$m$ is concentrated on $P(\mathscr{G})_{wan}$ if and only if*

$$\sum_{n=0}^{\infty} A(\beta)^n_{v,v} < \infty$$

*for one, and hence all $v \in NW_{\mathscr{G}}$.*

*Proof* Consider an $e^{\beta F}$-conformal measure $m$ on $P(\mathscr{G})$. For every $\mu \in P_f(\mathscr{G})$, set

$$Z(\mu)P(NW_{\mathscr{G}}) = \left\{(p_i)_{i=1}^{\infty} \in Z(\mu) : (p_i)_{i=|\mu|+1}^{\infty} \in P(NW_{\mathscr{G}})\right\} .$$

Note that since $m$ is $e^{\beta F}$-conformal,

$$m\left(Z(\mu)P(NW_{\mathscr{G}})\right) = e^{-\beta F(\mu)} m\left(P(NW_{\mathscr{G}}) \cap Z(r(\mu))\right) .$$

Combined with the observation that

$$P(\mathscr{G}) = \bigcup_{\mu \in P_f(\mathscr{G})} Z(\mu)P(NW_{\mathscr{G}}) \tag{4}$$

by Proposition 4.3 in [19], it follows that $m(P(NW_{\mathscr{G}})) \neq 0$. In short, no $e^{\beta F}$-conformal measure annihilates $P(NW_{\mathscr{G}})$. Therefore, all conclusions follow from (4) above and Theorem 4.10 in [19].

It follows that when $\mathscr{G}$ is cofinal every minimal $A(\beta)$-harmonic vector is either

- a proper almost $A(\beta)$-harmonic vector,
- a conservative $A(\beta)$-harmonic vector, or
- a dissipative $A(\beta)$-harmonic vector.

In this paper, we focus on the conservative minimal $A(\beta)$-harmonic vectors. We call the corresponding $\beta$-KMS-weights *conservative*. The adjective "minimal" is superfluous in connection with conservative $\beta$-KMS-weights because of the following.

**Theorem 4** *Assume that $C^*(\mathscr{G})$ is simple. There is a conservative $A(\beta)$-harmonic vector if and only if*

- *$NW_{\mathscr{G}} \neq \emptyset$,*
- *$\sum_{n=0}^{\infty} A(\beta)^n_{v,v} = \infty$, and*
- *$\limsup_n \left(A(\beta)^n_{v,v}\right)^{\frac{1}{n}} = 1$*

*for one and hence all $v \in NW_{\mathscr{G}}$. When it exists it is unique up to multiplication by scalars and it is minimal.*

*Proof* As observed above, there can not be any conservative $e^{\beta F}$-conformal measure on $P(\mathscr{G})$ unless $NW_{\mathscr{G}} \neq \emptyset$. Therefore, all statements follow by combining Theorem 3 above with Proposition 4.9 and Theorem 4.14 in [19].

In particular, when $C^*(\mathscr{G})$ is simple the existence of a conservative KMS-weight for $\alpha^F$ depends only on the restriction of $F$ to the strongly connected subgraph $NW_{\mathscr{G}}$. In Appendix 10 in [32], it is shown that there is at most one value of $\beta$ for which the conditions of Theorem 4 hold. When $\mathscr{G}$ is strongly connected and $F = 1$, in which case $\alpha^F$ is the gauge action, this value is the Gurevich entropy of $\mathscr{G}$ and the conditions will hold if and only if $\mathscr{G}$ is recurrent in the sense of Ruette, [33].

## 3 The Factor Type of a Conservative $\beta$-KMS-Weight

In the setting of Section 2.2, assume that $NW_{\mathscr{G}} \neq \emptyset$ and pick a vertex $v$ in $NW_{\mathscr{G}}$. Then

$$\left\{\beta F(\mu) - \beta F(\mu') : \ \mu, \mu' \in P_f(\mathscr{G}), \ r(\mu) = r(\mu') = s(\mu) = s(\mu') = v\right\}$$

is a subgroup of $\mathbb{R}$ which does not depend on the vertex $v \in NW_{\mathscr{G}}$. Let $R_{\mathscr{G},F}$ be the closure in $\mathbb{R}$ of this subgroup.

**Lemma 4** *Assume that $\mathscr{G}$ is cofinal and that $NW_{\mathscr{G}} \neq \emptyset$. Let $\psi$ be an extremal $\beta$-KMS weight for $\alpha^F$. Then $\pi_\psi(C^*(\mathscr{G}))''$ is a hyperfinite factor and*

$$\Gamma\left(\pi_\psi(C^*(\mathscr{G}))''\right) \subseteq R_{\mathscr{G},F} . \tag{5}$$

*Proof* $M = \pi_\psi(C^*(\mathscr{G}))''$ is hyperfinite by Lemma 3. In the following proof, we suppress $\pi_\psi$ in the notation and consider $C^*(\mathscr{G})$ as a subalgebra of $M$. We will show that $\mathbb{R}\backslash R_{\mathscr{G},F} \subseteq \mathbb{R}\backslash\Gamma(M)$. Let, therefore, $r \in \mathbb{R}\backslash R_{\mathscr{G},F}$ and choose a function $f \in L^1(\mathbb{R})$ whose Fourier transform $\hat{f}$ satisfies that $\hat{f}(t) = 0$ for all $t \in R_{\mathscr{G},F}$ and $\hat{f}(r) \neq 0$. Fix a vertex $v \in NW_{\mathscr{G}}$ and let $\mu, \nu \in P_f(\mathscr{G})$ be finite paths with $s(\mu) = s(\nu) = v$. We assume that $r(\mu) = r(\nu)$ since $S_\mu S_\nu^*$ is zero otherwise. Since $v$ is wandering and $NW_{\mathscr{G}}$ is strongly connected there is a finite path $\delta$ in $\mathscr{G}$ such that $s(\delta) = r(\mu) = r(\nu)$ and $r(\delta) = v$. Then

$$\beta F(\mu) - \beta F(\nu) = \beta F(\mu\delta) - \beta F(\nu\delta) \in R_{\mathscr{G},F} .$$

It follows that

$$\theta_f(S_\mu S_\nu^*) = \int_{\mathbb{R}} f(t)\theta_t(S_\mu S_\nu^*)\, dt = \hat{f}(\beta(F(\mu) - F(\nu)))S_\mu S_\nu^* = 0$$

because $\beta(F(\mu) - F(\nu)) \in R_{\mathscr{G},F}$. Since the elements of the form $S_\mu S_\nu^*$ for some $\mu, \nu \in P_f(\mathscr{G})$ with $s(\mu) = s(\nu) = v$ span a strongly dense subspace of $P_v M P_v$ we conclude that $\theta_f(P_v M P_v) = \{0\}$. Since $\hat{f}(r) \neq 0$ we conclude that $r \notin \mathrm{Sp}(P_v M P_v)$, and hence $r \notin \Gamma(M)$.

**Lemma 5** *Assume $C^*(\mathcal{G})$ is simple and that there is a $\beta$-KMS-weight for $\alpha^F$. Let $\mu, \nu, \delta$ be finite paths in $\mathcal{G}$ such that $|\delta| > \max\{|\mu|, |\nu|\}$ and $F(\mu) = F(\nu)$. Then*

$$P_\delta S_\mu S_\nu^* P_\delta = \begin{cases} 0, & \text{when } \mu \neq \nu \\ P_\delta P_\mu, & \text{when } \mu = \nu \ . \end{cases}$$

*Proof* If $|\mu| \neq |\nu|$ and $S_\delta^* S_\mu S_\nu^* S_\delta \neq 0$, the relations defining $C^*(\mathcal{G})$ imply that a piece of $\mu$ or a piece of $\nu$ will be a loop $\kappa$ of positive length in $\mathcal{G}$ such that $F(\kappa) = 0$. By Lemma 10.4 in [32], the existence of a $\beta$-KMS-weight rules out the existence of such a loop. It follows that $S_\delta^* S_\mu S_\nu^* S_\delta$ can only be nonzero when $|\mu| = |\nu|$. But $S_\delta^* S_\mu \neq 0$ implies that $\mu$ must be the initial piece of $\delta$ and similarly $S_\nu^* S_\delta \neq 0$ implies that $\nu$ must also be the initial piece of $\delta$. Therefore $S_\delta^* S_\mu S_\nu^* S_\delta \neq 0$ implies that $\mu = \nu$, in which case $P_\delta S_\mu S_\nu^* P_\delta = S_\delta S_\delta^* S_\mu S_\mu^* S_\delta S_\delta^* = P_\delta P_\mu$.

**Theorem 5** *Assume that $C^*(\mathcal{G})$ is simple and $NW_\mathcal{G}$ not empty. Let $\psi$ be a conservative $\beta$-KMS- weight for the generalized gauge action $\alpha^F$. Then*

$$\Gamma(\pi_\psi(C^*(\mathcal{G}))'') = R_{\mathcal{G},F} \ .$$

*Proof* The proof is a further development of the proofs of Proposition 4.11 in [14] and Theorem 4.1 in [16]. As in the proof of Lemma 4 we suppress $\pi_\psi$ in the notation and consider $C^*(\mathcal{G})$ as a subalgebra of $M$. The modular automorphism group $\theta$ on $M$ defined by $\tilde{\psi}$ is given by

$$\theta_t = \tilde{\alpha^F}_{-\beta t} \ ,$$

where $\tilde{\alpha^F}$ is the $\sigma$-weakly continuous extension of $\alpha^F$. Note that $\beta \neq 0$ by Proposition 2.4 in [32]. Let $N \subseteq M$ be the fixed point algebra for $\theta$ and consider a vertex $v \in NW_\mathcal{G}$. By Lemma 5, it suffices to show that $R_{\mathcal{G},F} \subseteq \Gamma(M)$, and from Lemme 2.3.3 and Proposition 2.2.2 in [24] we see that it suffices for this to consider a nonzero central projection $q \in P_v N P_v$ for some vertex $v \in NW_\mathcal{G}$, and show that $\beta F(l) \subseteq \mathrm{Sp}(qMq)$ when $l$ is a loop in $\mathcal{G}$ such that $s(l) = r(l) = v$. Let $\omega$ be the state on $P_v M P_v$ given by $\omega(a) = \psi(P_v)^{-1}\tilde{\psi}(a)$. Then $\omega$ is a faithful normal state which is a trace on $P_v N P_v$ and we consider the corresponding 2-norm

$$\|a\|_v = \sqrt{\omega(a^*a)} \ .$$

By Kaplansky's density theorem, there is an element $f \in P_v \mathcal{A} P_v$ such that $0 \leq f \leq 1$ and $\|q - f\|_v$ is as small as we want. Then $f$ is a linear combination of elements of the form $S_\mu S_\nu^*$ where $\mu, \nu \in P_f(\mathcal{G})$ and $s(\mu) = s(\nu) = v$. Note that

$$\lim_{R\to\infty} \frac{1}{R}\int_0^R \theta_t(S_\mu S_\nu^*)\, dt = \begin{cases} S_\mu S_\nu^* & \text{when } F(\mu) = F(\nu) \\ 0 & \text{when } F(\mu) \neq F(\nu) \ , \end{cases}$$

with convergence in norm, and that

$$\left\| q - \frac{1}{R}\int_0^R \theta_t(f)\, dt \right\|_v = \left\| \frac{1}{R}\int_0^R \theta_t(q-f)\, dt \right\|_v \leq \|q - f\|_v$$

by Kadisons Schwarz inequality. By exchanging $\lim_{R\to\infty} \frac{1}{R}\int_0^R \theta_t(f)\, dt$ for $f$, we may assume that

$$f = \sum_{i=1}^{N} \lambda_i S_{\mu_i} S_{\nu_i}^* ,$$

where $0 \leq \lambda_i \leq 1$ and $F(\mu_i) = F(\nu_i)$ for all $i$. Let $m_\psi$ be the $e^{\beta F}$-conformal measure on $P(\mathcal{G})$ defined by $\psi$ and let $m$ be the restriction of the measure $m_\psi(Z(v))^{-1} m_\psi$ to $Z(v)$. Then $m$ is a Borel probability measure on $Z(v) \subseteq P(\mathcal{G})$ such that

$$\omega\left(P_\mu\right) = m(Z(\mu))$$

for all $\mu \in P_f(\mathcal{G})$ with $s(\mu) = v$. For each $k \in \mathbb{N}$, we let $M_k$ be the set of paths $\delta$ in $\mathcal{G}$ such that $s(\delta) = v$ and $|\delta| = k$. Then

$$\sum_{\delta \in M_k} \omega(P_\delta) = \sum_{\delta\in M_k} m(Z(\delta)) = 1 ,$$

which implies that $1 = \sum_{\delta\in M_k} P_\delta$, where the sum converges with respect to the 2-norm and hence also in the strong operator topology. It follows that we can define $Q_k : P_v M P_v \to P_v M P_v$ such that

$$Q_k(m) = \sum_{\delta\in M_k} P_\delta m P_\delta .$$

Then $Q_k$ is a positive linear map of norm one and $Q_k(q) = q$. When $k > \max\{|\mu|, |\nu|\}$ it follows from Lemma 5 that

$$Q_k\left(S_\mu S_\nu^*\right) = \begin{cases} P_\mu, & \text{when } \mu = \nu \\ 0, & \text{when } \mu \neq \nu . \end{cases}$$

Thus, for some $k$ large enough, we have that $Q_k(f)$ is a linear combination

$$Q_k(f) = \sum_{i=1}^{N'} \lambda_i P_{\mu_i} ,$$

where $0 \leq \lambda_i \leq 1$ for all $i$. Using Kadisons Schwarz inequality again we find that

$$\|q - Q_k(f)\|_v = \|Q_k(q-f)\|_v \leq \sqrt{\omega\left(Q_k\left((q-f)^2\right)\right)} = \|q - f\|_v$$

since $\omega \circ Q_k = \omega$ for all $k$. By exchanging $f$ with $Q_k(f)$ for some $k$ large enough we may assume that

$$f = \sum_{i=1}^{N'} \lambda_i P_{\mu_i} . \tag{6}$$

Now, observe that since $\psi$ is conservative by assumption it follows that $m$ is concentrated on

$$\{p \in Z(v) : \ s(p_i) = v \ \text{ for infinitely many } i\} .$$

Fix one of the paths $\mu_i$ appearing in (6). Let $H_j$ denote the set of finite paths $\delta$ of length $j$ such that $s(\delta) = r(\mu_i), \ r(\delta) = v$. Then, up to an $m$-null set,

$$\bigcup_{j=1}^{\infty} \left\{Z(\mu_i\delta) : \ \delta \in H_j\right\} = Z(\mu_i) .$$

We can, therefore, find a finite set $K_i \subseteq \bigcup_{j=1}^{\infty} H_j$ such that

$$m\left(Z(\mu_i)\backslash \sqcup_{\delta\in K_i} Z(\mu_i\delta)\right)$$

is as small as we want. Note that

$$\left\| P_{\mu_i} - \sum_{\delta\in K_i} P_{\mu_i\delta} \right\|_v = \sqrt{m\left(Z(\mu_i)\backslash \sqcup_{\delta\in K_i} Z(\mu_i\delta)\right)}$$

is then also small. Therefore, by exchanging $\sum_{\delta\in K_i} P_{\mu_i\delta}$ for $P_{\mu_i}$ we may assume that $s(\mu_i) = r(\mu_i) = v$ for all $v$. Finally, since $q$ is a projection, a standard argument, as in the proof of Lemma 12.2.3 in [34], allows us to select a subset of the $\mu_i$'s and arrange, after a renumbering, that

$$p = \sum_{i=1}^{M} P_{\mu_i} \tag{7}$$

is a projection in $P_v \mathscr{A} P_v$ such that

$$\|q - p\|_v \leq \varepsilon , \tag{8}$$

where $\varepsilon > 0$ can be chosen as small as we need. We choose $\varepsilon > 0$ so small that

$$\omega(q) - e^{F(l)\beta}\varepsilon - 3\varepsilon \ > \ 0 .$$

For each $\mu_i$ from (7) we let $w_i \in P_v \mathscr{A} P_v$ be the elements

$$w_i = S_{\mu_i} S^*_{\mu_i l} .$$

Each $w_i$ is a partial isometry such that

(a) $w_i {w_i}^* = P_{\mu_i}$ ,
(b) ${w_i}^* w_i = P_{\mu_i l} \leq P_{\mu_i}$, and
(c) $\alpha^F_t(w_i) = e^{-iF(l)t} w_i$ for all $t \in \mathbb{R}$.

Set $w = \sum_{i=1}^M w_i$ and note that $w$ is a partial isometry. It follows from (b) that $wp = w$ and therefore from (8), (c), (b), and (a) that

$$\begin{aligned}
&\omega(qwqw^*q) \geq \omega(wqw^*) - 2\varepsilon = e^{\beta F(l)}\omega(qw^*w) - 2\varepsilon \\
&\geq e^{\beta F(l)}\omega(pw^*w) - e^{\beta F(l)}\varepsilon - 2\varepsilon = e^{\beta F(l)}\omega(w^*w) - e^{\beta F(l)}\varepsilon - 2\varepsilon \\
&= \omega(ww^*) - e^{\beta F(l)}\varepsilon - 2\varepsilon = \omega(p) - e^{\beta F(l)}\varepsilon - 2\varepsilon \geq \omega(q) - e^{\beta F(l)}\varepsilon - 3\varepsilon .
\end{aligned}$$

The choice of $\varepsilon$ ensures that $\omega(qwqw^*q) > 0$ and hence that $qwq \neq 0$. Since $\theta_t(qwq) = e^{-it\beta F(l)} qwq$ for all $t \in \mathbb{R}$, it follows from Lemme 2.3.6 in [24] that $\beta F(l) \in \mathrm{Sp}(qMq)$, as desired.

In combination with Corollary 1, we get the following

**Corollary 2** *Assume that $C^*(\mathcal{G})$ is simple and that $NW_{\mathcal{G}} \neq \emptyset$. Let $\varphi$ be a $\beta$-KMS-state for the restriction of $\alpha^F$ to $P_v C^*(\mathcal{G}) P_v$ for some $v \in \mathcal{G}_V$. When $\sum_{n=0}^{\infty} A(\beta)^n_{w,w} = \infty$ for some (and hence all) $w \in NW_{\mathcal{G}}$, the Connes invariant of $\pi_\varphi(P_v C^*(\mathcal{G}) P_v)''$ is $\Gamma\left(\pi_\varphi(P_v C^*(\mathcal{G}) P_v)''\right) = R_{\mathcal{G},F}$.*

In an Appendix in [32], it is shown that in the setting of Theorem 5 the subgroup $R_{\mathcal{G},F}$ is never $\{0\}$. Therefore, thanks to Connes' classification in [29] and Haagerup's result in [35], we get the following.

**Corollary 3** *In the setting of Theorem 5 and Corollary 2 the factors $\pi_\psi(C^*(\mathcal{G}))''$ and $\pi_\varphi(P_v C^*(\mathcal{G}) P_v)''$ are isomorphic; they are both isomorphic to the hyperfinite factor of type $III_\lambda$ for some $0 < \lambda \leq 1$.*

*Example 1* (*Generalized gauge actions on $O_\infty$.*) The Cuntz algebra $O_\infty$, [36], is the graph $C^*$-algebra $C^*(\mathcal{G})$ when $\mathcal{G}$ is the countable digraph with one vertex and infinitely many arrows, $a_n$, $n = 1, 2, 3, \cdots$. Since $O_\infty$ is unital all proper weights are bounded and can be normalized to states. Let $\{t_n\}_{n=1}^{\infty}$ be a sequence of real numbers and define $F : \mathcal{G}_{Ar} \to \mathbb{R}$ such that $F(a_n) = t_n$ for all $n$. The gauge action, where $t_n = 1$ for all $n$, was considered by Olesen and Pedersen who showed in [37] that there are no KMS-states for the gauge action. In general, it follows from [19] that a $\beta$-KMS-state exists iff $\sum_{n=1}^{\infty} e^{-\beta t_n} \leq 1$ and that it is unique. There is a conservative $\beta_0$-KMS-state for $\alpha^F$ iff

$$\sum_{n=1}^{\infty} e^{-\beta_0 t_n} = 1 .$$

By Theorem 5 the Connes invariant of the corresponding factor is the closed subgroup of $\mathbb{R}$ generated by the numbers $\{\beta_0 t_i\}_{i=1}^{\infty}$. It follows that the factor is always of type $III$ and never of type $III_0$ while all hyperfinite factors of type $III_\lambda$ for $0 < \lambda \leq 1$ can occur by varying the sequence $\{t_n\}$. The KMS-states for $\alpha^F$ that are not conservative occur for values of $\beta$ for which $\sum_{n=1}^{\infty} e^{-\beta t_n} < 1$ and they correspond to minimal proper almost $A(\beta)$-harmonic vectors, albeit of a somewhat degenerate kind since the vectors only have one coordinate.

**Acknowledgements** I am grateful to Johannes Christensen for discussions and help to eliminate mistakes. The work was supported by the DFF-Research Project 2 'Automorphisms and Invariants of Operator Algebras', no. 7014-00145B.

## References

1. Bratteli, O., Robinson, D.W.: Operator Algebras and Quantum Statistical Mechanics I + II. Texts and Monographs in Physics. Springer, New York (1979, 1981)
2. Haag, R., Winnink, M., Hugenholtz, N.M.: On the equilibrium states in quantum statistical mechanics. Commun. Math. Phys. **5**, 215–236 (1967)
3. Enomoto, M., Fujii, M., Watatani, Y.: KMS states for gauge action on $O_A$. Math. Jpn. **29**, 607–619 (1984)
4. Bost, J.-B., Connes, A.: Hecke algebras, type III factors and phase transitions with spontaneous symmetry breaking in number theory. Selecta Math. (N.S.) **1**, 411–457 (1995)
5. Okayasu, R.: Type III factors arising from Cuntz-Krieger algebras. Proc. Am. Math. Soc. **131**, 2145–2153 (2003)
6. Barreto, S.D., Fidaleo, F.: On the structure of KMS states of disordered systems. Commun. Math. Phys. **250**, 1–21 (2004)
7. Izumi, M., Kajiwara, T., Watatani, Y.: KMS states and branched points. Ergod. Theory Dyn. Syst. **27**, 1887–1918 (2007)
8. Neshveyev, S.: von Neumann algebras arising from Bost-Connes type systems. Int. Math. Res. Not. IMRN **2011**, 217–236 (2011)
9. Laca, M., Neshveyev, S.: Type $III_1$ equilibrium states of the Toeplitz algebra of the affine semigroup over the natural numbere. J. Funct. Anal. **261**, 169–187 (2011)
10. Yang, D.: Type III von Neumann algebras associated with 2-graphs. Bull. Lond. Math. Soc. **44**, 675–686 (2012)
11. Laca, M., Larsen, N., Neshveyev, S., Sims, A., Webster, S.B.G.: Von Neumann algebras of strongly connected higher-rank graphs. Math. Ann. **363**, 657–678 (2015)
12. Carey, A., Phillips, J., Putnam, I., Rennie, A.: Families of type III KMS states on a class of $C^*$-algebras containing $O_n$ and $Q_N$. J. Funct. Anal. **260**, 1637–1681 (2011)
13. Kajiwara, T., Watatani, Y.: KMS states on finite-graph $C^*$-algebras. Kyushu J. Math. **67**, 83–104 (2013)
14. Thomsen, K.: KMS weights on groupoid and graph $C^*$-algebras. J. Funct. Anal. **266**, 2959–2988 (2014)
15. Thomsen, K.: Exact circle maps and KMS states. Isr. J. Math. **205**, 397–420 (2015)
16. Thomsen, K.: Phase transition in $O_2$. Commun. Math. Phys. **349**, 481–492 (2017)
17. Yang, D.: Factoriality and type classification of k-graph von Neumann algebras. Proc. Edinb. Math. Soc. (2) **60**, 499–518 (2017)
18. Izumi, M.: The flow of weights and the Cuntz-Pimsner algebras. Commun. Math. Phys., To appear
19. Thomsen, K.: KMS weights on graph $C^*$-algebras. Adv. Math. **309**, 334–391 (2017)

20. Kustermans, J.: KMS-weights on $C^*$-algebras. arXiv:9704008v1
21. Kustermans, J., Vaes, S.: Weight theory for $C^*$-algebraic quantum groups. arXiv:990163
22. Kustermans, J., Vaes, S.: Locally compact quantum groups. Ann. Scient. Éc. Norm. Sup. **33**, 837–934 (2000)
23. Combes, F.: Poids associé à une algèbre hilbertienne à gauche. Compos. Math. **23**, 49–77 (1971)
24. Connes, A.: Une classification des facteurs de type III. Ann. Sci. Ecole Norm. Sup. **6**, 133–252 (1973)
25. Bates, T., Hong, J.H., Raeburn, I., Szymanski, W.: The ideal structure of the $C^*$-algebras of infinite graphs. Illinois J. Math. **46**, 1159–1176 (2002)
26. Kumjian, A., Pask, D.: $C^*$-algebras of directed graphs and group actions. Ergod. Theory Dyn. Syst. **19**, 1503–1519 (1999)
27. Drinen, D., Tomforde, M.: The $C^*$-algebras of arbitrary graphs. Rocky Mt. J. Math. **35**, 105–135 (2005)
28. Choi, M.D., Effros, E.G.: Separable nuclear $C^*$-algebras and injectivity. Duke Math. J. **43**, 309–322 (1976)
29. Connes, A.: Classification of injective factors. Cases $II_1$, $II_\infty$, $III_\lambda$, $\lambda \neq 1$. Ann. Math. (2) **104**, 73–115 (1976)
30. Szymanski, W.: Simplicity of Cuntz-Krieger algebras of infinite matrices. Pac. J. Math. **122**, 249–256 (2001)
31. Christensen, J., Thomsen, K.: Diagonality of actions and KMS weights. J. Oper. Theory **76**, 449–471 (2016)
32. Thomsen, K.: KMS weights, conformal measures and ends in digraphs. arXiv:1612.04716v2
33. Ruette, S.: On the Vere-Jones classification and existence of maximal measure for countable topological Markov chains. Pac. J. Math. **209**, 365–380 (2003)
34. Kadison, R.V., Ringrose, J.R.: Fundamentals of the Theory of Operator Algebras II. Academic, London (1986)
35. Haagerup, U.: Connes' bicentralizer problem and uniqueness of the injective factor of type $III_1$. Acta Math. **158**, 95–148 (1987)
36. Cuntz, J.: Simple $C^*$-algebras generated by isometries. Commun. Math. Phys. **57**, 173–185 (1977)
37. Olesen, D., Pedersen, G.K.: Some $C^*$-dynamical systems with a single KMS-state. Math. Scand. **42**, 111–118 (1978)

# Trotter–Kato Product Formulae in Dixmier Ideal

**Valentin A. Zagrebnov**

*On the occasion of the 100th birthday of Tosio Kato*

**Abstract** It is shown that for a certain class of the Kato functions, the Trotter–Kato product formulae converge in Dixmier ideal $\mathscr{C}_{1,\infty}$ in topology, which is defined by the $\|\cdot\|_{1,\infty}$-norm. Moreover, the rate of convergence in this topology inherits the error-bound estimate for the corresponding operator-norm convergence.

## 1 Preliminaries. Symmetrically-Normed Ideals

Let $\mathscr{H}$ be a separable Hilbert space. For the first time, the Trotter–Kato product formulae in Dixmier ideal $\mathscr{C}_{1,\infty}(\mathscr{H})$ were shortly discussed in conclusion of the paper [19]. This remark was a programme addressed to extension of results, which were known for the von Neumann–Schatten ideals $\mathscr{C}_p(\mathscr{H})$, $p \geq 1$ since [14, 24].

Note that a subtle point of this programme is the question about the rate of convergence in the corresponding topology. Since the limit of the Trotter–Kato product formula is a strongly continuous semigroup, for the von Neumann–Schatten ideals, this topology is the trace norm $\|\cdot\|_1$ on the trace-class ideal $\mathscr{C}_1(\mathscr{H})$. In this case, the limit is a Gibbs semigroup [25].

For self-adjoint Gibbs semigroups, the rate of convergence was estimated for the first time in [7, 9]. The authors considered the case of the Gibbs–Schrödinger semigroups. They scrutinised, in these papers, a dependence of the rate of conver-

V. A. Zagrebnov (✉)
Institut de Mathématiques de Marseille (UMR 7373) - AMU, Centre de Mathématiques et Informatique - Technopôle Château-Gombert,
39, rue F. Joliot Curie, 13453 Marseille Cedex 13, France
e-mail: Valentin.Zagrebnov@univ-amu.fr

© Springer Nature Switzerland AG 2019

T. M. Rassias and V. A. Zagrebnov (eds.), *Analysis and Operator Theory*,
Springer Optimization and Its Applications 146,
https://doi.org/10.1007/978-3-030-12661-2_18

gence for the (exponential) Trotter formula on the smoothness of the potential in the Schrödinger generator.

The first abstract result in this direction was due to [19]. In this paper, a general scheme of *lifting* the operator-norm rate convergence for the Trotter–Kato product formulae was proposed and advocated for estimating the rate of the trace-norm convergence. This scheme was then improved and extended in [2] to the case of *nonself-adjoint* Gibbs semigroups.

The aim of the present note is to elucidate the question about the existence of other than the von Neumann–Schatten proper two-sided ideals $\mathfrak{I}(\mathscr{H})$ of $\mathscr{L}(\mathscr{H})$ and then to prove the (non-exponential) Trotter–Kato product formula in topology of these ideals together with an estimate of the corresponding rate of convergence. Here, a particular case of the Dixmier ideal $\mathscr{C}_{1,\infty}(\mathscr{H})$ [4, 6] is considered. To specify this ideal we recall in Section 2, the notion of *singular* trace and then of the Dixmier trace [3, 5] in Section 3. Main results about the Trotter–Kato product formulae in the Dixmier ideal $\mathscr{C}_{1,\infty}(\mathscr{H})$ are collected in Section 4. There the arguments based on the lifting scheme [19] (Theorem 5.1) are refined for proving the Trotter–Kato product formulae convergence in the $\|\cdot\|_{1,\infty}$-topology with the rate, which is *inherited* from the operator-norm convergence.

To this end, in the rest of the present section, we recall an important auxiliary tool: the concept of *symmetrically-normed* ideals, see e.g. [8, 22].

Let $c_0 \subset l^\infty(\mathbb{N})$ be the subspace of bounded sequences $\xi = \{\xi_j\}_{j=1}^\infty \in l^\infty(\mathbb{N})$ of real numbers, which tend to *zero*. We denote by $c_f$ the subspace of $c_0$ consisting of all sequences with *finite* number of non-zero terms (*finite sequences*).

**Definition 1** A real-valued function $\phi : \xi \mapsto \phi(\xi)$ defined on $c_f$ is called a *norming function* if it has the following properties:

$$\phi(\xi) > 0, \qquad \forall \xi \in c_f, \quad \xi \neq 0, \tag{1.1}$$

$$\phi(\alpha\xi) = |\alpha|\phi(\xi), \qquad \forall \xi \in c_f, \quad \forall \alpha \in \mathbb{R}, \tag{1.2}$$

$$\phi(\xi + \eta) \leq \phi(\xi) + \phi(\eta), \qquad \forall \xi, \eta \in c_f, \tag{1.3}$$

$$\phi(1, 0, \ldots) = 1. \tag{1.4}$$

A norming function $\phi$ is called to be *symmetric* if it has the additional property

$$\phi(\xi_1, \xi_2, ..., \xi_n, 0, 0, \ldots) = \phi(|\xi_{j_1}|, |\xi_{j_2}|, ..., |\xi_{j_n}|, 0, 0, \ldots) \tag{1.5}$$

for any $\xi \in c_f$ and any permutation $j_1, j_2, \ldots, j_n$ of integers $1, 2, \ldots, n$.

It turns out that for any *symmetric norming* function and for any elements $\xi, \eta$ from the positive *cone* $c^+$ of non-negative, non-increasing sequences such that $\xi, \eta \in c_f$ obey $\xi_1 \geq \xi_2 \geq \ldots \geq 0$, $\eta_1 \geq \eta_2 \geq \ldots \geq 0$ and

$$\sum_{j=1}^{n} \xi_j \leq \sum_{j=1}^{n} \eta_j, \qquad n = 1, 2, \ldots, \tag{1.6}$$

one gets the Ky Fan inequality [8] (Sect. 3, §3) :

$$\phi(\xi) \leq \phi(\eta). \tag{1.7}$$

Moreover, (1.7) together with the properties, (1.1), (1.2) and (1.4) imply

$$\xi_1 \leq \phi(\xi) \leq \sum_{j=1}^{\infty} \xi_j, \qquad \xi \in c_f^+ := c_f \cap c^+. \tag{1.8}$$

Note that the left- and right-hand sides of (1.8) are the simplest examples of symmetric norming functions on domain $c_f^+$:

$$\phi_\infty(\xi) := \xi_1 \quad \text{and} \quad \phi_1(\xi) := \sum_{j=1}^{\infty} \xi_j \,. \tag{1.9}$$

By Definition 1, the observations (1.8) and (1.9) yield

$$\begin{aligned} &\phi_\infty(\xi) := \max_{j\geq 1} |\xi_j| \,, \quad \phi_1(\xi) := \sum_{j=1}^{\infty} |\xi_j| \,, \\ &\phi_\infty(\xi) \leq \phi(\xi) \leq \phi_1(\xi) \,, \quad \text{for all} \quad \xi \in c_f \,. \end{aligned} \tag{1.10}$$

We denote by $\xi^* := \{\xi_1^*, \xi_2^*, \dots\}$ a decreasing *rearrangement*: $\xi_1^* = \sup_{j\geq 1} |\xi_j|$, $\xi_1^* + \xi_2^* = \sup_{i\neq j}\{|\xi_i| + |\xi_j|\}, \dots$, of the sequence of absolute values $\{|\xi_n|\}_{n\geq 1}$, i.e. $\xi_1^* \geq \xi_2^* \geq \dots$. Then, $\xi \in c_f$ implies $\xi^* \in c_f$ and by (1.5) one obtains also that

$$\phi(\xi) = \phi(\xi^*), \qquad \xi \in c_f \,. \tag{1.11}$$

Therefore, any symmetric norming function $\phi$ is uniquely defined by its values on the positive cone $c^+$.

Now, let $\xi = \{\xi_1, \xi_2, \dots\} \in c_0$. We define

$$\xi^{(n)} := \{\xi_1, \xi_2, \dots, \xi_n, 0, 0, \dots\} \in c_f \,.$$

Then, if $\phi$ is a symmetric norming function, we set

$$c_\phi := \{\xi \in c_0 : \sup_n \phi(\xi^{(n)}) < +\infty\}. \tag{1.12}$$

Therefore, one gets

$$c_f \subseteq c_\phi \subseteq c_0.$$

Note that by (1.5)–(1.7) and (1.12), one gets

$$\phi(\xi^{(n)}) \leq \phi(\xi^{(n+1)}) \leq \sup_n \phi(\xi^{(n)}) \, , \text{ for any } \xi \in c_\phi \, .$$

Then, the limit

$$\phi(\xi) := \lim_{n\to\infty} \phi(\xi^{(n)}) \, , \qquad \xi \in c_\phi,$$

exists and $\phi(\xi) = \sup_n \phi(\xi^{(n)})$, i.e. the symmetric norming function $\phi$ is a *normal* functional on the set $c_\phi$ (1.12), which is a linear space over $\mathbb{R}$.

By (1.3) and (1.10) one also gets that any symmetric norming function is *continuous* on $c_f$:

$$|\phi(\xi) - \phi(\eta)| \leq \phi(\xi - \eta) \leq \phi_1(\xi - \eta) \, , \ \forall \xi, \eta \in c_f \, .$$

Suppose that $X$ is a compact operator, i.e. $X \in \mathscr{C}_\infty(\mathscr{H})$. Then by

$$s(X) := \{s_1(X), s_2(X), \dots \},$$

we denote the sequence of *singular values* of $X$ counting multiplicities. We always assume that

$$s_1(X) \geq s_2(X) \geq \dots \geq s_n(X) \geq \dots \, .$$

To define *symmetrically-normed* ideals of the compact operators $\mathscr{C}_\infty(\mathscr{H})$, we introduce the notion of a symmetric norm.

**Definition 2** Let $\mathfrak{J}$ be a two-sided ideal of $\mathscr{C}_\infty(\mathscr{H})$. A functional $\| \cdot \|_{sym} : \mathfrak{J} \to \mathbb{R}_0^+$ is called a *symmetric norm* if besides the usual properties of the operator norm $\| \cdot \|$:

$$\begin{aligned} &\|X\|_{sym} > 0, \qquad \forall X \in \mathfrak{J}, \quad X \neq 0, \\ &\|\alpha X\|_{sym} = |\alpha| \|X\|_{sym} \, , \qquad \forall X \in \mathfrak{J}, \quad \forall \alpha \in \mathbb{C}, \\ &\|X + Y\|_{sym} \leq \|X\|_{sym} + \|Y\|_{sym} \, , \ \forall X, Y \in \mathfrak{J}, \end{aligned}$$

it verifies the following additional properties:

$$\|AXB\|_{sym} \leq \|A\| \|X\|_{sym} \|B\|, \ X \in \mathfrak{J}, \ A, B \in \mathscr{L}(\mathscr{H}), \tag{1.13}$$

$$\|\alpha X\|_{sym} = |\alpha| \|X\| = |\alpha| \, s_1(X), \ \text{for any one} - \text{rank operator } X \in \mathfrak{J}. \tag{1.14}$$

If the condition (1.13) is replaced by

$$\begin{aligned} \|UX\|_{sym} = \|XU\|_{sym} = \|X\|_{sym} \, , \ X \in \mathfrak{J} \, , \\ \text{for any unitary operator } U \text{ on } \mathscr{H} \, , \end{aligned} \tag{1.15}$$

then, instead of the symmetric norm, one gets definition of an *invariant norm* $\| \cdot \|_{inv}$.

First, we note that the ordinary operator norm $\| \cdot \|$ on any ideal $\mathfrak{J} \subseteq \mathscr{C}_\infty(\mathscr{H})$ is evidently a symmetric norm as well as an invariant norm.

Second, we observe that, in fact, every symmetric norm is invariant. Indeed, for any unitary operators $U$ and $V$ one gets by (1.13) that

$$\|UXV\|_{sym} \leq \|X\|_{sym} \ , \ \ X \in \mathfrak{J} \, . \tag{1.16}$$

Since $X = U^{-1}UXVV^{-1}$, we also get $\|X\|_{sym} \leq \|UXV\|_{sym}$, which together with (1.16) yield (1.15).

Third, we claim that $\|X\|_{sym} = \|X^*\|_{sym}$. To this aim, let $X = U|X|$ be the polar representation of the operator $X \in \mathfrak{J}$. Since $U^*X = |X|$, then by (1.13), we obtain $\|X\|_{sym} = \||X|\|_{sym}$. The same line of reasoning applied to the adjoint operator $X^* = |X|U^*$ yields $\|X^*\|_{sym} = \||X|\|_{sym}$, which proves the claim.

Now, we can apply the concept of the symmetric norming functions to describe the symmetrically-normed ideals of the unital algebra of bounded operators $\mathscr{L}(\mathscr{H})$, or in general, the symmetrically-normed ideals generated by symmetric norming functions. Recall that any *proper* two-sided ideal $\mathfrak{J}(\mathscr{H})$ of $\mathscr{L}(\mathscr{H})$ is contained in compact operators $\mathscr{C}_\infty(\mathscr{H})$ and contains the set $\mathscr{K}(\mathscr{H})$ of finite-rank operators, see e.g. [20, 22]:

$$\mathscr{K}(\mathscr{H}) \subseteq \mathfrak{J}(\mathscr{H}) \subseteq \mathscr{C}_\infty(\mathscr{H}). \tag{1.17}$$

To clarify the relation between symmetric norming functions and the symmetrically-normed ideals, we mention that there is an obvious one-to-one correspondence between functions $\phi$ (Definition 1) on the cone $c^+$ and the symmetric norms $\|\cdot\|_{sym}$ on $\mathscr{K}(\mathscr{H})$. To proceed with a general setting, one needs definition of the following relation.

**Definition 3** Let $c_\phi$ be the set of vectors (1.12) generated by a symmetric norming function $\phi$. We associate with $c_\phi$ a subset of compact operators

$$\mathscr{C}_\phi(\mathscr{H}) := \{X \in \mathscr{C}_\infty(\mathscr{H}) : s(X) \in c_\phi\} \, . \tag{1.18}$$

This definition implies that the set $\mathscr{C}_\phi(\mathscr{H})$ is a proper two-sided ideal of the algebra $\mathscr{L}(\mathscr{H})$ of all bounded operators on $\mathscr{H}$. Setting

$$\|X\|_\phi := \phi(s(X)) \, , \qquad X \in \mathscr{C}_\phi(\mathscr{H}) \, , \tag{1.19}$$

one obtains a *symmetric norm*: $\|\cdot\|_{sym} = \|\cdot\|_\phi$, on the ideal $\mathfrak{J} = \mathscr{C}_\phi(\mathscr{H})$ (Definition 2) such that this symmetrically-normed ideal becomes a Banach space. Then, in accordance with (1.17) and (1.18), we obtain by (1.10) that

$$\mathscr{K}(\mathscr{H}) \subseteq \mathscr{C}_1(\mathscr{H}) \subseteq \mathscr{C}_\phi(\mathscr{H}) \subseteq \mathscr{C}_\infty(\mathscr{H}) \, . \tag{1.20}$$

Here, the trace-class operators $\mathscr{C}_1(\mathscr{H}) := \mathscr{C}_{\phi_1}(\mathscr{H})$, where the symmetric norming function $\phi_1$ is defined in (1.9), and

$$\|X\|_\phi \leq \|X\|_1 \, , \qquad X \in \mathscr{C}_1(\mathscr{H}) \, .$$

*Remark 1* By virtue of inequality (1.7) and by definition of symmetric norm (1.19), the so-called *dominance* property holds: if $X \in \mathscr{C}_\phi(\mathscr{H})$, $Y \in \mathscr{C}_\infty(\mathscr{H})$ and

$$\sum_{j=1}^{n} s_j(Y) \le \sum_{j=1}^{n} s_j(X) , \qquad n = 1, 2, \dots ,$$

then $Y \in \mathscr{C}_\phi(\mathscr{H})$ and $\|Y\|_\phi \le \|X\|_\phi$.

*Remark 2* To distinguish in (1.20) *nontrivial* ideals $\mathscr{C}_\phi$ one needs some criteria based on the properties of $\phi$ or $\|\cdot\|_\phi$. For example, any symmetric norming function (1.11) defined by

$$\phi^{(r)}(\xi) := \sum_{j=1}^{r} \xi_j^* , \qquad \xi \in c_f ,$$

generates for arbitrary fixed $r \in \mathbb{N}$, the symmetrically-normed ideals, which are trivial in the sense that all $\mathscr{C}_{\phi^{(r)}}(\mathscr{H}) = \mathscr{C}_\infty(\mathscr{H})$. Criterion for an operator $A$ to belong to a nontrivial ideal $\mathscr{C}_\phi$ is

$$M = \sup_{m \ge 1} \|P_m A P_m\|_\phi < \infty , \tag{1.21}$$

where $\{P_m\}_{m\ge1}$ is a monotonously increasing sequence of the finite-dimensional orthogonal projectors on $\mathscr{H}$ strongly convergent to the identity operator [8]. Note that for $A \in \mathscr{C}_\infty$, the condition (1.21) is trivial.

We consider now a couple of examples to elucidate the concept of the symmetrically-normed ideals $\mathscr{C}_\phi(\mathscr{H})$ generated by the symmetric norming functions $\phi$ and the rôle of the functional *trace* on these ideals.

*Example 1* The von Neumann–Schatten ideals $\mathscr{C}_p(\mathscr{H})$ [21]. These ideals of $\mathscr{C}_\infty(\mathscr{H})$ are generated by symmetric norming functions $\phi(\xi) := \|\xi\|_p$, where

$$\|\xi\|_p = \left(\sum_{j=1}^{\infty} |\xi_j|^p\right)^{1/p} , \qquad \xi \in c_f,$$

for $1 \le p < +\infty$, and by

$$\|\xi\|_\infty = \sup_j |\xi_j|, \qquad \xi \in c_f,$$

for $p = +\infty$. Indeed, if we set $\{\xi_j^* := s_j(X)\}_{j\ge1}$, for $X \in \mathscr{C}_\infty(\mathscr{H})$, then the symmetric norm $\|X\|_\phi = \|s(X)\|_p$ coincides with $\|X\|_p$ and the corresponding symmetrically-normed ideal $\mathscr{C}_\phi(\mathscr{H})$ is identical to the von Neumann–Schatten class $\mathscr{C}_p(\mathscr{H})$.

By definition, for any $X \in \mathscr{C}_p(\mathscr{H})$ the trace: $|X| \mapsto \mathrm{Tr}|X| = \sum_{j\ge1} s_j(X) \ge 0$. The trace norm $\|X\|_1 = \mathrm{Tr}|X|$ is finite on the trace-class operators $\mathscr{C}_1(\mathscr{H})$ and it is *infinite* for $X \in \mathscr{C}_{p>1}(\mathscr{H})$. We say that for $p > 1$, the von Neumann–Schatten ideals admit *no* trace, whereas for $p = 1$ the map: $X \mapsto \operatorname{Tr} X$ exists and it is continuous in the $\|\cdot\|_1$-topology.

Note that by virtute of the Tr-linearity the trace norm: $\mathscr{C}_{1,+}(\mathscr{H}) \ni X \mapsto \|X\|_1$ is *linear* on the positive cone $\mathscr{C}_{1,+}(\mathscr{H})$ of the trace-class operators.

*Example 2* Now, we consider symmetrically-normed ideals $\mathscr{C}_\Pi(\mathscr{H})$. To this aim let $\Pi = \{\pi_j\}_{j=1}^\infty \in c^+$ be a non-increasing sequence of positive numbers with $\pi_1 = 1$. We associate with $\Pi$ the function

$$\phi_\Pi(\xi) = \sup_n \left\{ \frac{1}{\sum_{j=1}^n \pi_j} \sum_{j=1}^n \xi_j^* \right\}, \qquad \xi \in c_f. \tag{1.22}$$

It turns out that $\phi_\Pi$ is a symmetric norming function. Then, the corresponding (1.12) set $c_{\phi_\Pi}$ is defined by

$$c_{\phi_\Pi} := \left\{ \xi \in c_f : \sup_n \frac{1}{\sum_{j=1}^n \pi_j} \sum_{j=1}^n \xi_j^* < +\infty \right\}.$$

Hence, the two-sided symmetrically-normed ideal $\mathscr{C}_\Pi(\mathscr{H}) := \mathscr{C}_{\phi_\Pi}(\mathscr{H})$ generated by symmetric norming function (1.22) consists of all those compact operators $X$ that

$$\|X\|_{\phi_\Pi} := \sup_n \frac{1}{\sum_{j=1}^n \pi_j} \sum_{j=1}^n s_j(X) < +\infty\,. \tag{1.23}$$

This equation defines a symmetric norm $\|X\|_{sym} = \|X\|_{\phi_\Pi}$ on the ideal $\mathscr{C}_\Pi(\mathscr{H})$, see Definition 2.

Now, let $\Pi = \{\pi_j\}_{j=1}^\infty$, with $\pi_1 = 1$, satisfy

$$\sum_{j=1}^\infty \pi_j = +\infty \qquad \text{and} \qquad \lim_{j\to\infty} \pi_j = 0\,. \tag{1.24}$$

Then, the ideal $\mathscr{C}_\Pi(\mathscr{H})$ is *nontrivial*: $\mathscr{C}_\Pi(\mathscr{H}) \neq \mathscr{C}_\infty(\mathscr{H})$ and $\mathscr{C}_\Pi(\mathscr{H}) \neq \mathscr{C}_1(\mathscr{H})$, see Remark 2, and one has

$$\mathscr{C}_1(\mathscr{H}) \subset \mathscr{C}_\Pi(\mathscr{H}) \subset \mathscr{C}_\infty(\mathscr{H}). \tag{1.25}$$

If in addition to (1.24), the sequence $\Pi = \{\pi_j\}_{j=1}^\infty$ is *regular*, i.e. it obeys

$$\sum_{j=1}^n \pi_j = O(n\pi_n)\,, \qquad n \to \infty\,, \tag{1.26}$$

then $X \in \mathscr{C}_\Pi(\mathscr{H})$, if and only if

$$s_n(X) = O(\pi_n) \,, \qquad n \to \infty \,, \tag{1.27}$$

cf. condition (1.21). On the other hand, the asymptotics

$$s_n(X) = o(\pi_n) \,, \qquad n \to \infty \,,$$

implies that $X$ belongs to some ideal in between of $\mathscr{C}_1(\mathscr{H})$ and $\mathscr{C}_{1+\varepsilon}(\mathscr{H})$.

*Remark 3* A natural choice of the sequence $\{\pi_j\}_{j=1}^{\infty}$ that satisfies (1.24) is $\pi_j = j^{-\alpha}$, $0 < \alpha \leq 1$. Note that if $0 < \alpha < 1$, then the sequence $\Pi = \{\pi_j\}_{j=1}^{\infty}$ satisfies (1.26), i.e. it is regular for $\varepsilon = 1 - \alpha$. Therefore, the two-sided symmetrically-normed ideal $\mathscr{C}_\Pi(\mathscr{H})$ generated by symmetric norming function (1.22) consists of all those compact operators $X$, which singular values obey (1.27):

$$s_n(X) = O(n^{-\alpha}), \quad 0 < \alpha < 1, \quad n \to \infty \,. \tag{1.28}$$

Let $\alpha = 1/p$ , $p > 1$. Then, the corresponding to (1.28) symmetrically-normed ideal is defined by

$$\mathscr{C}_{p,\infty}(\mathscr{H}) := \{X \in \mathscr{C}_\infty(\mathscr{H}) : s_n(X) = O(n^{-1/p}), \ p > 1\} \,,$$

which is known as the *weak-$\mathscr{C}_p$* ideal [20, 22].

Whilst by virtue of (1.28) the weak-$\mathscr{C}_p$ ideal admit no trace, definition (1.23) implies that for the regular case $p > 1$ a symmetric norm on $\mathscr{C}_{p,\infty}(\mathscr{H})$ is equivalent to

$$\|X\|_{p,\infty} = \sup_n \frac{1}{n^{1-1/p}} \sum_{j=1}^{n} s_j(X) \,, \tag{1.29}$$

and $\mathscr{C}_1(\mathscr{H}) \subset \mathscr{C}_{p,\infty}(\mathscr{H}) \subset \mathscr{C}_\infty(\mathscr{H})$.

## 2 Singular Traces

Note that (1.29) implies: $\mathscr{C}_1(\mathscr{H}) \ni A \mapsto \|A\|_{p,\infty} < \infty$, but any related linear, positive and unitarily invariant functional (*trace*), which is nontrivial on the ideal $\mathscr{C}_{p,\infty}(\mathscr{H})$ is *zero* on the set of finite-rank operators $\mathscr{K}(\mathscr{H})$. We remind that these *non-normal* traces:

$$\mathrm{Tr}_\omega(X) := \omega(\{n^{-1+1/p} \sum_{j=1}^{n} s_j(X)\}_{n=1}^{\infty}) \,, \tag{2.1}$$

are called *singular*, [5, 13]. Here, $\omega$ is an *appropriate* linear positive normalised functional (*state*) on the Banach space $l^\infty(\mathbb{N})$ of bounded sequences. Recall that the set of these states $\mathcal{S}(l^\infty(\mathbb{N})) \subset (l^\infty(\mathbb{N}))^*$, where $(l^\infty(\mathbb{N}))^*$ is dual of the Banach space $l^\infty(\mathbb{N})$. Note that the singular trace (2.1) is continuous in topology defined by the norm (1.29).

*Remark 4* (a) The *weak-$\mathcal{C}_p$* ideal, which is defined for $p = 1$ by

$$\mathcal{C}_{1,\infty}(\mathcal{H}) := \{X \in \mathcal{C}_\infty(\mathcal{H}) : \sum_{j=1}^{n} s_j(X) = O(\ln(n)),\ n \to \infty\}, \qquad (2.2)$$

has a special interest. Note that since $\Pi = \{j^{-1}\}_{j=1}^\infty$ does not satisfy (1.26), the characterisation $s_n(X) = O(n^{-1})$ is *not* true, see (1.27), (1.28). In this case, the equivalent norm can be defined on the ideal (2.2) as

$$\|X\|_{1,\infty} := \sup_{n\in\mathbb{N}} \frac{1}{1+\ln(n)} \sum_{j=1}^{n} s_j(X). \qquad (2.3)$$

By virtue of (1.25) and Remark 3 one gets that $\mathcal{C}_1(\mathcal{H}) \subset \mathcal{C}_{1,\infty}(\mathcal{H}) \subset \mathcal{C}_{1+\varepsilon}(\mathcal{H})$ for any $\varepsilon > 0$ and that $\mathcal{C}_1(\mathcal{H}) \ni A \mapsto \|A\|_{1,\infty} = \|A\|_1 \geq \|A\|_{1+\varepsilon}$.

(b) In contrast to the trace norm $\|\cdot\|_1$ on the positive cone $\mathcal{C}_{1,+}(\mathcal{H})$, see Example 1, the map $X \mapsto \|X\|_{1,\infty}$ on the positive cone $\mathcal{C}_{1,\infty,+}(\mathcal{H})$ is *not* linear although homogeneous. On the space $l^\infty(\mathbb{N})$, there exists a state $\omega \in \mathcal{S}(l^\infty(\mathbb{N}))$ such that the map

$$X \mapsto \mathrm{Tr}_\omega(X) := \omega(\{(1+\ln(n))^{-1} \sum_{j=1}^{n} s_j(X)\}_{n=1}^\infty), \qquad (2.4)$$

is *linear* and verifies the properties of the (singular) *trace* for any $X \subset \mathcal{C}_{1,\infty}(\mathcal{H})$. We construct $\omega$ in Section 3. This particular choice of the state $\omega$ defines the *Dixmier trace* on the space $\mathcal{C}_{1,\infty}(\mathcal{H})$, which is called, in turn, the *Dixmier ideal*, see e.g. [3, 4]. The Dixmier trace (2.4) is obviously continuous in topology defined by the norm (2.3). This last property is basic for discussion in Section 4 of the Trotter–Kato product formula in the $\|\cdot\|_{p,\infty}$-topology, for $p \geq 1$.

*Example 3* With non-increasing sequence of positive numbers $\pi = \{\pi_j\}_{j=1}^\infty, \pi_1 = 1$, one can associate the symmetric norming function $\phi_\pi$ given by

$$\phi_\pi(\xi) := \sum_{j=1}^{\infty} \pi_j \xi_j^*, \qquad \xi \in c_f.$$

The corresponding symmetrically-normed ideal is denoted by $\mathcal{C}_\pi(\mathcal{H}) := \mathcal{C}_{\phi_\pi}(\mathcal{H})$.

If the sequence $\pi$ satisfies (1.24), then ideal $\mathcal{C}_\pi(\mathcal{H})$ does not coincide neither with $\mathcal{C}_\infty(\mathcal{H})$ nor with $\mathcal{C}_1(\mathcal{H})$. If, in particular, $\pi_j = j^{-\alpha}$, $j = 1, 2, \ldots$, for $0 < \alpha \leq 1$, then the corresponding ideal is denoted by $\mathscr{C}_{\infty,p}(\mathscr{H})$, $p = 1/\alpha$. The norm on this ideal is given by

$$\|X\|_{\infty,p} := \sum_{j=1}^{\infty} j^{-1/p}\, s_j(X)\,, \quad p \in [1, \infty).$$

The symmetrically-normed ideal $\mathscr{C}_{\infty,1}(\mathscr{H})$ is called the *Macaev ideal* [8]. It turns out that the Dixmier ideal $\mathscr{C}_{1,\infty}(\mathscr{H})$ is dual of the Macaev ideal: $\mathscr{C}_{1,\infty}(\mathscr{H}) = \mathscr{C}_{\infty,1}(\mathscr{H})^*$.

**Proposition 2.1** *The space $\mathscr{C}_{1,\infty}(\mathscr{H})$ endowed with the norm $\|\cdot\|_{1,\infty}$ is a Banach space.*

The proof is quite standard although tedious and long. We address the readers to the corresponding references, e.g. [8].

**Proposition 2.2** *The space $\mathscr{C}_{1,\infty}(\mathscr{H})$ endowed with the norm $\|\cdot\|_{1,\infty}$ is a Banach ideal in the algebra of bounded operators $\mathscr{L}(\mathscr{H})$.*

*Proof* To this end, it is sufficient to prove that if $A$ and $C$ are bounded operators, then $B \in \mathscr{C}_{1,\infty}(\mathscr{H})$ implies $ABC \in \mathscr{C}_{1,\infty}(\mathscr{H})$. Recall that singular values of the operator $ABC$ verify the estimate $s_j(ABC) \leq \|A\|\|C\| s_j(B)$. By (2.3), it yields

$$\|ABC\|_{1,\infty} = \sup_{n\in\mathbb{N}} \frac{1}{1+\ln(n)} \sum_{j=1}^{n} s_j(ABC) \leq \tag{2.5}$$
$$\|A\|\|C\| \sup_{n\in\mathbb{N}} \frac{1}{1+\ln(n)} \sum_{j=1}^{n} s_j(B) = \|A\|\|C\|\|B\|_{1,\infty}\,,$$

and consequently the proof of the assertion. □

Recall that for any $A \in \mathscr{L}(\mathscr{H})$ and all $B \in \mathscr{C}_1(\mathscr{H})$ one can define a linear functional on $\mathscr{C}_1(\mathscr{H})$ given by $\mathrm{Tr}_{\mathscr{H}}(AB)$. The set of these functionals $\{\mathrm{Tr}_{\mathscr{H}}(A\cdot)\}_{A\in\mathscr{L}(\mathscr{H})}$ is just the *dual* space $\mathscr{C}_1(\mathscr{H})^*$ of $\mathscr{C}_1(\mathscr{H})$ with the operator-norm topology. In other words, $\mathscr{L}(\mathscr{H}) = \mathscr{C}_1(\mathscr{H})^*$, in the sense that the map $A \mapsto \mathrm{Tr}_{\mathscr{H}}(A\cdot)$ is the isometric isomorphism of $\mathscr{L}(\mathscr{H})$ onto $\mathscr{C}_1(\mathscr{H})^*$.

With the help of the *duality* relation

$$\langle A|B\rangle := \mathrm{Tr}_{\mathscr{H}}(AB)\,, \tag{2.6}$$

one can also describe the space $\mathscr{C}_1(\mathscr{H})_*$, which is a *predual* of $\mathscr{C}_1(\mathscr{H})$, i.e. its dual $(\mathscr{C}_1(\mathscr{H})_*)^* = \mathscr{C}_1(\mathscr{H})$. To this aim for each *fixed* $B \in \mathscr{C}_1(\mathscr{H})$, we consider the functionals $A \mapsto \mathrm{Tr}_{\mathscr{H}}(AB)$ on $\mathscr{L}(\mathscr{H})$. It is known that they are not *all* continuous linear functional on bounded operators $\mathscr{L}(\mathscr{H})$, i.e. $\mathscr{C}_1(\mathscr{H}) \subset \mathscr{L}(\mathscr{H})^*$, but they yield the entire dual only of compact operators, i.e. $\mathscr{C}_1(\mathscr{H}) = \mathscr{C}_\infty(\mathscr{H})^*$. Hence, $\mathscr{C}_1(\mathscr{H})_* = \mathscr{C}_\infty(\mathscr{H})$.

Now, we note that under duality relation (2.6), the Dixmier ideal $\mathscr{C}_{1,\infty}(\mathscr{H})$ is the dual of the Macaev ideal: $\mathscr{C}_{1,\infty}(\mathscr{H}) = \mathscr{C}_{\infty,1}(\mathscr{H})^*$, where

$$\mathscr{C}_{\infty,1}(\mathscr{H}) = \{A \in \mathscr{C}_\infty(\mathscr{H}) : \sum_{n\geq 1} \frac{1}{n}\, s_n(A) < \infty\}\,, \tag{2.7}$$

see Example 3. By the same duality relation and by similar calculations one also obtains that the *predual* of $\mathscr{C}_{\infty,1}(\mathscr{H})$ is the ideal $\mathscr{C}_{\infty,1}(\mathscr{H})_* = \mathscr{C}_{1,\infty}^{(0)}(\mathscr{H})$, defined by

$$\mathscr{C}_{1,\infty}^{(0)}(\mathscr{H}) := \{A \in \mathscr{C}_\infty(\mathscr{H}) : \sum_{j\geq 1}^{n} s_j(A) = o(\ln(n)),\ n \to \infty\}\,. \tag{2.8}$$

By virtue of (2.2) (see Remark 4), the ideal (2.8) is not self-dual since

$$\mathscr{C}_{1,\infty}^{(0)}(\mathscr{H})^{**} = \mathscr{C}_{1,\infty}(\mathscr{H}) \supset \mathscr{C}_{1,\infty}^{(0)}(\mathscr{H}).$$

The problem which has motivated construction of the Dixmier trace in [5] was related to the question of a general definition of the *trace*, i.e. a linear, positive and unitarily invariant functional on a *proper* Banach ideal $\mathfrak{I}(\mathscr{H})$ of the unital algebra of bounded operators $\mathscr{L}(\mathscr{H})$. Since any proper two-sided ideal $\mathfrak{I}(\mathscr{H})$ of $\mathscr{L}(\mathscr{H})$ is contained in compact operators $\mathscr{C}_\infty(\mathscr{H})$ and contains the set $\mathscr{K}(\mathscr{H})$ of finite-rank operators ((1.17), Section 1), *domain* of definition of the trace has to coincide with the ideal $\mathfrak{I}(\mathscr{H})$.

*Remark 5* The *canonical* trace $\mathrm{Tr}_{\mathscr{H}}(\cdot)$ is nontrivial only on domain, which is the trace-class ideal $\mathscr{C}_1(\mathscr{H})$, see Example 1. We recall that it is characterised by the property of *normality*: $\mathrm{Tr}_{\mathscr{H}}(\sup_\alpha B_\alpha) = \sup_\alpha \mathrm{Tr}_{\mathscr{H}}(B_\alpha)$, for every directed increasing bounded family $\{B_\alpha\}_{\alpha\in\Delta}$ of positive operators from $\mathscr{C}_{1,+}(\mathscr{H})$.

Note that every nontrivial *normal* trace on $\mathscr{L}(\mathscr{H})$ is proportional to the canonical trace $\mathrm{Tr}_{\mathscr{H}}(\cdot)$, see e.g. [6, 20]. Therefore, the Dixmier trace (2.4) : $\mathscr{C}_{1,\infty} \ni X \mapsto \mathrm{Tr}_\omega(X)$, is *not* normal.

**Definition 4** A *trace* on the proper Banach ideal $\mathfrak{I}(\mathscr{H}) \subset \mathscr{L}(\mathscr{H})$ is called *singular* if it vanishes on the set $\mathscr{K}(\mathscr{H})$.

Since a singular trace is defined up to trace-class operators $\mathscr{C}_1(\mathscr{H})$, then by Remark 5 it is obviously *not* normal.

## 3 Dixmier Trace

Recall that only the ideal of trace-class operators has the property that on its positive cone $\mathscr{C}_{1,+}(\mathscr{H}) := \{A \in \mathscr{C}_1(\mathscr{H}) : A \geq 0\}$ the trace norm is *linear* since $\|A + B\|_1 = \mathrm{Tr}\,(A + B) = \mathrm{Tr}\,(A) + \mathrm{Tr}\,(B) = \|A\|_1 + \|B\|_1$ for $A, B \in \mathscr{C}_{1,+}(\mathscr{H})$, see

Example 1. Then, the uniqueness of the trace norm allows to extend the trace to the whole linear space $\mathscr{C}_1(\mathscr{H})$. Imitation of this idea *fails* for other symmetrically-normed ideals.

This problem motivates the Dixmier trace construction as a certain limiting procedure involving the $\|\cdot\|_{1,\infty}$-norm. Let $\mathscr{C}_{1,\infty,+}(\mathscr{H})$ be a positive cone of the Dixmier ideal. One can try to construct on $\mathscr{C}_{1,\infty,+}(\mathscr{H})$ a *linear, positive* and *unitarily* invariant functional (called *trace* $\mathscr{T}$) via *extension* of the limit (called Lim) of the sequence of properly normalised finite sums of the operator $X$ singular values:

$$\mathscr{T}(X) := \mathrm{Lim}_{n\to\infty} \frac{1}{1+\ln(n)} \sum_{j=1}^{n} s_j(X) \, , \quad X \in \mathscr{C}_{1,\infty,+}(\mathscr{H}). \tag{3.1}$$

First, we note that since for any unitary $U : \mathscr{H} \to U$, the singular values of $X \in \mathscr{C}_\infty(\mathscr{H})$ are invariant: $s_j(X) = s_j(U\,X\,U^*)$, it is also true for the sequence

$$\sigma_n(X) := \sum_{j=1}^{n} s_j(X) \, , \quad n \in \mathbb{N} \, . \tag{3.2}$$

Then, the Lim in (3.1) (if it exists) inherits the property of *unitarity*.

Now, we comment that positivity: $X \geq 0$ implies the positivity of eigenvalues $\{\lambda_j(X)\}_{j\geq 1}$ and consequently: $\lambda_j(X) = s_j(X)$. Therefore, $\sigma_n(X) \geq 0$ and the Lim in (3.1) is a *positive* mapping.

The next problem with the formula for $\mathscr{T}(X)$ is its *linearity*. To proceed, we recall that if $P : \mathscr{H} \to P(\mathscr{H})$ is an orthogonal projection on a finite-dimensional subspace with $\dim P(\mathscr{H}) = n$, then for any bounded operator $X \geq 0$ the (3.2) gives

$$\sigma_n(X) = \sup_P \{\mathrm{Tr}_{\mathscr{H}}\,(XP) : \dim P(\mathscr{H}) = n\} \, . \tag{3.3}$$

As a corollary of (3.3), one obtains the Horn–Ky Fan inequality

$$\sigma_n(X+Y) \leq \sigma_n(X) + \sigma_n(Y) \, , \quad n \in \mathbb{N}, \tag{3.4}$$

valid, in particular, for any pair of bounded *positive* compact operators $X$ and $Y$. For $\dim P(\mathscr{H}) \leq 2n$, one similarly gets from (3.3) that

$$\sigma_{2n}(X+Y) \geq \sigma_n(X) + \sigma_n(Y) \, , \quad n \in \mathbb{N} \, . \tag{3.5}$$

Motivated by (3.1), we now introduce

$$\mathscr{T}_n(X) := \frac{1}{1+\ln(n)} \sigma_n(X) \, , \quad X \in \mathscr{C}_{1,\infty,+}(\mathscr{H}) \, , \tag{3.6}$$

and denote by $\mathrm{Lim}\{\mathscr{T}_n(X)\}_{n\in\mathbb{N}} := \mathrm{Lim}_{n\to\infty}\mathscr{T}_n(X)$ the right-hand side of the functional in (3.1). Note that by (3.6), the inequalities (3.4) and (3.5) yield for $n \in \mathbb{N}$

$$\mathscr{T}_n(X+Y) \leq \mathscr{T}_n(X) + \mathscr{T}_n(Y) \ , \quad \frac{1+\ln(2n)}{1+\ln(n)} \mathscr{T}_{2n}(X+Y) \geq \mathscr{T}_n(X) + \mathscr{T}_n(Y). \tag{3.7}$$

Since the functional Lim includes the limit $n \to \infty$, the inequalities (3.7) *would* give a desired linearity of the trace $\mathscr{T}$:

$$\mathscr{T}(X+Y) = \mathscr{T}(X) + \mathscr{T}(Y) \ , \tag{3.8}$$

*if* one proves that the $\mathrm{Lim}_{n\to\infty}$ in (3.1) exists and finite for $X, Y$ as well as for $X + Y$.

To this end, we note that if the right-hand side of (3.1) is finite, then one has (3.8), and hence the $\mathrm{Lim}\{\mathscr{T}_n(X)\}_{n\in\mathbb{N}}$ is a positive linear map $\mathrm{Lim} : l^\infty(\mathbb{N}) \to \mathbb{R}$, which defines a *state* $\omega \in \mathscr{S}(l^\infty(\mathbb{N}))$ on the Banach space of sequences $\{\mathscr{T}_n(\cdot)\}_{n\in\mathbb{N}} \in l^\infty(\mathbb{N})$. To make this definition more precise, we impose on the state $\omega$ the following conditions:

$$\begin{aligned} &(a)\ \ \omega(\eta) \geq 0 \ , \ \text{for } \ \forall\eta = \{\eta_n \geq 0\}_{n\in\mathbb{N}} \ , \\ &(b)\ \ \omega(\eta) = \mathrm{Lim}\{\eta_n\}_{n\in\mathbb{N}} = \lim_{n\to\infty} \eta_n \ , \ \text{ if } \{\eta_n \geq 0\}_{n\in\mathbb{N}} \text{ is convergent} \ . \end{aligned}$$

By virtue of (a) and (b), the definitions (3.1) and (3.6) imply that for $X, Y \in \mathscr{C}_{1,\infty,+}(\mathscr{H})$, one gets

$$\mathscr{T}(X) = \omega(\{\mathscr{T}_n(X)\}_{n\in\mathbb{N}}) = \lim_{n\to\infty} \mathscr{T}_n(X) \ , \tag{3.9}$$

$$\mathscr{T}(Y) = \omega(\{\mathscr{T}_n(Y)\}_{n\in\mathbb{N}}) = \lim_{n\to\infty} \mathscr{T}_n(Y) \ , \tag{3.10}$$

$$\mathscr{T}(X+Y) = \omega(\{\mathscr{T}_n(X+Y)\}_{n\in\mathbb{N}}) = \lim_{n\to\infty} \mathscr{T}_n(X+Y) \ , \tag{3.11}$$

if the limits in the right-hand sides of (3.9)–(3.11) exist.

Now, to ensure (3.8) one has to select $\omega$ in such a way that it allows to restore the equality in (3.7), when $n \to \infty$. To this aim, we impose on the state $\omega$ the condition of *dilation* $\mathfrak{D}_2$-invariance.

Let $\mathfrak{D}_2 : l^\infty(\mathbb{N}) \to l^\infty(\mathbb{N})$ be dilation mapping $\eta \mapsto \mathfrak{D}_2(\eta)$:

$$\mathfrak{D}_2 : (\eta_1, \eta_2, \ldots \eta_k, \ldots) \to (\eta_1, \eta_1, \eta_2, \eta_2, \ldots \eta_k, \eta_k, \ldots) \ , \ \forall\eta \in l^\infty(\mathbb{N}). \tag{3.12}$$

We say that $\omega$ is dilation $\mathfrak{D}_2$-*invariant* if for any $\eta \in l^\infty(\mathbb{N})$ it verifies the property

$$(c) \qquad \omega(\eta) = \omega(\mathfrak{D}_2(\eta)). \tag{3.13}$$

We shall discuss the question of *existence*, the dilation $\mathfrak{D}_2$-invariant states (the *invariant means*), on the Banach space $l^\infty(\mathbb{N})$ in Remark 6.

Let $X, Y \in \mathscr{C}_{1,\infty,+}(\mathscr{H})$. Then, applying the property (c) to the sequence $\eta = \{\xi_{2n} := \mathscr{T}_{2n}(X+Y)\}_{n=1}^\infty$, we obtain

$$\omega(\eta) = \omega(\mathfrak{D}_2(\eta)) = \omega(\xi_2, \xi_2, \xi_4, \xi_4, \xi_6, \xi_6, \ldots) . \tag{3.14}$$

Note that if $\xi = \{\xi_n = \mathscr{T}_n(X+Y)\}_{n=1}^\infty$, then the difference of the sequences:

$$\mathfrak{D}_2(\eta) - \xi = (\xi_2, \xi_2, \xi_4, \xi_4, \xi_6, \xi_6, \ldots) - (\xi_1, \xi_2, \xi_3, \xi_4, \xi_5, \xi_6, \ldots) ,$$

converges to *zero* if $\xi_{2n} - \xi_{2n-1} \to 0$ as $n \to \infty$. Then, by virtue of (3.11) and (3.14), this would imply

$$\omega(\{\mathscr{T}_{2n}(X+Y)\}_{n\in\mathbb{N}}) = \omega(\mathfrak{D}_2(\{\mathscr{T}_{2n}(X+Y)\}_{n\in\mathbb{N}})) = \omega(\{\mathscr{T}_n(X+Y)\}_{n\in\mathbb{N}}) ,$$

or by (3.11): $\lim_{n\to\infty} \mathscr{T}_{2n}(X+Y) = \lim_{n\to\infty} \mathscr{T}_n(X+Y)$, which by estimates (3.7) would also yield

$$\lim_{n\to\infty} \mathscr{T}_n(X+Y) = \lim_{n\to\infty} \mathscr{T}_n(X) + \lim_{n\to\infty} \mathscr{T}_n(Y). \tag{3.15}$$

Now, summarising (3.9), (3.10), (3.11) and (3.15), we obtain the linearity (3.8) of the limiting functional $\mathscr{T}$ on the positive cone $\mathscr{C}_{1,\infty,+}(\mathscr{H})$ if it is defined by the corresponding $\mathfrak{D}_2$-invariant state $\omega$ or a dilation-invariant mean.

Therefore, to finish the proof of linearity it rests only to check that $\lim_{n\to\infty}(\xi_{2n} - \xi_{2n-1}) = 0$. To this end, we note that by definitions (3.2) and (3.6), one gets

$$\begin{aligned}\xi_{2n} - \xi_{2n-1} &= \left[\frac{1}{\ln(2n)} - \frac{1}{\ln(2n-1)}\right] \sigma_{2n-1}(X+Y) \\ &+ \frac{1}{\ln(2n)} s_{2n}(X+Y).\end{aligned} \tag{3.16}$$

Since $X, Y \in \mathscr{C}_{1,\infty,+}(\mathscr{H})$, we obtain that $\lim_{n\to\infty} s_{2n}(X+Y) = 0$ and that $\sigma_{2n-1}(X+Y) = O(\ln(2n-1))$. Then taking into account that $(1/\ln(2n) - 1/\ln(2n-1)) = o(1/\ln(2n-1))$ one gets that for $n \to \infty$ the right-hand side of (3.16) converges to zero.

To conclude our construction of the trace $\mathscr{T}(\cdot)$ we note that by linearity (3.8) one can uniquely extend this functional from the positive cone $\mathscr{C}_{1,\infty,+}(\mathscr{H})$ to the real subspace of the Banach space $\mathscr{C}_{1,\infty}(\mathscr{H})$, and finally to the entire ideal $\mathscr{C}_{1,\infty}(\mathscr{H})$.

**Definition 5** The *Dixmier trace* $\mathrm{Tr}_\omega(X)$ of the operator $X \in \mathscr{C}_{1,\infty,+}(\mathscr{H})$ is the value of the linear functional (3.1):

$$\mathrm{Tr}_\omega(X) := \mathrm{Lim}_{n\to\infty} \frac{\sigma_n(X)}{1+\ln(n)} , \tag{3.17}$$

where $\mathrm{Lim}_{n\to\infty}$ is defined by a dilation-invariant state $\omega \in \mathscr{S}(l^\infty(\mathbb{N}))$ on $l^\infty(\mathbb{N})$ that satisfies the properties (a), (b) and (c). Since any self-adjoint operator $X \in \mathscr{C}_{1,\infty}(\mathscr{H})$ has the representation: $X = X_+ - X_-$, where $X_\pm \in \mathscr{C}_{1,\infty,+}(\mathscr{H})$, one gets $\mathrm{Tr}_\omega(X) = \mathrm{Tr}_\omega(X_+) - \mathrm{Tr}_\omega(X_-)$. Then, for arbitrary $Z \in \mathscr{C}_{1,\infty}(\mathscr{H})$, the Dixmier trace is $\mathrm{Tr}_\omega(Z) = \mathrm{Tr}_\omega(\mathrm{Re}Z) + i\mathrm{Tr}_\omega(\mathrm{Im}Z)$.

Note that if $X \in \mathscr{C}_{1,\infty,+}(\mathscr{H})$, then definition (3.17) of $\mathrm{Tr}_\omega(\cdot)$ together with definition of the norm $\|\cdot\|_{1,\infty}$ in (2.3) readily imply the estimate $\mathrm{Tr}_\omega(X) \leq \|X\|_{1,\infty}$, which, in turn, yields the inequality for arbitrary Z from the Dixmier ideal $\mathscr{C}_{1,\infty}(\mathscr{H})$:

$$|\mathrm{Tr}_\omega(Z)| \leq \|Z\|_{1,\infty}. \tag{3.18}$$

*Remark 6* A decisive for construction of the Dixmier trace $\mathrm{Tr}_\omega(\cdot)$ is the *existence* of the invariant mean $\omega \in \mathscr{S}(l^\infty(\mathbb{N})) \subset (l^\infty(\mathbb{N}))^*$. Here, the space $(l^\infty(\mathbb{N}))^*$ is *dual* to the Banach space of bounded sequences. Then, by the Banach–Alaoglu theorem, the convex set of states $\mathscr{S}(l^\infty(\mathbb{N}))$ is *compact* in $(l^\infty(\mathbb{N}))^*$ in the weak* topology. Now, for any $\phi \in \mathscr{S}(l^\infty(\mathbb{N}))$, the relation $\phi(\mathfrak{D}_2(\cdot)) =: (\mathfrak{D}_2^*\phi)(\cdot)$ defines the dual $\mathfrak{D}_2^*$-dilation on the set of states. By definition (3.12), this map is such that $\mathfrak{D}_2^* : \mathscr{S}(l^\infty(\mathbb{N})) \to \mathscr{S}(l^\infty(\mathbb{N}))$, as well as continuous and affine (in fact, linear). Then, by the Markov–Kakutani theorem, the dilation $\mathfrak{D}_2^*$ has a fix point $\omega \in \mathscr{S}(l^\infty(\mathbb{N}))$ : $\mathfrak{D}_2^*\omega = \omega$. This observation justifies the existence of the *invariant mean* (c) for $\mathfrak{D}_2$-dilation.

Note that Remark 6 has a straightforward extension to any $\mathfrak{D}_k$-dilation for $k > 2$, which is defined similar to (3.12). Since dilations for different $k \geq 2$ *commute*, the extension of the Markov–Kakutani theorem yields that the commutative family $\mathscr{F} = \{\mathfrak{D}_k^*\}_{k\geq 2}$ has in $\mathscr{S}(l^\infty(\mathbb{N}))$ the common fix point $\omega = \mathfrak{D}_2^*\omega$. Therefore, Definition 5 of the Dixmier trace does not depend on the degree $k \geq 2$ of dilation $\mathfrak{D}_k$.

For more details about different constructions of *invariant means* and the corresponding Dixmier trace on $\mathscr{C}_{1,\infty}(\mathscr{H})$, see e.g., [3, 13].

**Proposition 3.1** *The Dixmier trace has the following properties:*

*(a)* *For any bounded operator* $B \in \mathscr{L}(\mathscr{H})$ *and* $Z \in \mathscr{C}_{1,\infty}(\mathscr{H})$ *one has* $\mathrm{Tr}_\omega(ZB) = \mathrm{Tr}_\omega(BZ)$.
*(b)* $\mathrm{Tr}_\omega(C) = 0$ *for any operator* $C \in \mathscr{C}_1(\mathscr{H})$ *from the trace-class ideal, which is the closure of finite-rank operators* $\mathscr{K}(\mathscr{H})$ *for the* $\|\cdot\|_1$*-norm.*
*(c)* *The Dixmier trace* $\mathrm{Tr}_\omega : \mathscr{C}_{1,\infty}(\mathscr{H}) \to \mathbb{C}$ *is continuous in the* $\|\cdot\|_{1,\infty}$*-norm.*

*Proof* (a) Since every operator $B \in \mathscr{L}(\mathscr{H})$ is a linear combination of four unitary operators, it is sufficient to prove the equality $\mathrm{Tr}_\omega(ZU) = \mathrm{Tr}_\omega(UZ)$ for a unitary operator $U$ and moreover only for $Z \in \mathscr{C}_{1,\infty,+}(\mathscr{H})$. Then, the corresponding equality follows from the unitary invariance: $s_j(Z) = s_j(ZU) = s_j(UZ) = s_j(UZU^*)$ of singular values of the positive operator Z for all $j \geq 1$.

(b) Since $C \in \mathscr{C}_1(\mathscr{H})$ yields $\|C\|_1 < \infty$, definition (3.2) implies $\sigma_n(C) \leq \|C\|_1$ for any $n \geq 1$. Then, by Definition 5 one gets $\mathrm{Tr}_\omega(C) = 0$. Proof of the last part of the statement is standard.
(c) Since the ideal $\mathscr{C}_{1,\infty}(\mathscr{H})$ is a Banach space and $\mathrm{Tr}_\omega : \mathscr{C}_{1,\infty}(\mathscr{H}) \to \mathbb{C}$ a linear functional, it is sufficient to consider continuity at $X = 0$. Then, let the sequence $\{X_k\}_{k\geq 1} \subset \mathscr{C}_{1,\infty}(\mathscr{H})$ converges to $X = 0$ in $\|\cdot\|_{1,\infty}$-topology, i.e. by (2.3)

$$\lim_{k\to\infty} \|X_k\|_{1,\infty} = \lim_{k\to\infty} \sup_{n\in\mathbb{N}} \frac{1}{1+\ln(n)} \sigma_n(X_k) = 0\,. \tag{3.19}$$

Since (3.18) implies $|\mathrm{Tr}_\omega(X_k)| \leq \|X_k\|_{1,\infty}$ , the assertion follows from (3.19). □

Therefore, the Dixmier construction gives an example of a *singular* trace in the sense of Definition 4.

## 4 Trotter–Kato Product Formulae in the Dixmier Ideal

Let $A \geq 0$ and $B \geq 0$ be two non-negative self-adjoint operators in a separable Hilbert space $\mathscr{H}$ and let the subspace $\mathscr{H}_0 := \overline{\mathrm{dom}(A^{1/2}) \cap \mathrm{dom}(B^{1/2})}$. It may happen that $\mathrm{dom}(A) \cap \mathrm{dom}(B) = \{0\}$, but the form-sum of these operators: $H = A \dotplus B$ is well-defined in the subspace $\mathscr{H}_0 \subseteq \mathscr{H}$.

Kato proved in [12] that under these conditions the *Trotter product formula*

$$s - \lim_{n\to\infty} \left(e^{-tA/n}e^{-tB/n}\right)^n = e^{-tH} P_0, \qquad t \geq 0, \tag{4.1}$$

converges in the *strong* operator topology uniformly in $t \in [0, T]$, $0 < T < +\infty$ , where $P_0$ denotes the orthogonal projection from $\mathscr{H}$ onto $\mathscr{H}_0$. Moreover, in [12] it was also shown that the product formula is true not only for the *exponential* function $e^{-x}, x \geq 0$, but for a whole class of Borel measurable functions $f(\cdot)$ and $g(\cdot)$, which are defined on $\mathbb{R}_0^+ := [0, \infty)$ and satisfy the conditions:

$$0 \leq f(x) \leq 1, \qquad f(0) = 1, \qquad f'(+0) = -1, \tag{4.2}$$

$$0 \leq g(x) \leq 1, \qquad g(0) = 1, \qquad g'(+0) = -1. \tag{4.3}$$

Namely, the main result of [12] says that besides (4.1) one also gets convergence

$$\tau - \lim_{n\to\infty} (f(tA/n)g(tB/n))^n = e^{-tH} P_0, \qquad t \geq 0, \tag{4.4}$$

uniformly in $t \in [0, T]$, $0 < T < +\infty$, if topology $\tau = s$.

Product formulae of the type (4.4) are called the *Trotter–Kato product formulae* for functions (4.2), (4.3), which are called the *Kato functions* $\mathscr{K}$. Note that $\mathscr{K}$ is closed with respect to the *products* of Kato functions.

For some particular classes of the Kato functions we refer to [15, 25]. In the following, it is useful to consider instead of $f(x)g(x)$ two Kato functions: $g(x/2)f(x)g(x/2)$ and $f(x/2)g(x)f(x/2)$ that produce the self-adjoint operator families

$$F(t) := g(tB/2)f(tA)g(tB/2) \text{ and } T(t) := f(tA/2)g(tB)f(tA/2), \quad t \geq 0. \tag{4.5}$$

Since [14] it is known that the *lifting* of the topology of convergence in (4.4) to the *operator norm* $\tau = \|\cdot\|$ needs more conditions on operators $A$ and $B$ as well as on the key Kato functions $f, g \in \mathscr{K}$. One finds a discussion and more references on this subject in [25]. Here, we quote a result that will be used below for the Trotter–Kato product formulae in the Dixmier ideal $\mathscr{C}_{1,\infty}(\mathscr{H})$.

Consider the class $\mathscr{K}_\beta$ of Kato functions, which is defined in [10, 11] as given below:

(i) Measurable functions $0 \leq h \leq 1$ on $\mathbb{R}_0^+$, such that $h(0) = 1$ and $h'(+0) = -1$.
(ii) For $\varepsilon > 0$ there exists $\delta = \delta(\varepsilon) < 1$, such that $h(s) \leq 1 - \delta(\varepsilon)$ for $s \geq \varepsilon$ and

$$[h]_\beta := \sup_{s>0} \frac{|h(s) - 1 + s|}{s^\beta} < \infty\,, \quad \text{for } 1 < \beta \leq 2\,.$$

The standard examples are $h(s) = e^{-s}$ and $h(s) = (1 + a^{-1}s)^{-a}$ , $a > 0$.

Below we consider the class $\mathscr{K}_\beta$ and a particular case of generators $A$ and $B$, such that for the Trotter–Kato product formulae the estimate of the convergence rate is *optimal.*

**Proposition 4.1** ([11]) *Let $f, g \in \mathscr{K}_\beta$ with $\beta = 2$, and let $A$, $B$ be non-negative self-adjoint operators in $\mathscr{H}$ such that the operator sum $C := A + B$ is self-adjoint on domain $dom(C) := dom(A) \cap dom(B)$. Then, the Trotter–Kato product formulae converge for $n \to \infty$ in the operator norm:*

$$\|[f(tA/n)g(tB/n)]^n - e^{-tC}\| = O(n^{-1})\,, \; \|[g(tB/n)f(tA/n)]^n - e^{-tC}\| = O(n^{-1})\,,$$
$$\|F(t/n)^n - e^{-tC}\| = O(n^{-1})\,, \quad \|T(t/n)^n - e^{-tC}\| = O(n^{-1}).$$

*Note that corresponding to each formula, error bounds $O(n^{-1})$ are equal up to coefficients $\{\Gamma_j > 0\}_{j=1}^4$ and that each rate of convergence $\Gamma_j\,\varepsilon(n) = O(n^{-1})$, $j = 1, \ldots 4$, is optimal.*

The first *lifting* lemma yields sufficient conditions that allow to strengthen the *strong* operator convergence to the $\|\cdot\|_\phi$-norm convergence in the symmetrically-normed ideal $\mathscr{C}_\phi(\mathscr{H})$.

**Lemma 4.2** *Let self-adjoint operators: $X \in \mathscr{C}_\phi(\mathscr{H})$, $Y \in \mathscr{C}_\infty(\mathscr{H})$ and $Z \in \mathscr{L}(\mathscr{H})$. If $\{Z(t)\}_{t\geq 0}$ is a family of self-adjoint bounded operators such that*

$$s - \lim_{t \to +0} Z(t) = Z\,, \tag{4.6}$$

*then*

$$\lim_{r\to\infty}\sup_{t\in[0,\tau]}\|(Z(t/r)-Z)YX\|_\phi=\lim_{r\to\infty}\sup_{t\in[0,\tau]}\|XY(Z(t/r)-Z)\|_\phi=0\,,\quad(4.7)$$

*for any* $\tau\in(0,\infty)$.

*Proof* Note that (4.6) yields the strong operator convergence $s-\lim_{r\to\infty}Z(t/r)=Z$, uniformly in $t\in[0,\tau]$. Since $Y\in\mathscr{C}_\infty(\mathscr{H})$, this implies

$$\lim_{r\to\infty}\sup_{t\in[0,\tau]}\|(Z(t/r)-Z)Y\|=0\,.\quad(4.8)$$

Since $\mathscr{C}_\phi(\mathscr{H})$ is a Banach space with symmetric norm (1.13) that verifies $\|ZX\|_\phi\le\|Z\|\|X\|_\phi$, one gets the estimate

$$\|(Z(t/r)-Z)YX\|_\phi\le\|(Z(t/r)-Z)Y\|\|X\|_\phi\,,\quad(4.9)$$

which together with (4.8) give the proof of (4.7). □

The second *lifting* lemma allows to estimate the rate of convergence of the Trotter–Kato product formula in the norm (1.19) of symmetrically-normed ideal $\mathscr{C}_\phi(\mathscr{H})$ via the error bound $\varepsilon(n)$ in the operator norm due to Proposition 4.1.

**Lemma 4.3** *Let A and B be non-negative self-adjoint operators on the separable Hilbert space* $\mathscr{H}$ *that satisfy the conditions of Proposition 4.1. Let* $f,g\in\mathscr{K}_2$ *be such that* $F(t_0)\in\mathscr{C}_\phi(\mathscr{H})$ *for some* $t_0>0$.

*If* $\Gamma_{t_0}\varepsilon(n)$, $n\in\mathbb{N}$ *is the operator-norm error bound away from* $t_0>0$ *of the Trotter–Kato product formula for* $\{f(tA)g(tB)\}_{t\ge0}$, *then for some* $\Gamma^\phi_{2t_0}>0$ *the function* $\varepsilon_\phi(n):=\{\varepsilon([n/2])+\varepsilon([(n+1)/2])\}$, $n\in\mathbb{N}$ *defines the error bound away from* $2t_0$ *of the Trotter–Kato product formula in the ideal* $\mathscr{C}_\phi(\mathscr{H})$:

$$\|[f(tA/n)g(tB/n)]^n-e^{-tC}\|_\phi=\Gamma^\phi_{2t_0}\varepsilon_\phi(n)\,,\quad n\to\infty.\quad t\ge2t_0\,.\quad(4.10)$$

*Here* $[x]:=\max\{l\in\mathbb{N}_0:l\le x\}$, *for* $x\in\mathbb{R}_0^+$.

*Proof* To prove the assertion for the family $\{f(tA)g(tB)\}_{t\ge0}$, we use decompositions $n=k+m$, $k\in\mathbb{N}$ and $m=2,3,\ldots,n\ge3$, for representation

$$\begin{aligned}&(f(tA/n)g(tB/n))^n-e^{-tC}= \quad(4.11)\\&\left((f(tA/n)g(tB/n))^k-e^{-ktC/n}\right)(f(tA/n)g(tB/n))^m\\&+e^{-ktC/n}\left((f(tA/n)g(tB/n))^m-e^{-mtC/n}\right).\end{aligned}$$

Since by conditions of lemma $F(t_0)\in\mathscr{C}_\phi(\mathscr{H})$, definition (4.5) and representation $f(tA/n)g(tB/n))^m=f(tA/n)g(tB/n)^{1/2}F(t/n)^{m-1}g(tB)^{1/2}$ yield

$$\|(f(tA/n)g(tB/n))^m\|_\phi \leq \|F(t_0)\|_\phi\ , \tag{4.12}$$

for $t$ such that $t_0 \leq (m-1)t/n \leq (m-1)t_0$ and $m-1 \geq 1$.

Note that for self-adjoint operators $e^{-tC}$ and $F(t)$ by Araki's log-order inequality for compact operators [1], one gets for $kt/n \geq t_0$ the bound of $e^{-ktC/n}$ in the $\|\cdot\|_\phi$-norm:

$$\|e^{-ktC/n}\|_\phi \leq \|F(t_0)\|_\phi\ . \tag{4.13}$$

Since by Definitions 2 and 3 the ideal $\mathcal{C}_\phi(\mathcal{H})$ is a Banach space, and from (4.11)–(4.13), we obtain the estimate

$$\begin{aligned} &\|(f(tA/n)g(tB/n))^n - e^{-tC}\|_\phi \leq \\ &\|F(t_0)\|_\phi\ \|(f(tA/n)g(tB/n))^k - e^{-ktC/n}\| \\ &+\|F(t_0)\|_\phi\ \|(f(tA/n)g(tB/n))^m - e^{-mtC/n}\|\ , \end{aligned} \tag{4.14}$$

for $t$ such that $(1+(k+1)/(m-1))t_0 \leq t \leq nt_0$, $m \geq 2$ and $t \geq (1+m/k)t_0$.

Now, by conditions of lemma $\Gamma_{t_0}\varepsilon(\cdot)$ is the operator-norm error bound away from $t_0$, for any interval $[a,b] \subseteq (t_0,+\infty)$. Then, there exists $n_0 \in \mathbb{N}$ such that

$$\|(f(tA/n)g(tB/n))^k - e^{-ktC/n}\| \leq \Gamma_{t_0}\varepsilon(k)\ , \tag{4.15}$$

for $kt/n \in [a,b] \Leftrightarrow t \in [(1+m/k)a, (1+m/k)b]$ and

$$\|(f(tA/n)g(tB/n))^m - e^{-mtC/n}\| \leq \Gamma_{t_0}\varepsilon(m)\ , \tag{4.16}$$

for $mt/n \in [a,b] \Leftrightarrow t \in [(1+k/m)a, (1+k/m)b]$ for all $n > n_0$.

Setting $m := [(n+1)/2]$ and $k = [n/2]$, $n \geq 3$, we satisfy $n = k+m$ and $m \geq 2$, as well as, $\lim_{n\to\infty}(k+1)/(m-1) = 1$, $\lim_{n\to\infty} m/k = 1$ and $\lim_{n\to\infty} k/m = 1$. Hence, for any interval $[\tau_0,\tau] \subseteq (2t_0,+\infty)$, we find that $[\tau_0,\tau] \subseteq [(1+(k+1)/(m-1))t_0, nt_0]$ for sufficiently large $n$. Moreover, choosing $[\tau_0/2,\tau/2] \subseteq (a,b) \subseteq (t_0,+\infty)$, we satisfy $[\tau_0,\tau] \subseteq [(1+m/k)a, (1+m/k)b]$ and $[\tau_0,\tau] \subseteq [(1+k/m)a, (1+k/m)b]$ again for sufficiently large $n$.

Thus, for any interval $[\tau_0,\tau] \subseteq (2t_0,+\infty)$ there is $n_0 \in \mathbb{N}$ such that (4.14)–(4.16) hold for $t \in [\tau_0,\tau]$ and $n \geq n_0$. Therefore, (4.14) yields the estimate

$$\begin{aligned} &\|(f(tA/n)g(tB/n))^n - e^{-tC}\|_\phi \leq \\ &\qquad \Gamma_{t_0}\ \|F(t_0)\|_\phi\{\varepsilon([n/2]) + \varepsilon([(n+1)/2])\}\ , \end{aligned} \tag{4.17}$$

for $t \in [\tau_0,\tau] \subseteq (2t_0,+\infty)$ and $n \geq n_0$. Hence, $\Gamma^\phi_{2t_0} := \Gamma_{t_0}\ \|F(t_0)\|_\phi$ and $\Gamma^\phi_{2t_0}\varepsilon_\phi(\cdot)$ is an error bound in the Trotter–Kato product formula (4.10) away from $2t_0$ in $\mathcal{C}_\phi(\mathcal{H})$ for the family $\{f(tA)g(tB)\}_{t\geq 0}$.

The lifting Lemma 4.2 allows to extend the proofs for other approximants: $\{g(tB)f(tA)\}_{t\geq 0}$, $\{F(t)\}_{t\geq 0}$ and $\{T(t)\}_{t\geq 0}$. □

Now, we apply Lemma 4.3 in Dixmier ideal $\mathscr{C}_\phi(\mathscr{H}) = \mathscr{C}_{1,\infty}(\mathscr{H})$. This concerns the norm convergence (4.10), but also the estimate of the convergence rate for Dixmier traces:

$$|\mathrm{Tr}_\omega(e^{-tC}) - \mathrm{Tr}_\omega(F(t/n)^n)| \leq \Gamma^\omega \varepsilon_\omega(n). \tag{4.18}$$

In fact, it is the same (up to $\Gamma^\omega$) for all Trotter–Kato approximants: $\{T(t)\}_{t\geq 0}$, $\{f(t)g(t)\}_{t\geq 0}$ and $\{g(t)f(t)\}_{t\geq 0}$.

Indeed, since by inequality (3.18) and Lemma 4.3 for $t \in [\tau_0, \tau]$ and $n \geq n_0$, one has

$$|\mathrm{Tr}_\omega(e^{-tC}) - \mathrm{Tr}_\omega(F(t/n)^n)| \leq \|e^{-tC} - F(t/n)^n\|_{1,\infty} \leq \Gamma^\phi_{2t_0}\, \varepsilon_{1,\infty}(n)\ , \tag{4.19}$$

we obtain for the rate in (4.18): $\varepsilon_\omega(\cdot) = \varepsilon_{1,\infty}(\cdot)$. Therefore, the estimate of the convergence rate for Dixmier traces (4.18) and for $\|\cdot\|_{1,\infty}$-convergence in (4.19) are *entirely* defined by the operator-norm error bound $\varepsilon(\cdot)$ from Lemma 4.3 and have the following form:

$$\varepsilon_{1,\infty}(n) := \{\varepsilon([n/2]) + \varepsilon([(n+1)/2])\}\ ,\ n \in \mathbb{N}\ . \tag{4.20}$$

Note that for the particular case of Proposition 4.1, these arguments yield for (4.17) the explicit convergence rate asymptotics $O(n^{-1})$ for the Trotter–Kato formulae and consequently, the same asymptotics for convergence rates of the Trotter–Kato formulae for the Dixmier trace (4.18), (4.19).

Therefore, we proved in the Dixmier ideal $\mathscr{C}_{1,\infty}(\mathscr{H})$ the following assertion.

**Theorem 1** *Let $f, g \in \mathscr{K}_\beta$ with $\beta = 2$, and let $A$, $B$ be non-negative self-adjoint operators in $\mathscr{H}$ such that the operator sum $C := A + B$ is self-adjoint on domain $dom(C) := dom(A) \cap dom(B)$.*

*If $F(t_0) \in \mathscr{C}_{1,\infty}(\mathscr{H})$ for some $t_0 > 0$, then the Trotter–Kato product formulae converge for $n \to \infty$ in the $\|\cdot\|_{1,\infty}$-norm:*

$$\|[f(tA/n)g(tB/n)]^n - e^{-tC}\|_{1,\infty} = O(n^{-1})\ ,\ \|[g(tB/n)f(tA/n)]^n - e^{-tC}\|_{1,\infty} = O(n^{-1})\ ,$$
$$\|F(t/n)^n - e^{-tC}\|_{1,\infty} = O(n^{-1})\ ,\quad \|T(t/n)^n - e^{-tC}\|_{1,\infty} = O(n^{-1})\ ,$$

*away from $2t_0$. The rate $O(n^{-1})$ of convergence is optimal in the sense of* [11].

*By virtue of (4.19), the same asymptotics $O(n^{-1})$ of the convergence rate are valid for convergence of the Trotter–Kato formulae for the Dixmier trace:*

$$|\mathrm{Tr}_\omega([f(tA/n)g(tB/n)]^n) - \mathrm{Tr}_\omega(e^{-tC})| = O(n^{-1})\ ,$$
$$|\mathrm{Tr}_\omega(F(t/n)^n) - \mathrm{Tr}_\omega(e^{-tC})| = O(n^{-1})\ ,\quad |\mathrm{Tr}_\omega(T(t/n)^n) - \mathrm{Tr}_\omega(e^{-tC})| = O(n^{-1})\ ,$$

*away from $2t_0$.*

Optimality of the estimates in Theorem 1 is a heritage of the optimality in Proposition 4.1. Recall that, in particular, this means that in contrast to the Lie product formula for *bounded* generators $A$ and $B$, the *symmetrisation* of approximants $\{f(t)g(t)\}_{t\geq 0}$, and $\{g(t)f(t)\}_{t\geq 0}$ by $\{F(t)\}_{t\geq 0}$ and $\{T(t)\}_{t\geq 0}$, does not yield (in general) the improvement of the convergence rate, see [11, 23] and discussion in [26].

We resume that the *lifting* Lemma 4.2 and 4.3 are a general method to study the convergence in symmetrically-normed ideals $\mathcal{C}_\phi(\mathcal{H})$ as soon as it is established in $\mathcal{L}(\mathcal{H})$ in the operator-norm topology. The crucial is to check that for any of the *key* Kato functions (e.g. for $\{F(t)\}_{t\geq 0}$) there exists $t_0 > 0$ such that $F(t)|_{t\geq t_0} \in \mathcal{C}_\phi(\mathcal{H})$. Sufficient conditions for that one can be found in [16–18] or in [25].

**Acknowledgements** I am thankful to referee for useful remarks and suggestions.

## References

1. Araki, H.: On an inequality of Lieb and Thirring. Lett. Math. Phys. **19**, 167–170 (1990)
2. Cachia, V., Zagrebnov, V.A.: Trotter product formula for nonself-adjoint Gibbs semigroups. J. Lond. Math. Soc. **64**, 436–444 (2001)
3. Carey, A.L., Sukachev, F.A.: Dixmier traces and some applications in non-commutative geometry. Russ. Math. Surv. **61**(6), 1039–1099 (2006)
4. Connes, A.: Noncommutative Geometry. Academic, London (1994)
5. Dixmier, J.: Existence des traces non normales. C. R. Acad. Sci. Paris Sér. A **262**, 1107–1108 (1966)
6. Dixmier, J.: Von Neumann Algebras. North Holland, Amsterdam (1981)
7. Doumeki, A., Ichinose, T., Tamura, H.: Error bounds on exponential product formulas for Schrödinger operators. J. Math. Soc. Jpn. **50**, 359–377 (1998)
8. Gohberg, I.C., Kreĭn, M.G.: Introduction to the Theory of Linear Nonselfadjoint Operators in Hilbert Space. (Translated by A. Feinstein from the Russian Edition: "Nauka", Moscow, 1965) Translations of Mathematical Monographs, vol. 18. American Mathematical Society, Providence (1969)
9. Ichinose, T., Tamura, H.: Error bound in trace norm for Trotter-Kato product formula of Gibbs semigroups. Asymptot. Anal. **17**, 239–266 (1998)
10. Ichinose, T., Tamura, H.: The norm convergence of the Trotter-Kato product formula with error bound. Commun. Math. Phys. **217**, 489–502 (2001)
11. Ichinose, T., Tamura, Hideo, Tamura, Hiroshi, Zagrebnov, V.A.: Note on the paper "the norm convergence of the Trotter-Kato product formula with error bound" by Ichinose and Tamura. Commun. Math. Phys. **221**, 499–510 (2001)
12. Kato, T.: Trotter's product formula for an arbitrary pair of self-adjoint contraction semigroups. In: Gohberg, I., Kac, M. (eds.) Topics in Functional Analysis. Advances in Mathematics, Supplementary Studies, vol. 3, pp. 185–195. Academic, New York (1978)
13. Lord, S., Sukochev, F., Zanin, D.: Singular Traces. Theory and Applications. De Gruyter Studies in Mathematics, vol. 46. W. de Gruyer GmbH, Berlin (2013)
14. Neidhardt, H., Zagrebnov, V.A.: The Trotter product formula for Gibbs semigroup. Commun. Math. Phys. **131**, 333–346 (1990)
15. Neidhardt, H., Zagrebnov, V.A.: On error estimates for the Trotter-Kato product formula. Lett. Math. Phys. **44**, 169–186 (1998)
16. Neidhardt, H., Zagrebnov, V.A.: Fractional powers of self-adjoint operators and Trotter-Kato product formula. Integral Equ. Oper. Theory **35**, 209–231 (1999)
17. Neidhardt, H., Zagrebnov, V.A.: Trotter-Kato product formula and operator-norm convergence. Commun. Math. Phys. **205**, 129–159 (1999)
18. Neidhardt, H., Zagrebnov, V.A.: On the operator-norm convergence of the Trotter-Kato product formula. Oper. Theory Adv. Appl. **108**, 323–334 (1999)

19. Neidhardt, H., Zagrebnov, V.A.: Trotter-Kato product formula and symmetrically normed ideals. J. Funct. Anal. **167**, 113–167 (1999)
20. Pietsch, A.: Traces of operators and their history. Acta et Comm. Univer. Tartuensis de Math. **18**, 51–64 (2014)
21. Schatten, R.: Norm Ideals of Completely Continuous Operators. Springer, Berlin (1970)
22. Simon, B.: Trace Ideals and Their Applications. Cambridge University Press, Cambridge (1979)
23. Tamura, H.: A remark on operator-norm convergence of Trotter-Kato product formula. Integral Equ. Oper. Theory **37**, 350–356 (2000)
24. Zagrebnov, V.A.: The Trotter-Lie product formula for Gibbs semigroups. J. Math. Phys. **29**, 888–891 (1988)
25. Zagrebnov, V.A.: Topics in the Theory of Gibbs Semigroups. Leuven Notes in Mathematical and Theoretical Physics (Series A: Mathematical Physics), vol. 10. Leuven University Press, Leuven (2003)
26. Zagrebnov, V.A.: Trotter-Kato product formula: some recent results. In: Proceedings of the XIVth International Congress on Mathematical Physics, Lisbon (July 28–August 02, 2003), pp. 634–641. World Scientific, Singapore (2005)

The manufacturer's authorised representative in the EU is Springer Nature Customer Service Centre GmbH, Europaplatz 3, 69115 Heidelberg, Germany. If you have any concerns regarding our products, please contact ProductSafety@springernature.com

Printed and bound by CPI Group (UK) Ltd, Croydon, CR0 4YY
24/07/2026
02175951-0006